Decision Engineering

Series editor

Rajkumar Roy, Cranfield, Bedfordshire, UK

More information about this series at http://www.springer.com/series/5112

John Stark

Product Lifecycle Management (Volume 2)

The Devil is in the Details

Third Edition

John Stark
John Stark Associates
Geneva
Switzerland

ISSN 1619-5736   ISSN 2197-6589   (electronic)
Decision Engineering
ISBN 978-3-319-24434-1   ISBN 978-3-319-24436-5   (eBook)
DOI 10.1007/978-3-319-24436-5

Library of Congress Control Number: 2015936161

Springer Cham Heidelberg New York Dordrecht London

Printed on acid-free paper

Springer International Publishing AG Switzerland is part of Springer Science+Business Media
(www.springer.com)

Preface

This is the second volume of the third edition of *Product Lifecycle Management: Paradigm for 21st Century Product Realisation*.

Product Lifecycle Management (PLM) is the business activity of managing, in the most effective way, a company's products all the way across their lifecycles; from the very first idea for a product all the way through until it is retired and disposed of.

In the middle of the twentieth century, between 1945 and 1970, things changed little in the world of products. Companies, and their executives and employees worked out how to succeed in that environment. They had an accepted way of thinking, a paradigm, about the way products were managed.

Between 1970 and 2015, for various reasons, the product landscape changed rapidly and significantly. Facing so many changes, companies had to change to remain competitive. But change how? What's the new paradigm for managing products in the changed environment? Or, put another way, how should a company, its executives and employees be organised and work in this new environment? And, how should a company transition from the old paradigm to the new paradigm? Or, put another way, what set of actions will a company have to execute to achieve the change? What will be in the PLM Initiative? This book addresses these questions.

The new PLM paradigm emerged at the beginning of the twenty-first century and has been evolving since then. It was described in the first edition of this book, which was published in 2004. The second edition of the book was published in 2011. Since then, the paradigm has continued to evolve. There have been more changes in the technologies, products and the PLM environment. This third edition of the book addresses these changes, technological advances and the ever-increasing application of PLM.

As its name implies, Product Lifecycle Management enables companies to manage their products across their lifecycles; from the earliest idea for a product all the way through to the end of its life. This is one of the most important activities in any company that develops and supports products.

PLM is important because it enables a company to be in control of its products across their lifecycle. If a company loses control, the consequences can be serious. If it loses control during product development, the product may be late to market and exceed the targeted cost. The results of losing control during use of the product may be a frustration and a lack of satisfaction for the customer, or much worse, injury and death.

PLM addresses the heart of the company, its defining resource, the source of its wealth, its products. That is the role of PLM, which is why PLM is so important. Products define a company. Without its products, a company would not be the same. There is little in a company more important than its products and the way they will be developed and used. Without those products, there will be no customers and no revenues.

PLM is also important because it improves the activity of product development, without which a company will not survive. The source of future revenues for a company is the creation of new products and services. PLM is the activity that enables a company to grow revenues by improving innovation, reducing time-to-market for new products, and providing superb support and new services for existing products.

PLM is also important because it enables a company to reduce product-related costs. Product-related material and energy costs are fixed early in the product development process. PLM provides the tools and knowledge to minimise them. And PLM helps cut recall, warranty and recycling costs that come later in the product's life.

PLM provides a way to overcome problems with the use and support of existing products and with the development of new products. But PLM does not just have the potential to solve problems in the product lifecycle and in new product development. It also helps companies seize the many market opportunities for new products in the globalised environment of the early twenty-first century.

PLM is not easy to implement. It addresses areas previously considered separate, and managed separately. They include products, product data, business processes, applications, people and organisational structures. And PLM addresses them across the product's entire lifecycle, from cradle to grave. Most companies have, in their product portfolios, many products at different lifecycle stages. Managing product lifecycles in a global economy is a daunting proposition. PLM provides a framework in which all of a company's products can be managed together across their lifecycles.

PLM is cross-functional and, in the extended enterprise environment of the early twenty-first century, it is often cross-enterprise as well. Product lifecycle participants are often in different time zones, use different applications and work for different companies. The responsibility for the product may change at different phases of the lifecycle. At different times, it may be with marketing, engineering, product management, manufacturing, finance, marketing, sales and service groups in different companies. Getting agreement on a common approach among all these organisations can be time-consuming. PLM helps get everyone to work together effectively.

Implementation of PLM may take a long time. Clarifying and straightening out processes, data, organisational issues and applications can be time-consuming. Some of the processes and methodologies to propose, define, manufacture, support, upgrade, retire and recycle the product may not be aligned, or may even not exist. The knowledge about the product may be in different applications. The format in which data are created in one application may not correspond to the format in which it is needed in another application. In spite of these difficulties, companies must meet the increasing demands of their customers. They need to rapidly and continually improve their products and services. To achieve this they will turn to PLM. This book helps them understand and implement PLM. As for the previous editions, it draws on the extensive PLM consulting activities and experience of the author.

Contents

Chapter 1
Product Lifecycle Management

1.1 Overview

Product Lifecycle Management (PLM) is the business activity of managing, in the most effective way, a company's products all the way across their lifecycles; from the very first idea for a product all the way through until it is retired and disposed of.

At the highest level, the objective of PLM is to increase product revenues, reduce product-related costs, maximise the value of the product portfolio, and maximise the value of current and future products for both customers and shareholders.

The resources that can be organised in different ways to manage a product across its lifecycle are shown on the PLM Grid (Fig. 1.5). They include processes, applications, product data, people, products, facilities and equipment, methods and metrics.

PLM is focused on "the product". There's little in a company more important than its products, and the way that they'll be developed and used. Without those products, there'll be no customers and no revenues. Apart from PLM, there's no other approach to manage products across the lifecycle.

PLM solves many problems across the lifecycle. It provides many benefits across the lifecycle. The benefits include improved financial performance, reductions in time cycles, and quality improvements. Above all, PLM gets products under control across the lifecycle.

PLM emerged in about 2001. Before the 21st Century, a PLM approach wasn't sufficiently needed from a business point of view, and wasn't possible from the technical viewpoint.

PLM is used across a wide range of industries that develop, produce and support products. It's used in companies of all sizes.

People from all levels of the company, including top managers and executives such as the CEO and the CIO, are involved with PLM. And people from nearly all parts of the company are involved with PLM.

© Springer International Publishing Switzerland 2016
J. Stark, *Product Lifecycle Management (Volume 2)*,
Decision Engineering, DOI 10.1007/978-3-319-24436-5_1

PLM is usually implemented through a PLM Initiative, including a Feasibility Study, the development of the future PLM Strategy, and the development of the PLM Implementation Strategy and Plan.

Some companies involved in product development, manufacturing and/or support started a PLM Initiative between 2001 and the present. Most companies will have a PLM Initiative between now and 2025.

1.2   What Is PLM?

Product Lifecycle Management (PLM) is the business activity of managing, in the most effective way, a company's products all the way across their lifecycles; from the very first idea for a product all the way through until it is retired and disposed of.

PLM is the management system for the company's products. It doesn't just manage one of a company's products. It manages, in an integrated way, all of a company's parts and products, and the product portfolio. It manages the whole range, from individual part through individual product to the entire portfolio of products.

1.2.1   High-Level Objective of PLM

At the highest level, the objective of PLM is to increase product revenues, reduce product-related costs, maximise the value of the product portfolio, and maximise the value of current and future products for both customers and shareholders.

1.2.2   Activities of PLM

PLM is a high-level business activity. All of the lower-level product-related activities of a company are united under the PLM umbrella. Figure 1.1 shows some of these activities.

| managing a well-structured and valuable Product Portfolio |
| maximising the financial return from the Product Portfolio |
| managing products across the lifecycle |
| managing product development, support and disposal projects effectively |
| providing control and visibility over products throughout the lifecycle |
| managing feedback about products from customers, products, field engineers and the market |
| enabling collaborative work with design and supply chain partners, and with customers |
| managing product-related processes so that they are coherent, joined-up, effective and lean |
| capturing, securely managing, and maintaining the integrity of product definition information |
| making product data available where it's needed, when it's needed |
| knowing the exact technical and financial characteristics of a product throughout its lifecycle |

Fig. 1.1 Activities within the scope of PLM

1.2.3 A Joined-up, Holistic Approach

PLM is "joined-up" (Fig. 1.2). With PLM, the organisation manages the product in a coherent joined-up way across the lifecycle. PLM joins up many previously separate and independent processes, disciplines, functions and applications, each of which, though addressing the same product, previously had its own vocabulary, rules, culture and language. PLM brings together what was previously separate, for example, product development and product support.

PLM has a holistic approach to the management of a product (Fig. 1.3). It addresses many resources such as products, data, applications, processes, people, work methods and equipment.

1.2.4 Generic Product Lifecycle Phases

There are five phases in the generic product lifecycle (Fig. 1.4).

In each of these five phases, the product is in a different state. During the imagination phase, the product is just an idea in people's heads. During the definition phase, the ideas are being converted into a detailed description. By the end of

Fig. 1.2 A joined-up approach to the management of products

Fig. 1.3 PLM is holistic

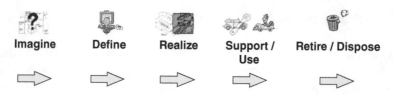

Fig. 1.4 The 5 phases of the generic product lifecycle

the realisation phase, the product exists in its final form (for example, as a car) in which it can be used by a customer. During the use/support phase, the product is with the customer who is using it. Eventually the product gets to a phase in which it's no longer useful. It's retired by the company, and disposed of by the customer.

The specific activities that take place across the lifecycle vary from one industry to another. As a result, companies in a particular industry may have a view of the product lifecycle that is specific to their industry. However, whatever the specifics of a particular company or industry, its activities can be mapped, in some way, to the five phases of the generic product lifecycle.

1.2.5 PLM, Managing the Product Across the Lifecycle

The product must be managed in all phases of the lifecycle to make sure that everything works well, and that the product makes good money for the company.

The product needs to be managed when it's an idea. Product ideas need to be managed to make sure, for example, that they are not lost or misunderstood.

The product needs to be managed when it's being defined. For example, a product development project has to be managed to be sure the product meets customer requirements.

The product needs to be managed when it's being realised. For example, it's important that the correct version of the definition is used during production.

The product needs to be managed when it's in use. For example, the product must be correctly maintained, taking account of its serial number, production date, previous upgrades, changes in the market and technical evolution.

The product needs to be managed at disposal time. Care has to be taken to make sure that poisonous components and toxic waste from the product don't get anywhere near sources of drinking water.

1.2.6 Managing the Product from Dawn to Dusk

It's sometimes said that PLM is about managing the product throughout its lifecycle, "from cradle to grave" of "from sunrise to sunset". However, both of these phrases miss the earliest part of the lifecycle. PLM manages the product "from dawn to dusk".

1.3 PLM—With What Resources?

PLM is the business activity of managing, in the most effective way, a company's products all the way across their lifecycles.

The resources that can be organised in different ways to manage a product across its lifecycle are shown in the PLM Grid (Fig. 1.5).

1.3.1 The PLM Grid

The PLM Grid is a simple 5 × 10 grid or matrix.

On the horizontal axis of the grid are the five phases of the product lifecycle.

On the vertical axis are the ten components (i.e., processes, applications, product data, etc.) that have to be addressed when managing a product across the lifecycle.

The PLM Grid helps show why the environment of the product can be so difficult to manage. The scope of the environment is broad. Many subjects are addressed, ranging from methods for identifying ideas for new products, through organisational structure, to end-of-life recycling equipment. The scope is wide, but that reflects the reality of managing products.

The Grid can be looked at in different ways. The many items and details may be seen as confirmation that the environment of products is complex and difficult to

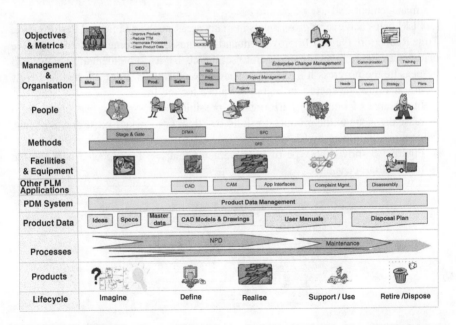

Fig. 1.5 The PLM grid

manage. Another way of looking at the PLM Grid is to see that the apparent complexity of the management of products can be summarised in a single diagram.

The environment is complex and difficult to manage, but once the complexity and difficulty have been understood, the environment becomes easier to manage.

The Grid gives companies that develop and support products a way of visualising PLM. It helps them organise and communicate about the many resources that are involved, and the related issues. It helps them visualise, quantify and communicate the status of their product-related resources, describe their current situation, and develop their future PLM Strategy.

1.3.2   Managing the Ten Components of the PLM Grid

PLM, the effective management of a company's products across the lifecycle, is achieved by managing the ten components shown on the PLM Grid.

1.3.2.1   Objectives and Metrics

The company's objectives for PLM drive all its PLM activities. The PLM objectives express at a high level what's expected from PLM. They express at a high level what must be achieved.

A PLM Strategy will be developed to describe the way to achieve the objectives. It defines how resources will be organised. It defines the policies that will apply for the management and use of resources.

PLM helps achieve improvements in many areas, such as Financial Performance, Time Reduction, Quality Improvement and Business Improvement. Metrics provide parameters that help an organisation to set targets for its implementation plans, and to measure the progress it's making towards the objectives. For each metric there's a current value and there can be target values for the future.

In the area of Financial Performance, possible metrics and targets could be to increase the value of the product portfolio by 20 %, or to reduce costs due to recalls, failures and liabilities by 75 %.

Figure 1.6 shows the type of improvements that can be expected between the current and target values.

Rate of introduction of new products	+ 100%
Revenues from extended product life	+ 25%
Part reuse factor	x 7
Costs due to recalls, failures, liabilities	- 75%
Development time for new products	- 50%
Cost of materials and energy	- 25%
Recycling of products	+ 90%
Product traceability	100%
Lifecycle control	100%
Lifecycle visibility	100%
Revenues from new services on existing products	+ 40%

Fig. 1.6 Improvement targets

It's not easy to define the objectives and set clear targets. Some of the potential problems are shown in Fig. 1.7.

1.3.2.2   Management and Organisation

All of the components of PLM are necessary to transform ideas into products which meet a company's objectives, meet customer requirements, comply with regulations, and meet environmental objectives. There are many resources to manage, and high volumes of many of these resources. And, as if the wide scope and high volumes didn't make it difficult enough, there are complex and changing relationships to manage between products, components, customers and suppliers.

To make all the resources and activities manageable, and to meet the objectives, organisational structures, strategies and plans must be defined. Many changes will need to be managed. The many resources in the PLM environment will need to be organised and aligned.

Effective organisation and management of resources is all-important for PLM. Just acquiring good resources, such as applications and people, doesn't automatically result in successful PLM. It's only when all the resources are organised to achieve the objectives of PLM, and managed to achieve the objectives of PLM, that the objectives can be met.

It's not easy to manage all the resources successfully. Some of the potential problems are shown in Fig. 1.8.

1.3.2.3   Processes

In every company, there's a lot of activity related to the product as it is developed, manufactured, supported and retired. The activity is organised into processes. In many companies, between 35 and 55 % of the company's processes are product-related. Many things have to happen if everything is to work well with the product. Processes are important. A process is something a company does. The

conflicting objectives	lack of training about metrics
unclear objectives, metrics not clear	poor reporting of the values of metrics
unbalanced metrics	lack of follow-up on reports of the values of metrics
inconsistent metrics	difficulty to relate metrics to performance
ineffective metrics	difficulty of measuring progress to objectives
not enough metrics (or too many)	multiple interpretations of objectives and metrics
metrics that are not relevant	lack of documentation of objectives and metrics

Fig. 1.7   Potential issues with objectives and metrics

each resource organised independently	unclear scope of resources	some resources without owners
some resources not organised	responsibilities not clear	lines of authority not clear

Fig. 1.8   Potential issues with organisation and management

Product Development process too slow	superfluous tasks, adding no value, are carried out
poorly defined processes lead to wasted resources	use of processes without metrics or quality control
Engineering Change process takes too long	duplication of the same activity in several processes
time lost in processes waiting for information	no process for managing the portfolio of development projects
no version control for processes	process ownership unclear

Fig. 1.9 Potential process-related problems

company has a choice. It can put in place good processes, and do the right things well. Or it can do things badly.

It's not easy to manage processes successfully. Some of the potential problems are shown in Fig. 1.9.

1.3.2.4 People

It takes many people to develop and support a product throughout its lifecycle. People are all-important. No product is made or managed without people. The company has a choice. It can hire highly-skilled people, motivate them and train them to do things the best way, or it can do the opposite. Throughout the product lifecycle, people are all-important. They define the requirements for new products, develop products to meet the requirements, produce high-quality products, and support them in the field.

It's not easy to manage people. Many people will fail to understand PLM, and will not want to change their daily activities and behaviour. Figure 1.10 shows some other potential issues with people.

1.3.2.5 Product Data

Product data defines and describes the product, and the product is the source of company revenues. A company's product data represents its collective know-how. As such, it's a major asset and should be used as profitably as possible. If there's something wrong with product data, then there'll be problems with the product and money will be lost. Throughout the product lifecycle, product data is all-important. It has to be available, whenever it's needed, wherever it's needed, by whoever needs it, throughout the product lifecycle. Product data is a strategic resource. Getting it organised, and keeping it organised, are major challenges.

Whatever the product made by a company, an enormous volume and variety of product data is needed to develop, produce and support the product throughout the

not the right skill mix	not sufficiently innovative	people don't learn new skills
not the right experience	not sufficiently audacious	slow response to new technology
not enough commitment	personnel costs too high	people aren't disciplined enough
not adaptive enough	people don't feel empowered	people don't learn new skills

Fig. 1.10 Potential issues with people

confidential data lost	difficulties to exchange product data with partners
multiple overlapping databases	unreleased versions of data mistakenly used
inability to access legacy data	product data stolen, and used by a competitor
ownership of product data unclear	duplicate data leading to confusion and errors
knowledge in heads not documented	difficulty of managing data at several locations
incorrect product data sent to a customer	conflicting different copies of the same data

Fig. 1.11 Potential issues with product data

lifecycle. Product data doesn't look after itself. If it's not looked after, then, like anything that's not properly organised and maintained, it won't perform as required. Over time, it will slide into chaos and decay. However this has to be avoided. The slightest error with product data can have serious consequences for the product and those associated with it. Figure 1.11 shows some of the potential problems with product data.

1.3.2.6   Product Data Management Systems

A PDM system is a very specific type of PLM application. It has the primary purpose of managing product data. A PDM system is one of the most important elements of a PLM solution. It can manage all the product data created and used throughout the product lifecycle. It can provide exactly the right information at exactly the right time.

Whatever a company's PLM Strategy, it's probable that a PDM system will be a major constituent. PDM gets product data under control, and, unless the product data in the product lifecycle is under control, it will be difficult to get the product under control.

For most of the product lifecycle, information is all-important. The PDM system gets this strategic resource under control, making it available, whenever it's needed, wherever it's needed, by whoever needs it, throughout the product lifecycle.

It's not easy to organise PDM effectively. Many problems may arise (Fig. 1.12).

1.3.2.7   Other PLM Applications

Just as there are many processes, and many types of product data, in the scope of PLM, there are also many application systems. Even in a medium size company, there may be as many as fifty different applications in use. PLM applications help

multiple PDM systems in the company	too much customisation required
duplicate PDM functionality is wasteful	insufficient user support
too much product data managed outside PDM	inappropriate data model
deficiencies in the PDM system	limited workflow functionality
poorly designed implementation of PDM	high costs for support
too much bureaucracy in application use	slow response time

Fig. 1.12 Potential issues with product data management systems

poorly used applications	too many applications	slow response to new technology
under-used applications	too much customisation	not integrated with the business
high costs of applications	too many interfaces	wildcat applications
insufficient user support	too many changes	duplicate applications

Fig. 1.13 Potential issues with PLM applications

people develop and support products. Without these applications, it's unlikely that so many complex and precise products could be developed, produced and supported. PLM applications enable people to achieve performance levels that would be impossible by manual means alone.

One of the challenges of PLM for a particular company is to identify the applications that are most relevant to the activities on which the company wants to focus its efforts.

PLM applications need to be managed, and that's not easy. Figure 1.13 shows some of the issues with PLM applications.

1.3.2.8   Facilities and Equipment

Facilities and equipment are used in every phase of the product lifecycle. They are needed to develop the product, to produce it, to maintain and service it, and to dispose of it. They affect the quality of the product, its cost and the time to develop and produce it.

In total, there are thousands of different machines and tools in the PLM environment. One of the challenges is to identify the facilities and equipment that are most relevant to the activities on which the company wants to focus its efforts. Companies are always looking for high quality equipment and output. They want flexibility, fast response and short set-up time. But a lot of equipment has a high capital cost. Its purchase has to be rigorously justified. Once installed, it has to be used as much as possible to ensure it meets financial targets and pays back the investment.

A company may face many problems when managing facilities and equipment (Fig. 1.14).

1.3.2.9   Methods

To improve performance across the lifecycle in terms of parameters such as product development time, product cost, service cost, product development cost, product

unreliable equipment	problems with cost-justification	lack of repeatability of machine behaviour
under-achievement	under-used equipment	difficult to integrate with other equipment
lack of standards	lack of support from the supplier	automated equipment not linked to company systems
lack of appropriate skills	high maintenance costs	difficult to upgrade the equipment

Fig. 1.14 Potential issues with facilities and equipment

difficulty to know which methods are most important	duplication of methods
overlap of one method with other methods	unclear relationship with business processes
difficulty to integrate methods with other PLM components	lack of knowledge about methods
lack of support for methods	conflicting methods

Fig. 1.15 Potential issues with methods

quality and disassembly costs, many methods and techniques have been proposed. Examples include Business Process Management (BPM), Concurrent Engineering, Design for Assembly (DFA), Early Manufacturing Involvement (EMI), Lean Production, Life Cycle Design (LCD), Open Innovation, Six Sigma, and Total Quality Management (TQM). Benefits typically proposed for these methods include: reduced time to market; improved quality; reduced manufacturing costs; improved service; reduced cycle time; and reduced Cost of Quality.

One of the methods-related challenges for a particular company is the identification of the methods that are most relevant to its objectives. Another is the effective management of their use. There are several potential issues with methods (Fig. 1.15).

1.3.2.10 Products

A company's products are one of its most important resources. A company gets its revenues from an on-going stream of innovative new and upgraded products. Great products make it the leader in its industry sector. Great products lead to great profitability. Whether it's a chair, a beverage, an aircraft or an anaesthetic, it's the product, and perhaps some related services, that the customer wants. The product is the source of company revenues. Without a product, the company doesn't need to exist and won't have any customers. Without a product, there won't be any related services. Products must be managed in all phases of the lifecycle to make sure that everything works well, and that the product makes good money for the company.

But successfully managing a product isn't easy. There are many potential issues with products (Fig. 1.16).

financial losses due to counterfeits	multiple incoherent product numbering systems causing confusion
complaints about products. Product Recalls	customers request better product support in Middle-of-Life
regulators require pollution-free End-of-Life	customers ask for customisation that is too costly
incorrect product classification	competitors offer products with better performance at lower cost
waste and rework due to quality problems	misinterpretation of customer requirements

Fig. 1.16 Potential issues with products

1.4 Why PLM?

1.4.1 There Is no Alternative

PLM is focused on "the product" . It addresses the heart of the company, its defining resource, the source of its wealth, its products. That's the role of PLM, which is why PLM is so important. Products define a company. The company's products are what the customer buys. They are the source of a company's revenues. Without its products, a company wouldn't be the same. There's little in a company more important than its products, and the way that they're developed and used. Without those products, there will be no customers and no revenues. As Fig. 1.17 shows, the product is at the heart of the PLM environment.

PLM gets products under control across the lifecycle. As a result, managers face less risk and fire-fighting. They can spend more time on preparing an outstanding future with awesome products.

PLM improves the activity of product development, without which a company won't survive. The source of future revenues for a company is the creation of new products and services. PLM is the activity that enables a company to grow revenues by improving innovation, reducing time-to-market for new products, and providing superb support and new services for existing products. PLM helps bring new products to market faster. It's important for a company to bring a product to market quickly. Otherwise the customer will choose a competitor's product before the company's product gets to market.

PLM helps companies to develop and produce products at different sites. It enables collaboration across the design chain and supply chain. PLM helps manage Intellectual Property. It helps maximise reuse of product knowledge. It helps bring together the management of products and processes, and to get processes such as engineering change management under control. It helps ensure compliance with regulations.

PLM enables a company to reduce product-related costs. It's important to reduce product costs. Otherwise the customer will choose a competitor's product that costs less than the company's product. Product-related material and energy costs are fixed

Fig. 1.17 The product is at the heart of the PLM environment

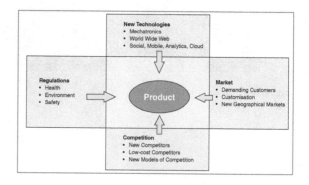

early in the product development process. PLM provides the tools and knowledge to minimise them. And PLM helps cut recall, warranty and recycling costs that come later in the product's life.

PLM gives transparency about what's happening over the product lifecycle. It offers managers visibility about what's really happening with products and with product development, modification and retirement projects. Without PLM, they are often faced by a huge mass of conflicting information about a product. PLM gives them the opportunity to manage better. With access to the right information, they can make better decisions.

PLM enables better support of customers' use of products. It's important for a company to support customers' use of its product. Otherwise they may stop using the company's product. They may start to use a competitor's product instead.

PLM enables the value of a product to be maximised over its lifecycle. With accurate, consolidated information about mature products available, low-cost ways can be found to extend their revenue-generating lifetimes.

For a company, there's nothing comparable to PLM. There's no alternative to PLM.

1.4.2 The Complex Environment of Products

One of the reasons that PLM emerged in the early years of the 21st Century is that the environment in which products were managed became increasingly complex. And to make matters worse, the environment underwent frequent changes.

Not only do companies find themselves in a complex environment, of many dimensions, that is affected by many different changes, but the changes are often intertwined. As a result, the product environment is becoming increasingly complex with many interwoven components and numerous interdependencies being affected by many overlapping changes. The resulting environment is so complex that it's often difficult to see what the changes are, and what is really driving them, or to understand how they will affect a particular company and a particular product. PLM enables a company to respond to the changes.

The changes in the complex environment of products are of different types and come from different sources. Macroeconomic and geopolitical changes can affect companies developing, producing and supporting products (Fig. 1.18).

Other changes affecting product companies have environmental sources such as accidents, waste, pollution, regulations and laws (Fig. 1.19).

globalisation	geopolitical developments	fluctuating commodity prices and exchange rates
ageing population	rise of service industries	shareholder demands to increase value
deregulation	influence of financial markets	multi-cultural, multi-lingual environments

Fig. 1.18 Some macro-economic changes facing product companies

Another type of change facing product companies is linked to changes in the overall corporate environment (Fig. 1.20).

Another type of change facing product companies relates to new technologies (Fig. 1.21).

Figure 1.22 shows some of the changes that are very closely related to the product itself.

Among the changes, some create opportunities, some create problems, some lead to the need to change, some are the source of more changes. The resulting changes can drive other changes. Unexpected events resulting from changes can be a source of further change. All these changes snowball, making it difficult to know how to respond.

In addition, changes have associated risks. Changes in one area may lead to enhanced risks in another area.

If the risks were only related to one component or change in the environment, it might be easy to manage them. Unfortunately though, they are often related to many changes, making their management difficult.

The complex, risky, continually changing, uncertain, highly competitive product environment makes life difficult for companies that develop, produce and support

sustainable development	recycling directives	regulatory requirements
accidents affecting the environment	product traceability	Pollution
nuclear waste	resource depletion	greenhouse gases
waste disposal	conservation	environmental laws

Fig. 1.19 Changes related to the environment

outsourcing, offshoring	improved supply chain	corporate restructuring
multi-site activities	Knowledge Management	retirement of knowledge workers
Partnerships	large volumes of data	product lifecycle focus
Change Management	multiple versions of processes	increased complexity of business
process reengineering	high cost of training new employees	changing business models

Fig. 1.20 Changes in the corporate environment

IS evolution and complexity	mobile communications	social networks
toolbox information systems	World Wide Web and Internet	Nanotechnology
Open Source software	Internet of Things	cloud computing
Big Data	smart machines	Analytics

Fig. 1.21 Technological changes

increasing product/solution/service complexity	products with very long lives	global products
parallel world of software development	new customer requirements	customisation
working effectively with partners	fast changing products	mechatronics

Fig. 1.22 Product-related changes

Imagine	Define	Realise	Support/Use	Retire/Recycle
Ideas pirated	Projects late/ failing	Pollution costs	Upgrades ignored	Incorrect identification
Lack of ideas	Costs too high	Poor factory layout	Missing applications	Poor documentation
Uncontrollable	Uncontrolled changes	Scrap	Poor communication	Low recycle rate
Suppression of ideas	Unclear processes	Rework	Data out of control	Materials wasted
Missing applications	Needs not clear	Costly prototypes	Culture of risk	High disposal costs
Culture of sterility	Design faults	Supplier problems	Customers lost	Fines
Failure punished	Application Islands	High material costs	Liability costs	No training
Bureaucracy	Long time to market	Excess inventory	Missing services	Lack of control
Priority, #1 CYA	Data silos	Limited part re-use	High service costs	Missing applications
Unknown cost	IP lost/missing	Slow ramp-up	Processes unclear	Processes undefined
No training	Project status vague	Safety problems	Product recalls	Lack of procedures
No process defined	Standards ignored	Wrong data versions	Product failures	Costly disassembly

Fig. 1.23 Problems across the product lifecycle

products. In such an environment, they need to have great products that leave competitors far behind. They need a great product deployment capability. They need to be continually in control of their products. If they aren't, and for one reason or another, they take their eye off the ball, unpleasant consequences can occur (Fig. 1.23).

1.4.3   Horror Story

Without PLM, problems can show up in many ways. One company summarised the resulting situation as "a nightmare" . In another, it was called "a horror story".

The problems are usually related to one or more of the resources in the pre-PLM product environment. Figure 1.24 shows some typical organisational issues.

Product data is an important resource. Many issues can arise with product data in the pre-PLM product environment. Some examples are shown in Fig. 1.25.

Figure 1.26 shows some typical issues with products in the pre-PLM product environment.

communication silos	not enough focus on products	insufficient training
misalignment of expectations	wasted development resources	lack of up-front planning
poor co-ordination with suppliers	high service costs	poor scheduling of projects
Sales/Engineering disconnects	projects coming in late	product development costs rising
departmental mentality	cycle times lengthening	service costs rising, performance dropping

Fig. 1.24 Typical organisational issues

redundant part numbers	discrepancies in product specifications used by R&D and Sales
multiple names for the same project	inconsistencies between data in Engineering and Production
data silos, a department's data unavailable to others	no history of maintenance tests on a particular product
unsure of usage of material across products	spreadsheets with conflicting information about a product
product labelling not corresponding to the product	conflicting lists of product configuration at a customer site

Fig. 1.25 Typical issues with product data

rework	using obsolete components in a new design
poor product quality	technical problems with products in the field
quality problems	optimising product performance, but fragmenting the supply chain
errors in product definition records	optimising product layout at the expense of longer delivery cycles
new products not performing as planned	interruptions and delays as new technologies become available

Fig. 1.26 Typical issues with products

slow engineering changes	increasing rework and engineering changes
inadequate customer service	not enough re-use of existing parts, reinvention of the wheel
product release delayed	products meeting specifications but failing to meet customer requirements
bureaucratic business processes	product labelling not corresponding to labelling regulations
uncoordinated changes to product data	increasing rework and engineering changes

Fig. 1.27 Typical issues with processes

Processes are an important resource. Figure 1.27 shows some typical issues with processes in the pre-PLM product environment.

1.4.4 Opportunities

PLM enables companies to take advantage of the many product-related opportunities available at the beginning of the 21st Century. Some of these opportunities are the result of new technologies. Others are due to social and environmental changes, or to macroeconomic forces such as globalisation.

Globalisation led to huge opportunities. Billions of people can now benefit from products to which they previously had no access. Companies can offer products to a global market of more than 7 billion customers and users. The resulting opportunities for sales and profits are enormous. So are the potential risks.

For most companies it's only recently that such opportunities have been available. In the 1990s, although many companies were international, or multi-national, only a few were able to offer a product throughout the world. Others were limited, for one reason or another, to smaller markets. As a result of the changes, the potential market for most companies is no longer a few hundred million customers for the product in a local regional market, but over 7 billion customers worldwide. Which means that, for many companies, the potential market is already more than 20 times larger than before. And the market is expected to grow to 8 billion by 2024, and 9 billion by 2040.

The number of opportunities opening up in the 21st Century seems boundless. Perhaps it was too risky to pursue them when the product development process was out of control, production runs in faraway countries had unexpected problems, and customers complained continually about product problems. But that was before PLM. Now PLM's here, allowing companies to develop and support tiptop services and products across the lifecycle. PLM can bring benefits across the lifecycle (Fig. 1.28).

Imagine	Define	Realise	Support/Use	Retire/Recycle
Best ideas selected	Projects on time	Reduced energy use	Fewer failures	Less waste
No ideas lost	Fast time to market	Trained workers	Better customer info.	Safer recycling
Fast innovation	Data under control	Efficient machine use	Add-on modules	Reduced pollution
Clear organisation	Clear processes	Less rework	More customers	Recycling costs cut
Support applications	IP under control	Green logistics	Happy customers	Lean processes
Supportive culture	Motivated people	Green production	Refurbishment	Re-usable materials
IP under control	Clear requirements	Optimal shop layout	Services revenues up	Re-usable parts
No bureaucracy	Customisation	Less inventory	More services	Environment-correct
Clear process	Clear decisions	More part re-use	Lower service costs	New applications
Breakthrough ideas	Reduced costs	Less scrap	In-service upgrades	Disassembly time cut
Imaginative people	#1 product family	Strategic suppliers	Liability costs cut	Fewer fines
More ideas	Standards adherence	Lower material costs	Warranty costs cut	Better compliance

Fig. 1.28 Benefits from PLM in each phase of the lifecycle

1.4.5   Benefits

PLM provides benefits throughout the product lifecycle. Examples include getting products to market faster, providing better support for their use, and managing the end of their life better.

With its focus on the product, companies are looking for PLM to provide benefits in the areas of financial performance, time reduction, quality improvement and business improvement (Fig. 1.29).

There are many ways in which PLM can increase revenues and reduce costs (Fig. 1.30).

Financial Performance	earlier market intro/increase revenue
	reduce development costs
	extend product life/increase revenue
	reduce recall costs
Time Reduction	reduce project time overrun
	reduce engineering change time
	reduce time to market
	reduce time to profitability
Quality Improvement	reduce manufacturing process defects
	reduce returns
	reduce customer complaints
	reduce scrap
Business Improvement	increase new product release rate
	increase the part reuse factor
	increase product traceability
	ensure 100% configuration conformity

Fig. 1.29 Targeted improvements with PLM

Sources of increased revenues	Sources of reduced costs
increased number of customers	reduced energy costs
increased range of products	reduced development costs
increased sales of new products	reduced material costs
increased sales of mature products	reduced liability costs
increased product prices	reduced prototyping costs
increased range of services	reduced rework costs
increased service prices	reduced documentation costs
increased service revenues	reduced warranty costs

Fig. 1.30 The benefits of PLM translate into increased revenues and reduced costs

The benefits of PLM are measurable and visible on the bottom line. Typical current targets for PLM are to increase product revenues by 30 % and decrease product maintenance costs by 50 %.

1.5   When PLM?

1.5.1   When Did PLM Emerge?

PLM emerged in about 2001. Before the 21st Century, a PLM approach wasn't sufficiently needed from a business point of view, and wasn't possible from the technical viewpoint. From a technical point of view, for example, database technology and Web technology weren't sufficiently advanced to enable PLM.

1.5.2   When Did Companies Get Started with PLM?

A few companies got started with PLM in 2001. Others started in the following years. Many still haven't started.

Before 2001, companies didn't have an approach that managed a product continuously and coherently throughout the lifecycle. Products were managed in one way in early stages of their life. Then in a different way during their development. Often the company didn't manage the product during its use, and partially or totally lost control of the product. Sometimes the company managed the product again when the product was due for disposal. Sometimes it didn't.

Before 2001, companies implicitly managed products across their lifecycles. But they didn't manage them, even conceptually, in an explicit, "joined-up", continuous way. Instead they managed separately department-by-department, for example, in Marketing, R&D, Manufacturing and Support. And other departments, such as IS and Quality, took product-related decisions separately.

Perhaps, before 2001, companies didn't manage the product as well as they could have done, but, of course, to some extent they managed it. Some managers made sure that products were sold, making money for shareholders, and enabling employees and suppliers to be paid. And in other parts of the organisation, other managers made sure that new products were developed and brought to market.

Many of the elements of what is now called PLM were put in place departmentally. Examples are Computer Aided Design (CAD), Product Portfolio Management (PPM), Product Data Management (PDM), Configuration Management (CM), Product Recall, Customer Complaint Management, Product Warranty Management and Engineering Change Management (ECM). To improve productivity, most companies started to address such product-related activities long ago. However they did this in a bit-by-bit way, with the result that the product was managed in different unconnected ways at different times in the lifecycle with different approaches, processes and applications. But that wasn't PLM. Use of the term PLM implies that the activity of managing products across the lifecycle is clearly-defined, well-documented, proactive, and carried out according to a particular design. It's carried out to meet specific objectives of increasing product revenues, reducing product-related costs, maximising the value of the product portfolio, and maximising the value of current and future products for both customers and shareholders.

1.5.3 When Do Companies Start a PLM Initiative?

A PLM Initiative (sometimes called a PLM Program) is an initiative to put in place, or improve, the capability to manage products across their lifecycle. Some companies involved in product development, manufacturing and/or support started a PLM Initiative between 2001 and the present. Most companies will have a PLM Initiative between now and 2025.

1.6 Where PLM?

1.6.1 Where Is PLM Used, in Which Industries?

PLM is focused on "the product", and is applicable across a wide range of industries that develop, produce and support products. Currently, PLM is being used in a wide range of industries (Fig. 1.31). It's used in discrete manufacturing, process manufacturing, distribution and service industries, as well as in research, education, military and other governmental organisations. There are many differences between these industries, and they have different needs and priorities. As a result, although PLM is used in many industries, it's implemented and used differently in different industries.

aerospace	apparel	automotive	beverage	chemical
consumer goods	construction equipment	defence	electrical engineering	electronics
financial services	food	furniture	life sciences	machine tool
machinery	medical equipment	mechanical engineering	petrochemical	pharmaceutical
plastics	plant engineering	rubber	shipbuilding	shoe
software	transportation	turbine	utility	watch

Fig. 1.31 Industries using PLM

1.6.2   Where Is PLM Used, in What Size of Company?

PLM is used in all sizes of companies ranging from large multinational corporations to small and medium enterprises. The particular PLM requirements of companies of different sizes may differ, but the fundamental requirements do not. In companies of all sizes, products have to be managed, product data has to be managed, product development and support processes have to be managed, and product data has to be exchanged with other organisations.

1.6.3   Where Is PLM Used, for What Type of Products?

The list of products for which PLM is used is long. It includes agricultural machinery, aircraft, beverages, cars, chemicals, computers, consumer electronics, electrical equipment, electricity, elevators, escalators, food, furniture, gas, insurance policies, machine tools, machines, medical equipment, medicines, mobile phones, mortgages, office equipment, offshore structures, pharmaceutical products, power plants, power transmission belts, processed food, refrigerators, rockets, ships, shoes, software, telecommunications equipment, telecommunications products, telephones, toys, trains, turbines, washing machines, watches, water and windows.

1.6.4   Where Is PLM Used, in What Type of Company?

PLM applies to all types of company that work with products. It applies to companies making many identical, or similar, products such as cars, machines and electronic equipment. It also applies to companies making one-of-a-kind products, and for companies such as "job shops" in which every product is customised to the customer's requirements. PLM is vital to a job shop because it provides control and visibility over each individual product. The configuration management features of PLM make sure all the information about the product is under control. And PLM keeps track of what was ordered and what was delivered, and what was done to the product after delivery to the customer.

1.6.5   Where Did Companies Start with PLM?

The starting place for PLM varied from one company to another. In some companies, one function was seen as a strategically important function, and able to act fairly independently. Initial PLM activities could be targeted on this function.

In some companies, the selection of the initial activities of PLM depended on the requirements of other corporate initiatives. In other companies, it resulted from particular needs, or particular problems.

1.6.6 Where Will Companies Continue with PLM?

The choice of starting point in the future is likely to depend on similar considerations as in the past. The main difference is likely to be that PLM will be increasingly seen as strategic. It will require an upfront activity to investigate the options, select the preferred option and develop a plan.

1.7 Who PLM?

The scope of PLM is broad. It addresses all products across the lifecycle. Not surprisingly, it has a widespread impact. It impacts everyone whose job in some way relates to the company's products and their performance. PLM maximises control, reduces risk, and provides an integrated view of what's happening with the company's products at all times. With such promise, it's of interest to all. However, the benefits of PLM are seen differently by people in different positions in the company.

1.7.1 Top Management Role

PLM is a top management issue. Top managers define the objectives and metrics of product performance, and the way the related resources are managed. Top management defines the PLM-related business objectives. The business objectives provide a clear business focus for PLM. There could be a need to reduce lead times significantly, to improve product quality, or to increase revenues. There could be specific issues that have to be avoided, or relationships with powerful customers that need to be improved. There may be the intention to suppress some product lines, or to develop new, or improved, products. There could be plans to change the way that clients and markets are addressed, or the way that work is carried out with development and support partners. Management may want to focus PLM on reducing product cost. Management may have specific targets in mind, for example to reduce the lead time in the engineering department by 50 %, and reduce recalls by 80 %.

At the highest level, the Chief Executive Officer (CEO) expects PLM to increase revenues and earnings by bringing better products to market faster, and extending the lives of mature products. CEOs look to PLM to maximise product value over the lifecycle, to maximise the value of the product portfolio, and to reduce risk. CEOs look to PLM to provide visibility and control over products, ensuring there are no unwanted surprises. With PLM, the CEO knows the value of the Product Portfolio and can see how it will evolve. The CEO can make decisions based on reality, rather than guesses. With PLM, it's easier for the CEO to take account of potential risks. With PLM, the organisational structure becomes clearer. The CEO can assign overall responsibility for all the products, which will be visible and under control, to one person, the Chief Product Officer (CPO).

PLM gives the CPO the ability to know the exact status of every part, and the exact status and structure of every product. The CPO can take control, both during product development and at later stages in the product lifecycle. The risks associated with products are better understood and managed. With uncertainties reduced, more reliable financial projections can be made for shareholders. The value of the Product Portfolio can be monitored, and scenarios built to understand the effect of a variety of possible circumstances such as reduced lifetimes, increased competition and acquisitions throughout the world. PLM helps the CPO get an integrated view over all product development projects. The CPO can develop and implement strategies for faster development and introduction of new products, and for better support of products across their lifecycles. PLM puts the product once again at the heart of business strategy.

With PLM, the Chief Financial Officer (CFO) can see the real financial figures for a product across its lifecycle, so knows precisely how much it cost and how much it has earned. Costs can be reduced as activities that don't add enough value become more visible. Better estimates can be made for the financial figures related to developing, supporting and retiring each product.

With PLM, the Chief Information Officer (CIO) can align IS applications to help bring competitive products to market faster and to support better their use. PLM provides the CIO the opportunity to carry out a wide range of necessary activities to clean up the company's product-related applications, processes and data. This will solve many everyday problems that currently occupy the time of IS professionals, enabling a reduction in IS costs and a redirection of IS effort to activities that add more value.

1.7.2 Everybody in the Company with a Product-Related Activity

PLM gives Product Managers an integrated view over their products and their product development projects. The exact financial and technical status of every part, product and project is known. With better information available, it's easier for them to take account of potential product and project risks. Product Managers can make decisions based on reality. Unwanted product-related costs, resulting from rework and scrap, warranty and liability claims, and returns and recalls will be reduced, and then eliminated. PLM enables secure collaboration with partners wherever they are located.

PLM helps Marketing Managers make better and faster responses to Requests For Proposal, and re-use material from successful proposals. It helps them price proposals realistically and competitively. It helps get better feedback about product use. With all the product data and customer specification information available, it will be much easier for Marketing Managers to evaluate future opportunities.

PLM helps Manufacturing operate more effectively. Manufacturing personnel can be involved increasingly in product development activities, advising against difficult to manufacture designs, reducing scrap and rework.

PLM helps Engineering Managers develop new products faster, and increase the success rate of new product introduction. It will help them reduce product costs during the development phase, when most of the product costs are defined. It will help them make sure existing designs are reused, or slightly modified, rather than creating new designs from scratch. It will help them control engineering changes and maintain exact product configurations.

PLM helps Product Support Managers know the configuration of each product that they're required to service.

1.7.3   The PLM Initiative Team

PLM is so important to the future of a company that PLM activities and, in particular, a PLM Initiative should be led, or sponsored, by the CEO or the CPO. Due to the enterprise-wide scope of PLM, it's at the C-level that action has to be taken if the expectations of PLM are to become reality.

The PLM Initiative Team will be a cross-functional team with members from the various departments within the company's scope of PLM. Some team members will come from departments such as Marketing, R&D, Production and Support. A Product Manager is often included in the team. And usually there are team members with cross-functional views of the product lifecycle. Examples are a project planner, someone from IS, someone from the organisation that manages product data and documents, and someone from the quality or process organisation.

Team members should be knowledgeable about the company and its products, respected, and have time to participate in the project.

The PLM Initiative Team Leader should be selected with care by top management. The Initiative's activities will be cross-functional, and involve working at many levels in the company. The PLM Initiative Team Leader needs to be a good manager, knowledgeable, diplomatic and respected. Many people in the company will find it difficult to address such a wide scope. Usually people are focused on a particular area, and want to improve what they understand. IS specialists want to implement IS solutions; process experts want to implement process solutions. However, in a PLM project, it's often necessary to make interrelated improvements simultaneously in different parts of the company.

1.8 How PLM?

The new paradigm of PLM leads to changes in organisational structures, processes and applications. It requires a different approach to the definition and structure of products and product data. The changes are important. The effort to implement them is considerable. Most companies implement them through a formally-defined PLM Project, PLM Program or PLM Initiative.

A PLM Project, or PLM Initiative, is a project to put in place, or improve, the capability to manage products across their lifecycle. Most companies involved in product development, manufacturing and/or support will have a PLM Project between 2015 and 2025. This could be for one of many reasons. One of these, looking backwards, is that there were so many changes between 1990 and 2015, that it's time to overhaul the company's operations to be sure they correspond to the new environment. Another one, looking forwards, is that the company wants to position itself better for the future. To take advantage of the opportunities of globalisation, it may need to improve its product deployment capability and its capability to deploy products globally. Another reason could be a recognition that the company has been informally managing products across the lifecycle, but it's now time to do it formally.

Whatever the reason for the project, one of the objectives of the PLM Project, or PLM Initiative, will be to improve PLM performance. However, the scope of PLM is wide. It addresses many subjects such as products, data, applications, processes, organisational issues, people, work methods, and equipment. It addresses many functions such as Marketing, Development, Production and Support. It's to be expected that implementation of PLM will take a long time.

PLM isn't the responsibility of just one function or department. PLM can't be the responsibility of an Engineering department, because Engineering isn't responsible for the product all the way across the lifecycle, for example, when it's in the field. Similarly, PLM can't be the responsibility of the Service department, because Service isn't responsible for the product all the way across the lifecycle, for example, when it's under development. PLM can't be the responsibility of the Information Systems Department, because IS isn't responsible for a company's products at any time. PLM can't be the responsibility of the Finance Department, because Finance isn't responsible for a company's products at any time.

It can be difficult for middle-level managers to start a PLM Project. They are aware, from their everyday activities, of the need for PLM. However, usually they don't have the required authority or responsibility. In addition, they're often overloaded with other activities and projects that have higher priority and are already running. The result is that they are able to make little or no headway with

PLM. As a result decisions about next steps are delayed, and PLM progress is slowed. Product developers and managers are frustrated. The company falls behind competitors.

It can also be difficult for business executives to get started with PLM. In a highly competitive global environment, many executives are overloaded with responsibility and work. Often they've been given additional responsibilities extending beyond their usual areas. For example, they may be responsible for integrating newly acquired companies or for overseeing operations in India, Russia, South Africa, Brazil or China. With little time available, they may not want to get involved with a subject such as PLM that can seem unclear in scope, duration and potential benefit.

Due to the unclear responsibility and enterprise-wide scope of PLM, it can be difficult to get a PLM Initiative started. Most companies aren't organised for such a project. They don't have organisational structures that operate across such a wide area.

In view of the above characteristics of PLM, how should a company proceed once it has decided to implement PLM?

Companies often face a dilemma over PLM. On one hand, it's clear that PLM makes sense and that it's necessary. On the other hand, it's not clear what to do about it, how to do it, or who should take action.

However, it's clear that, at some stage, the person who will have to decide and take action is a top-level business executive with the authority and responsibility to address a subject that's enterprise-wide and addresses so many resources.

1.8.1   The PLM Initiative

PLM Initiatives will be different in different companies because companies will be in very different situations when they start the Initiative (Fig. 1.32).

The PLM Initiative of a particular company will depend on a range of factors such as its existing PLM status, its financial health, its competitive environment, and the management skills it has available.

As a result, all PLM Initiatives are different. There are thousands of different PLM Initiatives in thousands of companies. However, although all PLM Initiatives are different, it's possible to learn some lessons from other companies' experience with them.

Full achievement of PLM will take a lot of effort and a long time. A PLM Initiative will run for several years. As a single, huge, multi-year PLM project is

companies have different products, such as aircraft and chocolates	companies have different supply chain positions
companies can have very different business objectives	companies are at different PLM maturity levels
companies have different levels of resources available	companies have different PLM awareness levels

Fig. 1.32 Different situations when starting a PLM initiative

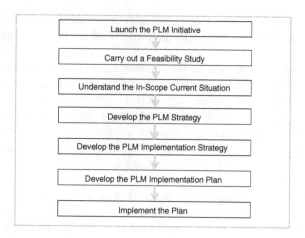

Fig. 1.33 Main steps in the PLM initiative

likely to end in disaster, PLM Initiatives should be made up of many smaller, shorter, more focused projects of various sizes.

Figure 1.33 shows the main steps in the PLM Initiative.

1.8.2 Different Approach, Different Result

Different approaches can be taken to improving performance. Figure 1.34 shows results from surveys aimed at identifying types and results of different approaches to improving performance. It can be seen that the results of different approaches are very different. For example, cherry-picking projects give results fast, but the results are limited in their effect.

Similarly, different approaches in a PLM Initiative will give different results. The PLM Initiative of a particular company may fall anywhere between "supremely strategic" and "totally tactical" (Fig. 1.35). The corresponding benefits can be expected to be anywhere between "high" and "low".

	Time period	Productivity change	Development cycle change	Product cost change
Company approach to improving performance				
Uncoordinated cherry-picking and lemon-squeezing	6 months	+4%	-3%	-3%
A short-term plan	1 year	+12%	-10%	-9%
A three-year Strategy and Plan	3 years	+40%	-39%	-28%
Integrated Vision, Strategy and Plan	5 years	+100%	-80%	-41%

Fig. 1.34 Different approaches, different results

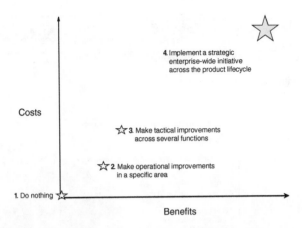

Fig. 1.35 Different approaches have different costs and different benefits

When starting a PLM Initiative, it's important to examine the different approaches. This will make it clear to everybody concerned what the PLM Initiative is likely to cost, and what it's expected to achieve. It's important to make sure people don't expect huge strategic benefits from a tactical approach and a small investment. If they do, they're likely to be very disappointed.

1.8.3 The Feasibility Study

Usually the first steps in a PLM Initiative are to understand what it means for the company, what it's about, its objectives, its scope, and the steps towards successful PLM. Often, a feasibility study will be carried out to find out which type of approach, and which level of response, are appropriate (Fig. 1.36).

There are similar Feasibility Study activities when investigating each approach (Fig. 1.37).

The results of the study should be documented in a Feasibility Study report. The typical contents of the Feasibility Study are shown in Fig. 1.38.

The next steps after the Feasibility Study will depend very much on the conclusions of the study. At one extreme, if an enterprise-wide initiative is proposed, then the next steps could be to develop and communicate a PLM Vision and a PLM Strategy for the future environment. These will help everyone understand where

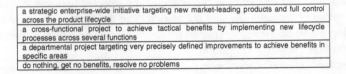

Fig. 1.36 Projects with different approaches

| document the objectives and the scope of the option |
| identify the benefits of achieving the objectives, and estimate their financial value |
| identify the activities and effort required to achieve the objectives, and estimate their cost |
| create the business case |
| create an outline plan for implementation of the activities identified |

Fig. 1.37 Activities for each approach

```
Feasibility Study : Options for our PLM Activity

Table of Contents

1. Executive Summary and Recommendations

2 Introduction
2.1 Background to the Study
2.2 Approach for the Study

3 Current Situation
3.1 Business Objectives
3.2 Current PLM Activity

4 Description of the Options
4.1 Approach A   Do Nothing
4.2 Approach B   Departmental improvements
4.3 Approach C   Cross-functional approach
4.4 Approach D   Strategic enterprise-wide initiative

5 SWOT Analysis

6 Conclusions and Proposed Next Steps

Appendix   Detailed Information
```

Fig. 1.38 The report from the feasibility study

they are going. The following step could be to define an Implementation Strategy to achieve the Vision. Then a plan could be developed to implement the strategy. Once the plan has been implemented, the benefits can be harvested.

However, if the Feasibility Study doesn't show the need for an enterprise-wide initiative, then the Vision, and even the Strategy, may not be needed. In all cases though, a plan will be needed to show what has to happen, when it should happen, and who does what to make it happen.

1.8.4   Understanding the In-Scope Current Situation

The Feasibility Study should lead to the selection of one of the approaches investigated.

The next step is usually to get a very good understanding of the current situation of the activities and resources in the scope of the selected approach. A very good understanding of the activities and the resources in the product lifecycle is needed to develop the PLM Strategy. This understanding must be based on factual information, not on guesses and opinions.

Understanding and documenting the current situation is a 13-step activity (Fig. 1.39).

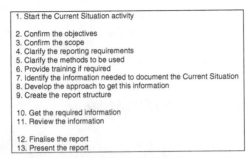

Fig. 1.39 13 steps to understand and document the current situation

Two documents should result from the activity of describing the current situation, the Current Situation Report and a PowerPoint Presentation addressing the current situation.

One of the dangers when describing the current situation is that important information can be drowned by the huge amount of data that's collected. As a result, it's useful to define the shape of the report before starting to collect data. The typical content of the report is shown in Fig. 1.40. The main findings about the current situation should be documented in a PowerPoint presentation that the Team Leader can present to top management and others.

Many methods and techniques are used to describe the current situation (Fig. 1.41).

The scope of the approach selected in the Feasibility Study defines the areas of the PLM Grid for which the current situation must be understood.

Once the information for the description of the current situation has been obtained and documented, it should be reviewed and cross-checked. In particular, it

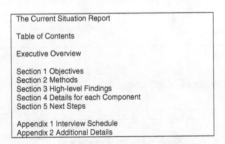

Fig. 1.40 Example of the content of the report on the current situation

creating and sending out questionnaires	holding meetings of study groups	mapping
reviewing responses to questionnaires	reviewing documentation	modelling
documenting real-life examples	carrying out interviews	walking the processes

Fig. 1.41 Methods and techniques to describe the current situation

should be shown to the people who provided it, to confirm that it really does correspond to the current situation. If it doesn't, corrections should be made and missing details added.

1.8.5 Developing the Future PLM Strategy

The future PLM Strategy describes how PLM resources will be organised, managed and used to achieve the objectives. As Fig. 1.42 shows, there is a PLM Strategy showing how PLM resources are currently used. There is also a PLM Strategy showing how they will be used in the future.

The future PLM Strategy isn't an independent stand-alone entity. It has to fit with the company's overall vision of its future, its mission and its objectives. Upstream of the PLM Strategy are the company's mission, objectives, vision, strategies and plans. Downstream, once the future PLM Strategy has been agreed, a suitable PLM Implementation Strategy has to be developed to achieve it.

As Fig. 1.43 shows, developing the future PLM Strategy is a 9-step activity.

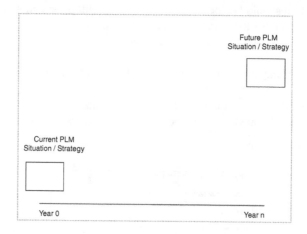

Fig. 1.42 Current and future PLM strategies

1. Start the activity to develop the PLM Strategy

2. Clarify the reporting requirements
3. Create the report structure
4. Gather information about the Future Situation
5. Develop Candidate Strategies (scenarios)
6. Select the Preferred Strategy

7. Finalise the report
8. Present the report

9. Communicate the Strategy

Fig. 1.43 9 steps to develop and communicate the future PLM strategy

A very good understanding of the expected future activities and resources in the product lifecycle is needed to develop the PLM Strategy. However, the description of the future situation will be much more concise than that of the current situation. Since the future situation doesn't exist, it's difficult to describe it in a lot of detail. However, the intent is only to describe the expected future situation at a high level. Main characteristics and desired performance levels can be described, and requirements can be prioritised.

The approaches, or methods, used to describe the future situation will depend on the scope and objectives. Probably some will focus on the company itself. And probably some will look outside the company.

Among the activities focusing on the company itself could be a review of existing project information, an analysis of the current situation, and identification of improvement suggestions from people working in the product lifecycle. Among the activities looking outside the company could be benchmarking, visits to other companies, reviewing maturity models, reading technical literature, and attending conferences.

It's useful to define the shape of the PLM Strategy Report before starting to develop the Strategy. Team members will then be aware of what they have to achieve.

In the next step of strategy development, several potential strategies are developed, formulated and described in terms of the organisation and policies to be applied to the resources. It's always useful to identify and describe several possible strategies. This will improve the chances of finding the best strategy since the most obvious strategies aren't necessarily the most appropriate.

In the third step of strategy development, potential strategies are tested, and the most appropriate strategy is selected and detailed. It will be useful to investigate three or four alternative strategies. This should lead to an in-depth understanding of the possible strategies. The strengths and weaknesses of a particular strategy often become clear when examining the strengths and weaknesses of other strategies.

In the following step of the strategy development process, the chosen strategy is communicated to the people who will be affected by it, or involved in its implementation. Communication of the strategy is essential. A strategy is useless unless the people who are going to be involved are fully aware of it, can understand it and can implement it.

1.8.6 Developing the PLM Implementation Strategy and Plan

The PLM Implementation Strategy shows how resources and activities will be organised to achieve the future PLM Strategy. In Fig. 1.44, three of the many potential PLM Implementation Strategies are outlined (Path 1, Path 2, Path 3). Path 2 shows an implementation strategy of the "Big Bang" type, with everything

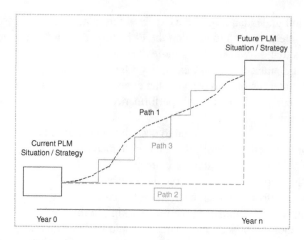

Fig. 1.44 PLM implementation strategy

changing in Year n. Path 1 shows an implementation strategy of the "Continuous Improvement" type, with many changes being made one after the other. Path 3 shows a phased approach with a small set of changes being introduced in each phase.

As Fig. 1.45 shows, developing the PLM Implementation Strategy and Plan is a 10-step activity.

It's useful to define the shape of the report before starting to develop the Strategy and Plan. Team members will then be aware of what they have to achieve. The typical content of the report is shown in Fig. 1.46.

The PLM Implementation Strategy defines the activities that have to be carried out to get from the current use of PLM resources in a company to the future use of PLM resources. To be able to develop the Implementation Strategy, it's necessary to have a basic understanding of both the current situation and the future situation in the company.

A starting point for developing the Implementation Strategy is to understand the gap between the current situation and the future situation. That's why it will be helpful if the future situation was described using the same structure as that used for

```
1. Start the activity to develop the PLM Implementation Strategy and Plan

2. Clarify the reporting requirements
3. Create the report structure
4. Gather information about the Current Situation and the Future Situation
5. Understand the factors that may influence timing and priorities

6. Develop Candidate PLM Implementation Strategies (scenarios)
7. Select the Preferred PLM Implementation Strategy

8. Detail the PLM Implementation Plan

9. Finalise the report
10. Present the report
```

Fig. 1.45 10 steps to develop the PLM implementation strategy and plan

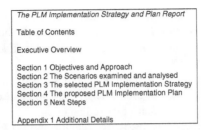

Fig. 1.46 Typical content of the report about the PLM implementation strategy and plan

the documentation of the current situation. That will make it relatively easy to compare what is needed with what exists. For example, the description of the future situation may call for a single PDM system, but there may be multiple PDM systems in the current situation. There may not be an obsolescence process in the current situation, but an obsolescence process may be required in the future situation.

The gaps should be listed and described in a Gap Description Matrix (Fig. 1.47).

With the gaps between the current and future situations identified, it's time to look for ways to close them. Several ways should be proposed to eliminate each gap. They should be described in a Gap Elimination Matrix (Fig. 1.48).

Once the Initiative team has completed the above steps, it should be in a position to identify possible candidates (or scenarios) for the PLM Implementation Strategy.

The PLM Implementation Strategy will show how to get from the current use of PLM resources to the future use of PLM resources. There are many ways to do this, and the likelihood of finding the most appropriate at the first attempt is low.

Resource	Gap Name	Current Situation	Future Situation
Processes			
	Product Idea	no process	single company-wide process
	Obsolescence	no process	single company-wide process
Data			
	Product Numbering	multiple numbering systems	single company-wide system
Applications			
	PDM System	3 PDM systems used	single company-wide system

Fig. 1.47 Excerpt from a gap description matrix

Gap Name	Potential Gap Elimination Approaches
Product Idea	develop the process based on currently perceived needs
	develop the process with the help of Business Process Management consultants
	use the process proposed by the vendor of whichever application will be used
	use the best of the current in-house approaches
	purchase a new best-in-class application
	add customised functionality to the current PDM system
	add customised functionality to the current ERP application

Fig. 1.48 Excerpt from a gap elimination matrix

Several potential scenarios should be identified and documented. Each scenario will show a different way to reach the future situation. It should also show related ways to reach the future situation for each component. Each scenario should be described in detail.

After the scenarios have been identified and described, they should be analysed. The strengths and weaknesses of each scenario should be described. This helps get an in-depth understanding of each proposed solution.

Analysis of the scenarios leads to identification of the preferred PLM Implementation Strategy. This should be documented in detail and described in the report.

1.8.7 Documenting the PLM Implementation Plan

The PLM Implementation Plan should show how the strategy will be achieved over the length of the total implementation (for example, 5 years).

There should also be a more detailed Plan for the first year. Different views of the Plan will be needed, with different levels of detail. The first view could be a block diagram showing in which years each Phase will take place (Fig. 1.49).

Other views of the plan will show more details of the activities (Fig. 1.50). They'll be needed for people who participate in, and manage, the activities.

The Implementation Plan should address the long term and the short term. For the long term, it provides management with the information necessary to understand

Phase Activity	Y1	Y2	Y3	Y4	Y5
Prepare Phase 1					
Execute Phase 1 activities					
Prepare Phase 2					
Execute Phase 2 activities					
Prepare Phase 3					
Execute Phase 3 activities					
Prepare Phase 4					
Execute Phase 4 activities					
Prepare Phase 5					
Execute Phase 5 activities					

Fig. 1.49 Timing of phases

Activity	M1	M2	M3	M4	M5	M6
Detail the plan for Phase 1 activities						
Manage the Phase 1 activities						
Carry out activities related to product structure						
Carry out activities related to processes						
Carry out activities related to product data						
Carry out activities related to PDM						
Carry out Portfolio Management activities						
Finalise deliverables. Prepare report						
Report Phase 1 activities						

Fig. 1.50 Lower-level, more-detailed implementation plan

activities, resources and timelines. The more specific the plan is, the better. It should define an overall implementation timetable. It should show how the PLM implementation will be split into manageable phases. The plan may be a phased plan, for example with Phase 1, Phase 2 and Phase 3. Or the intention may be to have a PLM Program, with several projects. Or the intention may be to have a PLM Project, with many sub-projects.

The short-term plan should show management which actions need to be taken initially. The plan is more likely to be accepted if it includes some actions that will lead to short-term savings and other short-term benefits.

Chapter 2
Product Lifecycle Management

2.1 Product Lifecycle Management (PLM)

PLM is the business activity of managing, in the most effective way, a company's products all the way across their lifecycles; from the very first idea for a product all the way through until it is retired and disposed of.

PLM manages both individual products and the Product Portfolio, the collection of all of a company's products.
PLM manages products from the beginning of their life, including development, through growth and maturity, to the end of life.

The objective of PLM is to increase product revenues, reduce product-related costs, maximise the value of the product portfolio, and maximise the value of current and future products for both customers and shareholders.

2.2 Managing the Product Isn't Easy

There are five phases in a product's lifecycle (Fig. 1.4). In each of the five phases, the product is in a different state. During the imagination phase, the product is just an idea in people's heads. During the definition phase, the ideas are being converted into a detailed description. By the end of the realisation phase, the product exists in its final form (for example, as a car) in which it can be used by a customer. During the use/support phase, the product is with the customer who is using it. Eventually the product gets to a phase in which it's no longer useful. It's retired by the company, and disposed of by the customer.

The product must be managed in all these phases to make sure that everything works well, and that the product makes good money for the company. That means managing the product throughout its lifecycle, "from cradle to grave".

© Springer International Publishing Switzerland 2016
J. Stark, *Product Lifecycle Management (Volume 2)*,
Decision Engineering, DOI 10.1007/978-3-319-24436-5_2

Managing the product across its lifecycle isn't easy. During the development of a product, it doesn't physically exist. Not surprisingly, during that phase of life it's difficult to control. Once a product does exist, it should be used at a customer location, where again, it's difficult for a company to keep control of it.

Within a company, the responsibility for the product is often different at different phases of the lifecycle. At one time it may be with Marketing, at other times with Engineering or Service. Maintaining a common coherent approach among these organisations, which may have different objectives, working methods and applications, can be difficult and time-consuming.

It becomes even more challenging in the Extended Enterprise environment. The issues are then no longer just cross-functional but also cross-enterprise. And it becomes even more challenging when a company works in different Extended Enterprises for different products. At different times the responsibility for the product may then be with different Marketing, Engineering, Manufacturing, Product Management, Finance, Marketing, Sales and Service groups in different companies. They may be on several continents, in different time zones and speaking different languages.

2.3 Loss of Control

In such an environment, it's easy for companies that develop, produce and support products to lose control over a product. But, if a company loses control, the consequences can be serious. If it loses control during product development, the product may be late to market and exceed the targeted cost. The results of losing control during use of the product may be frustration and a lack of satisfaction for the customer, or much worse, injury and death. For the company, the results may be damage to the company's image and loss of customers concerned about product problems. They could also include loss of revenues to companies that bring products to market faster, and reduced profit due to costs of recalls and legal liabilities resulting from product use.

An example of a product that was late to market is the Airbus A380. Delivery of the first A380 was originally planned for the last quarter of 2005. It was eventually delivered in the second half of 2007, two years late. The cost of late delivery was estimated to be $6 billion. Another example is the Airbus A400M program, which was launched in 2003. The development cost was estimated initially at about €20 billion, with first delivery planned for 2009. By 2009, the potential development cost had risen to about €30 billion. The first delivery was made in 2013.

The problem with the A380 occurred well into the development project. However, problems with products can occur even earlier in their lives, for example during their specification. At the time of the commercial launch of the Airbus A350 in December 2004, it was expected to enter service in 2010. The initial specification was based on an extension to an existing aircraft. That implied rapid availability and a relatively low development cost. However, in view of limited interest from potential customers, an aircraft with a new design, the A350 XWB (Extra Wide Body), was proposed in 2006. Entry into service was announced for 2013, three years later than previously expected. The first commercial flight took place in January 2015.

Problems can also occur during product manufacture. For example, in 2006, computer makers such as Apple Computer, Dell, Hitachi, Lenovo and Toshiba announced the replacement of Sony-made lithium-ion batteries that could overheat in certain circumstances and pose a safety risk.

In January 2013, after problems with lithium-ion batteries on JAL and ANA Boeing 787s, the Federal Aviation Administration (FAA) ordered all 787s grounded. The order was lifted in April 2013 after battery and containment systems had been redesigned.

Problems can also occur during product use. On 25 July 2000, the crew of an Air France Concorde noticed a loss of power and a fire under the left wing soon after take-off from Paris. The aircraft went out of control and crashed onto a hotel. Two years earlier, on 2 September 1998, not long after take-off from New York, the flight crew of Swissair Flight 111, an MD-11, noticed an abnormal odour in the cockpit. Their attention was drawn to an area behind and above them, but whatever it was apparently then disappeared. They decided it was smoke and decided to land, unaware of a fire above the ceiling in the front area of the aircraft. The fire spread, degrading aircraft systems and the cockpit environment. The aircraft crashed into the Atlantic Ocean near Halifax, Nova Scotia.

Other problems with planes include disappearance (MH 370 in 2014) and deliberate crash (Germanwings Flight 9525 in 2015).

Problems with products can involve big numbers. In October 2003, Nissan Motor Company said it would recall 2.55 million cars at an estimated cost of 15–16 billion yen ($138–148 million) due to an engine defect. In a few months in late 2009 and early 2010, Toyota announced recalls of more than eight million cars due to concerns over accelerator pedals and floor mats. The cost was estimated at $2 bn. In January 2010, Honda announced the recall of more than 600,000 cars to fix a switch defect that could lead, in some cases, to a fire. In June 2010, GM recalled over a million vehicles due to thermal incidents with heated washer fluid systems. GM listed 84 recalls affecting 30,433,365 vehicles on its "GM 2014 year-to-date North American recalls including exports" web page. In April 2010, an explosion on the Deepwater Horizon drilling rig led to the death of 11 people. The blowout preventer failed to activate correctly. For months, tens of thousands of barrels of oil spilled daily into the Gulf of Mexico, totalling perhaps a hundred million gallons.

Problems can also occur at product end-of-life. For example, the French Ministry of Defence had problems in 2005 and 2006 with Q790, previously known as the aircraft carrier Clemenceau. With hundreds of tons of asbestos on board, dismantling the hull for scrap was never going to be easy. A failed attempt to dismantle Q790 in Turkey was followed by a decision to dismantle it in India. Q790 left Toulon in France at the end of 2005 to be broken up at Alang in India. After being refused entry to India, it was towed 10,000 miles back to France.

Merck voluntarily withdrew VIOXX, an arthritis and acute pain medication, in September 2004 because a trial had shown an increased relative risk for cardio-vascular events. There were millions of users worldwide. VIOXX had been launched in 1999 and marketed in more than 80 countries.

If products don't meet the rules and regulations laid down by government and international authorities, there can also be problems. In 2001, authorities in the Netherlands found that some peripherals for a game console contained cadmium levels above the Dutch limits. Sony Corp. temporarily halted shipment. The estimated impact on sales was about 100 million euros.

Counterfeiting can be another result of loss of control. Companies making products as different as software, clothing, DVDs and pharmaceuticals suffer from product counterfeiting and product pirating. A 2009 report from the Organisation for Economic Co-operation and Development indicated that international trade in counterfeit and pirated products could have been up to $250 billion in 2007.

Another type of product-related problem was highlighted in 2006 when it was announced that the FBI had thwarted an attempt to steal and sell Coca-Cola's trade secrets, apparently including information about a new product.

Problems are not limited to high profile products and companies. Each month the website of the U.S. Consumer Product Safety Commission lists about 30 recalls of products such as drinking glasses that can break during use, cameras that can overheat, stools that can become unstable, lawn sprinklers that can crack, candle packaging that can ignite, and sweatshirt hood drawstrings that pose a strangulation hazard to children. Other products recalled include hair dryers that can pose an electrocution hazard, window blind cords that can pose a strangulation hazard for small children, and bicycle fenders that can break, posing a fall hazard to the rider.

Similarly, each month the U.S. Food and Drug Administration lists about 20 recalls, market withdrawals and safety alerts of products such as frozen strawberries, eye drops, herring in sourcream, teriyaki salmon jerky, atracurium besylate injection doses, blood glucose test strips, wet wipes and pharmaceutical drugs.

And, each month, the Office of Defects Investigation of the National Highway Traffic Safety Administration lists about 20 Vehicle Recall Reports addressing parts such as automatic transmissions, fuel tanks, wiper motors, airbags, brake hose assemblies, front passenger seat occupant detection mats, hoses, connectors, nuts and bolts.

2.4 Sources of Problems

Companies don't want to have such problems with their products. They can cost a lot of money. If a problem does occur, a company will do everything it can to understand the source, and to prevent the problem happening again.

In pre-emptive mode, it's also useful to identify and understand potential problems with a view to preventing them occurring. This isn't as easy as it may seem. There's a lot of information available about how to do things right. Much less

Problem Area	Issue(s)
Products	incorrectly, or unclearly, defined products
Product data	data out of control; data in silos; different definitions of data
Processes	processes not defined; unclear processes; conflicting processes
Applications	Islands of Automation; missing applications, ineffective interfaces
Projects	project status vague; unclear project objectives; too many projects
Equipment	machines and software licences under-utilised or not used
People	specific skills missing; lack of training
Organisation	working methods not defined; different methods used on different sites

Fig. 2.1 Some reasons for things going wrong with products

about how they are done wrong. Companies usually don't like to talk about their problems with products. However, sometimes the information becomes public. The Press may take an interest. Accident reports may be published. Technical journals publish case studies and other articles. And of course, people working with many companies, such as consultants, get to see the inside story. From these sources, conclusions can be drawn as to why things go wrong (Fig. 2.1).

2.5 Opportunities of Globalisation

The above issues can occur when a company operates in one country. However, globalisation has led many companies to operate in many countries. This has made it even more difficult to keep control of products. It's led to new ways to lose control of products.

Globalisation can have many effects on a company, even a small one. One positive effect is that, because of globalisation, it has the opportunity to sell its products and services worldwide. It has the opportunity to find many new customers and increase sales. Another effect of globalisation is that even small and medium-sized companies have competitors all over the world. And they may find that these competitors bring out similar products, but with better cost/performance than their own models. The result of the increased competition is that companies have to be more innovative, develop better products, develop them faster and develop them at lower cost.

Globalisation also implies that companies have to be close to customers in many places, and to understand customer requirements and sell products in many environments. However, the situation in different countries is different. Companies have to understand and take account of these differences. For example, they have to get pricing right in many different environments. They also have to provide technical information, parts, products and service in many locations. They must meet regulations in many countries. They have to coordinate the launch of new and modified products for the global marketplace.

The opportunities for sales and profits resulting from globalisation are enormous. As a result of the changes, the potential market for most companies is no longer a few hundred million customers for the product in a local regional market, but over 7 billion customers and users worldwide.

The opportunities are enormous, but so are the difficulties and potential risks. Many questions have to be answered. For which geographical markets should we offer our products? The whole world? One continent? Several continents? Just a few countries? If so, which ones? Should we introduce a new product everywhere in the world at the same time, or introduce it first in one market, then in the others? Do we understand these markets well enough? Should we have one product for customers throughout the world? Where will we develop our products? In a single location where we can bring our best people together and give them the best tools in the world?

Which business processes should we use? Which IS applications? Should we use the same processes and applications everywhere in the world? If not, what must be global, what can be local? Should we use a set of IS applications from just one vendor, and hope that will eliminate integration problems between applications in different application areas? Or should we use best-in-class applications in each area, even if they are from different vendors and do not integrate well? Where should we store the product data that defines our products? And how can we keep it safe from envious prying eyes?

Such questions, and their answers, are part of PLM, the management of a product across its lifecycle.

2.6   The Environment Before PLM

In the environment before the emergence of PLM, the paradigm for managing a product across the lifecycle was piecemeal. There was an Alphabet Soup of many activities and approaches, most known by a Three Letter Acronym (TLA) (Fig. 2.2). Each of these helped manage a product at a different moment in its life. Each had technical objectives, not business objectives. With this piecemeal approach, organisations didn't manage products in a joined-up way across the product lifecycle. For example, product development and product support were often carried out in different parts of the organisation even though they addressed the same products. Because they were addressed in different parts of the organisation, the activities were carried out by different groups of people with different managers. Each group created its own processes, defined its own data and

JIT	- Just In Time	CAD	- Computer Aided Design	ESI	- Early Supplier Involvement
VA	- Value Analysis	ABC	- Activity Based Costing	CAE	- Computer Aided Engineering
GT	- Group Technology	DFE	- Design For Environment	QFD	- Quality Function Deployment
LCD	- Life Cycle Design	NPD	- New Product Development	CAM	- Computer Aided Manufacturing
VE	- Value Engineering	DFM	- Design For Manufacturing	ERP	- Enterprise Resource Planning
DFA	- Design For Assembly	NPI	- New Product Introduction	BPR	- Business Process Reengineering
LCA	- Life Cycle Analysis	TQM	- Total Quality Management	EMI	- Early Manufacturing Involvement

Fig. 2.2 Alphabet soup for managing a product

document structures, and selected its own IT applications. Each group solved its own problems as best it could, adding an application here, a document there. Each group optimised its own activities, even though this might mean reducing overall effectiveness.

All these activities resulted, to some extent, in the company managing its products throughout the lifecycle. However, the way they managed them didn't result from a clear, deliberate, documented plan, but from the way the company organised other activities. The subject of how products were managed across the lifecycle hadn't been explicitly addressed by company management. It wasn't planned. It wasn't documented. In such a situation, often nobody in the company could describe in detail how the products were managed throughout the lifecycle. The resulting environment was one of all sorts of gaps, contradictory versions of the same data, information silos, islands of automation, overlapping networks, duplicate processes, redundant data functionality, ineffective fixes and product recalls. The end result was reduced revenues and higher costs.

2.7   PLM Paradigm

The PLM Paradigm emerged, as a way to avoid such problems, in the early 21st Century. It was driven by changes in the business environment that required better management of products. Improvements in technology made its emergence possible. The PLM Paradigm sees PLM as one major business activity with business objectives. It differs in many ways from the previous paradigm.

For example, PLM has a holistic approach (Fig. 1.3) to the management of a product. It addresses resources such as products, data, applications, processes, people, work methods and equipment. This holistic approach distinguishes it from the environment before PLM, in which activities, such as Product Data Management (PDM) and Business Process Management (BPM), focused on one particular resource.

PLM is "joined-up". With PLM, the organisation manages the product in a coherent joined-up way across the lifecycle. PLM brings together what was previously separate, for example, product development and product support (Fig. 1.2). PLM joins up many previously separate and independent processes, disciplines, functions and applications, each of which, though addressing the same product, had its own vocabulary, rules, culture and language.

Use of the term PLM implies that the activity of managing products across the lifecycle is clearly-defined, well-documented, proactive, and carried out according to a particular design. It's carried out to meet specific objectives of increasing product revenues, reducing product-related costs, maximising the value of the product portfolio, and maximising the value of current and future products for both customers and shareholders.

2.8   PLM Grid

On the horizontal axis of the PLM Grid (Fig. 1.5) are the five phases of the product lifecycle. On the vertical axis are the resources that have to be addressed when managing a product.

A simple 5 × 10 grid might seem too small to be of any use to a company's PLM efforts. However, it's often said that a picture is worth a thousand words. A lot of information can be communicated in a simple picture. For example, a small plaque on the side of the Pioneer 10 spacecraft, launched in 1972, is intended to give information (such as source and sender) to whatever form of life might find it. Communication was lost with Pioneer 10 in 2003. By 2015, it should have been nearly 11 billion miles from Earth, heading towards the Taurus constellation. The small plaque includes five graphics, and measures about 6 in. by 9 in. Small as it is, it provides enough information to introduce a different form of life, so it's not surprising that a 5 × 10 grid can be very useful for communicating about PLM.

The Grid gives companies that develop and support products a way of visualising PLM.

The PLM Grid helps show why the environment of the product can be so difficult to manage. The scope of the environment is broad. Many subjects are addressed, ranging from methods for identifying ideas for new products, through organisational structure, to end-of-life recycling equipment. The scope is wide, but that reflects the reality of managing products.

The PLM Grid is useful in many PLM activities. The most basic of these is communication of the scope of PLM. The Grid can also be used in many other circumstances such as for: increasing PLM awareness; discussing with PLM project team members; communicating with business executives; documenting the current situation of PLM; defining the PLM Vision, Strategy and Plan; and discussing with vendors of PLM products and services (Fig. 2.3).

2.9   Starting the PLM Initiative

When starting a PLM Initiative, it's important to make sure that management understands PLM. Make sure that PLM is brought to the attention of the most important participants, and make sure that they buy into the idea. The Initiative has

communicate about PLM	document the status of a PLM implementation
discuss the scope of PLM in a company	document the current situation of PLM
document the scope of PLM in a company	document and communicate a company's PLM Vision
communicate the scope of PLM in a company	communicate a company's PLM plan
communicate the contents of PLM in a company	provide a basis for cross-functional discussion
define PLM in a company	communicate a checklist for PLM activities
explain PLM in a company	set the basis for talking to vendors of PLM products
provide a basis for comparison	document the status of a PLM implementation

Fig. 2.3 Applications of the PLM grid

unclear justification for introducing PLM	doubts about migration paths for existing systems
incorrect definition of needs	uncertainty about PLM Return On Investment (ROI)
lack of skills and knowledge	departments disagreeing about working methods
lack of implementation support tools	fear of starting an ERP-style enterprise-wide mega-project
lack of understanding of available solutions	underestimate of management and training requirements
underestimate of the required investment	difficulty of objectively identifying the benefits of PLM systems
concern about high costs	lack of clarity about the scope of PLM
lack of interest after initial setbacks	difficulty of defining responsibilities of system vendors
lack of clarity about what to integrate	difficulty of defining responsibilities of system integrators

Fig. 2.4 Reasons for lack of success in PLM projects

to come at the right time for the company, and it has to show that PLM is relevant and applicable in the company's particular situation. Participants need to see how the Initiative meets their needs and how it will be of benefit for them as individuals. They need to believe that it's feasible to implement PLM, and to see how to achieve it step-by-step. The Initiative needs to be packaged in such a way that participants can understand it easily, and can spread the message to those around them.

Be aware that, as with any other improvement initiative that can offer a high return on investment, a project to implement PLM can be risky. Although PLM provides a solution to the challenges of the changing environment for product development, manufacturing and support, its implementation can be complex and have many repercussions. For example, what may appear at first as the simple purchase of a technical document management system, or the development of an online product catalogue—both of which could be components of a PLM solution —soon raises questions about the way it will be used and maintained. Other questions will be asked about how it fits with other systems, what training is needed, how to manage the data, which working methods to use, and how to communicate with suppliers and customers. And as more and more components of PLM are addressed, the complexity increases.

As a result, there is a high failure rate for PLM initiatives. Many overrun, many don't meet business objectives. A survey among PLM users and potential users showed many reasons for lack of success (Fig. 2.4).

Chapter 3
Complex and Changing Environment

3.1 Changes and Interconnections

One of the reasons that Product Lifecycle Management emerged in the early years of the 21st Century is that the environment in which products were managed became increasingly complex. And to make matters worse, the environment underwent frequent changes.

A lot of companies would be happy if there were no changes anywhere in the world in the product environment. They could then organise themselves, as well as possible, to provide customers, as efficiently as possible, the same product, day after day, year after year. They wouldn't need new products. Over time, they would probably be able to eliminate most of the problems with existing products. They could plan exactly how many products to produce and sell. Everyone would be happy. There would be no need for anything to change. As the global population increased, market sizes would increase. As companies went down the experience curve, they would reduce costs and increase profits.

However, the situation in the twenty-first century isn't anything like that. The world environment for products started to change significantly in about 1965, and then changed tremendously in the 25 years after 1985. The changed environment continues to change.

One company that I work with renewed about 75 % of its products in 2009. In the early 1990s, it annually renewed less than 10 %. Another company that I work with had operations in 6 countries in 2000. In 2010 it had operations in 26 countries.

In an environment of such change, it's best to have a clear understanding of objectives and capabilities, otherwise the result can be disastrous.

© Springer International Publishing Switzerland 2016
J. Stark, *Product Lifecycle Management (Volume 2)*,
Decision Engineering, DOI 10.1007/978-3-319-24436-5_3

3.1.1 Interconnections

Not only are companies in a complex environment, of many dimensions, that is affected by many different changes, but the changes are often intertwined. As a result, the product environment is becoming increasingly complex with many interwoven components and numerous interdependencies being affected by many overlapping changes. The resulting environment is so complex that it's often difficult to see what the changes are, and what's really driving them, or to understand how they will affect a particular company and a particular product. Some of the changes are primarily macroeconomic and geopolitical (Fig. 1.18), some primarily environmental and social (Fig. 1.19). Some of the changes are within the company itself (Fig. 1.20). Others are due to new technologies (Fig. 1.21) and new products (Fig. 1.22).

Among the changes, some create opportunities, some create problems, some lead to the need to change, some are the source of more changes. The resulting changes can drive other changes. Unexpected events resulting from changes can be a source of further change. All these changes snowball, making it difficult to know how to respond.

In addition, changes have associated risks. Changes in one area may lead to enhanced risks in another area. If the risks were only related to one component or change in the environment, it might be easy to manage them. Unfortunately though, they are often related to many changes, making their management difficult.

3.2 Macroeconomic and Geopolitical Changes

3.2.1 Globalisation

Globalisation has been a key factor for change. Globalisation can have many effects on a company, even a small one. One positive effect is that it can sell its products and services worldwide. It can find many new customers and increase sales. Another effect of globalisation is that even small and medium-sized companies are faced with competitors all over the world. And they may find that these competitors bring out similar products, but with better cost-performance than their own models. As a result, they have to be more innovative, develop better products, develop them faster and develop them at lower cost.

Globalisation also implies companies have to be close to customers in many places. They have to understand customer requirements and sell products meeting these requirements in many environments. A presence in many countries may be necessary. It may be achieved through the acquisition of foreign companies, or by setting up joint ventures. Companies have to take account of different requirements and different regulations in different countries. They have to get pricing right in many different environments. They also have to provide technical information, parts, products and service in many locations. They have to coordinate the launch of new and modified products for the global marketplace.

The complex reality of the globalised environment for companies developing, selling, manufacturing and supporting products (such as a machine, a car, a computer) becomes clear in the DASAMASA (Design Anywhere, Sell Anywhere, Manufacture Anywhere, Support Anywhere) acronym. It's becoming necessary for many companies to develop, to sell, to manufacture and to support their products anywhere in the world.

Globalisation is a change that has affected many products. The G-word describes the increasing economic interdependence of countries. Harvard professor Theodore Levitt used it in 1983 in an article called "Globalization of Markets". In the 1990s, globalisation became noticeable well beyond academic circles. A wave of imports from low-cost countries led to the price of goods dropping in advanced industrial countries. Globalisation isn't something that happened with a single "Big Bang". If it had, it might be easier to address. Everything would have changed, and then everything would have settled down. Instead, globalisation is an on-going long-term process, with new effects continuing to appear, and nobody being sure what will happen next.

It's sometimes difficult to clarify if a particular change is a driver of change, or an effect of change, or both. Often, the changes described in this chapter may be seen as reasons for change or as effects of change. For example, increased competition could be seen as a reason for change, or as an effect. For a particular company, increased competition may be seen as an effect of globalisation. However, for that company, increased competition may also be seen as a reason for changing the way it operates.

One change that has occurred in many industrially advanced countries over recent decades is a decrease in the percentage of GDP generated by manufacturing industry. According to the U.S. Department of Commerce's Bureau of Economic Analysis, in 2001, manufacturing accounted for only 14 % of the US Gross Domestic Product by industry. Finance, insurance, and real estate accounted for 20 %, and services accounted for 22 %. And, by 2008, manufacturing accounted for only 11.5 % of US GDP. In 2011, by value added, manufacturing industry accounted for 12.2 % of US GDP. "Finance, insurance, real estate, rental, and leasing" accounted for 19.9 %."Professional and business services" accounted for 12.6 %. Such changes have many knock-on effects. Whole regions that had an identity as a manufacturing community lose their way as manufacturing jobs are lost. Manufacturing often provided manual "metal-bashing" jobs reserved almost exclusively for men. As they are replaced by service jobs and knowledge-based manufacturing jobs, male unemployment rises and female employment rises.

By 2011, onshoring had started. Work that had been offshored to less-developed low-cost countries was brought back to high-cost countries such as the US. In 2012, for example, GM opened its first Information Technology Innovation Center in Austin, TX. In 2014 it opened, in Chandler, AZ, its fourth IT Innovation Center.

Part of the loss in manufacturing employment occurs because work is outsourced to currently lower-cost locations such as China, Mexico and India. The long-term effect of this is unknown. Countries that outsource jobs effectively export potential work for their populations, running the danger of high unemployment and social unrest. And from a production point of view, unless outsourcing is very well managed, it runs the danger of losing the associated know-how. For example, moving heat-treatment operations for a product from the US to China may eventually result in nobody in the US having the knowledge to define future heat-treatment work. When the next product is developed, not only production, but also the design work, may have to be done in China. For the following product, with all the development and production experience in China, it may make sense to move marketing and management there as well.

3.2.2 Geopolitical Developments

Geopolitical changes, for example those resulting from the end of the Soviet Union, affect the product environment. The end of the Cold War led to many countries taking different roles in the global economy. In the 1980s, most of Poland's exports went to Warsaw Pact countries. In 2009, Poland's main trading partner was Germany. In the years leading up to President Nixon's 1973 visit to China, there was little trade between China and the US. In 2008, the US was China's main trading partner.

Political change in China, its huge market potential, and its high availability of low-cost workers, have led to many changes. China has emerged as a large market for consumer and capital goods, a major manufacturing country and a major exporter of manufactured goods. For example, the leading steel-producing countries worldwide in 2008 were China (about 500 million metric tons), followed by Japan (about 100). In 2014, steel production was about 800 million tonnes in China, about 110 million tonnes in Japan and about 90 in the US. China's use of steel rose from about 120 mmt in 1999 to over 500 mmt in 2013. By 2009, China was the largest exporter worldwide, and the largest automotive market. In 2014, by number of billionaires, China was second only to the US.

Russia has become a leading producer of oil and gas. In November 2012, Russia produced 10.9 million barrels of crude per day, a million more than Saudi Arabia. Russia is the world leader in gas reserves. Russia's Gazprom had sales over $100 billion in 2008. In early 2007, only Exxon Mobil and General Electric had larger market capitalisations than Gazprom. In 2014, by number of billionaires, Russia ranked #5 worldwide.

India has emerged as a leading producer of software, software developers and IS companies. For example, Infosys, headquartered in Bangalore, has become a global IT solutions company with revenues over $4 billion in 2009. In 2015, revenues exceeded $8 billion. In 2014, by number of billionaires, India ranked #4 worldwide.

The end of the Cold War in 1991 enabled many countries that were in the Warsaw Pact to withdraw and join the European Union. By 2010, the European Union had expanded to include 27 countries, with an internal market of over 500 million people. By 2015, according to Eurostat, the population of the 28 EU member countries was about 506 million.

By 2015, the population of China was 1.4 billion, that of India 1.2 billion. More than 1.1 billion people lived on the continent of Africa. Indonesia and Brazil had populations of more than 200 million. Pakistan, Bangladesh, Nigeria, Russia, Japan, Mexico and Philippines all had populations of more than 100 million. These key future markets provide the opportunity of a lot of customers for some providers of global products. However, wherever you're located, many of the countries appear to be in faraway locations.

3.2.3 New Customer Requirements

Consumers want a product that corresponds to their requirements. They don't want the standard product imagined by a marketing specialist, or a customer focus group, on another continent. This leads to increasing pressure for mass customisation. This is the provision at a mass production price of products and services meeting the specific requirements of individual customers. Mass customisation provides a company the opportunity to increase the number of satisfied customers. However, customised products are more difficult to develop, sell and support than standard products. For mass customisation to become a reality, processes and applications have to be adapted to meet the new customer requirement.

Some customers also want more services offered along with the product. Sometimes it seems as if the services are more important than the product. Developing and supporting these services often requires additional skills. Companies that only used to sell products may not have these skills.

Consumer market segmentation and the prioritisation of target segments becomes more complex as the potential market becomes larger and more diverse. The demands of New York's Baby Boomers and Generation X, as they search for a deeply satisfying connected new experience from a customised product with the latest design from a globally recognised brand, may be relatively easy to understand. But what about the need of Grey Wolves, Generation Y and the Millenium Generation for digital pets, smartphones, wearables and domestic robots in London, Mumbai, Chongqing, Tokyo, Auckland, Soweto, Rio de Janeiro and Mexico City?

Consumers want to identify with their sport, fashion, music and screen heroes, from wherever they come. As these activities and industries become increasingly global, brands and products also become global. Crossover products emerge. They intertwine features and functionality previously only available in separate product areas.

a manufactured product
developed and engineered in many locations
assembled from materials and parts manufactured in many locations
can be purchased and used worldwide
is maintained and supported worldwide

Fig. 3.1 Characteristics of global products

trains	tractors	aircraft	food	clothes
office equipment	cars	tyres	beverages	watches
cookers	refrigerators	pharmaceuticals	machine tools	soap
toothbrushes	computer software	insurance products	PCs	television sets
smartphones	computer games	consumer products	ships	CDs, DVDs

Fig. 3.2 A wide range of global products

3.2.4   The Emergence of Global Products

A Global Product (Fig. 3.1) is usually available in many options or variants, for example, in different colours and different sizes. New versions are launched frequently. There may be a new version each year or each quarter. There may, as in the fashion industry, be a new version each season. There may be special versions for special events such as the Olympic Games and the World Cup.

Global Products may be consumer products, in which case they usually have a brand name known to consumers worldwide, or they may be industrial goods (also known as capital goods) such as civil aircraft, machinery, telecom equipment, power plants and chemical products.

If they are industrial goods, their brand names are known throughout the world to companies and other organisations in their particular industrial sector. And the customers and users of these organisations' products and services may also know their brand name. For example, aircraft brand names are known to their customers (the airlines) and to air travellers (the customers of the airlines).

The range of Global Products is very wide (Fig. 3.2). Global Products offer the opportunity of billions of customers, greatly increased sales and vastly increased profits.

3.2.5   Shareholder Value

In the 1990s, there was a strong trend for companies to increase shareholder value and thereby appear more attractive to current and potential shareholders. The desire to increase the pay-out to shareholders usually led to pressure to reduce costs. Cost reduction was usually achieved through headcount reduction, outsourcing and offshoring in the expectation that profits would increase as costs dropped and revenues held steady. The effect on products was a secondary consideration.

3.2.6 Market Mentality

In the market economy, the value of anything at any particular time is defined by what a buyer is prepared to pay for it. In 2006, the market value of a barrel of oil ranged between $44 and $76, even though it contained exactly the same product and the same volume. In 2008 it peaked at $146. In 2009 it fell below $40 a barrel, only to rise to $110 in 2013, and fall to $50 in 2014. Market values of other natural resources used in products, such as cocoa, steel, and platinum, are also subject to change. Between 2002 and 2005, the value of $1 fluctuated between 0.75 euros and 1.13 euros. Between 2010 and early 2015, it ranged between 1.48 in May 2011 and 1.06 in April 2015. The continual changes in the relative values of currencies, and in the prices of raw materials and semi-finished products, make it difficult for a company to know how much it'll have to pay for these in the future, and what price it should propose to customers. It can hedge the risks of changes, but this adds an additional cost to the business, and hence to the product. As hedges are hedged, and hedged hedges hedged, huge volumes of financial transactions are generated. In 2005, exports of goods and services of the 30 OECD member countries amounted to $8.5 trillion, yet annual global currency dealings were estimated to be close to $500 trillion. In 2012, exports of goods from the 34 OECD members amounted to $10.2 trillion. According to the Bank for International Settlements, trading in foreign exchange markets averaged $5.3 trillion per day in April 2013. On an annual basis, that's more than $1 quadrillion.

The need for shareholders and traders to know the value of a company as precisely as possible leads to companies producing their financial results within a few days or weeks of the end of the financial quarter. If the company fails to meet its guidance on earnings, the stock price may drop sharply in a few minutes. To avoid this, a lot of effort goes into setting up a company's systems to collect and manage financial data. A lot of management time goes into working on the figures to ensure they meet market expectations. A Stock Exchange mentality can develop, with managers more interested in quarterly results than in the long-term well-being of their products and services. The rewards for getting the figures right have increased in recent years. According to the Economic Policy Institute's report, "The State of Working America 2008/2009", in 2007, US CEOs of major companies earned 275 times more than an average worker. In 2011, the ratio was 231. In 1965, US CEOs in major companies earned 24 times more than an average worker.

Within companies, with financial figures being so important, product-related activities have been getting increasingly low priority. The main focus of most business managers has been the financial processes and the money flowing through them. Rightly so, some would say, since without positive cash flow, nobody will be paid for long, and workers will lose their jobs. Second priority goes to the sales process. This results in a cash inflow from customers in the short term. Then come the production processes, as customers often won't actually part with their hard-earned money, or their hard-earned credit, until they are sure the product exists. As for product development, this has a much lower priority. Most products can't be developed in a few weeks, so are unlikely to affect the top line, or the bottom line, in the next quarterly report. In the rear are the product support activities.

Financial markets want transparency about a company and its products so that investors can take decisions on the basis of full and current information. However, the company's long-term stakeholders may consider much of the information in a company, particularly about its products, to be confidential and a source of value, and will want to keep it secret. Intellectual property management becomes increasingly important at the same time as pressures rise for increased disclosure by companies.

3.2.7 Deregulation

In 1979, Margaret Thatcher became UK Prime Minister. Aiming to reduce the role of government and increase individual self-reliance, her programme included privatisation, deregulation, and the introduction of market mechanisms into education and health. Deregulation led to the break-up of large organisations. These often had well-defined responsibilities, but bureaucratic and inefficient behaviour, and offered poor service to customers. After the break-up, they were replaced by many companies, contractors and subcontractors with unclear relationships.

3.2.8 Regulation and Compliance

Companies are faced with an increasing number of regulatory requirements. These are aimed at protecting customers and other people. Regulators need proof that their requirements have been met. The proof comes in the form of documents. They include documents about product characteristics, documents about analysis of the product, and documents concerning tests of the product. Other documents, for example, process descriptions, describe the way that work is carried out. Regulations are often voluminous and liable to frequent changes. Just managing the regulations, and relating them to different products and services in different countries, is a time-consuming task for a company.

Regulations lead to requirements for analysis, auditing and reporting of everything from food and beverages to cosmetics and chemicals. Regulations are often introduced with the intention of doing good for mankind. The European Union, for example, introduced the Restriction of Hazardous Substances (RoHS) directive to address use of lead, mercury, cadmium, hexavalent chromium, polybrominated biphenyls and polybrominated diphenyl ether. The EU's Waste Electrical and Electronic Equipment (WEEE) directive was aimed at managing waste electrical and electronic equipment. The EU's End of Life Vehicle directive is aimed at getting manufacturers to dispose of vehicles in an environmentally sensitive way.

In 2006, the European Commission enacted legislation, known as Reach (Registration, Evaluation and Authorisation of Chemicals), to force companies to disclose basic data on the chemicals they produce. The long-term effect of many chemicals is unknown. A study on 7500 men in Aberdeen, Scotland between 1989 and 2002 showed a 29 % fall in the average sperm count from nearly 87 million sperm/ml to just over 62 million sperm/ml. In a study of 26,609 men in France between 1989 and 2005, average concentration went from 73.6 million sperm/ml in 1989 to 49.9 million sperm/ml in 2005. Pesticides, chemicals and radioactive material are likely causes.

CFCs (chlorofluorocarbons) were thought for many years to be safe refrigerants and solvents. In the 1970s it became clear that they create holes in the Earth's ozone layer, especially over Antarctica. Stable in the lower atmosphere, they are broken down higher up, releasing chlorine that depletes the ozone layer. Reductions in ozone levels in the upper atmosphere lead to more Ultraviolet B (UVB) getting through to the Earth's surface. UVB causes nonmelanoma skin cancer and has a role in the development of malignant melanomas.

In the 1930s, DDT (dichlorodiphenyltrichloroethane) was seen as a good insecticide, particularly effective against malaria-spreading mosquitoes. The World Health Organisation estimates it saved tens of millions of lives. However, by the 1950s, problems were appearing. Many insects developed resistance to DDT, and it was found to be highly toxic for fish. DDT has a half-life of about eight years, so it stays in the body for a long time. In the early 1970s, countries such as Sweden and the US banned its use. It's now thought to be carcinogenous, and to damage the liver, the nervous system and the reproductive system.

Asbestos is another material that industry used a lot before becoming aware of its dangers. It has interesting properties. It has long fibres, it's strong, and it's resistant to heat and fire. It was widely used in the early 20th Century in products such as roofing shingles, floor tiles, ceiling materials, cement compounds, textile products, and automotive parts. Use declined after it became clear that inhaling it was dangerous, and could lead to mesothelioma and other asbestos-related diseases.

Lead has been known to be harmful to people and the environment for years. However, it was used in many products including paint and pipes for drinking water. And, traditional solders, used for example to solder electronic components, were based on alloys of tin and lead.

Accidents can lead to new regulations that change the environment of manufactured products. In 1989, the Exxon Valdez oil tanker struck Bligh Reef in Prince William Sound, Alaska, spilling over 10 million gallons of crude oil. In the aftermath of the accident, Congress passed the Oil Pollution Act of 1990, leading to the phase-out and replacement of single-hulled oil tankers navigating in U.S. waters by double-hulled tankers.

In 1976, an explosion occurred in a reactor in a chemical plant about 20 km north of Milan, Italy. A toxic cloud of dioxin was accidentally released into the atmosphere and contaminated an area of about four square miles. The Seveso disaster, named after the town most affected, led to many changes in regulations.

Regulations often add costs. In 2006, the ex-Oriskany, a decommissioned aircraft carrier, became the largest ship intentionally sunk as an artificial reef. The US Navy spent $13.29 million to complete the environmental preparations and scuttling in conformance with Environmental Protection Agency (EPA) guidance.

3.2.9 Traceability

There are increasing demands for product traceability from regulators and consumers to provide and assure safety. Product traceability is important in industries ranging from food and pharmaceutical to automotive and offshore. If an airbag fails, a car manufacturer wants to find all the others from the same batch as soon as possible. If an oil rig collapses, any steel parts at fault need to be identified so that similar problems can be avoided on other rigs. Organisations that can successfully track products and parts are at an advantage compared to competitors that can't. Recalls of millions of parts, or millions of products, are very expensive, and may cost millions, or even billions, of dollars.

3.2.10 Education and Training

Historically, universities had roles of storing knowledge (the library), transmitting knowledge to the next generation (teaching), certifying that a student has reached a certain level (examination), and carrying out research. Many universities have existed for centuries; the University of Bologna was founded in 1088, Harvard in 1636, William and Mary in 1696, Yale in 1701.

In the 21st Century, universities are faced with a new world. The Web provides more knowledge than even the best traditional library. Google provides instant search. E-learning applications such as course management, interactive assessment, role-playing and simulation take over many administrative and teaching tasks (Fig. 3.3). The best teachers in the world can give courses to students throughout the world from collaborative virtual classrooms, located on the Web.

In the past, universities offered a once-in-a-lifetime teaching environment. The university staff taught the student for three years at the university. Then the student left, and usually never returned. Today, a lifelong learning environment is needed in which the student learns to learn, and then continues to learn and develop knowledge assets.

ease of use	possibilities to rerun lectures	unlimited number of students in a class
24/7 availability	365 availability	both human and virtual tutors
self-paced	lower environmental impact	staff and students can participate from any location

Fig. 3.3 Benefits of e-learning

email	communities	instant messaging
blogs	video-conferencing	virtual private networks (VPN)
conference calling	collaborative workspaces	Voice over IP (VoIP)
fax	e-commerce websites	personal digital assistants (PDA)
groupware	cloud computing	multifunction printers

Fig. 3.4 E-working tools

In many companies, there's a feeling that, because they've been forced to respond quickly to global changes, or go out of business, their know-how is now years ahead of that available in many universities, where there hasn't been such pressure. Some large companies have even set up their own "Universities".

Historically, the industrial revolution brought thousands, or tens of thousands, of workers together in factories. Later, hundreds, or thousands, of white-collar workers were brought together in offices. By 2010, as a result of the Communication and Computer Revolutions, many people worked alone, or in a small group, in a "Small or Home Office (SOHO)" environment. They could work for a large company, or as an individual, or for a small company. They needed products such as portable computers and smartphones to be effective, and were supported by e-working tools (Fig. 3.4).

3.2.11 Workforce Age Distribution

At the beginning of the 20th Century, life expectancy in the US and the UK was about 47 years. By the middle of the century it was about 68 years. At the beginning of the 21st Century, life expectancy in the US and the UK was about 80 years.

In the mid-20th Century, many countries in the West introduced a retirement age of 65 years. At the time, life expectancy wasn't much longer. Families were large, most work was manual, and health care was basic. The cost of retirement pensions was met by monthly contributions from workers.

By the end of the 20th Century, average life expectancy had increased by more than ten years, and birth rates had fallen, resulting in fewer young workers to fund pensions. However many countries find it difficult to change the retirement age. It's difficult to get people to think differently and accept abrupt change. For example, Germany's plan to raise the retirement age from 65 to 67 will be phased in gradually between 2012 and 2029. The UK will raise it from 65 to 66 by 2020, to 67 in 2028, and to 68 between 2044 and 2046.

3.2.12 Free Trade

Starting in the late 20th Century, there were many actions to enable the free trade of products and services. The World Trade Organisation was established (1995). The

North American Free Trade Area, linking Canada, Mexico and the United States, was established (1994). The Eurasian Economic Union, including Belarus, Kazakhstan and Russia was established (2015). Other free trade regions have expanded, and there have been reductions in trade barriers. This helps companies to offer products and services worldwide.

3.3 Environmental and Social Changes

3.3.1 Social and Health Issues

Perhaps it would be easier to sell products worldwide if so many potential customers weren't so far away. And, perhaps it would be easier to sell more products if so many people didn't live in poverty. In 1999, according to Human Development Reports from the United Nations Development Programme, 2.8 billion people lived on less than $2 a day and 1.2 billion lived on less than $1 a day. In 2008, about 1.3 billion people lived on less than $1.25 per day. In 2015, Oxfam published the "Wealth: Having it all and wanting more" report. It showed that the wealth of the 80 richest people in the world ($1.9tn) was the same as that of the bottom 50 % of the global population (3.5 billion people).

And, in the West, the rich are getting richer, and the poor are getting poorer. And the middle classes are becoming relatively poorer. The situation is similar in many ways to that at the end of the 19th Century and the beginning of the 20th Century, with people sleeping and begging on the streets of many Western capitals.

The world population is expected to rise from 7.3 billion in 2015, reaching 8.1 billion in 2025 and 9.6 billion in 2050. Worldwide, the average age is expected to increase by over 50 % by 2050. By then, it's expected there will be more people over 60 than under 14. As the world's population grows and ages, the demand for healthcare increases. In 2008, about 11 % of Germany's GDP was spent on health, and the percentage is expected to grow. Population ageing will lead to a demand for new types of products in areas such as medical equipment, home help robots, pharmaceutical drugs, replacement body parts and geriatric cosmetics. According to a 2015 Lancet Commission on Global Surgery report, 5 billion people in the world don't have access to safe, affordable surgical care. An opportunity for providers of surgical care.

More people now live in cities than in rural areas. As the world's population grows, and even more people move to cities, the need for decent housing will increase. In 2003, about a billion people lived in slums, and according to current trends, the number will rise to 2 billion by 2030. There may be 3 billion slum dwellers by 2050. An opportunity for providers of homes and home products.

AIDS was first recognised in 1981. In 1983, its cause, the HIV retrovirus, was identified. By 2008, it was estimated that 25 million people had died of AIDS. According to an UNAIDS report, 33.4 million people worldwide suffered from HIV in 2008. It's estimated that, by 2012, 36 million people had died of AIDS-related illnesses, and 35.3 million people were living with HIV. Many of these people live on less than $1 a day, and can't afford high-cost pharmaceutical drugs. An opportunity for providers of low-cost drugs.

3.3.2   Environmental and Sustainable Development

Since the 1960s, in response to the rising recognition of the potential dangers of products and production to mankind and the planet, politicians and ecologists have influenced business behaviour, forcing companies to think about environmental issues, waste products and recycling.

As global consumption increases, supplies of oil and water, and of elements such as iron, are put under pressure. There's widespread concern, for example, about the future lack of petroleum products. In 2004, the world used about 25 billion barrels of oil per year, and total world reserves of oil were estimated at about 1000 billion barrels. At that consumption rate, there would be none left in 40 years. By 2013, consumption had grown to about 34 billion barrels a year. Oil reserves had reached about 1600 billion barrels.

In 1987, the Brundtland Commission defined Sustainable Development as development that meets the needs of the present without compromising the ability of future generations to meet their own needs. It's a holistic concept that aims to unite economic growth, social equity, and environmental management. Sustainable development and related ecological/environment activities represent a major business opportunity. There are many product-related activities that can help achieve sustainable development targets (Fig. 3.5).

Aiming for sustainable development often implies a 90 % reduction in the use of new resources for a product. Sustainable development and related activities represent a major business opportunity that can provide opportunities for faster growth and profitability through improved current products and services, and innovation of new products and services. There are many potential areas to address (Fig. 3.6).

In 1997 at Kyoto, Japan, delegates from all over the world agreed on the need to reduce emissions of greenhouse gases, especially carbon dioxide. A lot of these

reduce the amount of energy and materials used by a product	enable use of low-carbon technologies
reduce the amount of energy and materials used in production	enable re-use of materials at End-of-Life
reduce the amount of energy used at End-of-Life	enable recycling of materials at End-of-Life

Fig. 3.5 Product development activities supporting sustainable development

monitor pollution	improve insulation of plants	reduce emission of greenhouse gases
recycle waste	manage the use of energy	improve the energy efficiency of machines
treat waste water	reduce the use of energy	restore contaminated soil and groundwater

Fig. 3.6 Production-related activities to support sustainable development

emissions come from products such as cars, aircraft and power plants driven by petroleum products. A report on the cost of global warming published in 2006 by economist Sir Nicholas Stern suggested that, if action is not taken on emissions, global warming could shrink the global economy by 20 %. That would affect the sales of many products. The report suggested 1 % of global gross domestic product should be spent on tackling climate change. That could be an opportunity for many global products.

The 2009 United Nations Climate Change Conference was held in Copenhagen, Denmark. Delegates at the ten-day event, attended by representatives of hundreds of countries and organisations, were unable to agree on action to address climate change. The 2011 conference, held in Durban South Africa, agreed to a legally binding agreement to be prepared by 2015.

3.3.3 Role of Women in Business

The role of women in the 19th Century is described in the German phrase Kinder, Küche, Kirche (children, kitchen, church). Women were excluded from many occupations. In the 20th Century, change came slowly. In 1910, a woman was awarded a medical degree in the United States. In 1960, the first female head of government was appointed. But it wasn't until 1981 that the US appointed a woman to the Supreme Court. It wasn't until 1999 that a woman, Carly Fiorina, was appointed CEO of a leading corporation (Hewlett-Packard). It wasn't until 2013 that a woman, Mary Barra, was appointed CEO of a major global automaker (General Motors).

In 1980, Sweden changed its succession rules, making the monarch's first-born child heir to the throne, regardless of gender. The UK's "Succession to the Crown Act 2013" made a similar change, replacing male-preference primogeniture with absolute primogeniture.

3.4 Corporate Changes

3.4.1 Changing Business Models

The changing environment provides opportunities for new business models to be developed. This can make life difficult for companies with more traditional models.

Some companies no longer manufacture their products, but outsource all production so that they can concentrate on product marketing, development and sales.

Some companies lease their equipment and facilities rather than purchase them.

Some companies look to the producer of a product to operate it as well.

Some companies offer their products for lease rather than for purchase. Aircraft, trains and cars can be leased. Some software is offered on a pay-for-use basis over the Web rather than for purchase.

Some companies offer guaranteed product performance. They may guarantee that their products will run for a certain number of hours per month. Or that a certain percentage of products will still be in service after 10 years.

Some companies cut out the traditional sales force by only selling over the Web. Others allow customers to set the price they will pay for a product at an online auction.

Some companies offer products free over the Web, with their income coming from Web advertising.

Low-cost product and service providers often cut out non-essential functions. They get customers to carry out some activities themselves, or to pay for them separately.

Fast-food eateries eat into the restaurant market by offering reduced choice, standard menus and no waiter service. Other companies compete by providing ready-to-eat food and drink products that are sold in shops for customers to eat on the street, or next to you on public transport.

Some pharmaceutical companies focus on providing low-cost generics that have the same effect as existing high-cost brand-name drugs.

New microfinance institutions are emerging, such as Grameen Bank, founded by Muhammad Yunus. Their products have billions of potential customers.

Non-governmental organisations and non-profit foundations are starting to develop and produce products, and own their intellectual property. This enables such products to get to the billions of people in the world who can't afford the prices demanded by multi-nationals based in high-cost countries.

3.4.2   New Company Structures

Company structures have changed in response to the changing environment. Many products are now the result of the concerted effort of an Extended Enterprise made up of a manufacturer and a chain of suppliers. The automotive industry has been accustomed for decades to these chains of car manufacturer, Tier 1 supplier, Tier 2 supplier, Tier 3 supplier, and so on. The concept became widespread in the electronics industry in the 1990s. In the aircraft industry, Boeing adopted it for the 787 Dreamliner. The concept is now also found in most other industries.

In the automotive industry, parts and products are developed round the world. As an example, in 2008, Ford Motor Company gave Gold and Silver World Excellence Awards to 37 suppliers. Among the award winners were companies from Brazil, Czech Republic, Germany, Italy, Japan, Mexico, Poland, South Korea, Spain, the UK and the US. In 2013, Ford gave Gold and Silver World Excellence Awards to 39 suppliers. Among the award winners were companies from Brazil,

Canada, China, Germany, Hungary, India, Japan, Korea, Mexico, Poland, Slovakia, Spain, Turkey, Ukraine, the UK and the US.

In the 1990s, suppliers of parts and components used to be referred to as Original Equipment Manufacturers (OEM), as they were the original manufacturers of the equipment included by another company in its products. In the following years, the meaning changed, and the company that incorporates parts and components from suppliers is now referred to as the OEM. Many OEMs put in place different design and supply chains for different products, or even for different product derivatives. Participants in the chain may be located in different time zones, with different business processes and different application systems. The long, widely dispersed chain necessitates better planning and control, better communications and well-defined business processes and information use.

In the early 1990s, globalisation led to imports from low-cost countries causing the price of goods in industrialised countries to drop. In response, companies in industrialised countries reduced their costs by outsourcing production to low-cost countries. In addition to outsourcing of production, companies then started outsourcing product development and service. Outsourcing provides the opportunity for a company to focus its efforts on the product lifecycle activities it considers most important and/or provide its competitive advantage. The companies to which it outsources activities can usually do these activities either better or at lower cost. In other cases, the outsourcing company may not even have the resources to carry out these activities.

Outsourcing affects a company's flow of information and materials. This leads to the need to realign processes and applications. Product development and support activities are complex and difficult to control even when they're in one company and on one site. When they're spread over many companies in many locations, their complexity increases, and so does the danger of loss of control. Outsourcing has led to long design, supply and support chains with the result that product development, manufacturing and support activities are spread out over different organisations, often over different continents. Managing them when they were in one company in one location was difficult enough. Managing them across an extended enterprise is many times more difficult.

Since the mid-1990s, many development and manufacturing organisations have moved away from the model of a single R&D or Product Engineering department in one location. For various reasons, they found this wasn't the most effective approach. In particular, companies have found that research is often carried out faster by small organisations operating without the overheads and bureaucracy of large organisations. As a result, in some cases, a company's researchers and developers may now be located in several places round the world. For example, Boeing has R&D centres in Australia, Brazil, China, India, Spain and Russia. In 2003, General Motors set up a $21-million automotive research laboratory in Bangalore, India. In 2012, GM opened a research centre in China. In some companies, much of the R&D work is outsourced to suppliers. In others, research is carried out through partnerships with other companies. Carrying out research on multiple sites offers an opportunity to bring new products to market faster.

However, relocating R&D activities changes the organisation of work. New approaches are needed to manage and work effectively in the new environment of networked and fragmented research, development and support.

The Extended Enterprise may sometimes appear as a convenient way for large companies to reduce headcount and increase shareholder returns. However, it also provides small and medium companies with the opportunity to increase their sales and to grow their profits. Initially, a small company may just supply one component to a large company. Later it can develop and produce more components for that company, supply it in more regions, and then supply similar products to other large companies. It can also move from supplying individual components to supplying sub-assemblies and assemblies.

Skilled professionals are sought by companies regardless of their nationality or culture. Their diverse skills are seen as an opportunity by companies looking to provide innovative new products. However, as product-related tasks are increasingly carried out by people from different countries working on different sites of different companies in different countries on different continents, the potential for misunderstanding due to different understandings of words, phrases, processes and behaviour increases. Translation increases costs, and the result may not convey 100 % of the meaning. Product development projects can slow down as cultural differences between different sites lead to difficulties in finding common solutions.

Many of the first generation of product developers that worked with computers, and implicitly or explicitly defined their companies' information and activity structures and elements, reached retiring age in the early years of the 21st Century. Born between 1945 and 1955, these Baby Boomers were among the first users of computers at the end of the 1960s and the beginning of the 1970s. By the year 2000 they were in management positions at the heart of their companies' product environments. By 2015, many had retired, taking with them the knowledge of why and how many activities in their organisations are carried out, and why particular design and other decisions were taken for specific products.

Most companies developing products have the two basic processes of New Product Development and Product Modification. The New Product Development (or New Product Introduction) process is fairly similar in every company, with the same input (requirements for a new product) and the same output (the new product). The Product Modification process is fairly similar in every company, with the same input (requirements to change an existing product) and the same output (the modified product). However, each company has developed its own processes separately, with the result that all companies have different processes. Managing a company-specific process takes a lot of effort. Implementing an IS application to support a company-specific process takes a lot of customisation effort. And, because there are no standard processes of New Product Development and Product Modification, universities and technical schools can't teach students a standard process. So when students join a company, a lot of time is wasted as they learn and understand all the details of the company's specific processes. And time is wasted when an OEM wants to work with several new suppliers, each with its own activities, applications and documents.

In 1873, Levi Strauss and Jacob Davis received a patent for the process of riveting men's pants. They began to make jeans in San Francisco. In the following years, the Levi's brand became an American icon, and Levi Strauss & Co. became one of the world's leading branded apparel companies, marketing its products in more than 100 countries world-wide. In September 2003, it announced that it would close its remaining manufacturing and finishing plants in North America as part of the shift away from owned-and-operated manufacturing. Production was shifted to its global sourcing network. According to chief executive officer Phil Marineau, "In order to remain competitive, we need to focus our resources on product design and development, sales and marketing and our retail customer relationships".

Corporate cultures change frequently with some companies empowering workers to enable them to make better use of resources to meet customer requirements. Other employers offer increasingly insecure part time and flexible employment conditions leading to workforces with little knowledge or understanding of the company they work for. And little knowledge of its products.

3.4.3 Business Process Reengineering

Since the end of the 1980s, companies have focused more and more on mastering their business processes. Major changes occur as they reengineer processes to remove wasteful activities, and to ensure they work the most effective way.

3.4.4 Corporate Theories

As companies strive for competitive advantage, the relative importance of business activities changes, and business theories change with them. In the late 1980's, quality was a key issue. Companies invested heavily in improving quality levels, implementing TQM (Total Quality Management) programs and creating VP TQM positions. As companies improved quality, it stopped being a differentiator. "Customer" became the new buzzword, accompanied by a whole family of slogans such as Customer Focus, Customer Service, Customer First and Customer Satisfaction. Companies created VP Customer Service, VP CRM, and VP Customer Relationship Management positions. As enterprises became more extended, logistics and lean supply chain management became more important. Positions such as VP Supply Chain Management and VP Logistics appeared. Now the pendulum is swinging to the product, the thing the customer wants to use. Whereas SCM addresses the logistics of parts and materials, and CRM focuses on customers, PLM focuses on the product. Without the product, a company doesn't have a customer. Without a product it doesn't need to think about logistics. Customers are very interested in using a product. They are much less interested in the details of the supply chain.

3.4.5 Standards

Standards, like regulations, are often voluminous, complex, and frequently changing. Many informal and formal standards may need to be respected during the activities in the lifecycle. Some may be company-specific, some may be industry-specific, some may be globally applicable.

There are industry-specific standards such as the Petroleum Industry Data Exchange (PIDEX) standards. There are the ISO 10303 standards which provide a representation of product information along with the necessary mechanisms and definitions to enable product data to be exchanged. ISO 14306:2012 provides the description of the structure and content for a binary file having the extension of jt. ISO 14739-1:2014 describes PRC 10001 of a product representation compact (PRC) file format for three dimensional (3D) content data. There are the ISO 9000 international standards for quality management which help demonstrate what the company does to meet customer and regulatory requirements, while enhancing customer satisfaction and continuously improving performance. There are the ISO 14000 environmental management standards which show what a company does to minimise harmful effects to the environment, and to achieve continuous improvement of its environmental performance. Some companies will use ISO 10007:2003 which gives guidance on the use of configuration management within an organisation. Some will use ISO/TS 16949:2009 "Particular requirements for the application of ISO 9001:2008 for automotive production and relevant service part organisations". Other companies will use ISO 13485:2003 "Medical devices—Quality management systems—Requirements for regulatory purposes". OHSAS 18001, which addresses quality, environmental and occupational health and safety management issues, may be applicable. Some companies need to comply with the U.S. Occupational Safety & Health Administration's OSHA 1910 and Code of Federal Regulations (CFR) 29. Some need to comply with the Food and Drug Administration's CFR Title 21 Part 11 regulations for electronic records and signatures, or CFR 21 Part 820 quality system regulations including Section 820.30, Design controls. Other companies must comply with the Federal Aviation Administration's Part 23 airworthiness standards.

3.4.6 Low-Cost and Lean

In the late 20th and early 21st Century, many manufacturing companies ran programs to become "Lean". Lean manufacturing is a management theory focused on creating value for the customer, eliminating any wasteful activities that don't create such value. There are seven groups of wasteful activities (Fig. 3.7).

transportation	inventory	motion	waiting	overproduction	non-value-adding processing	defects

Fig. 3.7 Seven groups of waste

3.4.7 Intellectual Property Management

Product data/information (product know-how) is one of the most valuable resources in a company. In the 1980s, it was usually on paper, difficult to access, difficult to transport. By the year 2000, most product data was electronic, increasingly easy to find and communicate anywhere. To protect it, in the face of increasing global competition and the potential risks from terrorism and economic espionage, companies need an "Intellectual Property Vault".

3.4.8 The Aftermarket

Since the end of the 20th Century, new opportunities have opened up in the aftermarket for long-life complex engineered products such as aircraft, heavy vehicles, trains, industrial equipment and power plants.

Owner/operators of such high-cost products are looking to improve performance in various ways. This is understandable. Downtime of such products, resulting in the inability to provide products and services for customers, can cost hundreds of thousands of dollars per day. Product operators are increasingly requesting the manufacturers of such products to associate service contract guarantees on product availability and reliability.

The OEMs are well placed to exploit aftermarket opportunities. They have deep technical knowledge of the product and through their access to the product design, can do most to reduce support costs and downtime. They can provide customers with product upgrades and other value-added services throughout the product lifecycle (Fig. 3.8).

Providing more product support is also a way to innovate. Knowing how the products are used in the real world is an asset when designing the next generation of products. Presence in the aftermarket shows that the OEM is there for the long run, offering a single source of service, and providing the customer with the benefits of clear responsibility and commitment to service over the longer term. Product support can become a more important revenue and profit generator than the initial sale of the product (Fig. 3.9).

manage product retirement	recycle and dispose of products in an environmentally-friendly way
disassemble the product quickly and at low cost	take informed reuse/recycling decisions at the end of life
effectively recycle and dispose of products	comply with EOL regulations

Fig. 3.8 Opportunities for OEMs in the aftermarket

minimise the response time to any problems	analyse and resolve issues that arise during the lifecycle
provide version control and history. Manage configurations	monitor product progress during the lifecycle
schedule maintenance based on actual product use	upgrade products in the field
reduce spares stocks by better knowledge of spares use	replace components before they fail, not after

Fig. 3.9 Product support activities in the aftermarket

3.5 Technological Changes

3.5.1 Improved Travel, Transport and Telecommunications

The possibilities for transporting information, parts and products all progressed significantly in the last decades of the 20th Century.

Improved travel services offer the opportunity to be closer to customers and suppliers in faraway locations.

Improved freight services ease part and product transport, providing the opportunity to be closer to customers and suppliers in faraway locations.

Improved telecommunications offer the opportunity to interact closely with customers and suppliers without being in the same room.

3.5.2 Revolutionary New Technologies

New technologies have appeared and caused such massive change that they are frequently referred to as revolutions. Examples include the Digital Revolution, the Electronics Revolution, the Computer Revolution, the Communication Revolution, the Biotechnology Revolution, and the Internet Revolution.

Each one of these revolutions leads to change and opportunities. For example, mobile telephony has provided a variety of opportunities to carry out activities in new ways. Service workers can connect to a central database from the customer site where they are working. On-the-move patients involved in trials of new drugs can send performance data rapidly to researchers. Designers of fashion goods can travel worldwide, yet be creative and deliver new designs within minutes of their conception. Billions of SMS text messages are sent each day. Companies blog and twitter. Social networks expand.

3.5.3 New IS Applications

Many new Information Systems have been brought to market. They've provided companies the opportunity to work more effectively internally, and, externally, to get closer to customers, suppliers and partners.

Application systems have evolved. More and more of the activities related to product marketing, development, sales and support have been automated. The cost of functionality has dropped. Applications with functionality that used to sell for hundreds of thousands of dollars now sell for hundreds of dollars. Database management systems have evolved to manage product data distributed on multiple sites in different countries around the world.

In the 1990's, improvement of the supply chain was a major focus for many manufacturing companies. New processes and applications were developed and implemented. The Internet provided new ways and sales channels (B2C, B2B, trading exchanges) to remove waste from the supply chain and get products to customers faster.

In the 1990's, application systems evolved. MRP2 (Manufacturing Resource Planning) evolved to ERP. As the scope of MRP2/ERP applications was extended, they covered more and more of the product lifecycle. In the 1990's, CAD and PDM functionality became a commodity. Systems with a lot of functionality sold for a few hundred dollars. CAD and PDM functionality could be bundled and included in other applications. Like ERP applications, CAD and PDM developed to cover more and more of the product lifecycle. The functionality of SCM (Supply Chain Management) and CRM (Customer Relationship Management) applications also expanded. This on-going evolution of applications raises questions for IS Departments about which applications to use for which business activities.

IS applications are acquired with the objective of improving productivity in the product environment. Large companies may have hundreds of these applications, medium size companies as many as 50. New applications are often implemented as Islands of Automation that aren't closely integrated with other applications. Many aren't interoperable with existing applications. Often they duplicate or triplicate the functionality of existing applications. Often they have separate databases, duplicating information that's already in other databases and raising questions as to where the "master" is. Due to gaps between incompatible Islands, data is transferred manually between the applications. Time is lost, and errors introduced, as information is transferred manually from one application to another.

Once an application has been implemented in a company to address a particular requirement, it tends to expand and be used to address other activities of the product environment. This on-going expansion of applications raises questions for the CIO as to which applications should be used to support the company's activities.

Enterprise-wide IS solutions are becoming ever more complex and time-consuming to implement. These solutions can be so complex that no single vendor can supply all components. This leads to numerous partnerships among application vendors and system integrators. Many implementations of enterprise-wide IS solutions fail. Some surveys claim failure rates as high as 50 %. Increasingly, many small and medium manufacturing companies don't have the in-house resources to implement such complex enterprise-wide systems, so outsource these activities.

There's pressure in many companies to reduce the cost of software and to make it more widely available. One way to do this is through use of open source software such as Linux. Many organisations would rather use a free-of-charge open source application with good functionality than a proprietary application with very good functionality costing hundreds, or thousands, of dollars per user. IS organisations need to find ways to get the benefits of Open Source while maintaining those of proprietary applications.

IS applications used in R&D, engineering and product development are often developed to address the widest possible market. They tend to be "toolboxes". They're rich in technical functions and features, but not focused on the specific needs of a particular company or industry. This approach helps minimise their cost, but results in the need for each company to work out how best to use them in their particular industry environment.

The Internet and the World Wide Web (first released in 1990) have provided the basic technology to allow many activities to be carried out faster and/or at lower cost. They enable information to be transferred faster. For example, engineering drawings can now be sent by e-mail or over the Web, instead of by post. Product requirements can be collected through questionnaires on the Web. Cars and other products can be configured and ordered on the Web. Product Development data in a Web-based project workspace can be shared between workers on several sites. Developers at different locations can review product designs together. Maintenance staff working on customer sites can get product data over the Web. Web-based developments are often relatively simple to implement, and provide companies with new opportunities. However, due to the freedom provided by the Web and the Internet, it's not easy for companies to manage the use of such functionality. It can be difficult to make sure that information remains coherent, and that competitors aren't able to sneak past ineffective security and browse through confidential know-how, looking for new ideas and gaps to exploit.

With new IS applications, more products are modelled and analysed, more information is developed about product models, and more simulation is carried out. The volume of data defining and describing products has become enormous, and continues to grow. Whereas developers once thought in terms of megabytes of data, they now think more of gigabytes. Whereas companies once thought in terms of gigabytes of data, they now think in terms of terabytes, petabytes, exabytes and zettabytes (Fig. 3.10). The high volume of data is needed to develop better products. However, it also increases the difficulty of managing and keeping control of product data. When applications are upgraded, it's often difficult to know exactly which data does not need to be transferred to the new application. As a result, all the data is transferred, including that which is not needed. Moves towards the "Paperless Office" often have the opposite effect to that intended, with each worker creating an unofficial and poorly managed archive of thousands of pages of paper-based information.

As IS applications and databases have evolved, the value that users try to draw from data has increased. Information is more valuable than data, knowledge more valuable than information. It's no longer enough to store data. Companies want to

Unit of Measure	Number of bytes	Unit of Measure	Number of bytes
kilobyte (kB)	10^3	petabyte (PB)	10^{15}
megabyte (MB)	10^6	exabyte (EB)	10^{18}
gigabyte (GB)	10^9	zettabyte (ZB)	10^{21}
terabyte (TB)	10^{12}	yottabyte (YB)	10^{24}

Fig. 3.10 From kilobyte to yottabyte

make sure that the data that's entered into computers can be easily and correctly interpreted, and is available as knowledge.

3.5.4 Communities

Internet and the World Wide Web have enabled the development of many communities. These groups of people have, and share, knowledge and experience of a particular subject. Sometimes linking thousands of people, they have collective knowledge and experience greater than that of most companies. Such reservoirs of knowledge, experience and ideas didn't exist in the twentieth century. They offer companies many opportunities. Companies can work with communities to find ideas for new products and processes, and to identify ways to improve existing products and processes. Communities can also fund new products.

3.6 Product Changes

3.6.1 Products

Many new products are launched each day.

The functionality of products goes on increasing, complicating their development and support. In many industries, onboard electronics and embedded software are major areas for innovation. For example, in 2010, some cars had about a hundred onboard electronic control units, with tens of millions of lines of software. These devices provide a wide range of functions, for example, to help drivers find the right direction, park, steer and avoid other cars. The value of the electronic components in a car may represent more than 25 % of the total value.

Products are becoming increasingly complex with more and more parts and functions. Although more complex products are proposed, they still need to be easy to operate, otherwise customers won't buy them. Cars contain more and more electronics, but aren't more difficult to use. Cameras have much more functionality, but are easier to use. Since many people are unable even to operate the controller of their television, companies have to make products that are easy to use, even though they are actually more complex.

Many companies now offer complete solutions, rather than individual products. This adds a new layer of challenges. Solutions are more complex to develop and support than single products. It's also more difficult to sell a solution than a product. For a product, the price and features are usually clear, and the sale often involves mainly bargaining to find an acceptable price. For a solution sale, a key step in the process is to understand the specific functions, features and performance that a potential buyer is looking for in the solution.

Product Lifetime	Percentage of products with lifetime in this range
less than 1 year	10%
less than 2 years	20%
less than 5 years	50%
less than 10 years	75%

Fig. 3.11 Typical product lifetimes (2003)

Some products and solutions are getting so complex that no single person can understand them. Companies need to find ways to gain and maintain control over such products throughout the lifecycle.

The support of products with very long lifetimes, such as aircraft, power stations and telephone exchanges, is complicated by the many changes in data media and formats that occur during their lifetimes. The IS applications that create this data evolve through many versions. Application vendors mature and disappear. Even the company that made a product may disappear during the product's lifetime. For example, Concorde was developed by the British Aircraft Corporation and Aerospatiale but, by 2000, and the Paris crash, neither of these companies existed. However, customers and regulations may require companies to produce documentation about products they, or predecessor companies, developed 50 or more years ago.

The lifetimes of many products (for example, telephones and computers) are decreasing. Many products now have lifetimes of less than a year (Fig. 3.11). The lifetime of some products is now so short that the development of a future generation has to start before the development of the previous generation has been finished. On the other hand, though, lifetimes for some other products are approaching 100 years. The B-52, for example, first flew in 1952, and is expected to fly beyond the year 2040.

Since the 1960's, in response to the rising recognition of the potential dangers of products and production to mankind and the planet, the focus on the product lifecycle has steadily increased. Politicians and ecologists influence business behaviour, forcing companies to think about environmental issues, waste products and recycling. Issues concerning the end of a product's life are increasingly taken into account during the design stage. And, during a product's life there's an increasing need for design data. Many companies have increased their customer interaction beyond the sales activity, leading to increased customer-oriented activity both earlier and later in the product lifecycle.

3.6.2 Mechatronic Products

Many companies develop mechatronic products. These are products that contain a mixture of mechanical, electrical, electronic and software modules. Companies often develop mechanical, electrical and electronic components in a similar way, with similar processes and applications. However, the processes and applications used for software development are usually very different. Companies have to adapt to work effectively with mechatronic products.

3.7   The Result and the Requirements

The result of the many changes mentioned in this chapter is a complex, risky, continually changing, uncertain, highly competitive, global product environment. This is characterised by demanding customers, horizontal integration across the Extended Enterprise, many small and medium companies in the design, supply and support chains, few layers of management, ubiquitous computing, fast technological evolution, and small numbers of knowledge workers from different functions working together in collaborative teams.

There's growing competitive and legislative pressure, such as that concerned with product liability, deregulation, health, safety and the environment. There are technology issues to be faced, including the effect of the increasing amount of electronics and software in products, the possibilities offered by widespread communication networks, the increasing availability of productised software components, and the rapidly decreasing cost of computer power. Multi-technologies in the product are making things more complex.

To be successful in this environment, a company must be able to supply and support the products that customers require, at the time they require them. The company must have great products. It must have a great product deployment capability. Customer expectations are rising. With so many manufacturers, from so many countries, proposing products, why should a customer settle for a second-rate product or late delivery? Customer demands imply better products and services, a wider product range, customisation and market niches.

But there's also increasing consumer resistance to price increases. Product costs must be trimmed so that they correspond exactly to customer requirements. Product functionality must be improved to match these requirements. Customer service must be improved with on-time documentation delivery, reliable delivery times, prompt complaint handling, and easy product repairability.

Products must be brought to market faster. Technology is evolving fast and products are becoming obsolete sooner. The reduced time between product launch and product retirement erodes sales revenues. As product lifetimes get shorter, significant market share is lost if a product isn't brought to market at the earliest possible moment. A company that gets to market first can capitalise on late market entry by other companies.

To successfully meet, in a complex, changing environment, customer requirements for great products, companies need effective Product Lifecycle Management.

Chapter 4
Product Pain

4.1 Product Environment

Products, in particular global products, offer companies the opportunity of billions of dollars of sales revenues. However, without the right product and the right product deployment capability, the opportunity will be lost. Even worse, customers and other product users may be killed. Billions of dollars may be lost. Executive reputations will be tarnished. Company workers may lose their jobs.

The complex, risky, continually changing, uncertain, highly competitive product environment makes life difficult for companies that develop, produce and support products. In such an environment, they need to have great products that leave competitors far behind. They need a great product deployment capability. They need to be continually in control of their products. If they aren't, and for one reason or another, they take their eye off the ball, unpleasant consequences can occur (Fig. 1.23).

And, since it's never been easy to develop and support good products, and there are many reasons why it could get even more difficult, companies need to be sure they are operating as effectively as possible. They shouldn't be applying principles proposed in the early 20th Century by industrial management specialists such as Frederick Taylor, Carl Barth, Henry Gantt, Frank Gilbreth and Lillian Gilbreth. Back then, the world was a very different place. Many products came from a local farm. Electronics and biotechnology didn't exist. Telecommunications, cars and air travel were in their infancy. Customer choice was limited. People were willing to wait years to buy a product. Since then, the global economy has changed enormously. In industrialised nations, agriculture only represents a few percent of the economy. Today's customers have a huge choice. They can buy products almost anywhere in the world. They expect products and services to be available and to work first time. If they don't, it's easy for customers to switch to a more competitive product.

© Springer International Publishing Switzerland 2016
J. Stark, *Product Lifecycle Management (Volume 2)*,
Decision Engineering, DOI 10.1007/978-3-319-24436-5_4

4.2   Private Life Experience

4.2.1   Washing Machine

You can't get much closer to private life than your laundry. Early in the 21st Century, I bought a washing machine. It had a label showing it came from a "world-class manufacturer". After it was installed, my daughter gave it a test run with a small load. Surprisingly, the washing machine ran across the floor and pinned her against a wall. She switched it off before it crushed her. It turned out that it was shipped with a fixture to stop the tub being damaged during transport. The installation engineer didn't remove the fixture. When the machine got to the spin dry stage, the fixture caused such vibration that the machine moved several feet.

The first electric washing machines appeared in the first decade of the 20th Century. There's about a hundred years of experience and knowledge about their development and use. That makes you wonder what kind of organisation manufactured the machine I bought. Do you think they implemented Total Quality Management? What about their Customer Satisfaction activities? What kind of installation service process did they define? How are the installation engineers trained? Are the installation engineers getting the right information? Why was there no Feedback Form for me to provide valuable customer input? And shouldn't they have designed the machine so that it couldn't start if the fixture was still inside? Don't they know about Poka Yoke (mistake proofing)? What training do their design engineers get? Are the service engineers involved in the design? Do you think this world-renowned manufacturer has control of its products across their lifecycle?

4.2.2   Telephone

As you know, telephones have been with us for some time. Alexander Graham Bell is credited with inventing the telephone in the 1870s. Some time after I bought the washing machine, I wanted to upgrade my GSM mobile (cellular) phone. It had the name of a world-class manufacturer on it. I went to the shop and showed the Customer Services Agent the phone and told him its number. He took a quick look at a database and told me which upgrades were possible. I chose one. He gave it to me, told me to put the SIM (Subscriber Identity Module) card from the old handset in the new one, told me I would need to charge the battery for about an hour. Isn't that great? Modern technology is so wonderful! It's so easy to live in today's service economy!

Well, actually it isn't. After I'd taken the phone home and charged the battery for an hour, the phone didn't work. I took it back to the Customer Services Agent. I told him it didn't work. "Oh, that's because your old phone worked with a 5 V SIM, but the new one only works with a 3 V SIM card. You'll need to get your service provider to change your card." The whole process took me about five hours. It could have taken less than five minutes. There was no apology from the service provider or the Customer Services Agent. There was no Feedback Form to return to the manufacturer. Seems the Customer Services Agent hadn't been properly trained. Seems the database used by the Customer Services Agent doesn't have the right data in it. Do you think this world-class manufacturer has control of its products across their lifecycle? And, by the way, who's in charge in this Extended Enterprise? The manufacturer, the shop, or the service provider?

4.2.3 Escalator

A bit later, I made two return journeys to Oxford, England over a two-week period. The first time I went through London Gatwick Airport, I noticed an "up" escalator wasn't working. It had a maintenance sign on it. As I walked up the nearby stairs to the bookshop, I wondered why the partner "down" escalator hadn't been switched to "up" to help people with their luggage. Four days later, on my way home, it was in the same state. I walked up the stairs again. Four days later, going through the airport again, it was in the same state. I walked up the stairs again. And four days later it was in the same state. I walked up the stairs again. According to the nameplate, the escalator had been supplied by one of the world's leading global escalator manufacturers. I wondered why product maintenance took so long. By the way, like many other products we use today (Fig. 4.1), the escalator was invented more than 100 years ago. Escalators were invented in 1891. How many more hundred years before they work properly?

4.2.4 Train

I travelled by train from Gatwick to Oxford. Going through Reading Station, which is on the main railway line going West from London, I noticed the long black plumes of smoke rising from the diesel engines pulling the carriages in their resplendent corporate liveries. (Weren't those plumes of smoke meant to disappear along with steam locomotives?) The passengers in the carriage I travelled in had the doubtful pleasure of being accompanied by diesel fumes. One of them told me travelling conditions that day were good. Apparently, a few weeks earlier, it had got

Product	Invented	Product	Invented	Product	Invented
aeroplane	1900s	elevator	1870s	telephone	1870s
air brake	1860s	escalator	1890s	washing machine	1900s
automobile	1880s	jeans	1870s	zipper	1890s

Fig. 4.1 More than 100 years of experience with some products

so hot in a carriage that the passengers had broken the windows to get fresh air. Another passenger added that a few weeks earlier, he'd been in a train when it had stopped, all the air conditioning and lighting had failed, and the doors were locked for passenger safety. After a couple of hours, people started screaming and breaking the windows.

On the final journey back I was amazed to see the driver of a superbly coloured engine, pulling an even more superbly coloured set of carriages, get down from the locomotive after he'd stopped the train. He walked over to a bucket of water. He picked up a broom, and proceeded to wash the windscreen. He then got back in, and continued the journey to London.

The first successful rail journey by a steam locomotive was made by Trevithick's Penydarren in 1804. In September 1825, for the first time, a locomotive pulled a passenger train. How come, nearly 200 years later, diesel fumes get in the carriages, passengers have to break windows to get fresh air, and locomotives don't have effective windscreen wipers? Are manufacturers in total control of their products?

4.2.5 Private/Professional Experience

Advising companies world-wide, I travel a lot, and get the opportunity to see a lot of products at work. Quite a lot haven't worked well (Fig. 4.2).

4.3 Professional Experience

In case you think these incidents aren't typical, but just happened to one particularly unlucky individual, here are a couple of examples from companies I've worked with. One company wanted to buy a batch of machines. The order was worth a few million dollars, and it took more than a year for the project team to decide exactly what they wanted. Eventually they ordered. "Sorry", said the manufacturer, "we can't deliver from that 2 year old product catalogue. We adopted Japanese manufacturing techniques and have Kanban and Zero Stock of machines. We don't even have the parts to make the machines you want. We purchased them, and the supplier doesn't make them any more." Is that manufacturer really in control of the product lifecycle? It seems the CEO wasn't too happy when he found out they'd

hotel coffee machines that don't work	hotel elevators that don't elevate
electronic keys that don't open hotel room doors	sensor-operated doors that don't operate
train toilet-door locks that don't work	vending machines that don't vend
rental cars that unexpectedly stop working	aircraft that can't fly
aircraft that hit another object before take-off	jetways that don't extend
credit cards that don't give credit	taxis that break down en route
reclining aircraft seats that don't recline	cash dispensers that don't dispense

Fig. 4.2 Some products that didn't work as expected

lost an order for a few million dollars. He didn't know who to fire. The Sales Manager, the Customer Relationship Manager, the Manufacturing VP, or the VP R&D?

Another company I work with did receive the brand-new high-tech machine it wanted. But the machine didn't work. Somehow there had been a mix-up concerning the hardware components of the machine and the software controlling them. Apparently the version of the software that was delivered didn't work with the hardware that was delivered. By chance, in a restaurant one evening, I met the service engineer sent to fix the problem. He told me that he was on the road every week fixing similar problems. The customers wanted customised products, but the company didn't have the systems to make sure all the parts for a specific order fitted together. What worried him even more was that, when he filed an error report about a part, it took more than a year before the problem was fixed. During that time the plant went on making the wrong parts. Logistics delivered them to customers, and he had to go and fix them and pretend he didn't know what was wrong. Is that manufacturer in control of the product lifecycle?

Problems can occur at any time in the life of a product. Sometimes the problem occurs while the product is being developed, sometimes while the product is being produced, sometimes while it's being used. Sometimes the problem occurs while the product is just an idea. Sometimes it happens at the end of the product's life. Making sure that such problems don't occur is a major challenge.

4.4 Public Experience

Maybe you think it's not relevant if a few individuals and a few companies have a few problems. And, since these stories are so different from the usual hype about today's wonderful products, you may not even believe them. After all, in the last few years, you've probably read many glossy articles and case studies about the development and production of great new products that customers simply can't wait to get their hands on. Did any of them mention this kind of problem? Or did they lavish praise on particular products, companies and executives? Have you read any Failure Stories to match the Success Stories? In case you can't remember anything going wrong, then maybe, digging deep into the distant past, we can find one for you. Here are some you may have heard about.

4.4.1 Electricity

On August 14, 2003, there was a power cut in the North-East of North America. More than 50 million people in dozens of cities including New York, Detroit and Toronto went without electricity all night. Some were trapped in trains and lifts, others were forced to sleep on the streets. The reason for the problem with this product seems to have been "a combination of circumstances".

Two weeks later, in a good demonstration of globalisation, London was blacked out for 30 min. This time, a clearer reason was given. Of the four main power lines serving part of London, two were down for maintenance and one was down due to an alarm. Unfortunately a couple of years earlier, the wrong protection relay had been installed on the remaining, fourth line. Five times smaller then "as intended" it triggered when power surged onto this line, causing the loss of about 20 % of London's power supply. This example shows one of the features of today's product lifecycles. There are many organisations involved in getting the product to the customer. In this case, they included the protection relay manufacturer, the organisation installing the relay, the designer of the network, and the operators of the network. Unless the bits are put together very carefully, something will fall through the cracks.

Then, at 11 pm on September 28, 2003, more than 50 million people in Italy were blacked out until the following morning. Apparently an overhead power line had got too close to a tree in Switzerland. One theory behind this problem is that the line sagged too far down as the high power load it was carrying caused its length to increase. Why would there be an abnormally high load of this product at 11 pm when most of the population is asleep? From the point of view of most of the world's population, electricity is an important product for society, saving lives in hospitals, helping children learn in schools, making the streets safer for all of us. Many people yearn to have it. According to the International Energy Agency, about 25 % of the world's population doesn't have electricity. Yet, for others, it's just another commodity to speculate in. Don't tie up capital, buy low and sell high, and if possible, reduce the risk by trading in options and futures. At 11 pm, there's a low demand for electricity, so that's the time to buy it.

On March 27, 2015 a power cut hit Amsterdam. Flights to and from Schiphol airport were cancelled. On March 31, 2015, a 10-h power cut hit more than half of Turkey's 81 provinces. Rescue teams had to help people stuck in lifts and underground trains.

Electricity has been understood for more than two hundred years. Haven't electrical engineers had time yet to learn about the effect of electricity on a power line? Isn't it possible for them to design and operate equipment in such a way that one fault doesn't affect 50 million people?

You may think that electricity isn't a product, so the above examples aren't relevant. In which case, consider some examples involving cars, ships, aircraft, bridges and rockets.

4.4.2 Cars

In 1997, the Mercedes A-Class car and the two-seater Smart car failed the "elk test". The test involves driving fast around a sharp curve as if avoiding a moose on the road. Correcting the flaw lengthened the development time. For the Smart car, the delay was about 5 months. The cost was about $150 million.

4.4.3 Bridges

In November 2003, 15 visitors died when visiting the Queen Mary 2 in its construction dry-dock in Saint Nazaire, France. At the time, the QM2 was the largest passenger ship ever built, at an estimated cost of $800 million. The visitors died when a 30 ft gangway from the quay to the ship broke. It had been installed the day before by a subcontractor.

When the Millennium Bridge opened in June 2000, it was London's first new Thames crossing since Tower Bridge in 1894. The first day that people walked across, it wobbled and swayed, and had to be closed. In some parts, the bridge moved nearly 3 in. The engineers fixed the problems by adding 91 dampers. The bridge was reopened in February 2002. The modifications cost £5 m, compared to the original cost of £18 m.

The Pont du Gard aqueduct was built by the Romans in 193 BC to supply water to Nimes, France. Two thousand years later, it's still standing. It has three tiers, with the top tier, which is about 150 ft above the river, holding the water-course. The Romans didn't have computers and CAD systems. Building a bridge that doesn't wobble and sway isn't exactly rocket science, is it?

4.4.4 Aerospace Products

Oops. Rocket science. On June 4, 1996, the Ariane 5 rocket disintegrated about 30 s after lift-off on its maiden flight. The 4 satellites it was carrying were destroyed. The loss is estimated at a few hundred million dollars.

In November 2003, officials at the Japanese space programme said a rocket carrying two spy satellites had to be destroyed after take-off from Tanegashima island because of an unspecified technical failure. The two satellites cost an estimated 125 billion yen (about $1 billion).

In 1999, NASA's $125 million Mars Climate Orbiter got to close to Mars and burned up in its atmosphere. An investigation found that a contractor's spacecraft engineering team (in Colorado) supplied information about propulsion manoeuvres in Imperial units (inches and pounds) to the navigation team (in California) which was using metric units. An investigation found the confusion started with a subcontractor of the contractor. Was NASA in control of its product across its life-cycle? To what extent does it learn from experience? The need to take care with metric/Imperial conversion is well known. In 1983, a Canadian Boeing 767 ran out of fuel, and had to glide down to an emergency landing after someone used the wrong metric/Imperial conversion factor to calculate how much fuel it needed to get from Montreal to Edmonton.

The Hubble Space Telescope was a collaborative development of NASA and the European Space Agency. It was deployed in April 1990. Initial images were found to be unexpectedly hazy. Two months later, the telescope was found to suffer from spherical aberration of the primary mirror. In places, the mirror was 2 microns too flat. The problem was corrected with COSTAR (Corrective Optics Space Telescope Axial Replacement) during a service mission in 1993. An inquiry was held into the problem and a technical explanation found. There was a fault in the null corrector, an instrument used in the mirror's manufacturing and testing process. Management failures were also identified. There had been insufficient testing, and under cost and time pressure, contradictory tests results from other equipment were not sufficiently investigated. No formal certification had been required for the null corrector even though it played a crucial role. Project managers lacked the expertise required to correctly monitor activities and there was poor communication. COSTAR and the corrective mission are estimated to have cost more than $500 million. Hubble's cost at launch was estimated at about $1.5 billion.

Maybe you still think these are exceptional cases? Or you think they weren't investigated seriously to find the real causes? Among the events of this kind that have been most fully investigated are the Challenger and Columbia Space Shuttle accidents. On the morning of January 28, 1986, the Challenger Space Shuttle was destroyed 73 s after launch. The seven-member crew died. It included Christa McAuliffe, who was to have been the first teacher in space. On February 1, 2003 the Columbia Space Shuttle broke up during re-entry. The seven-member crew died. In both cases, there was, of course, a physical reason for the accident, but in both cases the investigators also found organisational problems.

4.4.5 Power Plants

Sometimes the result of losing control of a single product can have world-wide consequences. In April 1986, operators at the Chernobyl nuclear power plant started a simple test run that went wrong. It led to a chain reaction and explosions that blew the roof off the reactor, releasing radioactive products which then travelled round much of the world. Fire-fighters died, hundreds of thousands of people were evacuated, and the incidence of thyroid cancer in local children increased.

That wasn't the first time things had gone wrong with a nuclear power station. In 1979, at the Three Mile Island nuclear power station near Harrisburg, PA, a minor malfunction in a cooling circuit led to a temperature increase causing the reactor to shut down automatically. Unknown to the operators, a relief valve failed to close, much of the coolant drained away and the reactor core was damaged. The resulting investigation found the causes were deficient instrumentation and inadequate emergency response training.

4.4.6 Financial Products

A home mortgage is a simple financial product. It's a loan from a financial institution to help purchase a property. In return for the loan, the customer agrees to make payments to pay it off, and to use the property as security. The product has a few basic product characteristics, such as loan size, length of loan period, interest rate, and repayment schedule. What could go wrong with such a simple product and such security? Well, financial organisations could offer variable interest rate mortgages to people with no capital, low earnings, and a history of unemployment and loan repayment delinquency. The result was the 2007–2008 global financial and economic crisis. Financial organisations round the world lost trillions of dollars. Some went bankrupt. Many were saved by taxpayer bailouts. Stock markets slumped. People lost their houses, companies closed, global economic growth declined, countries' economies went into recession.

A Stock Exchange is a place where stocks are traded. Some Stock Exchanges have high capitalisations. For example, according to the World Bank, the market capitalisation of US listed companies in 2010 was about \$17 trillion. From the outside, it looks as if it should be easy for a Stock Exchange to define policies and rules to ensure smooth trading, know the current price of stocks, and calculate Market Indices. However, on May 6, 2010, US markets went out of control. In about 15 min that afternoon, stock prices dropped nearly 7 %, the equivalent of about a trillion dollars. According to the joint report issued by the U.S. Securities and Exchange Commission and the U.S. Commodity Futures Trading Commission, more than 20,000 trades were executed at prices over 60 % away from their values moments before. Some of these trades were executed at prices of a penny or less, or as high as \$100,000. The cause was probably related to high-frequency computer-based trades. One of the effects was the introduction of "circuit breakers" that halt trading in a stock if it experiences a sudden 10 % price change.

4.4.7 Other Products

Occasionally problems are reported on television about high-profile products such as aircraft, power plants, ships, and cars. But as well as these very visible problems there are also, on an almost daily basis, publicly announced recalls of all sorts of products (Fig. 4.3).

airbags	processed food	cars	infant car seats	baby food
water heaters	bicycles	bunk beds	slippers	jackets
toys	television stands	refuse bins	shave gel	door knobs
security phones	cheese spread	chicken salad	cider	cookies
chairs	candleholders	flashlights	refrigerators	rifles

Fig. 4.3 A wide range of recalled products

And, apart from the well-publicised problems, you may also have noticed that some of the products and services that you acquired didn't quite come up to the level you expected. Your plane took off late because the pilot found a mechanical fault? Your computer crashed for some unknown reason? Your car didn't start, even though it worked perfectly the day before? Your pen leaked ink over your best dress? Your washing machine leaked water all over your favourite carpet? You couldn't open the hotel door with the electronic key? You broke your fingernail trying to switch your phone on? You tried a new lipstick and your lips broke out in a rash? You ate some cheese and were laid low for days?

4.5 Product Development Is Important

The above examples focus on the use and operation of products outside the companies that develop and manufacture them. There are also all sorts of problems in the purely company-internal activity of developing new products.

Product development is an important activity in a product's life. It's where everything is defined, from components to costs.

For example, it's in product development that a manufacturer of laundry powders defines which ingredients to use. Will it include phosphates, or a more environmentally-friendly alternative? If it includes phosphates, they may eventually get into lakes and seas, and result in eutrophication and proliferation of seaweed.

It's in product development that a manufacturer of plastic bottles decides which plastic to use, and how much plastic is needed for a particular product. The choice will have environmental and cost impacts.

On average, 80 % of a product's cost is defined during its development, even though more than 80 % of its life is beyond the factory gate. The cost of product development varies widely from one product to another. Product development can be expensive. Figures of about $10 billion are quoted for the A380. Advanced combat systems often cost billions to develop. New cars cost less to develop, usually just a billion or two. And, in the pharmaceutical industry, a blockbuster drug may cost more than $500 million to develop.

Product development is time-consuming. Like product development cost, product development time also varies widely from one product to another. About one-third of products are developed in less than 6 months, and about two-thirds in less than 18 months.

Surveys of product development in industry often show that about 50 % of the spend on product development is wasted. Wasted because a product development project is terminated before the product gets to market, or because the product fails in the market.

The frequency of introducing new products has increased in recent years. In the early 1990s, one company I work with renewed less than 10 % of its products annually. It renewed about 75 % of its products in 2009. Of the 300 or so products it had in 2005, only about 70 were taken forward into 2006.

4.6 Product Development Is Hard

Part One of the report of the Columbia Accident Investigation Board starts with the sentence "Building rockets is hard." This is true, but the reasons for things going wrong with the space shuttles can't be classified as advanced rocket science beyond the understanding of NASA's engineers. And, in other situations, things that go wrong with products aren't usually due to a failing of people to understand Einstein's general theory of relativity, but to simple everyday things that employees should know about, and for which managers should have prepared.

Building any product is hard. Product development is a complex process involving many poorly understood variables, relationships and abstractions. It addresses a wide range of issues, and is carried out by a wide variety of people using a wide range of practices, methods and applications, working in a wide variety of environments. Converting a concept into a complex multi-technology product under these conditions is not easy. It requires a lot of effort, definition, analysis, investigation of physical processes, verification, trade-offs and other decisions.

4.7 Pain in Use

Building any product is hard. Foreseeing what can go wrong with its use is also hard.

In the late 1950s, a newly discovered drug, thalidomide, was found to be a good sedative. It was prescribed to pregnant women. Unfortunately it had unforeseen effects. They resulted in the birth of more than 10,000 children with major malformations.

Other drugs may seem safe, yet be very dangerous for a particular type of patient. A particular drug might normally be safe but become dangerous when taken with another drug. Thousands of people die each year from adverse drug reactions.

In the 1970s, tens of thousands of people were infected with Hepatitis C after being given contaminated blood-clotting concentrates.

In December 1984, water caused a reaction in a tank of methyl isocyanate at a plant in Bhopal, India. The state government reported that the resulting gas leak led to about 3,800 deaths. Tens or hundreds of thousands of people may have been injured.

In the 1990s, millions of animals were slaughtered as a result of a BSE (bovine spongiform encephalopathy) epidemic. The culprit product was animal feed. It's thought that the epidemic originated when an animal developed BSE in the 1970s. The carcase of the animal was mixed into cattle feed. (The protein that animal carcases contained was intended to make the feed more nourishing, and help development.) Animals that ate the feed were infected. In turn, their carcases were mixed into cattle feed and infected others.

Many countries have problems with old military equipment. In particular, old ships may be polluted with hundreds of tons of carcinogenic polychlorinated biphenyls (PCBs) and asbestos, as well as flammable and poisonous fuels and oils. An accident can lead to environmental damage, toxic spills and fires. Dismantling is dangerous for people and the environment.

In November 2002, the Prestige single-hulled oil tanker sank off Spain, spilling over 60,000 tonnes of oil. The clean-up and environmental cost is estimated at several billion dollars.

In April 2010, the Deepwater Horizon rig explosion led to the death of eleven workers. BP stopped paying dividends to its shareholders, and agreed to finance a $20bn clean-up and compensation fund. On one occasion, US President Obama was reported as saying he had visited the Louisiana coast, "so I know whose ass to kick".

A 2005 report by the Centers for Disease Control and Prevention estimated an annual U.S. total of 76 million cases of food-borne disease, caused by consuming contaminated foods or beverages. It estimated an annual 325,000 hospitalisations and 5,000 deaths related to food-borne diseases.

Melamine is a useful chemical, but harmful to humans. It has a high level of nitrogen, and if added to food, leads some quality tests to overestimate the level of protein. For example, if it's added to low-quality milk, the tests will show a higher level of protein. As a result, the milk can be sold at a higher price. In 2008, more than 250,000 people in China fell ill after drinking milk to which melamine had been added.

Other examples of difficulties with product use can be seen by talking a walk in any large town. You may see elderly people struggling to climb three or four steps to get into buses, and wonder who wrote the product specifications. Or you may see a young person with a pram struggling to climb those steps, and wonder about the home lives of design engineers. You may see the automatic doors of a bus closing on people, yet see another bus that has sensors to prevent injury. Or buses with engines under the floor, providing unwanted vibration for passengers. Or try and get a bus ticket from a machine and find the machine doesn't give change. If these are customer-oriented products, can you imagine products that aren't customer-oriented?

4.8   Effects

When a company loses control of its products and product-related activities, there can be effects in several areas (Fig. 4.4).

Problems with products can result in high costs. In October 2006, Sony announced details of a global voluntary replacement program for certain battery packs using Sony-manufactured lithium ion battery cells. The estimated cost to Sony, based on a potential 9.6 million battery packs, was about 51 billion yen (about $440 million).

Fig. 4.4 Some effects of
losing control of a product

Area	Effect
Customers	Deaths and injuries
	Loss of customers concerned about product problems
Financial	Financial losses due to damages resulting from product use
	Reduced profit due to costs of recalls and legal liabilities
	High cost of problem clean-up
	Revenues lost to low-cost competitors
Image	Negative publicity in the media
	Damage to the company's image
Environment	Pollution of the environment
Products	Products not behaving as expected
	Development projects finishing late
	New products not providing competitive advantage
	Resignation of top executives
	Management appearances in court

When, in July 2000, an Air France Concorde crashed soon after takeoff, 100 passengers and 9 crew members died. Compensation agreements are believed to have cost insurers over $100 m. Concorde was taken out of service, and although commercial flights were restarted in 2001, it was withdrawn from service in 2003. Potential ticket sales of tens of millions of dollars were lost.

All 215 passengers and 14 crew members died when Swissair Flight 111 crashed into the Atlantic Ocean near Halifax, Nova Scotia, on September 2, 1998. 227 passengers and 12 crew members disappeared with MH 370 on March 8, 2014. 144 passengers and 6 crew members were killed on Germanwings Flight 9525 on March 24, 2015.

Estimates for the cost of the A380 delivery delay range up to $6 billion.

The development cost of the A350 XWB may be as high as $14 billion, three times that of the initially proposed A350.

Merck voluntarily withdrew VIOXX in September 2004. Worldwide sales in 2003 were $2.5 billion.

The BSE crisis cost the UK more than $4bn in slaughtered cattle, compensation and lost exports. More than one hundred people died from the related Creutzfield-Jacob disease, possibly infected by eating contaminated beef.

In 2001, as a result of high tyre failure rates, Ford Motor Company announced it would replace all 13 million Firestone Wilderness AT tyres on its vehicles. It took a charge of $2.1 billion to cover the costs of replacing the tyres.

Toyota estimated that the recall of millions of cars in late 2009 and early 2010 could cost as much as $2 billion.

GM's fourth quarter 2014 earnings release showed that, in 2014, full-year net income was impacted unfavourably by recall-related pre-tax costs of $2.8 billion.

4.9   Causes

Often an enquiry will be held when there's a serious problem with a product. Usually it's found that there are causes of several types (Fig. 4.5).

Typical sources of the problems are shown in Fig. 4.6.

| physical causes | technical causes | organisational and cultural causes |

Fig. 4.5 Different types of cause

bureaucracy	design alternatives ignored	information lost, misunderstood, ignored
lack of training	customer needs misunderstood	inter-departmental communication problems
design faults	lack of prototypes and testing	informal decision-taking and change-making
culture of risk-taking	standards not suitable/not adhered to	management pressure overriding technical rules

Fig. 4.6 Typical sources of problems with products

| details understood, but not the overall picture | risks weren't fully analysed | information got lost |
| customer requirements misinterpreted | key relationships ignored | decisions not co-ordinated |

Fig. 4.7 Issues affecting products

Often it seems that everyone was doing their job the way they should have, but somehow, things fell through the cracks because something wasn't done (Fig. 4.7).

Although the physical effects of a major problem with a product may be the most visible, the principal causes are often organisational and technical. These causes have to be identified and understood so that measures can be taken to prevent their effects recurring.

4.9.1   Challenger

The Presidential Commission investigating the Challenger Space Shuttle accident found that the physical cause of the accident was the failure of the O-ring pressure seals in the aft field joint of the right Solid Rocket Booster. This was due to a faulty design overly sensitive to several factors. One of these was temperature. O-ring resiliency is directly related to temperature. A warm O-ring that's been compressed will return to its original shape quicker than a cold one when compression is relieved. The O-ring seals weren't certified to fly below 53 °F. The Commission found that, on the eve of the launch, NASA and the Booster builder debated whether to operate the Shuttle in the expected cold weather. (Overnight the temperature dropped to 19 °F and at launch time was 36 °F.) The engineers recommended a launch postponement. Under pressure from mid-level managers, they reversed the recommendation and gave the go-ahead to launch. The Commission found that higher-level NASA managers weren't informed of the late-night debate. The Commission looked at management practices and the command chain for launch commit decisions. It found a culture that had begun to accept escalating risk, and a safety program that was largely ineffective.

| reliance on past success as a substitute for sound engineering practices (such as testing) |
| organisational barriers preventing effective communication of information and stifling differences of opinion |
| lack of integrated management across program elements |
| an informal chain of command and decision-making processes |

Fig. 4.8 Practices detrimental to safety

4.9.2   Columbia

The Columbia Accident Investigation Board found that the physical cause for the Columbia Space Shuttle's break-up during re-entry was a breach in the thermal protection system on the left wing's leading edge. This was caused by insulating foam which separated from the External Tank 81.7 s after launch and struck the wing. During re-entry, this breach allowed superheated air to melt the aluminium structure of the wing, resulting in break-up. According to the Board's report, the organisational causes of the accident were rooted in Space Shuttle Program history and culture. Cultural traits and organisational practices detrimental to safety had developed (Fig. 4.8).

4.9.3   SR-111

The Canadian Transportation Safety Board investigation into the crash of the Swissair Flight 111 MD-11 found that the accident was probably caused by an arcing event on an in-flight entertainment network (IFEN) cable, which set alight nearby flammable material. The Board's report has a long list of "Findings as to Causes and Contributing Factors". The investigation found that aircraft certification standards for material flammability were inadequate. They allowed use of materials that could be ignited and propagate fire. And the type of circuit breakers used in the aircraft was not able to protect against all types of wire arcing events. The original design philosophy had been for "non-essential" passenger cabin equipment to be powered by one of eight cabin buses. These couldn't provide sufficient power for the IFEN system that was originally planned, so another bus was used. The new design didn't include a way to deactivate the IFEN system when the pilot switched off the cabin power. It didn't provide the pilots with a procedure to deactivate the IFEN system during an emergency. There were no built-in smoke and fire detection and suppression devices in the area where the fire started and propagated. And, in the deteriorating cockpit environment, the positioning and small size of standby instruments would have made it difficult for the pilots to transition to their use, and to continue to maintain the proper spatial orientation of the aircraft. On the organisational side, the investigation found that, in the past, Swissair had relied on its MD-11 maintenance provider, SR Technics, to manage modifications to its MD-11 s. However, after SAir Group was restructured, SR Technics became a separate business entity. For the IFEN project, Swissair chose another contractor for the design, certification, and integration services. It made a separate agreement with SR Technics to provide support to the contractor. The contractor subcontracted parts of the project, and the contractor's prime subcontractor further subcontracted some of the work.

This example shows again the complexity of the product lifecycle, and why it needs to be properly managed. In this case, it's apparent that even an industry certification standards organisation plays an important role in the product lifecycle. And the complexity of the extended enterprise is seen again. There's a contractor supporting another contractor which has contracted to a subcontractor which has contracted to a sub-subcontractor.

4.9.4   Ariane 5

The Ariane 501 Enquiry Board identified the chain of events corresponding to the technical causes for disintegration of the rocket. The rocket started to disintegrate because of high aerodynamic forces resulting from a sudden change to the direction of flight. This led to the rocket self-destructing. The change of flight direction was due to erroneous data transmitted by Ariane's inertial reference system (IRS). A software exception had occurred in the IRS unit while executing a data conversion from 64-bit floating point to 16-bit integer. The exception was detected, but inappropriately handled.

The Board found that the loss of guidance and attitude information was due to specification and design errors in the software of the IRS. And the reviews and tests carried out during Ariane 5 development hadn't included adequate analysis and testing of the IRS or of the complete flight control system, which could have detected the potential failure.

Recommendations from the Enquiry Board included: improved testing; software qualification reviews for each item of equipment incorporating software; making all critical software a Configuration Controlled Item; including external participants when reviewing specifications, code and justification documents; setting up a team to propose the procedure for qualifying software; and a more transparent organisation of the partners in the program.

4.9.5   Multiple Causes

There are usually multiple causes leading to a problem. However, people often have a tendency to look for a single root cause, a single cause that leads directly to the effect, or occurs at the beginning of the series of events that leads up to the problematic effect. Perhaps they hope that, when they have identified such a single cause, they will be back in control. And it will then be easy to identify the measures needed to prevent recurrence of the effect.

The approach of looking for a single root cause may be valid when applied in an environment that is well-structured and limited in scope. However, in an environment in which activities are carried out in parallel, and in other more complex environments, there are usually multiple causes. In these environments, which include the environment of global product development, production, use and support, there is usually not just a single cause, but a network of interrelated causes. They'll all have to be understood and addressed if recurrence of the effect is to be prevented.

4.10   Causes and Measures

When the organisational, cultural, physical and technical causes have been understood, corrective measures can be identified and taken to prevent repetition of the effects.

In September 2006, in connection with the battery problem, Sony Corporation explained that, on rare occasions, microscopic metal particles in battery cells could come into contact with other parts of the battery cell, and this could lead to a short circuit, which could lead to overheating and potentially flames. Sony announced it had introduced additional safeguards in its battery manufacturing process to address this condition and provide more safety and security.

After the attempt in 2006 to steal Coca-Cola's trade secrets, Coca-Cola carried out a thorough review of information protection policies, procedures and practices to ensure that its intellectual capital was safeguarded.

The investigation into the Concorde accident in 2000 found that, during takeoff, a tyre was damaged when it ran over a strip of metal which had fallen from another aircraft (a Continental Airlines DC-10). Tyre debris was projected against the left wing, and led to rupture of a fuel tank. Leaking fuel ignited. The investigators made numerous recommendations, including strengthened tyres and strengthened fuel tank linings for Concorde.

In October 2006, in connection with the A380 delay, Airbus announced that the amount of work to finalise the installation of electrical harnesses was underestimated. Airbus announced, in a Press Release, "Beyond the complexity of the cable installation, the root cause of the problem is the fact that the 3D Digital Mock up, which facilitates the design of the electrical harnesses installation, was implemented late and that the people working on it were in their learning curve. Under the leadership of the new Airbus President and CEO Christian Streiff, strong measures have been taken, which, in addition to management changes, include the implementation of the same proven tools on all sites, as well as the creation of multi-national teams to better use the best skills available. Simultaneously, training is being organised to swiftly bring the employees using those tools to the optimum level. With the right tools, the right people, the right training and the right oversight and management being put in place, the issue is now addressed at its root, although it will take time until these measures bear fruit."

4.11   Pre-Emptive Measures and PLM

When there is a serious problem with a product, an enquiry is usually held to understand in detail what happened, to identify the causes, and to take measures to prevent the problem recurring. The same approach can be helpful even if there hasn't been a serious problem. The product environment can be reviewed to find and eliminate potential problems.

Ideally, of course, a company would want to identify potential problems with a product and take preventive action before a problem occurs. Often, though, this isn't as easy as it may seem.

In many organisations there are cultural barriers to admitting the existence of potential problems. Executives prefer to be seen to be running a problem-free, well-oiled, high-performing organisation. Not one riddled with problems waiting to happen. Engineers don't want to draw attention to potential problems, as they don't want to be seen as the source, or the cause, of the problem. Or as a troublemaker.

If potential problems can be discussed, care needs to be taken in the way they're presented. Investigations into problems with products often find many causes. An investigation into a potential problem may show that people could do a lot better. Although this shows opportunities for the company to improve, it also implies that people were not performing as well as possible in the past.

4.12   Current and Future Nightmare

4.12.1   It's a Nightmare

Most companies don't have products that cause disasters and get to be front-page news. However, that doesn't necessarily mean that they don't have the occasional problem. Most of the companies I work with haven't suffered from disasters to their products. Usually, they're just looking to improve the business, and make more money for shareholders. When we look in detail at the product environment, we often see the same kind of issues that are identified in accident investigations. There are organisational issues (Fig. 1.24), issues with data (Fig. 1.25), issues with products (Fig. 1.26), issues with processes (Fig. 1.27), and other issues such as IS issues (Fig. 4.9).

In one company, the CEO summarised the situation as "a nightmare". In another company, they called it "a horror story".

Islands of Automation	many unused parts maintained in databases
lack of good product developers	not possible to migrate data from an old to a new application
equipment under-utilised or over-booked	unwillingness to benefit from external developments, NIH syndrome

Fig. 4.9   Other issues in the product environment

Over the years, I've worked with more than 200 companies. I've seen issues like these in companies of all sizes and in all industries. Many of these companies are highly successful, with some great products and a strong five-year financial track record.

Usually, taken singly, the issues listed don't lead to major problems. However, cumulatively they can result in, at best, unnecessarily long lead times, increased product costs and reduced product quality, and, at worst, in disaster.

4.13 Global Growing Pains

The type of issues mentioned above can occur when a company only has operations on one site, and only sells to customers in a local market. When the company goes global, it's faced by additional issues. If you work for a company that develops, markets, produces and/or supports products, you can probably think of a few questions that your company needs to answer if it's looking at the opportunity of Global Products.

For which geographical markets could we offer such products? The whole world? One continent? Several continents? Just a few countries? If so, which ones? Would we introduce a new Global Product everywhere in the world at the same time? Or introduce it first in one market, then in the others?

Should we sell direct to the customer everywhere? Or should we sell through third parties? Should we sell direct in some countries, and through third parties in others? Should we provide support directly to customers everywhere? Or should we provide support through third parties? Should we sell over the Web?

Should we have the same price everywhere? Or adjust the price to each market? Should the price be quoted in our Head Office currency? Or that of the customer? If we have the same price everywhere, should we quote it in dollars, or euros, or yen, or yuan? And what happens when exchange rates change? Which prices do we change?

Should we have one product for customers throughout the world? Or should we have a different product for each continent, or even a different product for each country? Maybe we know what a potential customer in Columbus, Ohio wants, but how about customers in Seoul and Bogota? Will the same product satisfy customers in Vostok, where the temperature can drop to −129 °F, and in El Azizia, where it can rise to 136 °F? Should we have one product for everybody? Or different products for women and for men? And different products for people of different religions and cultures?

What architecture should we have for our Global Products? Should the product be modular? If so, how do we decide on the modules? How do we define the interfaces between modules? Will interfaces be country-specific? Will we have product platforms? How do platforms relate to modules? Should we have a core product that we can sell world-wide with local customisations? If we are able to make a product that we can sell world-wide, how can we retain market leadership over other companies in the world that, presumably, can do the same thing? Which of our competencies really set us apart from competitors? Which of our product features and functions set us apart?

Where will we develop our Global Products? In a single location where we can bring our best people together and give them the best tools in the world? Or, to be closer to the market, should we develop in several regional locations, even though this implies limited resources at each location? Should we develop the product in one location and then offer the same version worldwide? Or should we develop in one location, and then localise that development in different locations round the world? Or should all the locations work together to develop a common product that can then be produced with local variations? How will we know what to develop for customers in faraway places? How will we know on which development projects we should work? How will we manage development projects that involve companies in different locations with different management structures?

Will we manufacture in-house? Or should we just assemble in-house? Should we move all our manufacturing to a new subsidiary that we build up in a low-cost country? Should we outsource manufacturing? Should we always work with our "preferred suppliers"? Or should we always select on the basis of lowest-cost? And what happens if, as a result of exchange rate changes, another supplier becomes lower-cost than a previously preferred low-cost supplier? And what about design? And development? And marketing? And IS? And finance? Should we outsource them? What should we outsource? What should we keep in-house? What should we offshore? What should we keep at home? What should we insource?

How will we inform customers around the world about our products? In which language? Over the Web? On television, in magazines, in journals, in newspapers, on billboards? Should we have the same message in all countries?

How will we address regulatory issues? Should we specifically aim to meet regulations country by country? Or should we aim to have a product that will meet the toughest regulations in all countries so that we are sure we can meet all country-specific regulations?

Which business processes should we use? Which IS applications? Should we use the same processes and applications everywhere in the world? If not, what must be global, what can be local? Should we use a set of IS applications from just one vendor, and hope that will eliminate integration problems between applications in different application areas? Or should we use best-in-class applications in each area, even if they're from different vendors and don't integrate well? And where should we store the data that defines our products? How can we keep it safe from envious prying eyes?

How will we train our people? Should everybody get the same training, or should training be country-specific? Should we speak the same language everywhere?

Global Products offer the opportunity of billions of customers, greatly increased sales and vastly increased profits. But, from the above questions, it appears that developing and supporting products worldwide may not be as easy as talking about billions of dollars. There's an awesome number of questions to answer. And, apart from all those questions, can we take the risk of something going wrong?

If something does go wrong, people could die, large numbers of people could lose their jobs, billions of potential customers could hear about it, and billions of dollars could be lost.

4.14   No Silver Bullet

Sometimes it feels as if the number of difficulties appearing in product development and support is continuously increasing. And that unless a revolutionary new approach to solving them is invented, companies will grind to a halt. But such a magical solution is unlikely. Progress very rarely happens like that. Even when what may appear as a sudden breakthrough occurs, it's usually based on a succession of improvements made by different people over a long period of time. Watt's invention of the steam engine built on the work of Newcomen, Savery and others. The Wright brothers built on the research of Samuel Langley, Otto Lilienthal and others. As Isaac Newton wrote in 1676, "If I have seen farther than others, it is because I was standing on the shoulders of giants".

As it's highly unlikely that a revolutionary new approach to developing and supporting products is going to appear in the next few years, managers need to make the most of what is currently available. PLM is available, and has a lot to offer.

Chapter 5
Emergence of PLM

5.1 Product

The word "product" has many meanings and implications in the context of PLM. There is the individual physical product that's in the hands of the customer. And there are various descriptions of that product in the company or companies that develop, produce and support it. These may be on paper and/or in electronic form.

The individual product used by the customer may be just one of an identical batch of thousands, or it may be a unique product. The product may be a successor or a derivative of another product. It may be one of a product range or a product line. These may be parts of a product family. In turn these all make up the company's product portfolio.

The product may be made of many assemblies and thousands of parts. An assembly may also be made of a large number of parts. A part used in one assembly may be used in other assemblies. An assembly used in one product may be used in other products.

At any given time, most companies will have products in all phases of the lifecycle. The products may have been developed by the company itself. They may be under development. Or they may have been acquired as a result of merger and acquisition (M&A) activity.

The product may be a tangible product, or an intangible product, such as a software product or an insurance policy. The "product" may actually be a service.

5.2 Lifecycle

PLM is the activity of managing a product throughout its lifecycle, "from cradle to grave", "from sunrise to sunset".

© Springer International Publishing Switzerland 2016
J. Stark, *Product Lifecycle Management (Volume 2)*,
Decision Engineering, DOI 10.1007/978-3-319-24436-5_5

There is nothing new in the concept of a lifecycle. In 1599, Shakespeare described a lifecycle when he wrote of the seven ages of man (the infant, school-boy, the lover, a soldier, the justice, the lean and slippered pantaloon, second childhood).

The concept of a product having a lifecycle has existed for a long time in many industries. It's frequently found in industries where products (such as aircraft and power plants) have long lives. Yet in other industries, many companies have tended to ignore what happens to their product once it's gone out the factory gate.

Sometimes it's not very clear what is meant by the lifecycle, as manufacturers and users of products may have different views of the product life and the product lifecycle.

From the Marketing viewpoint there are market-oriented lifecycles. A four-stage example is product introduction, growth, maturity and decline. A five-stage example is product development, market introduction, market growth, market maturity and sales decline. Different approaches to the product's identity, pricing and sales strategy may be taken in different stages.

Raymond Vernon of the Harvard Business School developed a four-stage international product lifecycle theory showing how a product's production location changes as the product goes through its lifecycle.

And, from the global resource viewpoint, there's an environmental product lifecycle. In this lifecycle, first a natural resource (such as an ore, or oil) is extracted from the Earth. Then the resource is processed, and the processed resource is used in the manufacturing of a product. The product is used. When the product is no longer needed, the resource/waste is managed. It may be reused, recycled or disposed of.

A user of a product may think of a product having a "life" from the moment they acquire it and start using it, to the moment they stop using it, or dispose of it (Fig. 5.1).

As seen by the user of the product, there are five phases in a product's lifecycle: imagination; definition; realisation; use; disposal (Fig. 5.2).

As seen by a manufacturer of a product, there are also five phases in a product's lifecycle: imagination; definition; realisation; support; retirement (Fig. 5.3).

The first three phases are the same for the manufacturer and the user, but the last two are different. When the user is using the product, the manufacturer will probably need to provide some kind of support. For example, a car has to be repaired or serviced. Later, for various reasons, the manufacturer will stop

first, there's an idea for the product (such as a car). At this stage the car may just be a dream in someone's head
then the car is defined in detail. In other words an exact description is created. In this stage, the physical product, the car, doesn't exist and can't be used
then the product is 'realised', for example, all the parts of the car are produced and assembled in a form in which it can be used
then the product is 'used' by someone, or maybe 'operated' on their behalf
and finally the product comes to the end of its life. Some parts of it may be reused, some recycled and some disposed of

Fig. 5.1 Activities in the lifecycle

A	B	C	D	E
Imagine	Define	Realise	Use/ Operate	Dispose/ Recycle

Fig. 5.2   Product user's view of the lifecycle

A	B	C	D	E
Imagine	Define	Realise	Support/ Service	Retire

Fig. 5.3   Product manufacturer's view of the lifecycle

producing the product. Later still, it will stop supporting the product. The company "retires" the product, gradually reducing support levels, and eventually no longer providing any service.

Not only are the last two phases different for the manufacturer and the user, but there is no simple chronological relation between them. A user may stop using a product (Phase D), but the manufacturer has to continue supporting it (Phase 4) for other users. A manufacturer may retire a product (Phase 5) well before a user disposes of it (Phase E). Or a user may dispose of the product (Phase E) before the manufacturer retires it (Phase 5). Further complicating the situation, the user may dispose of the product by returning it to the manufacturer, who then recycles it.

The phases seen by the user don't occur one after the other, they overlap (Fig. 5.4).

For example, while the user is using the car (Phase D), poorly performing parts may stop working and be disposed of (Phase E). They may be replaced by parts that have been redesigned (Phase B) and newly manufactured (Phase C).

Another way to describe the five phases is shown in Fig. 5.5.

While this representation neatly shows where "End-of-Life" fits, it masks the different views seen by the manufacturer and the user.

In reality, what happens in the End-of-Life phase is not very clear for many manufacturers. According to the report of the GGI 2000 Product Development Metrics Survey carried out in 2000 by the Goldense Group, Inc. (Needham, MA), only 19 % of companies had an active product obsolescence or product retirement

A	Imagine	XXX
B	Define	XXXXXXXXXXXX
C	Realise	XXXXXXXXXXXX
D	Use	XXXXXXXXXXXXXXXXXXXXX
E	Dispose	XXXXXXXXXXXXXXXXXXXXX

Fig. 5.4   Overlapping lifecycle phases

i	ii	iii	iv	v
Imagine	Define	Realise	Useful life	End-of-Life

Fig. 5.5   End-of-Life view

activity. In 81 % of companies, old products just faded away over time as fewer and fewer orders were placed for them.

For a customer, the "product life" usually relates to the particular product they use, as in "my car is 10 years old" or "over my car's life, total emissions of carbon dioxide will be more than 50 tonnes".

For a manufacturer, the "product lifetime" is usually the time period over which a particular product is produced. After this time period, a replacement product may be available. The Ford Model T had a lifetime of 18 years. It was in production from September 1908 to June 1927. It was replaced by the Ford Model A, which had a lifetime of 4 years. It was in production from October 1927 to August 1931. The Wright Model B Flyer had a lifetime of 2 years. It was produced from 1910 to 1912. The last Wright Model B flew in 1934. It had a life of more than 20 years. The Boeing 707 was in production from 1957 to 1978, a lifetime of 21 years. Some were still in service in 2015.

An individual customer's product with a long "product life" may still be in use after it's been retired by its manufacturer.

The activities that make up the lifecycle vary from one industry to another. And their relative importance changes from one industry to another. There are many activities. They include product screening, specification, design, sourcing, costing, development, testing, release, manufacturing, packaging, operation, deployment, maintenance, repair, refurbishment, service, decommissioning, dismantling, demolition, recycling and elimination. Whatever the industry, the activities fit into one of the five phases.

The many activities in the lifecycle enable current products and services to be produced and supported. In parallel, the product portfolio is maintained. Platform products are defined and built. Derivative products follow. Product lines, product groups, and product families are created and maintained. Plans are prepared for future products and services. Projects for new products are defined and carried out. Projects are defined and carried out to modify existing products and services.

5.3   Changing Views of Products

Years ago, nobody even thought about Product Lifecycle Management. The market for most companies was local. Customer Choice was limited, "any colour … so long as it's black". In the 1950s and 1960s, getting products to market was the main focus. What happened after that was secondary, people just wanted new products. Following on from the Second World War, those were golden decades for manufacturing industry. Demand exceeded production capability. Social trends away from rural life produced more city-dwellers. Prosperity led to new customers happy just to buy a widening range of consumer goods such as cars, washing machines, wirelesses, telephones and television sets. Factories were often vertically integrated.

Most industrial goods were sold in the country where they were designed and produced. Computers were in their infancy. Development of new products was left to the engineers. At the end of their life, products went to a tip, dump, field or the sea.

However, some people were beginning to ask questions about the way products were developed, and about their effects. In the late 1950s there was the thalidomide tragedy. And in 1962, Rachel Carson published "Silent Spring", warning about the use of chemical pesticides such as DDT. In 1965 Ralph Nader published "Unsafe at Any Speed: The Designed-In Dangers of the American Automobile" and started a movement towards consumer safety. In the 1950s, Iceland became concerned that the fish stocks around its coast were being depleted by trawlers from other countries, particularly the UK. Iceland's economy was dominated by fish and fish-related activities. In 1958, it extended its fishing limits from 4 miles to 12 miles.

The first Numerical Control (NC) programs were written in the 1950's, the first CAD programs in the 1960's. They ran on mainframe computers. The first minicomputers, such as Digital Equipment Corporation's 12-bit PDP-8 were introduced in the early 1960's. By the early 1970's, more powerful minicomputers, such as the PDP-11, were used for CAD.

In the 1970s, the Oil Shock and Nixon's scrapping of the Bretton Woods agreement led to inflation and currency fluctuations. Companies in the West were worried by the increasing presence and quantity of high-quality, low-cost Japanese cars and electronics. They eventually responded with JIT, ISO 9000 and TQM. Computer power increased, while computer prices decreased. CAD, Computer Aided Manufacturing (CAM) and Material Requirements Planning (MRP) developed, and were used more widely in industry.

In 1973, Joseph Harrington published "Computer-Integrated Manufacturing". Someone even came up with the equation CIM = CAD + CAM + MRP + FA (Factory Automation) which gives the impression that there's not much more to product development and manufacturing than connecting up the computers. There was no sign of the customer or product support in the equation.

During the 1970s, the development of new products was mainly carried out by young engineers. Product support was left to the older, more experienced engineers. As young engineers are often not all that interested in customer requirements, and like to work as individuals, customers received products with unwanted functionality and little support, the product developers communicating with neither the market nor the support engineers.

During this period, system development methodologies were introduced to manage software development. These methodologies divided projects into phases, and used deliverables and approvals to maintain control. In 1970, W.W. Royce published an article called "Managing the development of large software systems: concepts and techniques." This referred to the Waterfall system development methodology (requirements, analysis, design, coding, testing and maintenance). Other methodologies had different phases, for example: initial investigation; feasibility study; requirements definition; system design; coding; unit testing; integration and system testing; implementation and system maintenance.

In the 1970's, companies had departmental organisations. Many engineering activities were carried out in series. The Engineering Department did all its work alone, then "threw the design over the wall" to the Manufacturing Department. Manufacturing found all sorts of problems with the design, and sent it back to Engineering for improvement. In the 1980's, companies started to implement Concurrent Engineering and Simultaneous Engineering methodologies to overcome the problems of serial engineering activities.

In its December 1988 report "The Role of Concurrent Engineering in Weapons System Acquisition", Concurrent Engineering was defined by the Institute for Defense Analysis (IDA) as - "...a systematic approach to the integrated, concurrent design of products and their related processes, including manufacture and support. This approach is intended to cause the developers, from the outset, to consider all elements of the product life cycle from conception through disposal, including quality, cost, schedule, and user requirements."

In the 1980's, MRP evolved to MRP2. Minicomputers gave way to workstations and PCs. Networks became increasingly powerful. CAD and CAM systems were no longer centralised. As companies started to be swamped by engineering data, the first Engineering Data Management (EDM) systems appeared. The development of ISO 10303, an ISO standard for the representation and exchange of product data, started.

In 1983, Theodore Levitt wrote an article called 'Globalization of Markets'. In 1987, the Brundtland Commission reported on Sustainable Development, defining it as development that meets the needs of the present without compromising the ability of future generations to meet their own needs.

The end of the Cold War in 1991 altered the geopolitical equation, leading to major changes in many areas. Some of them affected the Aerospace and Defence (A&D) sector of manufacturing industry.

In the 1990's, globalisation hit. A wave of imports from low-cost countries led to the price of goods dropping in advanced industrial countries. In response, production was outsourced to low-cost countries. Then product development was outsourced.

In 1992, "Engineering Information Management Systems: Beyond CAD/CAM to Concurrent Engineering Support" was published. The link was made between CAD, management of product data, and business processes.

In 1993, Joseph Pine wrote "Mass Customization: The New Frontier of Business Competition". In 1994, Michael Hammer published "Reengineering the Corporation". Many companies started Business Process Reengineering initiatives, significantly streamlining many business activities. However, little effort went into reengineering the processes of the product lifecycle.

In the 1990's, the World Wide Web, e-commerce, B2B and trading exchanges appeared. MRP2 evolved to ERP. CAD functionality became a commodity. EDM systems were relabelled as PDM systems and brought some order to all the product data. Web-based sales configurators allowed customers all over the world to purchase the exact products they wanted.

With development outsourced to different locations, and developers needing to work closely together, Concurrent Engineering, the Web and CAD morphed into Collaborative Development. Air freight, DHL and FedEx enabled rapid transport of designs, parts and goods. A presence in other countries became necessary.

Ecologists became more vocal, questioning why more products weren't recycled, and why the production and distribution processes of manufacturing industry created so much pollution. Was the real cost of shipping parts and products being taken into account? Did it make sense for the components of your Sunday lunch to travel 50,000 miles before they got to your plate? Why is aviation fuel exempt from tax? Did it make sense for cows to be transported hundreds of miles from one place to another 10 times during their lifetime so that various subsidies could be collected? Questions were asked about the true cost of the pollution from all the planes and trucks. And how should Total Cost of Ownership and Total Cost of Use be defined?

In 1997 at Kyoto, Japan, delegates from all over the world agreed on the need to reduce emissions of greenhouse gases, especially carbon dioxide. A lot of these emissions come from products such as cars, aircraft and power plants driven by petroleum products.

During the late 1990's, trillions of dollars were created in the dot.com boom. Banks, pension funds, insurance companies and mutual funds demonstrated their skills at investing their customers' money. The Nasdaq and the New York Stock Exchange lost about 8 trillion dollars in the crash that followed.

In 2000, the first version of the Global Reporting Initiative's (GRI) Sustainability Reporting Guidelines was released.

In December 2001, after admitting accounting errors that inflated earnings, Enron Corporation filed for bankruptcy protection. Arthur Andersen, one of the world's leading accountancy firms, was found guilty of shredding key documents. In July 2002, WorldCom filed for bankruptcy after $11 bn of accounting irregularities were uncovered. The Sarbanes-Oxley Act, signed into law in 2002, introduced harsh penalties for anyone found guilty of corporate wrong-doing.

The term "Web 2.0" started to be used. Adoption of Software as a Service (SaaS) and Cloud computing started. Cloud-based CAD and PDM solutions appeared. The Internet of Things emerged.

In the early years of the new millennium, Develop Anywhere Sell Anywhere Manufacture Anywhere Support Anywhere (DASAMASA) became the aim of leading manufacturers. An example would be a car that was developed at 2 sites in the USA, 3 in Europe and 2 in Japan. It was sold on a web-site in Germany to a Swede living in Australia. It was manufactured in Poland, China, Vietnam, Singapore, Mexico, Taiwan, France, Canada, the USA and Italy. It was serviced in Brazil when the Swede moved there. And it was serviced later in Mexico, where it was resold. Before finally being scrapped and recycled in California. Not just a global car. A DASAMASA car (Fig. 5.6).

Fig. 5.6 Lifecycle of a
DASAMASA car

Imagine	Define	Realise	Use/ Operate	Dispose/ Recycle
Europe	Europe	Canada	Australia	USA
Japan	Japan	China	Brazil	
USA	USA	France	Mexico	
		Italy	USA	
		Mexico		
		Poland		
		Singapore		
		Taiwan		
		USA		
		Vietnam		

In a presentation in June 2001, the President of ArvinMeritor, Inc. said, "For example, at ArvinMeritor, we manufacture a manifold for the VW Beetle that was conceived in Warton, U.K., developed and produced as components in Finnentrop, Germany. Key components are shipped, assembled and incorporated into production in our exhaust plant in Mexico. The Beetle is then sold to customers in the U.S. and Canada, and, finally, to customers back in Germany."

In 2007, financial markets peaked, with bankers paying themselves bonuses of tens of billions of dollars. The following year, the sub-prime crisis led to the biggest financial collapse since the 1930s. Lehman Brothers, one of the world's leading financial firms, collapsed. Bankers begged for tens of billions of dollars of state aid. In 2009, General Motors sought bankruptcy protection.

Faced with the financial crisis, it was the G20, formed in 1999, that acted, and not the G7. The G7 (Canada, France, Germany, Italy, Japan, United Kingdom, and United States) had been formed in 1976. In addition to the G7 countries, the G20 group also includes Argentina, Australia, Brazil, China, India, Indonesia, Mexico, Russia, Saudi Arabia, South Africa, South Korea, Turkey and the European Union.

Meanwhile, the tallest buildings in the world were no longer being built in the US. The successors to the Sears Tower, Chicago (442 m, built in 1973) were Kuala Lumpur's Twin Towers (452 m, 1988), Taipei 101 (509 m, 2004) and Dubai's Burj Khalifa (828 m, 2010). In 2015, the second and third tallest buildings were the Shanghai Tower (632 m, 2013) and Mecca's Abraj Al-Bait Tower (601 m, 2012).

Huge container ships were built, capable of transporting more and more containers. By 2015, the largest could carry more than 19,000 containers.

In the 21st Century, the global economy with its uncertain rules, unanswered questions, intense competition, ever-changing consumer preferences and financial fluctuations asks a lot from companies making products. Cost, quality, revenues, shareholder value and market-share are continual concerns for CEOs. Should they look to outsource even more activities to progressively lower-cost countries, or should they keep them close to home and close to government protection and subsidies? Air freight, travel, telecommunications, video conferencing and the Web make it easy to work with people anywhere. With development, sales, manufacturing and support activities possible anywhere, how should CEOs manage functional roles and boundaries?

5.4   Emergence of PLM in the 21st Century

Before the 21st Century, a PLM approach wasn't sufficiently needed from a business point of view, and wasn't possible from the technical viewpoint.

PLM emerged in about 2001. Before then, companies implicitly managed products across their lifecycles. But they didn't do this, even conceptually, in an explicit, "joined-up", continuous way. Instead they managed separately department-by-department, for example, in Marketing, R&D, Manufacturing and Support. And other departments, such as IS and Quality, took product-related decisions separately. Without explicit, "joined-up", management of the products across the lifecycle, things fell through the cracks (Fig. 3.8). Even if everyone in the marketing, R&D, manufacturing and support chain had done their work correctly according to company procedures, the end result could be problems such as products getting to market late, and products not working properly in the field.

By bringing together previously disparate and fragmented activities, systems and processes, PLM helps overcome the many problems, such as these, that resulted from the old unconnected approach.

5.5   A New Paradigm

PLM is a holistic business activity addressing not only products but also organisational structure, working methods, processes, people, information structures and information systems. It's a new paradigm, a new way of looking at the world of products. (A paradigm is a conceptual structure that helps people think about a particular subject.) This new paradigm for a company's particular situation of products, market, customers, competitors and technology leads to new opportunities and new ways to organise resources to achieve benefits.

PLM manages each individual product across its lifecycle, from "cradle to grave"; from the very first idea for the product all the way through until it is retired and disposed of. Managing a product all the way across its lifecycle allows a company to take control of what happens to it.

But PLM doesn't just manage one of a company's products. It manages, in an integrated way, the parts, the products and the product portfolio. It manages the whole range, from individual part through individual product to the entire portfolio of products.

The portfolio of product development projects is the project portfolio. However it's sometimes referred to as the product portfolio. This is confusing since the term "product portfolio" is also used to refer to the company's portfolio of products.

The product development projects can be put in an Integrated Portfolio with the products, providing an overview of all products, whether developed or under development. In Fig. 5.7, Product 1 has been retired, Product 2 is no longer realised

Product/Project	Phase	Imagine	Define	Realise	Support	Retire
1						X
2					X	
3				X	X	
4				X	X	
5				X	X	
6			X			
7		X				

Fig. 5.7 An integrated portfolio

but is still supported. Products 3, 4 and 5 are realised and supported. Product 6 is in the definition phase, while Product 7 is in the idea phase.

5.6   Across the Lifecycle

PLM manages a company's projects to innovate and develop products, and their related services, all the way across the lifecycle. Without new products, company revenues will decline. Innovation activities are the source of growth and wealth generation in a company, and PLM makes them more effective. PLM helps a company get control of its products and services, and enables it to take responsibility for them across the lifecycle. Mastering the activities in the lifecycle makes it easier to provide reliable products, sell services on them, and even sell services on competitors' products.

PLM has a wide scope in terms of application across a company because it's used throughout the lifecycle of a product. Customer input into product design early in the lifecycle aids customer satisfaction and identifies the demand. PLM is needed at this stage, for example, to capture product requirements. Companies want to develop excellent products, so they need PLM during research and development, when they are discussing ideas and concepts, and defining the product. They need it when developing new parts and the new product, modifying existing parts, testing prototypes, introducing the new product to market, and retiring existing products. They want to sell excellent products to their customers, so they need PLM during the sales process. They want to provide excellent support to customers, so they need PLM during the use stage.

5.7   A New Way of Thinking

At first glance, PLM may just appear to address the management of products across the lifecycle. However, that's just the tip of the iceberg. PLM offers a new way of thinking about products and manufacturing industry (Fig. 5.8).

Before PLM	With PLM
Think Product Manufacturing	Think Product Lifecycle
Think vertically about the company	Think horizontal
Think functionally about the company	Think lifecycle
Think about one activity of the company	Think about several activities
Think product development	Think cradle-to-grave
Focus on the customer	Focus on the product, and then the customer
Listen to the Voice of the Customer	Listen to the Voice of the Product
Think going forward in time	Think forwards and backwards
Think Customer Survey	Think Customer Involvement
Think product portfolio & project portfolio	Think Integrated Portfolio
Think bottom-up, starting with a part	Think top-down, starting with the portfolio
Think about the product lifecycle bit-by-bit	Think about PLM in a joined-up, holistic way
Think PLM is for the techies	Think PLM is a top management issue
Think profit	Think profit and planet
Think our Processes	Think standard processes
Think our Data	Think standard information
Think our Applications	Think standard applications

Fig. 5.8 Before PLM and with PLM

5.7.1   Thinking About Manufacturing

Before PLM, companies aimed to make a product and get it in the hands of the customer. The activities of Manufacturing and Assembly created most of the value for the company.

Thinking PLM, it's clear that, in developed countries, companies' future profits won't come from Manufacturing and Assembly of commodity products. Companies in a country where wage costs are 10 % of those in the US will be able to carry out those Manufacturing and Assembly activities at a much lower cost. To survive, companies in developed countries have to add value and create revenues elsewhere in the lifecycle. Opportunities include innovating breakthrough products, developing ideas for new environment-friendly products, providing customised products, providing services to support product use, refurbishing existing products, and taking financial and environmental responsibility for products produced in low-cost countries.

5.7.2   Thinking About the Company

Before PLM, when thinking about a company in simple terms, people broke it up into functional departments. Typical departments included marketing, engineering, manufacturing, sales, and after-sales. Jobs were defined with reference to those functions. They included marketing director, design engineer, manufacturing worker, sales associate and after-sales service person. Information systems were implemented to help the functions in their work. There were CAD systems for engineering, MRP systems for manufacturing.

Thinking PLM, people view the company in terms of product lines. They identify the lifecycle for the product line. They look to build the best design chain and supply chain for that product line. They hire people to work in its product teams. They define lifecycle processes, and implement lifecycle PDM systems to manage the data across the lifecycle.

5.7.3 Thinking About a Function

Before PLM, people thought functionally about the company. A Marketing VP, an Engineering VP and a Manufacturing VP would report to the CEO. Managers of product lines would report in through a matrix.

Thinking PLM, people first think about the product lifecycle. A Chief Product Officer (CPO) has the responsibility for all the products across the lifecycle. The CPO reports to the CEO. So do the Chief Financial Officer (CFO) and the Chief Information Officer (CIO). Product Managers report to the CPO.

5.7.4 Thinking About an Activity

Before PLM, people would think about one activity in a company at a time. In the Engineering Department, without considering the needs of other people in the lifecycle, design engineers would buy new application software to design products faster. In the Recycling Department, recycling specialists scratched their heads and wondered what they were going to do with the freshly-arrived truck-load of old products too difficult to disassemble for recycling purposes.

With PLM, people think about more than one activity at a time. They think about the product across its lifecycle. Engineers designing a product take account of how it will be manufactured. And how it will be disassembled and recycled. The recycling specialists keep up-to-date with environmental laws and keep development engineers informed. Together, they work out how to design products that can be disassembled quickly, and how to re-use parts in new products.

5.7.5 Thinking About the Product Development Activity

Before PLM, people would think about one activity in a company at a time. Product development, in particular, was often seen as a separate island, somehow disconnected from the other activities. Design engineers seemed to have problems communicating with people in other functions, so were left to work alone.

With PLM, people think about the entire product lifecycle, from cradle to grave. Product development is one of many activities in the lifecycle, and is closely integrated with the others. Design engineers talk to people in other functions to find out as much as possible about product needs and behaviour, and to learn about experience with similar products.

5.7.6 Thinking About Focus

Before PLM, the rule was "focus on the customer".

With PLM, the rule is "first focus on the product", then focus on the customer. Customers buy great products. Companies can have all the knowledge in the world about their customers, and what the customers have said, but they won't get a sale without a competitive product.

5.7.7 Thinking About Voices

Before PLM, the rule was listen to the "Voice of the Customer".

With PLM, the rule is "listen to the Voice of the Product as soon as possible". Get the product to report back about how it is working. And, of course, don't forget to listen to the "Voice of the Customer".

5.7.8 Thinking About Time

Before PLM, people thought towards the future: first came product development, then manufacturing, then support. Time goes forward. One thing comes after another.

With PLM, not only are there flows, such as time, going forwards. There are also flows going backwards. Information comes back from product use and operation to be used in product development. Feedback about the use of one generation of a product helps improve future generations. Products that have reached the end of their life are disassembled, and some parts are reused in the start-of-life of new products.

5.7.9 Thinking About Customers

Before PLM, people would carry out a Customer Survey to find out what customers thought of existing and future products.

With PLM, people think Customer Involvement. Using technologies such as mobile telephony, GPS, Radio Frequency Identification (RFID) technology and the Internet of Things, they exchange information directly with a customer who's using the product. Getting feedback from a customer at the actual time of use provides more valuable information than a survey form.

5.7.10   Thinking About the Portfolio

Before PLM, people in Marketing and Sales would refer to the product portfolio. This was the portfolio of existing products. Meanwhile, people in Engineering would refer to the project portfolio. This was the portfolio of new products in the pipeline.

With PLM, everyone in the lifecycle refers to the Integrated Portfolio which contains both the existing products and those under development.

5.7.11   Thinking About the Product

Before PLM, people would think bottom-up, starting with parts and building up to the product. After parts were developed, it would be found that they didn't fit into assemblies. So they were redesigned. Assemblies were redesigned to fit together. Companies developed some parts in one CAD system, then found, because of differences in CAD data representations, that they didn't fit with parts developed in other CAD systems. Engineers developed data transfer software to address the problem. Engineers wasted time down at the level of bits and bytes, instead of focusing on the product.

With PLM, people start by thinking about the Integrated Portfolio, then work down through product families, platforms, and modules, to products, and then to parts. PDM systems manage the information about a product across its lifecycle. Maybe a company will outsource design engineering, and won't need a CAD system. Maybe a company will outsource Manufacturing. Maybe Assembly will be limited to putting together a few dozen major modules. Engineers focus on the product, which creates value, and not on the bits and bytes.

5.7.12   Thinking About the Product Lifecycle Approach

Before PLM, the many product-related issues weren't considered together. For example, Product Recall, Product Development and Product Liability would be addressed separately and independently.

With PLM, all the product-related issues are united under PLM and are addressed together in a joined-up way. The approach is holistic. PLM is seen as the way to address all the product-related issues.

5.7.13 Thinking About the Management Role

Before PLM, product-related issues weren't considered to be a subject for management.

With PLM, top managers understand and can formulate the need for effective product lifecycle management. They define the key metrics. And how the activity will be managed.

5.7.14 Thinking Profit or Planet

Before PLM, companies often put profit before the planet. They fouled the air, the water and the land. They burned wood, coal and oil as if there was no tomorrow. Major corporations boasted of their success while much of mankind lived in poverty and squalor, and died young of preventable hunger and disease.

With PLM, companies think profit and planet. They take more account of non-financial issues, such as the environment, social issues, health, education and sustainable development.

5.7.15 Thinking About Processes, Data, Applications

Before PLM, companies thought about ourProcesses, ourData, ourApplications. In the extended enterprise environment, each inter-organisation interface (process, data, or application) was a source of chaos, adding costs and slowing down life-cycle activities.

With PLM, companies think of the standard processes, standard data and standard systems that they, and their numerous suppliers, customers, and partners in the extended enterprise environment, can use to save an enormous amount of time and money.

Chapter 6
Opportunities and PLM

6.1 Opportunities of a Growing Market

Globalisation has increased the number of potential customers for many companies. The world headcount continues to grow by more than 100,000 per day, promising even more customers in the future. The world population is expected to rise from 7.3 billion in 2015 to 9.6 billion in 2050. That's an awesome number of customers, but most of them are in faraway locations. PLM will be needed to keep control of the products they acquire.

New markets create new opportunities. U.S. Census Bureau data shows that out of a 2015 US population of 320 million, 40 million were aged 65 and over. By 2050, of the projected 400 million population, 83 million are expected to be aged 65 and over. The 85 and over population was about 6 million in 2015, and is expected to rise to nearly 20 million by 2050. As the population of elderly people increases, so does the need for special health products and services.

The population of overweight and obese people has grown rapidly. About one-third of US adults are overweight, another third are obese. The percentage of children and adolescents defined as overweight doubled in the last quarter of the 20th Century. The market for weight reduction products is over $100 billion worldwide.

Concerns about health create new markets for pharmaceutical products and medical equipment in developed and developing countries. In the UK for example, it was estimated in 2003 that, on average, each person would take about 14,000 pills, tablets and other forms of medication in their life.

There's a growing market for drugs for domestic animals. For example, in the US, the market size for drugs that relieve arthritis in dogs is more than $100 million.

New markets create opportunities. As the average working time drops in developed countries, the leisure market expands. As populations move away from subsistence living, and disposable income increases, the market for products such as cosmetics expands.

© Springer International Publishing Switzerland 2016
J. Stark, *Product Lifecycle Management (Volume 2)*,
Decision Engineering, DOI 10.1007/978-3-319-24436-5_6

6.2   Technology Opportunities

New technologies open up new markets and lead to new products. The transistor, which was invented in the late 1940s, led to a seemingly endless stream of electronics and communication products throughout the second half of the 20th Century. The invention of the computer led to other new products such as software applications. Biotechnology appeared in the early 1970s, leading to countless new drugs.

Towards the end of the 20th Century, new technologies such as nanotechnology, the control of matter at the nanometre level (1 nm = 10**−9 m), appeared. They opened the way to even more new products. Down at the nano level of individual molecules, matter can be arranged and assembled to make products such as new drugs, plastics and electronic circuits.

Existing technologies such as electronics, computing, telecomms, robotics and biotechnology continue to offer scope for new products. For example, designer drugs will be developed to match an individual's particular genetic make-up. Intelligent clothes will change performance as the weather changes and the wearer's mood changes.

The Internet, the World Wide Web and the Grid all offer opportunities for new products and services, and new ways to develop, sell and support products. The Global Positioning System (GPS) underlies many new products. Mobile telephony offers new opportunities, as do portable computers and other portable devices. Further developments will create and meet needs that had not even been thought of before.

Cyborg technology (part human, part machine), with electrodes that pick up brain activity and control the machine part of the cyborg's body, offers new possibilities. Direct brain implants of memory and processing power will increase human performance. People will have their individual web address. Sensors implanted in the body will monitor organ performance. Results will be automatically transmitted and viewable in real-time on personal web-sites. Initially, uses will be passive, for example showing heart beats and brain activity. Later cyborg behaviour will be controlled directly by changing parameters on a web form or from a mobile device.

RFID technology allows products to be tagged with chips that can provide information about the product when they are scanned. This allows products to be tracked throughout their lifetime. RFID offers opportunities to get a better understanding of the way products behave over their lifecycle. Many products will have their own web address, with performance data and feedback being available online.

Environmental requirements and the desire for sustainable development will open up new opportunities, in particular for a product's end of life.

Business models will change, with more and more services being offered around products. PLM will enable the development and support of new products and services.

"seeing", "feeling", "reading" and monitoring	with various types of sensors
"speaking"	with a voice synthesiser
moving	with motors
locating	with GPS
showing information	on a display
"thinking and calculating"	with a microprocesser
remembering information	with a memory
self-identification	with a memory
sending information over a network	with a transmitter

Fig. 6.1 Smart functions

6.3 Smart Product Opportunity

Smart Products, also known as Intelligent Products, are products that can sense and communicate information about their condition and environment. In addition to their primary functionality, they have functionality to decide or communicate about their situation or environment. A washing machine has primary functionality to wash clothes. A Smart washing machine, equipped with a scanner, can read the labels on clothes, and select the most appropriate washing and drying cycle. Packaging can be Smart. For example, labels in transparent foil around meat products can change colour from blue to red when the temperature rises above the safety limit.

There are several types of smart functionality (Fig. 6.1).

Many different types of smart products and smart packaging are available and they carry out numerous tasks. For example, a smart lawn mower can be programmed to cut the grass for you. Its sensors see if there are any obstacles, identify the height of the grass, and switch on its motors to go down the garden and cut the grass. Smart vacuum cleaners have similar functionality to make life easier for their owners. A smart microwave oven can identify the food to be cooked, then set the timer and the temperature. A smart water softener can identify the hardness of incoming water, and treat it as required by its hardness and the intended use.

As well as working independently, smart products can also work together. For example, smart home appliances such as an electric blanket, a toaster, a coffee-maker, a bathroom scale and a blood pressure monitor can be networked together to make life better for their user.

Voice applications include the scales that speak your weight and the voice in your car that reminds you to fasten your safety belt. Smart security devices can listen to your voice and look into your eyes before giving you access to a secure zone.

Smart products have many applications. They can work in dangerous areas to render them safe. They can identify End-of-Life parts and materials that can be reprocessed or reused.

6.4   Opportunity of Global Products

Global Products provide huge opportunities. They allow billions of people to benefit from products to which they previously had no access. They allow companies to offer products to a global market of more than 7 billion customers and users. The resulting opportunities for sales and profits are enormous. So are the potential risks.

For most companies it's only recently that such opportunities have been available. In the 1990s, although many companies were international, or multi-national, only a few were able to offer a product throughout the world. Others were limited, for one reason or another, to smaller markets. There are several reasons for the changed situation (Fig. 6.2).

As a result of the changes, the potential market for most companies is no longer a few hundred million customers for the product in a local regional market, but over 7 billion customers worldwide. Which means that, for many companies, the potential market is already more than 20 times larger than before. And the market is expected to grow to 8 billion by 2025 and 9 billion by 2040.

The unit of measure for use of consumer products is now the billion. There are more than a billion PCs in use. There were over 4 billion mobile phone accounts in 2009, more than 5 billion in 2010, and more than 6 billion in 2014. In 2010, there were more than 2 billion Internet connections. In 2014, there were 3 billion. In 2006, the world's airlines carried more than 2 billion passengers. In 2014, they carried more than 3 billion. Billions of items of clothing and footwear are sold each year. There are more than a billion vehicles (cars, trucks, buses, motorcycles, and bicycles) on the world's roads. More than a billion people wear a watch. There are more than a billion copies of word processing software. Only founded in 2006, by the end of 2009 Twitter enabled more than a billion tweets per month.

Nearly 10 billion embedded systems were produced in 2009. These are the very small devices, usually with computing, control and communication capability, that are built into more and more products to provide new functionality. The number of devices connected to the Internet of Things in 2020 is estimated at more than 20 billion.

There's nothing to stop a company taking a slice of the billion-customer market. For example, in the early 1980s, US manufacturers dominated the world market for large civil aircraft. Yet 20 years later, Airbus, which is part of the Airbus Group, had taken a 50 % share. The most successful Airbus product family is the A320 family (which includes the A318, A319, A320 and A321). By 2013, more than

| the end of the Soviet Union in 1991 |
| economic reforms in China that started in 1978 |
| economic reforms in India that started in the 1980s |
| reduced trade barriers |
| improved travel, transport and telecommunications |

Fig. 6.2 Some reasons for the changed situation

10,000 aircraft of the A320 family had been ordered, and more than 5,000 delivered. With a catalogue price of about $50 million, that's trillions of dollars of sales.

In 2006, sales of the BMW brand, at nearly 1.2 million vehicles, were close to double the figure of about 700,000 recorded in 1998. In 2014, BMW sold 2.1 million vehicles. BMW's home is in Munich, Germany. It started producing vehicles in the US in 1992. By 2006 it manufactured at 22 sites in 12 countries on 4 continents. In 2014, it had a production network of 30 locations in 14 countries. And a research and innovation network of 12 locations in 5 countries.

Toyota's annual vehicle sales in the 1970s were about 3 million. They grew to 4 million in the 1980s, and 5 million in the 1990s. Production rose to 8 million vehicles in 2006, and 10 million in 2014. Toyota began exporting to the US in 1957. In 2006, Toyota had manufacturing companies in 28 countries, design and R&D centres in 5 market regions (North America, Europe, Australia, Asia and Japan), and marketed its vehicles in more than 170 countries and regions of the world.

There are similar examples from other industry sectors. In 2005, sales of The Coca-Cola Company reached 20.6 billion unit cases. That's nearly 500 billion servings. The Coca-Cola Company sold more than 400 brands in over 200 countries. The geographical sales split in 2005 was 28 % North America; 6 % Africa; 9 % East, South Asia and Pacific Rim; 16 % European Union; 25 % Latin America; 16 % North Asia, Eurasia and Middle East. In 2014, the company had 20 billion-dollar brands and sold 1.9 billion servings a day. 28.2 billion unit cases were sold worldwide. The geographical sales split was 29 % Latin America; 21 % North America; 21 % Pacific; 15 % Eurasia and Africa; 14 % Europe.

The opportunities of Global Products aren't limited to just a few large companies with thousands of employees. The opportunities are also there for small and medium companies, with tens or hundreds of employees. There are millions of these companies throughout the world. The smaller company may sell its product direct to end users and consumers worldwide. Alternatively it may supply its product to a larger company, operating worldwide, that will include it in the products it offers to its customers.

6.5 Social and Environmental Opportunities for Products

The world faces increasing social problems.

Society has always had an impact on manufacturing industry. Years ago, among the most highly visible effects of manufacturing industry were the factory chimneys and coal-burning fires that polluted cities. From the 1850s, London, England suffered from smog, a mixture of fog and smoke resulting from the combustion of coal. In 1952, a smog led to 4000 excess deaths. This was a key event in environmental history. Laws were passed requiring the use of cleaner fuels. Nevertheless, as late as 1962, London experienced a smog with 340 excess deaths. For a time, London then had cleaner air, but it now suffers from photochemical smog which occurs when sunlight acts on nitrogen oxides in vehicle exhaust gases to form ozone. In addition, incomplete combustion of fuel leads to the production of carbon monoxide, a colourless, odourless, poisonous gas. Having removed industrial pollution from their cities, advanced industrial countries now introduce laws concerning emissions from cars, disposal of cars, and disposal of electric and electronic goods. Other effects of manufacturing industry that are of concern to society include acid rain, global warming and the ozone hole. Initially, in the name of Progress, much is accepted, but eventually society catches up and legislates against dirty, poisonous products that kill and pollute. PLM will play a key role in addressing all these issues because it provides the opportunity to get control of products across their lifecycles.

Sustainable Development was defined as "development that meets the needs of the present without compromising the ability of future generations to meet their own needs" by the Brundtland Commission 1987. It's a holistic concept that aims to unite economic growth, social equity, and environmental management. The problems it addresses, and the ideas for their solution, are not new. Over the years, population growth, lack of disposal sites, and scarce natural resources have led to all sorts of reduction, reutilisation, recycling and recovery programs. As far back as 1958, Iceland started to extend the fishing exclusion zones around its coasts to prevent over-fishing, particularly by UK trawlers. In 1975, Iceland extended its exclusion limit to 200 miles from its coastline and a "Cod War" ensued. The UK deployed about forty ships, including more than twenty warships, to protect its forty trawlers. When Iceland threatened to close a NATO base, the UK agreed to Iceland's requirements. In 2003, the fishing grounds around Iceland were among the few in the world that weren't dangerously over-fished. World-wide, 90 % of major fish species are fished-out. It's easy to over-exploit natural resources such as oil, water, farmland, fishing grounds, forests and minerals. It requires more thought to use them in a sustainable way.

6.6 More Opportunities for Products

6.6.1 Unsolved Problems

The world has many problems in need of solutions (Fig. 6.3).

6.6.2 Future Changes

Although many problems have not been solved, new and enhanced approaches to product development and distribution have appeared. They include Shareware, Open Source, microcredit and microsavings, generic products, and Fair Trade for exports from developing countries to developed countries. The role of

Area	Problem
Environment	Global Warming, threatening the flooding of many cities and states
Poverty	The World Bank estimated that over 1 billion people live on $1.25 a day or less
Unclean Water	More than a billion people live without safe drinking water and electricity
Hunger	The United Nations Food and Agriculture Organization estimated that 805 million people don't have enough to eat. Every 10 seconds, a child dies from hunger-related diseases.
Slums	In 2000, about a billion people lived in slums. According to current trends the number will rise to 3 billion by 2050
Disease	Thousands of people die each day of curable diseases and illness. For example, diarrhoea, usually caused by lack of clean water, kills more than half a million children every year. In 2009, according to the World Health Organisation (WHO), about 1 billion people were affected by Neglected Tropical Diseases such as schistosomiasis and trypanosomiasis. Rabies kills more than 50,000 people each year. According to the WHO, by April 2015, more than 10,000 people had died in the Ebola outbreak that started in 2014.
Accidents	The WHO estimates 1.2 million people die in road crashes each year. Another 50 million are injured. In the US, about 100 people die each day. Estimates for the annual cost of motor vehicle crashes in the US range from about $75 billion to over $200 billion.

Fig. 6.3 Problems in need of solutions

non-governmental organisations (NGO), foundations and associations has greatly increased. For example, the Bill & Melinda Gates Foundation has committed billions of dollars in global health grants to organisations worldwide.

Many problems remain to be solved, creating many opportunities for product developers. It's to be expected that, in coming years, many other opportunities for product developers will result from technological advances and from customer demands. Of course, it's possible that the future won't be so rosy. World trade patterns could change radically, for example because of wars, revolutions, terrorism, epidemics, global warming, and/or increased radiation disrupting communication systems.

6.6.3   Balance of Power

In 2014, the G7 countries of Canada, France, Germany, Italy, Japan, the UK and the US accounted for about two-thirds of the world economy, and 10 % of the world population. Assuming though, that progress continues along a similar path as in recent years, by 2050 the seven countries with the largest economies could include China, India, Brazil, Mexico and Russia. They would have a combined population of nearly 4 billion, nearly 50 % of the world population. Meeting the needs of large populations of poor people may seem more important to their governments than sending profits to foreign companies. They may have the economic power to change the rules of world trade to better meet their needs. For example, they may change the rules protecting property rights that render many products unavailable at prices that people throughout the world can afford. They may consider that true innovation should be justly rewarded, but people should not die because of laws overly protective of past activities in other parts of the world. They may find that state-owned manufacturing companies can be as effective as privately-owned profit-driven companies at producing mature "human rights products" and commodity products.

In such a scenario, with NGOs and state-owned companies developing and manufacturing basic low-cost products for a large part of the world's population, foreign privately-owned companies could be led to focus on developing new, advanced, highly complex, high-value-adding products. They would still be offering products globally, and the need for PLM would be similar. There would still be the need to collaborate, for the development and manufacturing of products, with different types of organisation worldwide. And they would still need to be in total control of their Intellectual Property.

6.6.4   Increased Regulation

As products are so important to customers, governments and investors, it's likely that regulation, certification and audits will increase. Company performance in areas as diverse as community contribution, environmental performance, pollution, sustainability and R&D performance could be a target for new or enhanced regulation or certification.

Currently, many characteristics of a company, such as the financial figures, and the compliance with quality standards, are audited by external organisations. However the value of a company's products isn't audited in a standard way. As PLM becomes increasingly important, and more and more standard processes are used, the company's products will be seen as central to its value. They'll be audited for the good of users and investors. Auditors could report on the expected future value of a company's product portfolio, a useful figure for investors. PLM will play a key role in this valuation.

The different requirements and regulations in different countries and trade zones will complicate the activity of compliance. PLM will continue to play a role in keeping all the activities and documents under control.

6.6.5   Better Managed Product Companies

In the global product environment, the potential for problems with products is magnified. (For example Toyota recalled about 8 million vehicles in late 2009/early 2010. GM announced recalls of more than 30 million vehicles in 2014.) In response to the increased risks with global products, more emphasis will be put on risk management activities. They'll be needed to help avoid major losses and to assure major benefits are achieved. Product development projects will be managed better to be sure that products get to market on time. The Product Portfolio will be managed better to maximise product value.

In the 1990s and the early years of the twenty-first century, many companies made good progress with Lean Manufacturing in production. Increasingly they will look to make similar improvements in white-collar areas. Targets of 50 % reduction seem likely. Standard processes, applications and documents will be used wherever possible to cut down on waste.

Increasingly, society is expecting producers to take responsibility for their products. In coming years, companies can expect to have to take more responsibility, which will mean getting more and more control over the product lifecycle. The penalties for problems resulting from products, and their related processes and services, will probably increase.

In the early twenty-first century, there's little automated feedback from products. In coming years, products will send back much more information to their manufacturers. It will be used to support product use and disposal, and to develop new generations of products.

Increased counterfeiting is one of the effects of globalisation. Companies will manage their product IP better to reduce the risks of IP theft and counterfeit products. Unique identifiers for products will help improve traceability, IP protection and customer support.

6.6.6   Multitude of New Products

Bundling existing products together and offering them as an easy-to-use solution is one way to meet customer requirements. It can be a quick way to bring new products to market.

Tens of thousands of new products will be developed each year. With about a million engineers graduating each year from India and China, more and more products will be developed in Asia.

Companies in the West will have to improve their product innovation capabilities, and their potential for collaboration with companies in Asia.

6.6.7   More Web-Based Product-Related Services

A first phase of product-related services on the Web led to on-line sales and auctions of products and services. As the Internet of Things evolves, and becomes more and more ubiquitous, all the functions associated with a product will need to be available over the Web, leading to Web-based Product Lifecycle Management. Ideas for new products will be generated over the Web. New products will be developed over the Web. Product information will be available on the Web. The product will be serviced over the Web. And the Web will play a part when the product reaches the end of its life.

6.6.8   Breakthrough Computer Aided Product Development

With a need for faster innovation of products, applications to suggest new products will appear. Although products that will appear in five years may be unknown today, it's sure that they will exist. Companies need to find ways to identify them sooner.

With products becoming increasing complex, there'll be an increased need for simulation of a product's physical and financial performance before large sums are invested in its development.

6.7   So Much Opportunity

PLM has the potential to solve the problems in the product lifecycle and in new product development (Fig. 1.23). It also enables companies to seize the many market opportunities for new products in the early 21st Century.

The number of opportunities opening up in the 21st Century seems boundless. Perhaps it was too risky to pursue them when the product development process was out of control, production runs in faraway countries had unexpected problems, and customers complained continually about product problems. But that was before PLM. Now PLM's here, allowing companies to develop and support tiptop services and products across the lifecycle.

Surely, with so much opportunity, new products should be rolling off the production line at an ever-increasing rate. And revenues should be going through the roof.

6.8   Response to Opportunity

PLM opens up a huge number of opportunities and benefits. But resources will have to be applied to achieve those benefits. And they will have to be organised effectively.

Companies need to understand which opportunities and benefits are relevant for them. They have to make sure they don't miss out on a potential benefit, or aim for the same benefit twice. Their response to the opportunities has to be matched to the benefit they want to achieve. If they make no response, they'll get no benefits.

The opportunities are similar for all companies. The achievements of different companies will be very different, ranging from great success to the opposite.

Airbus sales grew in a market against competition from aircraft such as the Boeing 747, the McDonnell Douglas DC-10 and MD-11, and the Lockheed L-1011. In 1997 McDonnell Douglas became part of The Boeing Company. MD-11 production stopped in 2001. The L-1011 was the last large civil aircraft built by Lockheed. In 1995, Lockheed became part of Lockheed Martin. Boeing is now the only US producer of large civil aircraft.

Sales of BMW and Toyota vehicles more or less doubled in a decade. Ford Motor Company's worldwide vehicle unit sales of cars and trucks stood at 6.6 million in 1995, 6.8 million in 2005, and 6.3 million in 2014.

The Good News about PLM is that it offers companies the opportunity to address larger markets, to develop a great product, sell it to billions of customers and users, and rack up huge profits.

The Bad News about PLM is that developing and supporting products worldwide isn't easy. There are a lot of questions to be answered, a lot of choices to be made, and a lot of decisions to be taken. Once these decisions have been taken, all sorts of problems can occur. Products can take a lot longer to develop than planned. The wrong product can be developed. Problems can occur during manufacturing. When the product is in the market, a competitor may bring out a better product that will eat into sales. And problems can occur during use, all the way through until the product's end-of-life, where there can be even more problems waiting.

However the Best News, for some, is that their companies will understand the opportunities, questions and potential problems of PLM. And then they'll develop appropriate responses for their products. And for their capability to deploy those products to customers throughout the world.

6.9 From Opportunities to Detailed Benefits

It's great to talk of awesome opportunities on a global scale. It's equally important to be able to detail the resulting benefits for a particular company.

One way to understand the benefits of PLM is to focus on the revenue increases it can provide. Revenue increases can be achieved in many areas (Fig. 6.4).

Type of Increase	PLM Involvement
increase the number of customers	by developing and supporting new products
increase the product price	increasing product quality enables justifiable price increases. New functions and features can justify higher prices. Being first to market enables pricing premiums
increase the range of products that customers can buy	for example, by improving product structure management, PLM enables more customer-specific variants. It enables companies to expand the size of their product portfolios. It enables breakthrough products that can create new markets
increase the number of products of a particular type that a customer buys	by increasing product quality, PLM allows customers to dispense with second sourcing
increase the re-ordering percentage	by increasing product and service quality
increase the buying frequency	by getting products to market faster and more frequently
increase the service price	by using PLM to improve the quality of existing services
increase the range of services	by using PLM to develop and support additional services
get customers to pay sooner	by developing and delivering products faster
increase sales of new products	by introducing innovative new products
increase sales of mature products	by lengthening the life of a product. Make more frequent product enhancements, product derivatives, niche offerings, and add-ons to product platforms. Offer simple enhancements to mature products

Fig. 6.4 Revenue increases with PLM

Type of Reduction	PLM Involvement
reduce direct labour costs	Specialists across the lifecycle, for example in engineering, production and the field, waste a lot of time on data retrieval and management activities. PDM systems, which are components of PLM, can do this work for them, leaving them more time for value-adding activities. As a result, fewer specialists will be needed
reduce overhead costs	In an effective, "joined-up" PLM environment, a lot of the paper-shuffling, data re-entry, data formatting, and administrative work that is currently carried out by many people (such as administrators, supervisors, coordinators, assistants, gofers, checkers, filing staff, data entry clerks, inspectors, documenters and BOM conversion staff) across the lifecycle will be eliminated. As a result, fewer people will be needed for these tasks
reduce material and energy consumption costs	In the PLM environment, people will have better information, allowing them to take better design, engineering and purchasing decisions that will lead to reduced production costs. Digitally simulating production facilities and production processes for all the possible configurations of a new product will reduce costs
reduce the cost of purchased goods	More accurate and more detailed information will be available with PLM, allowing people to negotiate better prices for purchased products and services
reduce the cost of quality (COQ)	PLM will reduce the number of errors made along the marketing / engineering / production / delivery / service chain enabling a reduction in scrap and rework, non-compliance costs, penalty costs, warranty costs, recall parts, erroneous order and production of parts, obsolete parts, product liability costs
reduce costs of storing information	Information will be stored on low-cost, compact media rather than on space-eating paper
reduce costs of communicating information	Information will be transferred quickly and cheaply by electronic means rather than by the slow and expensive transport of paper documents
reduce the cost of space used	Once all the information currently stored on paper has been transferred to electronic media, the buildings, rooms, vaults, cupboards and drawers currently used for paper storage can be used more usefully. Once headcount has been reduced, fewer offices will be needed
reduce costs of holding finished inventory and work in progress	PLM applications can enforce use of standard parts and catalogues, enabling a reduction in stocks. Examination of the detailed descriptions of all the existing parts and products in use will show many similarities and a high potential benefit from rationalisation
reduce costs of equipment and tooling	Examination of detailed descriptions of all processes and machines in use will show many similarities and the potential for rationalisation
reduce IS costs	A PDM system can act as a central repository and infrastructure for the company's product data. As a result, some of the company's existing applications will no longer be needed. Many of the interfaces between these applications and other applications will no longer be needed
reduce costs of product development projects	PLM helps make project progress clearer. Action to prevent increased costs can be taken earlier
reduce costs of customer acquisition	Requirements Management applications, which are components of PLM, help clarify exact customer requirements and reduce costly changes to requirements

Fig. 6.5 Cost reductions with PLM

Another way of understanding the benefits of PLM is to focus on the ways it helps cut costs. Costs can be reduced in many areas (Fig. 6.5).

Chapter 7
Product

7.1 Product Importance, Range, Instance

7.1.1 Importance

The product is important. Whether it's a chair, a beverage, an aircraft or an anaesthetic, it's the product, and perhaps some related services, that the customer wants. The product is the source of company revenues. Without a product, the company doesn't need to exist and won't have any customers. Without a product, there won't be any related services.

The company generates revenues from an on-going stream of innovative new and upgraded products. Great products make it the leader in its industry sector. Great products lead to great profitability.

7.1.2 Range of Products

There's a huge range of products. There are tangible products, products you can touch, products such as a computer or a car. And there are intangible products such as software, insurance policies and mortgages. There are products as diverse as an Airbus A380 and a dollar bill, a book and a beverage.

Products come in all sorts of shapes and sizes. The movement of a Swiss watch may be little longer and wider than a postage stamp, and only a few millimetres in thickness. A postage stamp is even smaller. Many other products are much larger. For example an Airbus A380 is 73 m long, with a wingspan of nearly 80 m.

The contents of products are diverse. Some products contain mechanical, electronics and software components. Others contain components of animal or plant origin. Most of the tangible parts of this product, this book, came from forests.

© Springer International Publishing Switzerland 2016
J. Stark, *Product Lifecycle Management (Volume 2)*,
Decision Engineering, DOI 10.1007/978-3-319-24436-5_7

A product may actually be a service. A product can also be a package of services, or a bundle of products and services, or a solution containing several products, or a solution containing products and services.

7.1.3   More than the Product

The product is often more than what seems, at first glance, to be the product. Product packaging is often a part of the product. So is product labelling. The product may include wires, plugs and other components that connect it to the outside world. The product may include product literature, such as user documentation or regulatory documentation. The product may be a six-pack or a single can. If it's a six-pack, it may have additional packaging, but the product you drink is the same as if it's a single item. The delivery mechanism may be part of the product. Inside the packaging of an anaesthetic may be a sterile syringe.

7.1.4   Instance of a Product

At home, among your products, you may have a washing machine. The manufacturer may have produced 1,000 identical machines, yours and 999 others. Yours is an "instance" of a "series". If the manufacturer produces 100 identical machines every month for a year, then your machine is an instance from a batch which is part of a series.

The use of words in the PLM environment isn't standardised. For example, sometimes the word "product" is used to refer to an instance, sometimes to a batch, sometimes to a series.

7.1.5   Number of Products

There's a huge number of different products. Many companies have thousands of different products. In some industry sectors, there are hundreds of thousands of products. For example, even before the European Community regulation on the Registration, Evaluation, Authorisation and Restriction of Chemical substances (REACH) came into force in 2007, more than 140,000 substances had been pre-registered.

It's virtually impossible to know the exact number of different products, but there must be tens of millions, if not hundreds of millions. And the number of instances is probably in the trillions.

7.1.6   Commonality

Although there is such a huge number and wide range of products, from a product lifecycle management viewpoint, much is common between different products. Figure 7.1 shows some of the issues that need to be considered for most products, and will be addressed in this chapter.

7.2   Parts, Ingredients, Components, Assemblies

7.2.1   Range of Parts

Products are often made up of many "things". Depending on the type of product, one of these things may be called by a name such as a part, a piece, an item, a component, an element, a module, a sub-assembly, an assembly or an ingredient.

The things that make up a product can be very different. Ingredients of a shampoo could include Ammonium Laureth Sulfate, Ammonium Lauryl Sulfate, Sodium Chloride and Glycol Distearate. Ingredients of a deodorant could include Cyclomethicone, Stearyl Alcohol, Aluminium Zirconium Tetrachlorohydrex GLY, and PPG-14 Butyl Ether. A sandwich could contain flour, water, milk, eggs, salt and sugar. The parts that make up a Personal Computer could include a case, a screen, a keyboard, a battery, a processor, system memory, a hard drive, a Communications device and a power adapter.

7.2.2   Number of Parts

As Fig. 7.2 shows, many products contain a lot of parts.

Numbering, Naming	Product Portfolio	Performance
Content, Structure, Architecture	PRODUCT	Ownership, IP
Representations, Model	Requirements	Classification

Fig. 7.1 Subjects common to many products

Product	Typical number of parts
Deodorant	20
Sandwich	30
Shampoo	50
Watch	300
Machine tool	2000
Car	25000
Aircraft	400000
Space shuttle	2000000
Application software (lines of code)	20000000

Fig. 7.2 Typical number of parts in products

A company's product may be made of many assemblies and thousands of parts or components or constituents or ingredients depending on the type of product. An assembly may also be made of a large number of parts. These assemblies and parts could be made by the company itself, or could be the products of other companies, its suppliers. Many products contain industrial components (products) of various types, such as hardware, software, electrical, electronic and chemical. Many products also contain other types of components, such as agricultural, forestry and fishery products.

Just as the term "product" may refer to an instance of a product, or to all products of a certain type, the word "part" is also used in different ways. It may be said that "this product is made up of fifty parts". But that phrase could have several meanings. Sometimes, standard parts such as screws and bolts won't be included in such a count. And the number of parts will only refer to the parts specifically designed for the product. Sometimes parts such as wiring will be excluded from the part count. Sometimes the individual parts within a purchased part will be excluded from the part count, and the purchased part considered as one part. Sometimes the part count will refer to the number of different parts so, although there may only be fifty different parts, there may actually be a total of eighty parts.

7.2.3   Part and Product

Sometimes a "thing" will be referred to as a product in one context, but a part in another context. For example, if a bag to carry your PC is included in the original packaging, it may be seen as a part of the product you buy. However, if you buy the bag separately from the computer, then the bag is a separate product.

7.3   Identifier

7.3.1   Need for an Identifier

There are so many products, so many parts in products, and so many instances of a particular product, that special identifiers are needed to know exactly which "thing" is being referred to in situations as diverse as defining a product, assembling the product, controlling stock levels, ordering, billing, accounting, and handling complaints and returns. The number should be unambiguous so that it's clear to which thing it refers.

7.3.2   Name, Number

Products and parts often have several identifiers. An identifier may be a number and/or a name. A product or part may have a model number, or a series number or a type number. It may have an company-internal number and name. It may be known by other names and numbers to the rest of the world. It may have a code name or a project name during its development. It may have other names in production. As well as the name and number, it may also have a description. The manufacturer may have one description for the product. A retailer may have another description for the same product. And a customer yet another.

The product may have a series identifier (such as BMW Series 3) and an extended series identifier (such as Series 3 Sedan, or Series 3 Touring or Series 3 Convertible). Your instance of the product may have a specific batch number or serial number.

7.3.3   Internal, and Other, Names/Numbers

Many consumer products have SKU (Stock Keeping Unit) numbers. A SKU is an identifier used for tracking the product once it's been made and is in a warehouse, or is in a sales location. A product can have a SKU for when it's sold singly. And another SKU when it's sold as a group, for example, three bottles of shampoo wrapped in plastic, and sold for the price of two. Similarly, beverages will have different SKUs when sold individually, or as a six-pack.

7.3.4   Serial Numbers

Sometimes things are numbered according to a serial numbering system, sometimes according to a significant numbering system. The numbers used to identify products may be serial numbers or significant numbers.

A serial number is a unique number assigned to a product. It differs from the unique numbers assigned to other products of the same type by a multiple of a particular number. Often the particular number is one.

Banknotes are products that have numbers. For example, I have a 5 Euro note with the number X17150510036. From that note, I can't see whether 5 Euro notes have a serial number or a significant number. However, I have two Swiss 100 Franc notes. One has the number 04E1676337. The other is numbered 04E1676338. The last digits differ by one. It looks as if Swiss 100 Franc notes have serial numbers.

Serial numbers aren't intended to tell you anything about a product.

7.3.5   Significant Numbers

Significant numbers are also referred to as "speaking numbers", "intelligent numbers", "meaningful numbers" and "coded numbers". Like a serial number,   a significant number should be unique. Unlike a serial number, a significant number is meant to say something to somebody about a product or part. For example, L20-US-P1 could refer to a 20 inch product sold in the US in package 1. L20-JP-P4 could be the same product sold in Japan in package 4. L20-GE-P8 could be the same product sold in Germany in package 8.

Although, in some situations, significant numbers can be helpful to people who fully understand their significance, they can lead to various misunderstandings and problems. For example, L20-US-P1 may refer to a 20 inch lamp in the US, but in Germany the product may be measured in cm, so the equivalent of the L20-US-P1 in Germany may actually be the L51-GE-P8 and not the L20-GE-P8.

And when sold in Poland, perhaps the equivalent of the L20-US-P1 is identified as L51-PO-P8. But in Portugal, perhaps it's also identified as L51-PO-P8. When a L51-PO-P8 is sent in for upgrade from someone who has moved to France, it's difficult to know if it should be upgraded to Polish specifications or Portuguese specifications. Significant numbering systems can cause confusion.

Another issue is that a significant numbering system will eventually overflow. In a system that uses a single letter of the alphabet (such as L for lamp, T for table), once the 26 letters of the alphabet have been used up, the next product may be assigned an AA code. But then, instead of 7 significant digits, there will be 8. As more products are developed, it may be found that the code that should logically be assigned to a new product has already been assigned to an existing product. If the letter C is assigned to Chairs, what letter will be assigned for Cupboards? And if the company merges with another company using a similar but slightly different system, then more confusion can occur, with a particular product from one company being known by one name, but exactly the same product from the other company being known by another name.

7.3.6   Product Key

Product keys are another type of identifier used for some products. For example, when installing software it's often necessary to enter a long string of perhaps 25 numbers and/or letters that may be found on the packaging.

7.3.7   Naming Languages

The description of the product may need to be translated into several languages. The number of languages (such as English, Mandarin, Japanese and Spanish) needed for the product name and the product description is often higher than the number of numbering systems (for example, 0–9, A–Z) used for the number.

Multiple languages, and partial or unclear translations, can lead to confusion. An English-speaker may understand that the "L" in L20-US-P1 refers to a lamp. A potential customer who doesn't speak English may not.

7.3.8 Some Product and Part Identifiers

Working with product and part numbers isn't easy. There are many of them, and they are often very long. 10- and 20-digit numbers are common.

My computer has been used a lot, and some of the HP service tag on the underside is difficult to read. However, I can just about make out the content (Fig. 7.3).

The computer is an HP 2133 Netbook.

On the Web, HP PartSurfer shows that Product Number FU345EA corresponds to a description "HP 2133 Mini-Note PC", and that FU345EA matches part number FU345EA, description HP 2133 CM-7 8 2048/120 NB PC.

s/n is probably the Serial Number of my HP 2133, the number of this unique instance of the product.

The underside of the power adapter is covered with various icons, names and numbers in English and Mandarin. It has a HP part number, but that's too small to read with the naked eye. Easier to read are S/N F1-08095029370B and CT: WACLP0BL9WL7LX.

On a different product, a shampoo bottle, there are other numbers. There's a barcoded number on the back, 5 011321 833616. There's a label on the front with the number 98581815 and a label on the back with the number 95245768.

7.3.9 Product Name and Part Name

When a part may also be sold as a product, it may have a part name, part number and part description, and a different product name, different product number and different product description.

Going back to the PC, product number RR316AA has description HP Executive Leather Case. Part number 439427-001 has part description HP Executive Leather Case.

```
Product HP 2133
s/n CNU841OHIM
p/n FU345EA#UUZ
H2133VC7MI5W8N120BBNNN2CA
```

Fig. 7.3 Content of the service tag

7.3.10   Trade Mark

Another type of identifier that can be used to identify a product is a trademark. This helps potential customers to identify its source.

7.4   Requirements

Once a product exists, it's relatively easy to see what it is. Initially though, it doesn't exist. There are just some requirements and ideas for the product. They may include what a customer has requested, and/or what the person developing the product thinks customers may want, and/or what will be required to meet the regulations.

7.4.1   Customer Requirements

Customer Requirements may be very specific or they may be very vague. There may be many of them. Or there may be very few.

For example, I just had a few vague requirements for a computer. I wanted a small, light computer with easy access to Internet that I could use for my consulting work in different countries. Also, because I write a lot, I wanted a full-size keyboard. But, I didn't write down my requirements, didn't differentiate between "must-have" and "nice-to-have" requirements.

Often, customers aren't explicit about all their requirements. They think some requirements are implicitly obvious and don't need to be mentioned. In addition to the requirements for a computer mentioned above, I had some implicit requirements. For example, I expected it to work with mains electricity or with a battery. I expected it to work in a normal office environment (so in normal temperature and humidity ranges). As I don't like carrying extra equipment such as external modems, or internal modems that need special cards, I expected it to be robust and transportable in a small case.

Many customers aren't explicit about their requirements. On the other hand, some of the companies I work with have customers who know exactly what they want, and provide many pages of explicit, detailed requirements. They often differentiate between "must-have" and "nice-to-have". They use international standards to define everything as precisely as possible. For example, when asking for a particular colour, they wouldn't ask for a "kinda nice steel-blue", but specify LB5T on a standard colour coding system.

7.4.2 Requirements for Global Products

With globalisation, and more potential customers in more countries, the need to clarify requirements increases.

A potential customer of a global consumer product probably sees it first on the Web, or on television, or in a glossy magazine. It's probably being used by happy, healthy, beautiful people in some great situation. And the potential customer wants to share the experience. They want that product, and are going to pay good money for it. It goes without saying that they want a product they can use. They want a product that they can use where they live, in their country. And they want product documentation and instructions in a language they understand. They don't want, for example, a car built for people in a country where the average height is 20 cm less than in their country. They don't want a product, or product packaging, or product labelling, that is offensive to their religion or national culture. They don't want a product that they can't even use in their own country because it falls foul of government regulations. They want service close by, in a language they understand.

Often, country-specific implicit requirements will be difficult to identify from a faraway corporate headquarters. For example, one company I worked with eventually found that its product wasn't selling well in one particular country because people there thought it didn't have the smell they associated with a new product.

Similar issues arise with users of industrial products. Users expect them to be easy to use and safe. They want understandable operating instructions. They don't want to work all day at a machine that was built for a country where people are on average 20 cm taller. They don't want to strain their arm and leg muscles all day long to work the machine.

7.5 From Customer Requirement to Product Specification

I only had a few requirements when I bought my HP 2133. However, a few months after buying it, I had a question about its use. I looked on the Web for some information. I found a 20-page specifications document for a HP 2133 Mini-Note PC. On the first page, this showed front view, left view and right view drawings. There was a view of the underside, and a rear view. The drawings were clear and informative. They even showed features I had never noticed before.

There was a lot of information in those 20 pages, but I expect the developers had an even larger specification document. Very detailed specifications are needed to develop and assemble a product. That may seem obvious, but when products have hundreds or thousands of parts, a huge amount of product specification data is needed.

Going back to the HP 2133 document, I found that the HP 2133 has a spill-resistant keyboard which is 92 % of full size. It is 101/102-key compatible with isolated inverted-T cursor control keys, both left and right control and alt keys,

12 function keys, and hotkey combinations for audio volume, power conservation, brightness, and other features. US and international key layouts are available. For Internet access there's a Broadcom 4322AGN 802.11a/b/g/draft-n Wi-Fi Adapter. There's a C7-M Ultra Mobile Processor and Genuine Windows Vista Home Basic 32. As for dimensions it has H × W × D of 1.05 (at front) × 10.04 × 6.5 in. The weight starts at 2.8 lb (1.27 kg), including a 3-cell battery, 1 GB memory, 4 GB flash module, and the 802.11 wireless communications module. The operating temperature is 32° to 95 °F (0° to 35 °C), and so on, and so on, for 20 pages.

7.6 Identification Standards

There are so many numbers and names for parts and products, that standards have been proposed to make them easier to work with. Some of these are described below. There are many others that are not described here.

7.6.1 Global Trade Item Number

The GS1 organisation created a system of standards for use in the supply chain. These include eleven GS1 Identification Keys. Among these are GTIN (Global Trade Item Number) and GLN (Global Location Number).

There are several GTIN data structures, with different numbers of digits. They include GTIN-8, GTIN-12, GTIN-13 and GTIN-14. A quick look at the shampoo bottle reveals a barcode with a 13-digit GTIN, 5 011321 833616. In this structure, the 13 digits should be read in 4 sections. The first section, with two or three digits, is the GS1 prefix. The next four, five or six digits are for the Company Number. The next two to six digits are the Item Reference. The last digit is a checksum digit.

The GS1 prefixes (Fig. 7.4) are "country codes". They are used in each country to create GS1 identification keys for companies that apply. These companies may manufacture products anywhere in the world. So, the GTIN identifies the product, but the GS1 prefix doesn't identify the country of origin of a product.

The GS1 country code on the shampoo bottle, 501, is assigned to the UK. The Global Electronic Party Information Registry (GEPIR) shows the owner of 5

GS1 prefix, "Country Code"	Country
000-019	USA and Canada
400-440	Germany
490-499	Japan
500-509	United Kingdom
690-695	China
760-769	Switzerland
930-939	Australia

Fig. 7.4 Some GS1 country codes

011321 833616 to be the company with GLN 5000174000009, Procter & Gamble UK, Newcastle upon Tyne, United Kingdom.

7.6.2   International Standard Book Number

Books are products. The identifier for a book is the International Standard Book Number (ISBN). The ISBN is a special 13-digit GTIN code. The first 3 digits of the ISBN code are 978 (the GS1 "country code" for all books is 978), the next 1 to 5 digits identify the language of the book, the next 4 identify the publisher, the next 3 a particular item (book title). The last digit is a checksum digit. As an example, the ISBN identifier for the English-language version of a book published by Springer with the title "Product Lifecycle Management: 21st century Paradigm for Product Realisation" is 978-185233810-7. In this example, the 1 after 978-identifies the book as being in English.

7.6.3   International Mobile Equipment Identity

Another product identifier is the International Mobile Equipment Identity (IMEI) number. This is a 15-digit number. Each mobile equipment instance has a different IMEI. The number has a structure such as 35-098420-998049-0. For that instance, the initial 35, for the "Reporting Body", identifies BABT as the reporting body. The next six digits are the Type Identifier, and the following 6 digits are the Serial Number of the device.

7.6.4   International Standard Music Number

The International Standard Music Number (ISMN) is an identifier for printed music. It consists of 13 digits starting with 979-0, followed by a publisher identifier, an item identifier and a check digit.

7.6.5   CAS Registry Numbers

Chemical Abstracts Service (CAS) assigns identifiers (CAS registry numbers, CAS#s) to chemicals. CAS is a division of the American Chemical Society. There are more than 90 million substances in the CAS database. A CAS number can be up to 10 digits long. The CAS# of water, for example, is 7732-18-5. That of caffeine is 58-08-2.

7.7   Unique Identifier, Unique Key

The name of an author, for example John Smith, doesn't uniquely identify a book. The author could have written several books. There could be several authors with the same name.

However, the ISBN does uniquely identify a book (Fig. 7.5). Each book has a different ISBN. Two books don't have the same ISBN. The ISBN is an example of a unique key. It uniquely identifies the object (the book).

Another example of a unique key is a Social Security number. It uniquely identifies a person in a particular country.

7.8   Traceability

Even with the product identifiers described above, it may not be possible to uniquely identify a particular instance of a product. More numbers may be used, such as a batch number or a production date, to more precisely identify an instance, or a batch of instances.

For example, in addition to the numbers mentioned above, the shampoo bottle has another number, 9307484732 L12, stamped on it near the neck of the bottle. A more recently purchased bottle has a different number in that position, 9310484729 L12, but all the other numbers are the same.

7.9   Communication of Identifier

It's one thing to have an identifier, it's another to communicate it.

7.9.1   Type of Communication

In the case of Smart Products, the product may identify itself when it's near a sensor. Or it may identify itself when asked.

In the case of more traditional products, the identifier may be painted on the product, written on, etched in, printed on a label that is attached to the packaging, printed on a label that is sewn on the product, or be on a plate that is attached to the product.

ISBN	Book Title
978-044201075-1	Engineering Information Management Systems
978-185233810-7	Product Lifecycle Management: 21st century Paradigm for Product Realisation
978-184628914-9	Global Product: Strategy, Product Lifecycle Management and the Billion Customer Question

Fig. 7.5 The ISBN is a unique key

The identifier may be communicated in a coded form, or uncoded. An example of a coded form is a bar code.

7.9.2   UPC Barcode

Barcodes were invented in the 20th Century. Different organisations proposed different systems with different numbers and shapes of bars. One of these, the Universal Product Code (UPC), achieved wide acceptance for tracking trade items in North American stores. It has 12 decimal digits (for example, 0 39047 00513 6). Each decimal digit is encoded in 7 binary digits (where 0 is represented by white space, and 1 is represented by a black bar).

7.9.3   EAN-13

The EAN-13 barcode system was defined by GS1 for use worldwide. "EAN" originally stood for European Article Number, but now stands for International Article Number. It extends the 12-digit UPC system, and is compatible with UPC. A 13-digit GTIN number can be encoded in a EAN-13 barcode.

7.9.4   Two-Dimensional Barcodes

Two-dimensional barcodes started to be used at the end of the 20th Century. One of the most frequently used is the QR system. A QR barcode is made up of black squares on a square white background.

7.10   Product Classification

Classification is different from identification. Identification identifies one thing, for example, your instance of a washing machine.
   Classification enables grouping of similar objects.

7.10.1   Classification

Classification is a way of grouping similar objects according to some criteria. It differentiates an object from others that are not the same. This may be done to help recognise a product, or to name it, or to describe it, or to find it. For example, it could help show that a part on a list on the Web is a screw, and not a power pack. It could help show that in one aisle of a store there are biscuits, and in another aisle there are dairy products.

7.10.2 Advantages of Classification

Classification can help get a quick overview of things. For example, "in this group are all books in English about PLM". It can help physically sort products. For example, "all products with a red stripe are for the US, all those with a blue stripe are for the Rest of the World".

7.10.3 Classification Systems

Classification systems may be company-specific, industry-specific, specific to a continent, or global. And, there may be several hierarchical levels to a classification scheme.

Eurostat is the statistical office of the European Union. It provides statistics at the European level that enable comparisons between countries and regions.

In the Eurostat Central Product Classification, for example, classification codes 451 and 452 are at one level (Fig. 7.6). 4522 and 4523 are at the next level of detail, within 452. And 45221 and 45222 are at the next level of detail, within 4522.

7.11 Versions, Variants, Options

7.11.1 Lifecycle State

At different times in the lifecycle, a product will be in different states, for example, Preliminary, Prototype, Pilot, Production, Service-only and Obsolete.

Code			Products
451			Office and accounting machinery, and parts and accessories thereof
452			Computing machinery and parts and accessories thereof
	4522		Portable automatic data processing machines weighing not more than 10 kg, such as laptops, notebooks and sub-notebooks
		45221	Portable automatic data processing machines weighing not more than 10 kg, such as laptop and notebook computers
		45222	Personal digital assistants and similar computers
	4523		Automatic data processing machines, comprising in the same housing at least a central processing unit and an input and output unit, whether or not combined

Fig. 7.6 Classification example

7.11.2 Version, Iteration

Most products evolve. At a certain time, a product, for example a software product, may have certain features and functions, and be at Version 1.0. Later, with more features and functions, it can be at Version 2.0.

Alternatively, the name may change as the product evolves. For example, the HP 2133 was superseded by the HP Mini 2140 Notebook PC.

7.11.3 Variant, Option

The words variant and option are used in various ways.

Often a variant will be a variation of a basic working product. For example, the HP 2133 was available in one variant with Windows Vista, in another variant with SuSE Linux. There was a variant with a 8.9-inch WXGA display with 1280×768 resolution, another with a 8.9-inch WSVGA display with 1024×600 resolution.

Options are additions to a basic working product. For the HP 2133, there were various carrying case options such as the ultra-portable Carrying Case, the Basic Carrying Case, a Value Nylon Case, a Universal Nylon Case, and an Executive Leather Case. I don't have any of them. But that's not a problem. A product should work without an option part. However, a product often won't work without a variant part. The HP 2133 wouldn't be so useful with neither the 8.9-inch WXGA display nor the 8.9-inch WSVGA display.

7.12 Product Ownership

7.12.1 Rights

Ownership of a product is an important issue. Ownership gives certain rights over property.

You may be the owner of your washing machine (for example, if you bought it), but you are only the owner of that instance, and you have limited rights. Buying it probably gave you the right to use it, but didn't give you the right to clone it and to produce another one.

Ownership gives rights over property, but it also gives responsibilities. The responsibilities associated with parts and products have to be defined in detail, particularly when companies sell products containing parts that they haven't developed and/or manufactured.

7.12.2 *Intellectual Property*

Ownership gives rights over property, whether it's a tangible property or intellectual property. Intellectual Property (IP) includes copyright, patents and trade secrets.

Copyright refers to the rights, including the right to copy, granted to the author of an original work.

A patent is a set of rights granted to an inventor in exchange for public disclosure of an invention.

A trade secret is information, such as a recipe, that's not known outside the company, and enables a company to gain competitive advantage.

7.13   Product Structure and Architecture

7.13.1   *Structures*

During the product lifecycle, many people, such as designers, salespeople, customers and recyclers, will work with the product. They'll be involved in different activities. They'll want to work with the product in the most appropriate way for their activity. They'll want to work with the most appropriate structure. Some users may want to work with a list of what's in the product, others may just want a list of what has to be ordered to manufacture it.

A list is one-dimensional, either a column of entries (a vertical list) or a row of entries (a horizontal list).

A structure with more dimensions, for example a two-dimensional array of Number of Items and Item Names, can hold more information, and be easier to understand.

A structure is helpful and adds meaning. For example, from the list of 5 parts in Fig. 7.7, you might not know in which order to assemble the product.

The array in Fig. 7.8 is easier to understand.

Similarly, it's easier to understand a one-line entry in a two-dimensional array with "20 Ph8 screws" than a list of 20 lines each with "Ph8 screw".

Hierarchical structures can be used to model the components of a product.

| bottle | liquid shampoo | screw cap | front label | back label |

Fig. 7.7 A horizontal list

Assembly 1	bottle, front label, back label
Assembly 2	labelled shampoo bottle, liquid shampoo
Assembly 3	filled bottle, screw cap

Fig. 7.8 An array

Figure 7.9 shows the structure of a product, its main assemblies, and the relationships between them.

The entity at the top of the structure, in this case the car, is sometimes referred to as the root of the structure.

The entities at the next level down are known as its children.

Working top-down, each entity, unless it is at the lowest level, may have children. Working bottom-up, each entity, unless it is at the highest level, has a parent.

The information about a structure that is shown graphically (Fig. 7.10) can also be shown in array form (Fig. 7.11).

The information in the array in Fig. 7.11 is identical to that in the hierarchical graphical structure in Fig. 7.10.

7.13.2   Bill of Materials

A Bill of Materials (BOM) is a hierarchical structure showing the things that make up an end item (part or product). A BOM is hierarchical in that it shows the end

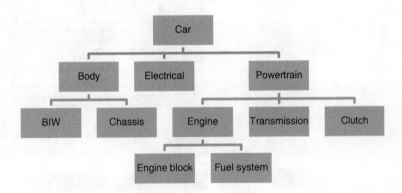

Fig. 7.9 A structure of a car

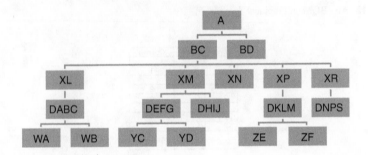

Fig. 7.10 A graphical structure

item, the things that make up the end item, the things that make up the things that
make up the end item, etc.

There can be several BOMs for the same product. The Engineering Bill of
Materials (eBOM or EBOM) shows the objects that make up the end item from a
design viewpoint. It can be shown in graphical (Fig. 7.12) or tabular form
(Fig. 7.13).

In addition to the things that make up the end item, and are shown in the eBOM,
a Manufacturing Bill of Materials (mBOM or MBOM) also shows the other things
(such as machine oil) that are needed to make the part or product.

In Fig. 7.12, C is made from an H and an I. B is made from 2 F's and 4 G's. One
end item, A, is made from a B, a C, 2 D's and 4 E's.

Fig. 7.11 The same structure
in array form

Level	Occurrence in sequence	Item
0	1	A
.1	1	BC
..2	1	XL
...3	1	DABC
....4	1	WA
....4	2	WB
..2	2	XM
...3	1	DEFG
....4	1	YC
....4	2	YD
...3	2	DHIJ
..2	3	XN
..2	4	XP
...3	1	DKLM
....4	1	ZE
....4	2	ZF
..2	5	XR
...3	1	DNPS
.1	2	BD

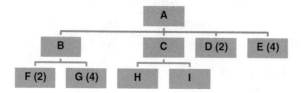

Fig. 7.12 An eBOM in graphical form

Level	Item	Quantity
0	A	1
.1	B	1
..2	F	2
..2	G	4
.1	C	1
..2	H	1
..2	I	1
.1	D	2
.1	E	4

Fig. 7.13 The same eBOM in array form

7.13.3   Product Architecture

Product architecture links the physical structure of the product to other character-istics of the product, such as functionality. These relationships are important throughout the lifecycle. Architecture is closely linked to design. It defines which function will be in which assembly. Architecture is closely related to production. It defines in which order the product will be assembled. Architecture is closely related to support. It defines how the product will be disassembled for service.

For example, if a product has components that need power, and also has a power supply, then should the power supply and the components using the power be in one assembly? Or should they be in separate, interfaced assemblies? If they are in the same assembly, then an interface won't be necessary. However, service engi-neers may prefer the power supply to be in a separate assembly, so that it can be replaced more easily. And customers may prefer the power supply to be in a separate assembly, as this may reduce the cost. However, the Manufacturing organisation, trying to reduce the number of suppliers, may prefer working with a preferred supplier that offers the power pack and the components in one assembly.

7.13.4   Product Portfolio

Most companies have more than one product. They have a portfolio of products. Usually, there are several well-defined groups of products in the portfolio. These groups may be referred to as product lines, or product families (Fig. 7.14). The products may be grouped as a function of various characteristics, for example, because they have similar functionality, or are sold in a particular industry, or are in the same price category, or have similar production processes.

The name of the product line may also be an identifier. Separate product cata-logues may be produced for each product line.

From a top-down viewpoint, the Portfolio is made up of product families, which are made up of assemblies, which are made up of products, which are made up of parts (Fig. 7.14). From a bottom-up viewpoint, all parts fit into higher-level entities such as assemblies or products, which fit into product families which make up the Portfolio (Fig. 7.15).

| Product Portfolio |
| Product Family, Product Line, Product Group |
| Assembly |
| Product |
| Part |

Fig. 7.14 From portfolio to part

Part
Product
Assembly
Product Family, Product Line, Product Group
Product Portfolio

Fig. 7.15 From part to portfolio

7.13.5   Product Model

A model is a simpler representation of something else. Because it's less complex, a product model can be built much quicker than the product itself. Once built, the model will help increase understanding, and reduce the risk of misunderstanding. Models may address many different properties, including physical structure and appearance, electrical behaviour and aerodynamic performance.

Before a building such as a skyscraper is built, it's normal practice for an architect to build a model for public display. The model shows what the building will look like, and how it will fit in the environment.

After the Great Fire of London in 1666, Christopher Wren created models for the new St Paul's Cathedral. Some of them can still be seen in the crypt.

Similarly, models of many products are created before the product is manufactured. They may not be put on public display, but they will be shown to many people, such as potential customers and suppliers. They will also be shown to people from the many parts of the company who will work with the product across its lifecycle. The models will help increase knowledge and understanding of the product.

Christopher Wren made models out of wood. Today, virtual (computer-based) models can be made.

7.14   Description, Definition and Representation

During the product lifecycle, the product will be described and represented in many ways. Different customers and users will need different representations. Different types of developers will need different representations.

For example, a potential customer might want to see a photograph, or a drawing, of a product before ordering it. Other customers might want to see a list of specifications. Another customer may want to browse through a catalogue containing a set of similar products, and then select one particular product. A customer who buys software will want to see the user manual. A customer may want a service manual to know how to repair the product if it breaks down.

Product developers will want to see other representations. On the shop floor, people will want to see manufacturing drawings. Maintenance engineers won't be interested in seeing how the product is manufactured, but will be interested in drawings showing how to service it as quickly as possible. Dismantling operatives will want to see yet other representations. For some, a 2D drawing could be enough, for others a 2D representation could be ambiguous and they would want to see a 3D model.

Although the representations mentioned above may all be different, and be used in different situations, they all represent the same product, and must be consistent.

7.15   From Customer Requirement to Performance

Most customers and users of products probably don't think about how the product was developed, manufactured and made available. That's not their concern. They just want it to perform the way they like. And they expect it to perform the way the manufacturer claims it will perform.

7.16   No Product Is an Island

No product is an island (Fig. 7.16), isolated from the rest of the world. All products are closely related with other PLM components. They are also influenced by other forces within the company, and outside the company.

As an example, in response to customer demand, electronic parts and software may be added to a product that before had none. Then, assuming nothing is out-sourced, the company will need to hire some people with skills in electronics and software. The company organisation may be changed by the addition of an electronics department. Processes will need to be created and changed to address electronics and software. There will be new data types, perhaps circuit diagrams and code lists. Part numbering and naming will have to be reviewed. The classification scheme may need to be extended. New applications will be needed to develop, simulate, analyse and test the new parts. New equipment will be needed in development and production. Depending on the product and its use, conformity may be needed with specific regulations. New performance indicators will be introduced to measure and improve performance of the new parts.

data	process	applications
people	**product**	organisation
metrics	methods	equipment

Fig. 7.16 No product is an island

| Need to reduce Time To Market |
| Customers look for more customisation |
| Waste and rework due to overlapping assembly structures |
| All available product numbers used up |
| Due to mergers, multiple incoherent product numbering systems |
| Overlap of significant numbers |
| Customers request better product support in Middle-of-Life |
| Losses due to counterfeits |
| Incorrect classification |
| Regulators require pollution-free End-of-Life |
| A key customer requirement was misinterpreted |
| Unclear lifecycles definition |

Fig. 7.17 Potential product-related challenges for a company

7.17 The Challenges

At a particular time, the specific product-related challenges that a particular company faces could come from several sources (Fig. 7.17).

7.18 The Way Forward

The issues described in this chapter all relate to the product, but they can't be solved in isolation from the other components of PLM. Their solution, within the PLM environment, is addressed from Chap. 21 onwards.

Chapter 8
Product Data

8.1 Product and Product Data

Whatever the product made by a company, an enormous volume and variety of product data is needed to develop, produce and support the product throughout the lifecycle. The term "product data" includes all data related both to a product and to the processes that are used to imagine it, to design it, to produce it, to use it, to support it, and to dispose of it.

Some of this data will describe the characteristics of the product, or a part of the product, or its packaging, or a label, or an identifier. Some will describe a structure such as a BOM or a list of ingredients. Some will describe a process related to the product, how something has to be done. Some of the information may describe a regulation that the product must comply with. Some may describe a best practice guideline developed by an industry organisation or an international standards organisation.

Product data is created and used throughout the product lifecycle. Some product data (such as part geometry, NC programs and Bills of Materials) is created in design engineering and manufacturing engineering, some is created elsewhere (such as feedback results from customers in field tests). Some of the data (such as results of stress analysis or circuit analysis) is used within design engineering, some (such as welding instructions) in production. Some of the data (such as installation instructions) is used on a customer site, some (such as disassembly instructions) at the end of the product life.

Product data doesn't look after itself. Like anything that's not properly organised and maintained, it won't perform as required. Over time, it will slide into chaos and decay. However, this has to be avoided as the slightest slip can have serious consequences for the product and those associated with it. Getting product data organised, and keeping it organised, are major challenges in PLM.

For much of the 20th Century, paper was the medium for storage, transfer and communication of most product data. The paper with a collection of data related to

© Springer International Publishing Switzerland 2016
J. Stark, *Product Lifecycle Management (Volume 2)*,
Decision Engineering, DOI 10.1007/978-3-319-24436-5_8

record	form	report	procedure	policy
directive	guideline	rule	list	standard
template	document	protocol	sheet	chart
drawing	file	folder	bill	instruction
plan	diagram	schedule	log	order

Fig. 8.1 Paper documents with product data

Analytic models	Analysis results	Assembly drawings	As-built configuration	Bill of Materials
CAD geometry	Consumables lists	Cutaways	Engineering drawings	Change data
Costing data	Customer requirements	Disposal lists	Design specifications	Cutsheets
Equipment logs	Equipment data sheet	Exploded views	Factory layouts	Failure reports
Flowcharts	Formulae	Functional specs	Label information	Ingredients list
Line lists	Machine libraries	Maintenance info	Material certification	Mounting data
NC programs	Packaging standards	Parts classifications	Parts lists	Patent reports
Photographs	Pipe specifications	Pneumatic diagram	Process model	Project flows
Project plans	Process plans	Purchasing data	QA records	Recipes
Regulatory rules	Results of calculations	Scanned drawings	Schedules	Service lists
Service manuals	Shop floor instructions	Simulation results	Sketches	Software
Spare part info	Specifications	Standards	Standard costs	Status logs
Test data files	Technical publications	Test results	Tool designs	User guides
User manual	Validation reports	Versioning data	Wiring diagram	Video files

Fig. 8.2 Examples of product data

a particular activity, or for a particular purpose, was given a variety of names
(Fig. 8.1). As often in the PLM environment, the meaning of these names wasn't
standardised. As a result, the collection of data that was referred to in one company
as a record would be referred to in another company as a report. A policy for one
company would be a procedure for another, and an instruction for a third.

8.2   Product Data Examples

In the typical company, a huge variety and volume of product data (Fig. 8.2) is
generated and used during the development, manufacturing and support of a
product across its lifecycle.

8.3   Product Data Issues

There are many types of product data, each with its individual characteristics.
However, although some characteristics are specific to just one type of data, many
issues are common to many or all types (Fig. 8.3). These common issues are
described below in alphabetical order. They're all potential sources of problems to
be addressed in PLM.

Volume, Variety, Vocabulary	Status, Change, Version	Users, Access, Availability
Definition, Content, Structure	PRODUCT DATA	Ownership, Value, Meaning
Representations, Media, Source	Numbering, Naming,	Security, Confidentiality

Fig. 8.3 Issues common to many types of product data

8.3.1   Access to Product Data

Access to product data has to be provided to people when they need it. However, access to product data has to be controlled so that only authorised people, with specific rights, can access the data. For example, a company may want customers to be able to access some information about a product over the Web, but not want competitors to be able to access details about the product.

Complicating access management, the access rights of a user of product data may change during the product lifecycle. At one time, a user may have the right to create and modify some product data. But, at another time, they may not even be allowed to see that data. And, at a particular time, a person may have different access rights on different projects and products.

Access to product data can be further complicated by the obsolescence of computers and the introduction of new applications. For example, new generation applications often have difficulty in using data produced for old products by old applications. And old applications (legacy applications) are often unable to work with new data structures created in new applications. Yet users may need to access old data (for old products), enriched data on old products, and data in new formats (for new products).

8.3.2   Applications

There are many applications in the PLM environment. Most of them create and store product data in different ways. This complicates access to, and management of, product data. It complicates transfer of data between applications.

There's usually redundant data in applications. Many applications store all the information they require, leading to duplication and overlap between the information stored by different applications. As an example, many applications will store lists of product names and part numbers. Should the structure of these change, the corresponding applications may have to be changed, and the files they use and create may have to be changed. This takes time and effort, and may introduce errors.

8.3.3   Archiving of Data

A lot of product data must be kept for a long time. Requirements vary by industry, but customers and regulators may require data to be kept for several decades.

8.3.4   Availability of Data

Product data needs to be available to users where they need it, when they need it. Product data may need to be available anywhere. For example, an aircraft or a ship may need to be repaired in any part of the world.

People may wait hours, even weeks, for the product data they need. The person who should send it may be overloaded with other work, and then be out of the office for several days visiting prospects. Work is held up because nobody knows who else has the authority to send it. When people do receive information, they may not be sure if they've received the correct version. Sometimes they just want to make a simple request for product data from an application, in the same company, that "belongs" to another department. And they have to wait several days to get it.

8.3.5   Change

Most product data undergoes change at various times in its lifetime. A change to one part often has knock-on effects. Adjacent parts may need to be changed. Some of the changes may lead to changes in materials, manufacturing processes and user documentation. Other changes may lead to changes in labels and regulatory documents. All these changes have to be managed: formally requested; evaluated; properly approved by all interested parties; publicised; and recorded. There's a huge volume of product data. In a fast-changing environment, that leads to a huge volume of changes to product data.

Changes to product data can be difficult to co-ordinate, with the result that unnecessary changes are introduced or required changes are held back. As a result, design cycles are lengthened, and unreleased versions of data are acquired by manufacturing, sales and support, causing confusion and waste. The time taken for raising, approving and implementing changes becomes much longer than necessary. The change process may take days, weeks or even months, whereas the actual processing time may only be a few minutes or hours. In large companies, it can cost thousands of dollars to process a change to product data.

Engineering change control systems are often bureaucratic, paper-intensive, complex and slow. A central engineering services group may have the responsibility, but not the tools, to push the changes through as quickly as possible. Many departments may be involved. It may take several months, and numerous forms, to get a proposed change approved and incorporated into a product design. Even when a change has been agreed and announced, many months may go by before the corresponding documentation gets to the field.

For some product data, there may not be a formal change control process. Some companies even have a formal ISO 9000-compliant change process, but management doesn't expect anyone to use it. Minor modifications to products and drawings are made without informing anyone. Components are substituted in end products without corresponding changes being made to test routines. People fail to maintain the trace of the exact ingredients in ever-smaller batches of products. Nobody notices until something goes wrong or another change has to be made. Then, extra effort will be needed to find out where the problem comes from. And additional support staff will be employed to try to prevent further problems.

8.3.6   Copies of Data

As soon as there's more than one copy of a file or document, there's a danger that someone will change one copy, but the other copies won't be updated. Procedures need to be in place to ensure that copies and changed copies can't lead to problems.

8.3.7   Confidentiality of Data

Product data is valuable. Much of it is confidential, and shouldn't be seen by people in other organisations, such as competitors.

8.3.8   Configuration

Configuration control may break down. Configuration documentation no longer corresponds to the actual product. Unexplained differences appear between as-designed, as-planned and as-built BOMs. Increased scrap, rework and stock result. Incomplete products are assembled and delivered. Field problems are difficult to resolve as nobody knows exactly which parts the product contains.

8.3.9   Definition of Data

Data needs to be clearly and correctly defined. When there's not a standard definition of the data associated with a particular part or product, each user (and application program) can have a different definition of the data, and all the definitions can be different. Different departments may even use different numbering systems.

If there's more than one definition of something, such as the name of a product, then people will get confused. Yet, the same item of product data is often defined differently in different parts of a company. And in different parts of an Extended Enterprise. This results in confusion and waste when product data is transferred from one part of the organisation to another, or when people in different parts of the organisation work together.

Multiple definitions lead to errors, and wasted time and money. Yet many companies have several different definitions of some data items. A CAD program may have one definition of a part. A part programmer may redefine it. A stress analysis program may use a third definition. In the Bill of Materials, the part may have another definition. It may be redefined for inspection, and again in assembly instructions. Unless the company has introduced strict procedures ensuring that all these definitions are equivalent, there'll probably be minor differences between them. These will lead to confusion when modifications are made to the part, or if an attempt is made to reuse the part in another product or design. Since the definitions aren't identical, the result of a modification to one definition may not be the same as the result of the same modification to another definition. Special software may have to be developed and maintained to allow each user to continue working with their own definition.

8.3.10 Duplicate and Redundant Data

A lot of product data will be held in duplicate. Most people don't fully trust "the system". To protect themselves, they keep their own copies of data, and try to stop other people accessing, and perhaps changing, "their" data. As a result, groups of users often have different copies of what should be the same data. A design engineer will have a copy of some of the information that's used in production planning. Copies of some of the information used to generate NC programs will be kept by both design engineers and manufacturing engineers. Maintenance engineers may want to keep "their" drawings close to hand so that they can respond quickly to urgent customer calls. Most of this information will also be stored elsewhere in the company.

Some data will be duplicated in many documents. For example, the name of a product will be entered on many documents. Similarly, the name of a project will be entered on many documents. The characteristics of a product will probably be entered many times. Creating duplicate data wastes time and can lead to errors.

8.3.11 Exchange of Data

Product data often needs to be moved from one place to another. It may be communicated from one person to another, or exchanged between one application (or representation or owner) and another. As product data may be represented in different ways in different applications and media, there may be problems when it's transferred from one representation to another.

Part descriptions and Bills of Materials developed with a CAD application may be manually transferred to an ERP application on a computer that's not linked to the CAD computer. The Manufacturing BOM may be different from the Engineering BOM. The CAD and ERP applications may be the responsibility of different organisations in the company. The change processes in the two organisations may be different and out-of-step. At a particular time, a given change may have been made in one application, but not in the other. As a result, some users may not have immediate access to the most up-to-date information.

8.3.12 File-Based Data

Computers have been used in the product development, manufacturing and support environment since the 1950s. The earliest applications of computers in this environment were point solutions, creating "Islands of Automation". The data that an Island of Automation required and generated led to the creation of an Island of Data. The users of applications are much more interested in real objects, such as products and parts, than in the structure and format of data in files. Only too often though, information on the real objects is only available after wading through, and understanding, many sets of files. Again, this represents a waste of time.

In the file-based environment, it's often difficult to maintain information on the almost limitless number of relationships between information in different files. For example, one file may contain analysis results on a part in a second file that was subjected to forces specified in a third file. The second file may also contain the geometry model of another part. A dimensioned drawing of this part may be stored in a fourth file. When the second part is modified, a fifth file may need to be created with the corresponding dimensioned drawing, but because the first part may not have been modified, it may not be necessary to repeat the stress calculation. Probably the file will contain no indication of whether or not the first part was modified, so the stress calculation will be repeated, generating a sixth file that's identical to the first file. In the real environment, in which there will be thousands of files, it's very difficult to maintain the correct information on relationships between files.

8.3.13   Formal Description

It's important to have a formal description of product data. Unless product data is clearly and consistently described, it will be difficult to manage and to improve. The formal description needs to be documented and communicated.

In many companies though, there isn't a formal description of product data.

8.3.14   History

At any time, it may be necessary, for any one of a variety of reasons, to look back at the design of a particular part or batch. A batch of biscuits may be inedible. The manufacturer will want to look back to see what the ingredients were, where they came from, how they were used. Based on this information, the company can take appropriate action. In other cases, a batch of airbags may be faulty, or a part may have failed on a forty year old aeroplane.

8.3.15   Identification and Classification Systems

There's so much product data that special identification and classification systems have to be used to keep track of it. Unique numbers are needed to identify every specification, drawing, list, test procedure, operating manual and other document which defines the functionality, physical construction, and/or performance of a solution, product, component, assembly, sub-assembly, or part.

Many issues can arise with numbering systems, for example, when a numbering system runs out of digits, or when two companies with different numbering systems merge. Or when a customer has special requirements for numbers. Or when a company receives one number for a part from a supplier, and another number for the same part from a sub-supplier.

8.3.16   Inconsistent Data

Product data from different sources may be inconsistent. For example, some data about product performance may come from the sales organisation. Other data may come from the service organisation. The two organisations may be using different data collection techniques and different applications, and focusing on different parameters. The resulting information may be contradictory. Some information may show increasing customer satisfaction. Other information may show decreasing customer satisfaction.

8.3.17   Incorrect Data

Mistakes with product data can be expensive. For example, if incorrect documentation is created, the wrong tooling may then be developed, and wrong orders placed with suppliers. The longer it takes to discover that the wrong information has been released, the more costs will be incurred and time wasted.

8.3.18   Informally Annotated Documents

Documents, both manual and electronic, are often annotated informally. Manual documents can be annotated using a pen or pencil, electronic documents by typing on a blank part of the document. The document author has an important comment to add, or perhaps wants to reference another document. However, in many cases, this important information will be lost forever. Nobody else will even know that it exists.

8.3.19   Informal Communication of Data

Informal communications are developed between departments to cope with the lack of suitable formal communications. Few records will be kept of this type of product data transfer. In the absence of a particular individual, it may be impossible to find any trace of important product-related information.

8.3.20   Input of Data

Data entry needs to be carefully controlled. It's easy to type the wrong character or copy the wrong file. Data is easily lost and may be impossible to retrieve. Then it's re-created and errors may be introduced. The wrong product goes to a customer. When a defective part is found in the field, unless data was well organised, many more products than necessary have to be recalled. Often, information about problems with product use isn't fed back to product developers. As a result, they design the same problems into the next generation of products. Design history isn't maintained, so it's difficult to build on previous experience.

8.3.21 Interoperability

In the context of PLM, interoperability is the ability of applications to meaningfully exchange product data without human intervention. The top management of manufacturing companies would like all applications to be interoperable, as this would increase freedom, choice and functionality. However, one way for a software vendor to control the market is to limit interoperability between its applications and those of other vendors. And hope that this will lead customers to choose only its applications.

8.3.22 Languages

Product data may need to be available in several languages. If you buy a sandwich in a closed wrapper, you'd like the name of the sandwich and the list of ingredients to be visible somewhere in a language you can understand. Otherwise, you won't know what you're buying, or what you're going to eat. However, there can be issues with exact translation, with translation of translated text, and with changes to texts in different languages. Many words don't have exact equivalents in all other languages.

8.3.23 Level of Detail

The level of data definition changes throughout a product's lifecycle. In early stages of the lifecycle, there's little data available. The level of detail increases as the product is developed and used. Once the product has been shipped to the customer, the level of detail may fall. The definition of the product doesn't have to be identical at all stages, but it does have to be consistent.

8.3.24 Library of Data

It can be useful to build up a library of standard data, for example on standard parts or ingredients. This will enable re-use of data. However, if some users don't like what's in the library, they may keep their data outside the library, or even make another library. To avoid this, it has to be agreed which data is to be stored in libraries, what format it will be in, when it's to be created, who can access it and perhaps most importantly, who can modify it and when.

Problems can arise with references made in old products to library data when that library data has to be updated. In some cases, it may be better to stay with the old library data for old products. In others, it may be better to bring old products into line with the updated version of the library.

8.3.25 Location of Data

There can be a lot of product data in a company (Fig. 8.4). It can be found in many locations. It can be in a database, in an application, in a file, in a drawer, on a piece of paper on someone's desk, in someone's head. The users of product data may be in the same building, or in the same plant, but they could also be in different locations in different countries, or even on different continents. Product data will be distributed over several locations. Copies of each individual data item may be kept in several storage places.

Some of the users of the data will be inside the company, others in other organisations (such as suppliers, partners, customers and regulatory bodies).

8.3.26 Long-Life Data

Product data supports the product across its lifecycle. In some cases, for products such as power plants and aircraft, the overall product life may be more than fifty years. During this time, there will be a huge volume of data generated, first to design and manufacture the product, and then to support its use.

Product data must be capable of outliving the people, applications and computers that generate and process it. It may even have to outlive the companies that generate and process it. This is essential particularly when the product has a long lifecycle, and/or can have long-term effects. Examples of such products include aeroplanes, offshore platforms, weapon systems, pharmaceutical and medical products, ships and process plants.

Analytic models	Analysis results	Assembly drawings	As-built configuration	Bill of Materials
CAD geometry	Consumables lists	Cutaways	Engineering drawings	Change data
Costing data	Customer requirements	Disposal lists	Design specifications	Cutsheets
Equipment logs	Equipment data sheet	Exploded views	Factory layouts	Failure reports
Flowcharts	Formulae	Functional specs	Label information	Ingredients list
Line lists	Machine libraries	Maintenance info	Material certification	Mounting data
NC programs	Packaging standards	Parts classifications	Parts lists	Patent reports
Photographs	Pipe specifications	Pneumatic diagram	Process model	Project flows
Project plans	Process plans	Purchasing data	QA records	Recipes
Regulatory rules	Results of calculations	Scanned drawings	Schedules	Service lists
Service manuals	Shop floor instructions	Simulation results	Sketches	Software
Spare part info	Specifications	Standards	Standard costs	Status logs
Test data files	Technical publications	Test results	Tool designs	User guides
User manual	Validation reports	Versioning data	Wiring diagram	Video files

Fig. 8.4 Examples of product data

For many products, the support cycle is longer than the design and manufacturing cycle, and may produce correspondingly more data. Technical manuals will have to be first produced and then kept up to date, perhaps for decades. Spare parts will have to be ordered and manufactured. For each maintenance job, information will need to be available on the handling tools and the required skills. Product specification data may need to be given to second sources. Data supporting repair and replacement will be needed. Field data needs to be managed. Performance data needs to be maintained so as to be able to plan preventive maintenance.

A lot of data will be produced during a long product lifecycle. Many users, perhaps in different companies, will want to access the data. Each will want the data to be available in the most suitable place and format. Different types of data will be produced and needed at different times. New data will be produced, existing data will be reused and perhaps modified. Over a long life, the product may be repaired or upgraded to such an extent that most of the original product will have been replaced.

8.3.27   Manuals

There may be specific problems with manuals. User Manuals may be difficult to understand, apparently written by people who haven't used or seen the product. Translated into other languages by people with little knowledge of the product, or its associated vocabulary, they become even more difficult to understand.

Technical manuals may not be updated sufficiently frequently, and become outdated. Logistics support data can get out of control. Inadequately documented configurations become more and more difficult to maintain. Spares replenishment becomes inaccurate, and inefficiencies occur in spare parts management.

8.3.28   Media

Product data are on a variety of media. Some of the data may be on traditional media such as paper and aperture cards. Some may be on electronic media. Some may be on Mylar (biaxially-oriented polyethylene terephthalate) or microfilm. Some data may be on magnetic tape. Some may be on CD-ROM or on DVD. Some may be in the Cloud. Data on different media have to be managed in different ways. Data on one medium, such as paper, usually has different management requirements to that of data on other media.

Companies have to keep product data that's on both traditional and electronic media for similar periods and similar purposes. Just as some traditional media deteriorate over time, some electronic media also deteriorate over time and aren't suitable for long-term storage.

8.3.29   Meaning of Data

Raw data often has little meaning, and is of limited use. Yet, in many cases, it's often only raw data that's kept. Typically, the data that's maintained on a product, such as a machine, will include CAD files, some drawings, Bills of Materials and some information on the manufacturing process. Little information will be maintained on user needs, the process that was used to develop the data, or the various choices and activities that took place during the design cycle. Most of that information will be lost and forgotten. The next time that similar activities are carried out, it won't be possible to benefit fully from past experience, and development will start again from scratch.

8.3.30   Missing Data

Part costs may be difficult to estimate. Often there's no relevant data available to provide a basis for calculation. When some data is found, it's seen to be inaccurate or out-dated. When data from different sources is brought together, it may conflict. Or it may not make sense.

8.3.31   Navigation to Data

From one piece of data, people want to be able to get to (to navigate to) related information. For example, knowing a product name, a recycler may want to find a list of dangerous components. Unless the navigation path has been defined, the list will be difficult to find.

8.3.32   Ownership of Data

Traditionally it's been made clear, through organisation charts, which human resources belonged to which part of a company's organisation. It's been less clear which information belonged to which part of a company's organisation. The real ownership of product data is often unclear. Even within a particular part of the organisation, such as the Engineering Department, it may not be clear who are really the owners of information. Designers, analysts, drafters, supervisors and managers will all have their own ideas as to ownership of information. They will probably be willing to defend what they see as their property if anyone suggests that ownership of the information should actually be assigned to somebody else. But, although they may want to enjoy some of the benefits of owning the data, they may be less willing to accept the responsibilities of maintaining it properly, and making it available in a suitable form when it's needed by others. Outside the company, similar ownership issues arise when product data is shared with partners, contractors, suppliers and customers.

8.3.33   Processing of Data

Data is processed in many ways. The best way for processing and managing any one of the many different types of data may not be the best way of processing and managing another. If users want optimum performance, they will suffer from not having a common approach. If they prefer a common approach, they won't have optimum performance.

8.3.34   Project Data

Project and resource management tools should be linked to product data, but often they aren't. Unintended overlap in data and project activities results, wasting time and money. To avoid this, the activities may be run in serial, lengthening project cycles. Design Rules and procedures may be ignored because there's no way to enforce them. Project planning exercises can't draw on real data from the past, but are based on over-optimistic estimates. Project managers find it difficult to keep up-to-date with the exact progress of work. As a result, they are unable to address slippage and other problems as soon as these occur.

8.3.35   Re-invention of Existing Data

Manual transfer of data introduces errors and wastes time. It can take so long that users may decide it's not worthwhile. Instead, they may work with out-of-date, or incomplete, data. For example, an engineer may not be able to get cost data directly from the product costing application, or product quality data from field support applications. As a result, they may develop a design without taking sufficient account of cost and quality information. The resulting design will probably be of lower quality than a design built on the basis of full understanding of cost and quality information.

If it's difficult for users to access data that's not readily available, then, rather than searching and waiting for it, they may prefer to re-invent it.

8.3.36   *Relationships Between Data*

There are many types of relationships in the product lifecycle. There are relationships, for example, between products and parts, between parts and data, between one part and another, and between parts and processes. There may be relationships between the parts of one product and the parts of another product. There may be relationships between parts and development projects. There may be relationships between apparently separate product development projects. There are hierarchical relationships linking parts to a product. Bills of Materials, parts lists, assembly drawings and where-used lists contain information on such relationships. The various types of data (such as specifications, drawings, models, test results) supporting a product need to be related. A component may only fit a particular variant of a particular product. An engineering drawing corresponds to a particular part. An NC program corresponds to a particular version of a part.

There are relationships between activities. In some cases, for example, the design of a particular part may only be started when the design of other, related parts has been completed. Data needs to be linked both to its source and to derived data. It also needs to be linked to the activities that create and use it. Users need to be able to navigate through the various relationships and links.

Information on the procedures used to develop a product needs to be related to the product and to the activities. Similarly, information on the procedures that should be followed for new products should be linked to the activities and the data.

Information needs to be maintained on the reasons why choices were made between competing designs. Relations need to be maintained between different designs, selection criteria and decisions. If the selection criteria can be retained, they can be used to help make better choices in the future. If the context within which a particular design was chosen is clear, it may be obvious how the part should be made.

In the absence of the right links between the information and the task that creates it, it's impossible to get real control of the environment. This is true from the technical point of view where information concerning relationships, such as selection criteria for alternatives, is lost. It's equally true from the management point of view, with the result that product development doesn't take place as efficiently as it could.

In addition to the wide variety of relationships, added complexity arises due to changes that occur in relationships as product development takes place. For example, an ingredient that was previously used in three recipes may, in future, be used in a fourth recipe. A part that was originally expected to be made of metal may instead be made of plastic. Similarly, the relationship between logic changes and Printed Circuit Board (PCB) designs needs to be maintained so that changes in the logic are reflected in the board. Parts of a design may be due not to structural or aesthetic reasons but to the manufacturing process that will be used. A part that has been designed to be cast, may have features that are completely unnecessary if, in future, the part is to be machined.

Information concerning the completeness of a design needs to be maintained from the viewpoints of functionality, constituent parts and necessary activities. If a particular part is replaced, it should be possible to identify what's needed to maintain completeness. A "missing" relationship may be as important as a "present" relationship. Comparative relationships such as "duplicate" are needed, for example, to prevent development of identical products.

8.3.37 Representations of Data

An entity may be represented in different ways. For example, a circle may be represented by three points on its circumference in one CAD application, but by its centre and radius in another CAD application. A line may be represented by a vector in a CAD application, but by a set of points in a rasterised representation. The same object may exist electronically in one representation in a spread-sheet, and in another representation on a piece of paper.

Different users will want to look at, and work with, different representations of data. Sometimes, the representations will make use of the same data type. At other times, even the data type will be different. In all cases, the representations must relate to the same underlying data. Modifications have to be made to the underlying data and not to the more superficial representation.

At different times, a user will want to work with different structures of data. Different users will want to work with different structures. Different users will want to work with different hierarchical levels within a given structure. Different users will want to include the same part in different functional or hierarchical structures. Engineering, Accounting, Production Management and Assembly may all have different requirements for Bill of Materials structures.

The structures and levels have to be consistent. At the lowest levels, the data will be in the form of bits and bytes. These may represent numbers, characters, sounds, and lines. At the next level up, these may represent geometric information, or information about a material or a colour. At a higher level, this information may actually represent a part, which in turn may be a component of an assembly such as a wing flap, which in turn belongs to a wing, which in turn belongs to an aircraft. Each level of the structure is of interest to particular users. The machinist drilling the holes for the bolts that attach the wings of an aircraft to the fuselage, wants to know the exact position of the holes and any deviations during drilling, and isn't interested in the aerodynamic qualities of the wing.

One user may need a bottom-up structure of the product, starting with nuts and bolts and small parts, and then working upwards through larger components and assemblies. Another user may require a top-down structure, starting with the complete product, and then working down through the major assemblies.

Some users will be happy to work with 2-dimensional data. An engineer laying out a single Printed Circuit Board may not need to take account of the third (thickness) dimension. On the other hand, a stylist defining the shape of a car wing will, at some time, want a full three-dimensional representation of data. Other users may need to work with both 2D and 3D representations, and want a modification to one of these to be reflected immediately in the other.

A related issue is that of a drawing of a part on paper that's electronically scanned, and then converted from raster format to CAD vector format. The three representations of the part are different representations of the same object. Each representation can play a useful role in the product lifecycle. An analyst may need to use the CAD model. A machinist may need a rasterised picture of a part on a shop floor terminal. A maintenance engineer may have to make a major repair on a customer site, and may need to take a paper drawing to the customer site. Procedures have to be in place so that the different representations can co-exist, and so that any necessary modifications can take place. Any modification made to one of the representations has to be reflected in the others.

Under their own data management functionality, different applications may store the same data in different formats. For example, one CAD application may represent a circle by its centre and radius. Another application may represent it by three points on its circumference. A third application might represent it by its centre and two points on its circumference. Even applications that use the same representation may physically store the data differently. One application may store it in the order x co-ordinate of centre, y co-ordinate of centre, radius, whereas another application may store it in the order radius, x co-ordinate of centre, y co-ordinate of centre.

For one application, 10-11-12 may mean 10 November 2012, for another it may mean 11 October 2012, or 12 November 2010, or 11 December 2010. And 10 November 2012 may be acceptable to one application, but not November 10 2012.

8.3.38   Rules Deficit

Years ago, in days of paper documents and life-long employment, there were supervisory staff to explain the rules to newcomers and check that their work kept to the rules. In the early 21st Century, new hires are expected to start work at their workstations and be productive immediately. There's nobody to explain whatever rules may govern their work. So they start working and do their best. Not knowing the rules, assuming there are some, they make basic mistakes, such as leaving a space in a product name where the company doesn't leave a space. Or capitalising where the company doesn't capitalise. Or using a table in Word, where the company usually uses Excel. Once small errors like this get into the system, they're unlikely to be corrected. However, they lead to a waste of time and effort later as other people try to find and use the information.

8.3.39 Searching for Data

People can easily waste a lot of time looking for product data. Sometimes they're even unable to find the data they need. And, if they do find it, it may not correspond to the actual state of the product. Developers may be unable to rapidly access a particular design among the mass of existing designs. To find specific information, they may have to search through many paper and electronic files. Studies show that design engineers spend up to 80 % of their time on administrative and information retrieval activities. Rather than searching for hours to find an existing design that they can reuse, they may develop a new design which is almost identical to an existing design. Then, unnecessary additional costs are generated as the new design is taken through all the activities necessary for manufacture, and then supported during use.

Within individual engineering activities, the percentage of time that individuals spend looking for, or transferring, information is high. For many, otherwise productive, individuals it may be 30 % or even 50 %. As time may also be taken up by management tasks, the time actually spent on the functional activity may only be about 30 %. In addition, in many companies there are technical "liaison" staff who spend 100 % of their time looking for information that should have been transmitted by other departments.

8.3.40 Security of Data

Product data is valuable and should be kept secure. For various reasons, such as its volume and its easy communication by e-mail or the Web, it's not easy to maintain product data secure. It would be easier if security rights were always the same, but they often change during a product lifecycle. In addition, users may have different rights on different projects and products. For example, some designers may not have the right to see some financial information, and some maintenance staff may have limited access rights to the detailed designs of particular parts.

At the level of the data itself, there's a need to provide better security, control, access and protection. Although data is so valuable, it's often very easy for users to mistakenly destroy, or lose, data. One company I worked with actually lost all its CAD files. A combination of very unusual events led to all files, copies, back-ups and archived files being lost. Another company lost the data it had created over the previous 2 months. When it tried to restore the data, it found that the back-up functionality only worked correctly in certain circumstances.

The need to provide full protection of data has to be addressed alongside the need to make data available immediately to authorised users. At different times in the product lifecycle, data should have different levels of protection. Until a design is released, a design engineer may be free to modify it. After release, manufacturing engineers shouldn't be allowed to make "improvements" without following the correct change procedures. At times, the situation will be complicated by the need to share data among several users each of whom has different access rights.

8.3.41   Sources of Data

Product data is created in many functions, and is used in many functions. Some data is created in one application, other data in other applications. Some product data is created during development. Some is created later. Some of the data is used during product development. Some is used elsewhere. Some of the data will be created in the company. Some will be created by suppliers and customers. Some data may be printed out on paper. Some of it may stay in electronic files and databases. Often, a lot of data will be in files created by word processing applications (such as MS Word), spread-sheet applications (such as MS Excel), and other applications such as CAE, CAD and CAM. Whatever source it comes from, it has to be managed.

8.3.42   Software

There may be specific problems with data representing a product's software. Onboard software may not be documented sufficiently. Modifications may be made to software without appropriate change control.

8.3.43   Standards for Data

There are standards for product data. The most well-known of these is the ISO 10303 set of standards (STEP, the Standard for the Exchange of Product Data). The STEP standard enables exchange of data between many applications. It includes many Application Protocols (AP), addressing the needs of particular industries and particular activities. For example, AP 215, AP 216 and AP 218 address product data for ships. AP 236 addresses product data for furniture. Other standards include ISO 14306:2012 "Industrial automation systems and integration—JT file format specification for 3D visualization", and EN 9300 standard for "Long Term Archiving and Retrieval of digital technical product documentation such as 3D, CAD and PDM data".

8.3.44   States of Data

Product data can be in various states. These include in-work, in-process, in-review, released, as-designed, as-planned, as-built, as-installed, as-maintained, and as-operated. The lifecycle of an object, such as a part or a drawing, can be described by these states and the transitions between them.

Different rules apply to access and modification of data in different states. In early stages of development, data is frequently modified. Once it's been released it's much more stable. Users may need to work at any time with product data that are in different states.

8.3.45   Structure of Data

A product, or a plant, may be made up of assemblies, sub-assemblies, components and parts. Another product will be made up of many ingredients. Product data, such as Bills of Materials, recipes and goes-into lists, describe the structure and relationships. This structural data has to be managed.

8.3.46   Tabulated Documents

Tabulated Documents make life easy for the person who creates them. For example, instead of making 12 separate drawings, each with its own name and identifier, of similar parts, they make a drawing of one of the parts. Then, on the same sheet, they create a table showing the characteristics that change from one part to another. The one drawing they make has one name and identifier. The document creator may have saved a few minutes by not making 12 separate drawings, but downstream users of the document waste hours, or even days, creating copies that they can work with. Downstream, they may believe that life is simpler if there's one drawing, with its own name and identifier, for each and every part. Maybe they'll photocopy the drawing 12 times so that they can have a drawing of each of the parts. Or, maybe someone will create the drawing 12 times in the CAD application they have at home. Soon after, the person who created the tabulated drawing leaves the company. A customer requests urgent changes to two of the parts. The newly hired replacement thinks it would be quicker, to avoid going through the company's time-consuming engineering change process, to create separate drawings for the two parts. Downstream, they make two new copies. Now they have the original 12 drawings and the 2 new drawings. Initially, maybe, that isn't a problem. But before long, for example, someone downstream leaves, and other customers start complaining that the parts they've been purchasing for years no longer work properly.

Then, more time, effort and money is wasted as everyone tries to work out why they're having problems with some of the parts.

8.3.47 Traceability of Data

Traceability is a requirement in many industries. For example, consumers increasingly want food to be traceable back to its farm of origin. Food producers see this as a key element in increasing consumer confidence in food product safety.

An audit trail needs to be kept so that it's possible to go back in time and see how, and why, a particular part or product was made. An audit trail shows which actions were taken on which data. It helps locate errors.

8.3.48 Training Deficit

The PLM environment is complex and continually changing. In such an environment, people need frequent training to keep up to date. Many companies though, feel they're under such pressure to respond to market changes, that they don't have time to provide such training. And anyway, they reason, professionals shouldn't need training, they should know it all. Well, the professionals do know a lot about their subject matter. But often that's not where they need training. More likely, they need training about the changes in the way the company operates as it responds to market changes.

8.3.49 Type and Format of Data

Product data exists in many different forms. There's text data (specifications, schedules, process plans, manuals, project plans), numeric data (descriptive geometry, formulae, results of analytic experiments and calculations, computer programs), graphics data (photographs, drawings, sketches) and voice data. Within each of these different types of data, there may be different sub-types. For example, some of the graphics will be in vector form, some will be in raster form. The materials on which data are stored will be of different sizes. Paper will be of various sizes. Paper from different countries may have different sizes such as ANSI A and B in the US, and ISO A3 and A4 in Europe. Electronic storage devices will be of various sizes.

Some of the product data in the PLM environment will actually be computer programs. Among the programs that may need to be managed are those that are components of products (onboard programs, programs developed to be used within the company's products). Other programs that may need to be managed are those that are used to define and support the product, such as CAD.

8.3.50   Update Frequency of Data

Different types of data may be updated with very different frequencies. Some data on existing products may not be changed for years. On the other hand, software under development may be changed several times per day.

8.3.51   Users of Data

There will be many users of product data. Product data will be used by many people in many different functions and at many different locations. They may be working on the company's premises. They may be working for a supplier, a partner, or they may even be the final customer of the company's product. Product data has to be made available to all these people. At the same time, product data must be protected against unauthorised access.

Product data will be used and shared by several departments and functions. A lot of product data will be created in R&D and Engineering Departments. But information will also be created and used in the Manufacturing, Marketing, Finance, Sales and Support Departments. Some of the data will be with design engineers, some with manufacturing engineers. Some will be with production planners, some on the shop floor. Some will be with service engineers. Some data will be with the customer. Some with suppliers. Some may be created by customers. Some may be created by suppliers. Wherever the product data is, and whoever it's with, it needs to be managed if it's to be used effectively.

8.3.52   Uses of Data

Across the product lifecycle, many people, such as machine operators, salespeople, customers and recyclers, will work with the product. They'll be involved in different activities. They'll want to work with the product in the most appropriate way for their activity. They'll want to work with the most appropriate structure of product data for their task. Some users may want to work with a list of what's in the product. Others may just want a list of what has to be ordered to manufacture the product. The different data structures have to be coordinated to avoid information loss when data is exchanged between them. Otherwise errors will occur.

Users of product data will be working on a variety of tasks. Depending on what they're doing, and their level of computer literacy, they will have different product data usage and product data management needs. Some will create data. Some will modify it. Some will delete it. Others will only want to reference data, perhaps for management purposes.

Some of the users will be working with advanced concepts such as the creation of data that will help take Man out of the Solar System. Others will need data to help them solve more down-to-earth problems like finishing a part before the shift ends.

While different users may have different requirements, they may also have some common requirements. For example, they may all want to make use of the same basic spread-sheet, text processing and electronic mail systems.

Some of the users will have an engineering or science background. Some of those who don't have such backgrounds may have backgrounds such as accountancy, human resource management, marketing and sales. Some of the users will be customers with completely different backgrounds. Some of the users may be schoolchildren, some may be retired farmhands.

8.3.53   Value of Data Unknown

A company's product data represents its collective know-how. As such it's a major asset and should be used as profitably as possible. Yet many companies ignore their product data. If $100,000 goes astray in their financial environment, there's a major panic. If $10,000,000 goes astray in their product data environment, there's generally no panic at all, since most people in the company are completely unaware of the loss. Many managers find it difficult to put a value on their product data. Top management, in particular, is rarely aware of the extent to which valuable product data is ignored and misused.

8.3.54   Variants and Options

Many products, such as cars, are available with a variety of options and variants. The definitions and descriptions of these have to be managed carefully. As product lifetimes decrease and customisation increases, the possible number of options and variants increases, as does the required effort to manage their definitions and descriptions.

8.3.55   Versions of Data

The product development environment is typified by many versions, iterations and alternatives. Products are made with different models, versions, options, variants, releases and alternatives. Throughout the development process, designs change, components are modified, products are restructured and project status is updated accordingly.

Different versions of the same information, for example the recipe of a soft drink, will exist. One person may need the latest version, but others may need earlier versions. People who need the latest version want to be sure that the version they receive really is the latest, and not an out-of-date, or superseded, version.

In many companies there are different versions of the document on which a particular type of product data is captured. For example, customer requirements may be captured on a standardised form for several years. Then it will be decided that more detailed customer data is needed, so the form will be modified. A new version of the form will be developed, but instances of the old form may continue to be used.

8.3.56 Versions of Applications

Application software is regularly upgraded by application vendors. Each time an application is upgraded, there's the risk that the addition of new functionality and a richer information content will make it impossible to access and use product data that was created in previous versions of the application.

8.3.57 Views of Data

Different users will want to see different views of product data, but many users will only want to see and work with one view of the data. For example, a manager may want to see current progress on all parts of a product development project, but not details of the product design itself. A project engineer may want to check an assembly of parts, but have no interest in the progress of stress or thermodynamic analysis activities. A drafter may only be interested in an individual part. A company may only want to give a supplier a very restricted view of its overall database. In all these cases, while users may want to see different views of the data, and the applications they use may be different, the underlying data must be the same.

The available views of a product change during its lifetime. In early stages of development, the product is defined in specifications describing its functionality and required performance. Later in the development cycle, a top-level design or architecture is prepared which assigns specific functions to specific parts of the product. Towards the end of the design process, the physical arrangement of these parts is fully described, through detailed models, drawings and parts lists. Information is generated to support the processes of production and test of the product. Information is also prepared to support the product in operational use.

Once the product has been manufactured, information on its actual performance in the field will be measured and recorded.

8.3.58   Vocabulary

People in different parts of a company often use different words to describe the same thing. Sometimes they use the same word to describe different things. Each department develops its own jargon that may be misunderstood by other departments, partners and customers. This can lead to mystification and confusion (Fig. 8.5) as people try to communicate information about an object that's described and defined in many different ways in different organisations.

8.3.59   Volume of Data

The sheer volume of product data makes it difficult to manage. There can be millions of objects, descriptions, numbers and words of product data to manage. Estimates for medium-to-large companies foresee data volumes exceeding 1 petabyte. Since the creation of several gigabytes of data only requires a few seconds, it doesn't take long, even in small companies with only a few users of computer-based product data, for manual data management techniques to break down. Users soon find that they're unable to efficiently and effectively locate data. In larger companies, with several hundred engineers, thousands of GB of product data may be created and accessed each week, and the volumes keep growing.

Usually there's a lot of product data on paper. And usually there's also a lot of electronic product data. Some companies have millions of paper documents and millions of electronic files.

Terms	Issues
Research, develop, discover, design, ideate, imagine, invent, innovate, conceive	Are they the same? Are the differences clear? In what order do they occur?
Product definition, description, specification	Are the differences clear?
Version, variant, release, option, model, revision	Are the differences clear?
Product life, lifetime, lifecycle	Are the differences clear?
Prototype, pilot, product	Are the differences clear?
Recycle, dispose, retire, reuse, upgrade, refurbish	Are the differences clear?
Project cost	Which costs are included in the cost of a project? Which overhead costs are included?
Number of parts in a product	Is it clear which parts are being counted? How are duplicate parts counted?
Portfolio Management	Which portfolio is being managed? A Product Portfolio? A Project Portfolio?
Date	When is 11/10/12? Is that October 11, 2012? Is it 10 November 2012? Is it October 12, 2011?

Fig. 8.5 Product and lifecycle mystifusion

As products are customised, the number of possible combinations of parts rises dramatically. For the manufacturer, the environment becomes increasingly complex and hard to manage. As the number of potential product configurations increases, it becomes harder to know which configurations are meaningful and can be produced, and which shouldn't even be proposed to a customer. It becomes harder to keep track of the real configurations of individual products. With more and more different products, it becomes harder to make sure that configuration documentation corresponds exactly to individual products. Yet the customer requires the same after-sales service on a product that's unique, as on a product that was produced as one of an identical batch of several thousand.

8.3.60   Workflow

All the activities along the product lifecycle create and/or use product data. They exist to provide the product data necessary to produce, use and support the product. Without product data, there would be no need for these activities.

Each step, or activity, in the activities has its own information needs, information input and information output. Within an activity, people use information. If information is not available, it may not be possible to complete the activity. Often, the end of an activity is characterised by information being prepared, signed off and released. Between activities, information is transferred. When an iteration or change occurs in the activity, corresponding information is produced. Information flow has to be synchronised with the workflow so that the information is available when and where it's needed.

In all but the smallest organisations, product data is created to be used by someone else. Presumably the creator of information knows for who it's being created. Once created, it should be moved on, it should flow, to the activity that's going to use it. Since product data and product activities are so closely linked, it's not possible to control one without becoming involved with the other.

8.3.61   Consequence

With all these issues related to product data, it's not surprising that product development and support activities are delayed for many apparently random and minor, but cumulatively significant reasons. And the resulting product and service quality is erratic despite vast investments in technology, and significant expenditure of management time.

8.4 Metadata

8.4.1 Data Fields in Paper Documents

On a paper document describing a product, usually one part of the document, at least, will have a particular structure reserved for information about the rest of the information on the document.

A paper drawing, for example, often has an informative "title block" with fields for information such as drawing title, identifier, scale, units and creation date.

A paper text document may have a "header" with fields for information such as document name, product name, creation date and author name.

8.4.2 Data About Data

The title block or header on a paper document contains metadata. Metadata is "data about data", "data describing other data". It's key information about a larger volume of data, such as its name, its status, its location, its owner. It's similar to the catalogue information of a book in a library. That might contain the book title, author name, book number and book location.

The amount of metadata is usually much, much smaller than the amount of data it describes. For example, there might be just 10 metadata for a book in a library catalogue. There might be just 20 metadata describing a 50 kB text document, or 40 metadata describing a 1 MB CAD object. Even though there's not much metadata, it's very helpful when identifying, managing and accessing product data.

8.4.3 Examples of Metadata

There are different types of metadata. For example, there's administrative data such as file size, type, creating application, and time stamps. There's descriptive metadata such as object type, object identifier, title, subject matter, and owner. There's structural metadata such as the link between a drawing and a part, or the links between sub-assemblies and a product.

The metadata of a file could include the name of the file, its title, its type, the application that it was created with, the location of the file, its size, and its lifecycle state (such as in-work, under review, in rework, approved, rejected, cancelled). It could include its creation date, the date last modified, the date last accessed, permissions (such as who can read it and who can modify it), the author, its status (checked in, or checked out), and by who it was last saved. With less than 20 metadata, it's possible to get a good overview of the file. Yet the file itself could, for

document title	subject	name	creator	type
reviewer	format	creation date	date last modified	access rights
description (1)	source	template	location	structure
description (2)	dependencies	classification	attachments	author
lifecycle state	usage	version	iteration	owner

Fig. 8.6 Some metadata of a word processed file

drawing number	type	title	date created	date modified
date reviewed	date released	page number	sheet number	drawing owner
revision level	reviewer	approval date	approver name	drawing scale
releaser	drawing format	contact name	drawing size	standards

Fig. 8.7 Some metadata of a CAD drawing file

part number	part superseding	part description	part superseded
drawing number	make/buy source	unit(s) of measure	cost
revision level	lead time	dependencies	certifier
creating system	status	price	colour

Fig. 8.8 Some metadata of a part

example, hold the results of 20,000 measurements of pressure and temperature at various positions on the product.

The metadata of a word processed document file could include a lot of useful information about the file (Fig. 8.6).

With less than 25 metadata, it's possible to get a good overview of the word processed file. Yet the file could contain dozens of pages of detailed description of a market segment or a new product.

Similarly the metadata of a CAD drawing file could include a lot of useful information about the file (Fig. 8.7).

And the metadata of a part (Fig. 8.8) gives a good overview of the part.

8.5   Models

8.5.1   Need for Models

The scope of the PLM environment is wide. It's a complex environment. It's difficult to understand. Simple models are needed to help people understand and communicate about it. A model acts as a common basis for discussion and communication. It helps people increase understanding and reach a common view.

Some models address just one component of the PLM environment, such as a model of a product. Other models address several components. For example, a model could show how product data is created and used in business processes by people from different departments using different applications.

8.5.2   Sub-models

An overall model may be built up using other models. Each of these could have sub-models. There's so much to be modelled in the PLM environment that a single model would be so large and complex that it would be very difficult to understand.

8.5.3   Different Models

Many different models can be developed. They can show the situation at different times, and from different viewpoints. Some of these models are complementary.

For example, companies often find that it's useful to create models of both the current ("as-is") situation and the future ("to-be") situation. These will eventually be related by a plan.

Another useful pair of models is the "top-down" model and the "bottom-up" model. The "top-down" model is developed from a business-oriented overview of the PLM environment, working down towards individual operations and detailed descriptions of data and activities. The complementary "bottom-up" approach starts from individual operations and detailed descriptions of data and activities. Then it links data and operations, and builds successively higher levels of information and processes.

8.5.4   Different Levels of a Model

The PLM environment, like any other environment, can be described in many different ways, for many different purposes, using many different types of representation. For example, in another environment, such as that of a new boat, a naval architect may make a drawing that shows a yacht as the promoter wants it to appear. Once an owner has been identified, the architect will make a more detailed drawing showing the required internal layout of walls, cabins and other spaces. Later, once this layout is clear, another drawing will be needed to show the detailed electrical layout. Another drawing will detail electrical components such as light switches. Another drawing will be needed by the installer of the wiring. And another for the person who eventually sails the yacht.

8.5.5   Iterative Approach

The development of a model is generally an iterative approach. The first attempt at a model will probably lack detail and be incorrect. However, it provides a starting point from which further refinement can take place.

8.5.6   Involvement in Modelling

Development of models usually involves many people in the company. Involvement in this activity helps them better understand the activities and entities in the PLM environment. The relationships between entities, activities, processes, documents, applications and people will become clearer. It will be possible to understand the events that link them, and to identify the major management milestones used to control them. Involvement of people at this stage will increase their commitment to the success of the future environment.

8.5.7   Modelling Tools

It's possible to make a model with simple tools such as a pencil and a piece of paper.

Software tools are also available to provide support for modelling. They can be of great benefit in modelling, since the amount of data generated by the modelling activity, and the frequent updates and changes, can be difficult to handle by purely manual means. However, the ease with which such software allows modelling to take place shouldn't be allowed to distract people from making sure that what they're modelling is correct and of use to the business. It's only too easy, when mapping and modelling the PLM environment, to lose sight of the target. A very detailed data model will be of little use if the PLM project stops before the model is completed.

8.5.8   Modelling Techniques

There are many different modelling techniques and languages. Often they use a mixture of text and graphics. They use symbols such as different types of parallelogram (such as square boxes, rectangles, diamonds and rhomboids) and curves (such as circles and ellipses). Depending on the language, each of these may represent an object or an activity. Arrows usually show a flow between the objects or activities represented by the symbols. Lines between symbols usually show relationships.

The various languages and techniques often have a number of different types of "diagram" (or "maps" or "domains") that show different aspects of an object or an activity. These include for example, the structure of objects and activities, their development, how they communicate, and how they're related to other objects and activities.

One of the languages used is Unified Modelling Language (UML). Maintained by the Object Management Group (OMG), it has many types of diagrams that are used to show information about objects and activities. Some of these are described and illustrated below. The types of diagram are divided into three categories: Structure Diagrams; Behaviour Diagrams; and Interaction Diagrams.

Structure Diagrams represent static structure and include the Class Diagram, Component Diagram and Object Diagram. There are 3 Behaviour Diagrams. They are the Activity Diagram, the State Machine Diagram and the Use Case Diagram. They represent various types of behaviour. The group of Interaction Diagrams represents interactions, and includes the Communication Diagram, Sequence Diagram, Timing Diagram, and Interaction Overview Diagram.

8.5.9 Characteristics of Models

Usually some of the people involved in the modelling activity have a good understanding of a particular modelling technique gained in academic or business courses. They know the rules and try to keep to them. Many people though, haven't been trained to use a particular model. They may not want to keep to any rules. The result is that models are often produced with elements of different techniques, and are only understood by those who produce them.

8.6 Product Data Models

The most frequently used models of product data are used to show its flow and its structure.

8.6.1 Data Flow

Figure 8.9 shows the flow of data from a Sales Engineer to a Design Engineer and then to a Manufacturing Engineer. The "Detail the design" activity is represented by a box characterised by the information on three arrows. The input arrow shows the input necessary to perform the activity, and the output arrow shows the output of the activity. The third arrow, the vertical mechanism arrow, shows the means by which the activity is accomplished. Activity boxes are linked on a diagram to show the overall environment. The output of one activity is the input for another.

The principles behind such a model are simple and easy to understand. Creation of a model for a small activity is quick and easy. But creation, change and management of models with hundreds of activities and participants takes a lot of time and effort.

Fig. 8.9 Data-flow diagram

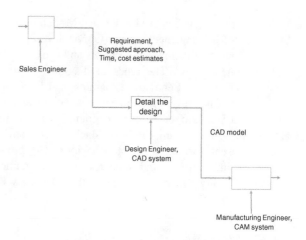

Data-flow models can show how data flows through the lifecycle, and by which activities it's processed and stored. Data-flow models relate the various activities of the PLM environment to the use and flow of data.

Typically, data-flow modelling is carried out as a top-down exercise with as many decompositions or hierarchical levels as necessary. The environment is first described at the top level, Level 0. Then each element of the top level is separately described in more detail.

8.6.2   Entity-Relationship Model

Entity-relationship models have been used since the 1970s. In this modelling technique, an entity is any physical or logical object of interest in the environment being modelled. Customers, products, suppliers, processes, locations, machines, money, documents and employees are all examples of entities. There are also many possible relationships. For example, one entity may use another entity. One entity may own other entities. One entity may contain another entity.

An entity doesn't represent an individual item (an instance), but a class of similar items, each of which can be characterised in the same way and will be used in the same way in the environment being modelled. Examples of entities in a particular environment could include aircraft, manufacturer and airport. The common properties and characteristics of an entity are referred to as its attributes. Examples of an aircraft's attributes could include colour, weight, fuselage length and wing span. Attributes of a manufacturer could include manufacturer name and assembly plant location. Relationships are associations that describe the link between entities, for example between an aircraft and a manufacturer.

Graphically, entities are often represented as rectangular boxes, attributes as circles or ellipses, and relationships as diamonds. Lines associate entities and

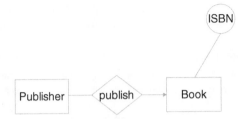

Fig. 8.10   Entity-relationship diagram

relationships to their attributes. Annotations indicate the nature of relationships: one-to-one, 1:1; one-to-many, 1:n; many-to-many, m:n. Alternatively, single-headed or double-headed arrows can be used to indicate the nature of the relationship. In Fig. 8.10, there would be a one-to-many relationship between publisher and book. A publisher publishes many books, a book is only published by one publisher. Each book has a unique ISBN.

8.6.3   Class Diagram

The Class Diagram is used to show the static structure of a system. It shows the system's classes, their attributes, and the static relationships between classes.

In the class diagram, a class is represented by a box with three parts (Fig. 8.11). The upper part shows the name of the class, the middle part shows the attributes of the class, and the bottom part shows the class's operations. These are the operations or methods performed by the class. As there may be many attributes and operations, sometimes some attributes aren't shown, and sometimes the operations aren't shown at all.

The "Person" Class would have attributes such as Name, Date of Birth, Height and Weight. Each person (member of the "Person" Class) has these attributes, but has their own value for each attribute.

A person with the Name of John Smith is a member of the Person Class, with a Date of Birth value of January 1, 1981, a Height value of 5 ft 10 in, and a Weight value of 220 lb.

All instances of a class have the same attributes and relationships. So, for example, all drawings of the Drawing Class would have an attribute of Number. But they would all have a different number.

Name of the Class
Attributes
Methods

Fig. 8.11   A class is represented in three parts

A line can represent the relationship between classes. An arrowhead indicates the role of the entity in the relationship. Numbers at each end of the line indicate the multiplicity.

The Part-Drawing Relationship (Fig. 8.12) shows that a Drawing can exist without a Part (0), but that a Drawing can be associated with many Parts (n). A Part doesn't necessarily have a Drawing (0); but a Part can be associated with many Drawings (n).

The Part-Colour relationship (Fig. 8.13) shows that a Colour can exist without a Part (0), but that a Colour can be associated with many Parts (n). A Part always has exactly one Colour (1).

The Part-Part relationship in the model in Fig. 8.14 shows that parent-child relationships are possible. A Part-Part relationship is used to show a part belongs to a sub-assembly, or a sub-assembly belongs to an assembly.

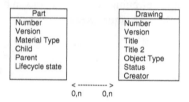

Fig. 8.12 Part-Drawing model

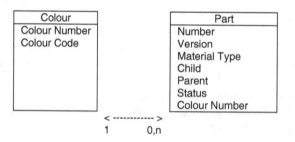

Fig. 8.13 Part-Colour model

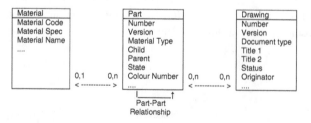

Fig. 8.14 Material-part-drawing model

Fig. 8.15 A state diagram

Object	Lifecycle	State 1	State 2	State 3	State 4	State 5
A		In Work	Completed	Cancelled		
B		In Work	Under Review	Released	Rejected	Cancelled
C		In Work	Under Review	Complete	Cancelled	
D		In Work	Under Review	Released	Rejected	Cancelled
E		In Work	Under Review	Released	Complete	Cancelled
F		In Work	Under Review	Approved	Accepted	Cancelled

Fig. 8.16 Object lifecycle state table

8.6.4 State Diagram

A state diagram shows the states that exist for an entity, and the allowable transition paths between them.

Figure 8.15 shows four possible states for an entity.

When the entity is In Work, the only allowable transition is to the Under Review state. When the entity is Under Review, the allowable transitions are to In Work, Released and Cancelled states. When the entity is in the Released state, the only allowable transition is to the Cancelled state.

Some entities in the PLM environment have the same lifecycle, some have different lifecycles. Figure 8.16 shows the lifecycle states of the objects A, B, C, D, E and F.

8.7 Product Data Is not an Island

Product data is not an island (Fig. 8.17), isolated from the rest of the world. Product data is closely related with other PLM components. It's also influenced by other forces within the company, and outside the company.

As an example, if a change is needed to a specific type of product data, the application that creates the data may need to be modified and the application that stores the data may need to be modified. To enable the change in the product data, the working method may need to be changed. The description of the process, showing how the product data is created and used will be changed. Users of the data

product	process	applications
people	**Product Data**	organisation
metrics	methods	equipment

Fig. 8.17 Product data is not an island

Problems accessing data on old magnetic tapes
Difficulties to exchange data with partners
Unreleased versions of data mistakenly used
Inability to access legacy data
Duplicate data
Incorrect data sent to a customer
Confidential data lost
Multiple overlapping databases
Difficulty to manage data at several locations
Conflicting different copies of the same data
Ownership of data unclear
Knowledge in heads not documented

Fig. 8.18 Potential product data-related challenges for a company

will need to be informed of the changes. The data may be used in a performance indicator, which may need to be adjusted. Downstream users of the data may need to make changes to equipment.

8.8   The Challenges

The specific product data-related challenges that a particular company faces could come from several sources (Fig. 8.18).

8.9   The Way Forward

The issues described in this chapter all relate to product data, but they can't be solved in isolation from the other components of PLM. Their solution, within the PLM environment, is addressed from Chap. 21 onwards.

Chapter 9
Process

9.1 Introduction and Definition

9.1.1 Action Across the Lifecycle

There's a lot of activity in a company as a product is developed, manufactured, supported and retired. Many things have to happen if everything is to work well with the product. Figure 9.1 shows some of the activities that take place across the lifecycle.

Although there are 35 activities in Fig. 9.1, far more activities occur in the product lifecycle. Figure 9.2 shows another 35 activities. And you can probably think of some more.

9.1.2 Organising the Action

For most of the 20th Century, companies were mainly organised by functional departments such as Marketing, Engineering, Manufacturing and Sales. People and activities were assigned to a department. People in Engineering worked the way the boss of Engineering told them to work. People in Manufacturing worked the way the boss of Manufacturing told them to work.

However, although a departmental structure may make it easy to organise people, and tell them what to do, it doesn't reflect the way that a company works. In reality, a lot of the activities that take place involve people from many departments. If some people are working one way (for example, the way the Engineering VP tells them to), and others are working another way (for example, the way the Manufacturing VP tells them to), then confusion will result and time will be wasted.

© Springer International Publishing Switzerland 2016
J. Stark, *Product Lifecycle Management (Volume 2)*,
Decision Engineering, DOI 10.1007/978-3-319-24436-5_9

Manage the Portfolio	Capture product ideas	Screen ideas	Evaluate proposals	Prioritise projects
Identify requirements	Specify products	Define BOMs	Define Design Rules	Design products
Cost products	Purchase parts	Simulate parts	Test parts	Manage orders
Configure products	Plan Manufacturing	Make parts	Assemble parts	Use products
Get feedback	Solve problems	Make changes	Replace parts	Maintain products
Refurbish products	Compare actual costs	Hire people	Upgrade equipment	Retire products
Disassemble products	Recycle parts	Train people	Report progress	Measure progress

Fig. 9.1 Examples of product-related activities in the lifecycle

alliance management	contract preparation	contract review	corrective action	delivery
risk management	design control	disposal	document control	service provision
change management	handling	inspection	leadership	operations
analysis	packaging	process control	supplier audit	integration
project management	prototyping	validation	quality assurance	quality control
equipment purchase	progress review	machine set-up	plant maintenance	verification
product modification	acquisition	project planning	part storage	disposal

Fig. 9.2 More product-related activities in the lifecycle

Towards the end of the 20th Century, companies began organising another way, grouping the activities into processes, and organising around processes. Since the 1980s, the ISO 9000 family of standards has underlined this process focus.

A process is as an organised set of activities, with clearly defined inputs and outputs, that creates business value.

There are many processes, of different size and importance, in a company. There are actually so many processes that companies often create a hierarchy of business processes, processes, sub-processes, sub-sub-processes and activities. Within each of the activities there are usually tasks, roles, responsibilities, checklists, milestones, deliverables and metrics that specify in detail the scope, nature, type, information needs, required skills and measurement of work.

Companies position the processes in a process architecture. A correctly-organised, coherent process architecture will enable effective working across the product lifecycle.

At the highest level of the hierarchy are the Business Processes. They are often laid out in a Business Process Architecture (Fig. 9.3). This is a top-level diagram of the company's processes. It's a common reference for everyone in the company when thinking about processes.

At the top level in Fig. 9.3 are the Management processes, three Main processes, and a set of Support processes. The three Main processes are Supply Chain Management, Customer Relationship Management, and Product Lifecycle Management. The Product Lifecycle Management process runs from Portfolio Management to Phase Out.

At this high level, a process is often identified as the "XYZ Process" (Fig. 9.4). However, at lower levels, since processes describe activities, descriptions need to start with a verb, for example, "Create a list of new ingredients", as in Fig. 9.1.

A lot of processes are product-related. In many companies that I've worked with, between 35 and 55 % of the company's processes are product-related.

Fig. 9.3 Example of a business process architecture

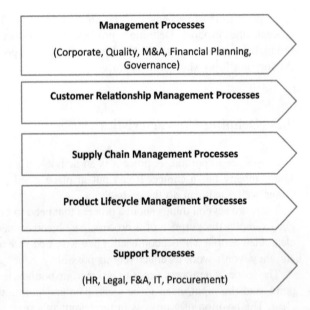

Fig. 9.4 Some processes in the product lifecycle

Assembly Process
Requirements Management Process
Test Process
Concept Development Process
Welding Process
Risk Management Process
Refurbishment Process
Calibration Process

There are so many processes, and it's so important to get them working right, that many companies have a special Process Group to define, maintain and improve processes.

And, to organise the processes, there are special processes for establishing, defining, documenting, maintaining and improving processes. These include sub-processes for planning, review, measurement, audit, monitoring, verification and validation.

9.1.3 Processes for PLM

Companies are in such different businesses that it's not possible for them to have exactly the same PLM processes. Product-oriented, project-oriented, and order-oriented companies, for example, would be expected to have different PLM processes as they obviously don't have the same activities. At a high level, though, there are six high-level processes that are found in most companies. Five of them correspond to the five phases of the product lifecycle. These are the Product Idea process, the Product Definition process, the Product Realisation process, the Product Support process, and the Project Phase Out process. In addition there's a Product Portfolio Management process.

9.2 Defining Characteristics

Processes are important. A process is an activity. It's something a company does. The company has a choice. It can put in place good processes, and do the right things well. Or it can do things badly.

There are several things about a process that need to be defined clearly (Fig. 9.5). For example, the purpose of a process, its objective, needs to be clear. If it's not clear, then people will get confused. They won't be sure about what they should do, and they won't work as effectively as possible.

The scope of a process needs to be clear. Its boundaries should be clear. The start point, and the input, need to be clear. The end point, and the output, need to be clear. The position of a process in the company's process architecture needs to be clear. It needs to have the right neighbours. Communication with its neighbouring processes needs to be clearly and correctly defined.

The activities of a process need to be clear, as do the participants in the process, the roles of the participants, the information they use and create, the tools they use, and the owner of the process. Anything that isn't clear will lead to hesitation and confusion. Time and money will be wasted.

Each process has a customer. If nobody's going to use the output of a process, then the process doesn't need to exist. The customer of a process may be an internal customer (inside the company) or an external customer (outside the company). The process must add value. Otherwise it's not needed.

Each process needs an owner who is responsible for its performance and improvement.

Processes are sometimes divided into three groups. These are operational processes, support processes and management processes. Operational processes create value for external customers. Support processes create value for internal customers.

Process metrics, or Key Performance Indicators (KPIs) are needed so that process performance can be measured, reported, analysed and improved.

Naming	Change, Version	Performance, Improvement
Definition, Purpose, Scope	PROCESS	Ownership, Training
Architecture, Structure	Value-adding, Lean	Representations, Model, Map

Fig. 9.5 Issues common to many processes

9.3   Unwanted Characteristics

Processes have many unwanted characteristics.

One of the unwanted characteristics is that there are few standards for processes. Every company has to develop its own processes. When developing processes it may be helpful to refer to proposals for processes such as those of ISO/IEC 15288 and CMMI/SEI.

9.3.1   Unclear Names

Companies are free to define the scope of their processes, and name them as they like. The many product-related activities and processes across the product lifecycle are given widely different names in different industries and different companies. Figure 9.6 shows some of these names.

Any one of these processes may be given several names in the same company. For example, one person in a company may call a process New Product Introduction (NPI), another may call it New Product Development (NPD). Someone else in the company may call it the Product Innovation process. Others may call it the Innovation process, or the Product Idea process, or the Product Creation process, or the Product Commercialisation Process. As all these people are from the same company, they are actually referring to the same process. When they work together, they work in the same process, whatever its name.

However, sometimes they may need to work with someone from a partner company. Then, for example, the person from the company who refers to this process as the Product Creation Process may meet someone from the partner company who also refers to a process called the Product Creation Process within their company. They may agree that they will work together, with one of the companies doing the first half of the process, and the other company doing the second half. This could lead to problems if this process has been defined differently in the two companies. In one company, for example, it could include the collection of ideas generated by customers and by people in the company. In the second company, the idea collection activity could be in another process, for example in the Idea Management process.

New Product Development	Product Idea Management	Product Maintenance	Portfolio Management
Product Quality Management	Product Change Management	Product Selection	Product Upgrade
Product Improvement	Engineering Change Management	Product Retirement	Program Management
Product Concept Management	Product Obsolescence Management	Product Feedback	Project Management
Requirements Management	Intellectual Property Management	Product Renewal	Concept Management
Product Failure Management	Resource & Capacity Management	Product Modification	Product Liability
Product Data Management	Configuration Management	Product Creation	Product Support

Fig. 9.6 Many similar names for processes

9.3.2 *Other Unwanted Characteristics*

There is often duplication or overlap of activities between different processes. Their interfaces may be unclear, or overlap. Boundaries between processes may be unclear.

Process ownership may not be clearly defined. As a result, perhaps nobody will feel responsible for monitoring or improving the process. Alternatively, perhaps several people will feel authorised to modify it.

As time passes, and the environment evolves, processes may change, resulting in several versions of the same process. But there may be no management system for processes. Some people may start to work with a new version of the process, while others continue to work with the old version of the process.

Processes may be poorly defined, and poorly documented.

There may be no metrics for some processes. In other cases, people may measure too much, or measure the wrong things. Or the same thing may be measured differently in different parts of the company.

There may be no training about processes. Or training may not be sufficiently detailed or relevant.

There may be no management commitment to ensure that processes are followed.

It's often difficult, when developing a process, to get away from a departmental focus. A process developer from one department will tend to include everything needed by their department, and ignore the needs of other departments. They may add extra steps (and cost and time) to address a specific activity that interests them, even though it may rarely be needed in practice.

It's easy to forget, when developing processes, that the business environment is changing all the time, and processes will need to change. It's important to be flexible when working with business processes and process models. They need to able to change to take account of these changes.

Process developers may focus on developing one process, and ignore its interactions with other processes. In the process that they are developing, they may include activities that are already in other processes. This can lead to redundant effort across the organisation.

To develop and document the process quickly, developers may use the words that they are familiar with, their jargon. That will make it more difficult for other people to understand what's happening. And because they are busy people, they may not have time to define the purpose of the process clearly, or its scope, or even make sure that it has an owner.

Without constant attention, processes become slow and bureaucratic. There's always someone who wants to add in extra steps, just to be sure. There's rarely anyone willing to take the risk of slimming the process down. There's always a danger of perhaps removing something that's really important. The result is that the processes take longer to execute than needed. They suffer from low quality, poor communications, a lack of management understanding, and a lack of structure.

Issue	Examples
Transportation	unnecessary movement of product, of product data
Inventory	unnecessary stores of product, of product data
Motion	unnecessary movement of people, unnecessary notifications
Waiting	waiting for information, waiting for processing
Overproduction	producing unneeded information, or information before it's needed
Over Processing	too precise definitions, too many tests, too many iterations, superfluous conversions, unnecessary reformatting
Defects	rework, wasting effort on unnecessary inspections to prevent defects, translation errors, ambiguous data

Fig. 9.7 Examples of waste in processes in the product lifecycle

9.3.3 Process Waste

Just as there's the opportunity for waste in any activity, there's the opportunity for waste in processes in the product lifecycle. Figure 9.7 shows some examples.

9.4 Application Workflow, Product Workflow

9.4.1 Application Workflow

An application workflow is a small set of connected actions that are frequently carried out, and that has been automated in a particular application.

An example of an application workflow is a workflow for document creation. Other examples are application workflows for document approval and for document change.

There are clearly defined steps and roles in an application workflow. Activities are carried out, using pre-defined documents, in a pre-defined order by people in those roles working according to pre-defined rules.

An application workflow overcomes some of the difficulties of carrying out the set of actions in a purely paper-based environment. Consistency is achieved though pre-defined rules, procedures, roles, documents and data types. Progress is easier to track. An audit trail of actions can be automatically created.

9.4.2 Product Workflow

The product workflow is the flow of work through the activities that create or use product data. It runs across the product lifecycle. It runs through many processes. It runs through many departments. Some of the activities in the workflow may take place inside the company, others outside.

In theory, product workflow starts with initial product specification, includes product use by the customer, and ends with product retirement and recycling. In practice, it's not so easy to describe. The product workflow is rarely linear, starting with one well-determined activity, and continuing serially through other well-determined activities, until it reaches a well-determined final activity. Usually, it's much more complex. Usually some activities run in series, and some run in parallel. Sometimes it's only when an activity has been completed, and its result is known, that the next activity will be known. In some cases, the result of one activity will mean that previous activities have to be repeated.

The product workflow depends on many parameters such as product/market features, costs, time, resources, number and frequency of iterations, number and frequency of changes, and extent of parallel activities. The workflow in a particular company will depend greatly on the type of product and the company's position in the value chain. Within process industries, product workflow for a producer of fine chemicals will differ from that of a producer of bulk chemicals. Similarly, in discrete manufacturing industries, the product workflow will be different for producers of airplanes, machine tools and durable consumer goods. Even between companies in the same sector, product workflow will be different. Differences will be due to different customer bases, product lifecycles, production runs, new product introduction rates, and special regulations, as well as to company size and organisational structure. Product workflow is company-specific.

It's important to understand where a company defines costs in the product workflow, and where costs are actually incurred. In the same way, it's necessary to understand where quality and time cycles are defined, and where they become reality. Studies have been made of the distribution of costs in the traditional product lifecycle. Typically they show some 60–75 % of costs defined (and 1–5 % of costs incurred) during conceptual design, and 85–95 % of costs defined (and 10–15 % incurred) before release to manufacture.

The incurred cost of changes during conceptual design is negligible compared to that of changes after the design is released. When design changes occur late in the product workflow, they can be extremely expensive, impacting the product, production equipment and the launch date. In the worst case, the customer will receive a defective product. Fixing a design problem when a product is in the field is often thousands of times more expensive than preventing it during initial design.

The above figures for product costs don't take account of product use by the customer. Normally this is the longest phase in a product's lifecycle. Some products require a great deal of support and maintenance. The cost of their maintenance and spare parts may exceed their purchase price. For example, airlines typically spend 2 or 3 times the purchase price of an engine on spare parts for the engine. Support and maintenance activities can significantly change the distribution of lifecycle costs, often reducing even further the percentage of costs incurred during the early, development phases.

Unlike the flow of materials on the shop floor, for which the workflow is defined in detail, many parts of the product workflow are poorly defined and poorly structured. Often there are no formal procedures for estimating the time and cost of activities. In the absence of any clear structure, it's difficult to optimise the product workflow, with the result that estimates of, for example, lead times, will invariably be longer than necessary. The lack of product workflow structure makes it difficult to manage the time and cost of a given project, with the result that product development projects are often late and over budget. In many cases, these problems are reduced by the skill and experience of the people involved. However, the opposite is often true in cases where management doesn't have a good understanding of the product workflow. More overhead is then added, in the form of unnecessary management reports, which are sent to people who neither understand their contents nor are capable of acting on them.

The product that the customer will eventually receive is designed and manufactured by the activities of the product workflow. This means that the quality and cost of the product are functions of the workflow. The elapsed time between the first idea for a product, and the moment that the first customer receives the product, depends on the efficiency of the workflow. Since the product workflow affects product cost and quality, and manufacturing and overall lead times, companies faced by competitors producing higher-quality products faster, and at lower cost, need to improve their product workflow. Before being able to do this, they must first understand the workflow, and bring it under control. Unless they take control of it, they can't hope to improve it, and, as a result, improve control over the product, and improve customer service.

It's not difficult for companies with a traditional product workflow made up of sequential independent steps (for example, customer requirement to conceptual design to preliminary design to analysis to detailed design to process planning to purchasing to production planning to production to quality control to field tests to support to retirement) to understand where the problems lie. Disjunctures, superfluous steps, and inefficient activities in the product workflow all contribute to unnecessarily extending lead times, increasing costs and reducing quality. With the traditional product workflow, it's impossible to accurately forecast costs and lead times at the beginning of the process, because nobody knows what will actually happen. In any given step of the workflow, people don't have all the necessary knowledge and experience, so they make assumptions, and get something wrong. Later in the sequence, a correction has to be made. The workflow loops back and time is lost.

In many companies, few people can describe the product workflow, and even fewer know why it takes the shape it does. In most cases, the flow results not from a reasoned design, but from a long series of minor reorganisations that resulted from changes in departmental structures, product characteristics and human resources.

Most companies are organised to manage individual steps in the product workflow, but the overall product workflow isn't managed throughout the lifecycle. Individual activities are at best managed on a departmental or functional basis, and may take account of neither the overall flow, nor the detailed needs of other groups. Often, no one individual has the responsibility for the overall workflow. There may be Process Owners but not a Product Lifecycle Owner. Problems are neither understood in detail, nor addressed from an overall point of view. As a result, they're not solved. The overall workflow remains inefficient, wasteful, and out of control. In this state it can't be optimised.

When overall product workflow is brought under control, lead times are reduced, quality goes up, and costs go down. Once the workflow is controlled, a good understanding of final product costs is achieved earlier in the product development process. Accurate estimates of lead times can be made.

9.5   Product Workflow and Information Flow

At first glance, it may appear that a key objective in a product development project is to get the information about the new product to flow smoothly (Fig. 9.8) from the idea, to the design, to production.

At second glance, it's apparent that it would be better if information flowed smoothly all the way down the product lifecycle (Fig. 9.9).

But the flow of information isn't enough, there must also be a flow of material (Fig. 9.10).

After further thought, it will be realised that the picture is nowhere near so simple. The product's lifecycle begins with the imagine phase. Usually information from the definition, realisation, use, support and retirement phases of similar products will be needed in this phase. And material from existing products may also be examined. Before the information for the new product flows from the imagine phase to the define phase, there has usually been a flow in the opposite direction (Fig. 9.11).

In the next phase, it will be the same. Previous definitions will be re-used. Information about previous realisations will be reviewed. Existing products may be reviewed. Information about the use of products, coming perhaps from the Internet

Information	>>>>>	>>>>	>>>>		
	Imagine	Define	Realise	Support	Retire

Fig. 9.8 Information flow from product idea to realisation

Information	>>>>>	>>>>	>>>>	>>>>	>>>>
	Imagine	Define	Realise	Support	Retire

Fig. 9.9 Information flow from product cradle to grave

Material	>>>>>	>>>>	>>>>	>>>>	>>>>
Information	>>>>>	>>>>	>>>>	>>>>	>>>>
	Imagine	Define	Realise	Support	Retire

Fig. 9.10 Information and material flow from product cradle to grave

Material	<<<<<	<<<<	<<<<	<<<<	<<<<
Information	<<<<<	<<<<	<<<<	<<<<	<<<<
	Imagine	Define	Realise	Support	Retire
Information	>>>>>	>>>>	>>>>	>>>>	>>>>
Material	>>>>>	>>>>	>>>>	>>>>	>>>>

Fig. 9.11 Information and material flow across the lifecycle

of Things, will be used. Thought will be given to the way the product will be disposed of. And so it goes on, getting more and more complex as each phase is examined, and the relationships with the other phases brought to light.

9.6   Process Mapping and Modelling

The term Process Mapping is often used to describe the activity of documenting an existing process. This activity is also sometimes referred to as Business Process Mapping, or Process Charting, or Process Flow Charting.

One of the reasons for documenting an existing process is to make it clear to everybody exactly what happens in a particular process. Understanding the process is a first step to improving it.

The term Process Modelling, or Business Process Modelling, is often used to describe the activity of creating models of future processes.

Business Process Management (BPM) is an overall approach to the improvement of a company's business processes. It includes process mapping, process modelling and process measurement.

Process measurement is an essential part of the activity of process management. Process measurement must be built into every process. Determining what to measure, and how to measure it, are key initial activities of process measurement. Some of the metrics that are chosen are likely to be process-oriented metrics, such as the number of documents in a process, or the number of steps in a process. Some metrics are likely to be business-related metrics, such as the time it takes to execute the process. And some metrics are likely to be product-related metrics such as the number of product defects resulting from a process.

9.7   Hierarchical Process Structure

Hierarchical decomposition is a frequently-used approach in process modelling. Starting at the top level (Level 0), the main "things" at the next level down (Level 1) are identified. Usually between 4 and 7 activities are identified. Then, the same

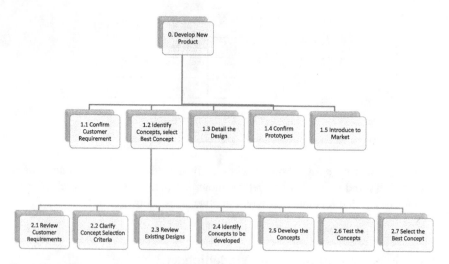

Fig. 9.12 Hierarchical process decomposition

technique is applied at the next level down. For each of the entities at Level 1, the main "things" at the next level down (Level 2) are identified. Again, usually between 4 and 7 activities are identified for each entity. In Fig. 9.12, "Develop New Product" is at Level 0. There are five main activities at the next level. One of these is "Identify Concepts, Select Best Concept". This is made up of seven activities at the next level down.

9.8 Activity Flow

Figure 9.12 is a model showing the structure of a process. Figure 9.13 is a map, or model, showing the flow of activities, including the activity of taking a decision.

Another way of modelling a flow of activities is to show them in "swimlanes" (Fig. 9.14). This approach allows information about the roles of participants ("actors") to be shown.

The figures above show models of the structure and the flow of process activities. Another type of model is a control-flow model. These models describe the allowed values and combinations for the inputs and outputs of functions, and show how these are related to detailed activities. An example of a control-flow model is the PetriNet. A programmer's flow chart is also an example of a control-flow model. It shows the exact sequence of instructions to be executed, and the points where flow is dependent on the value of particular variables.

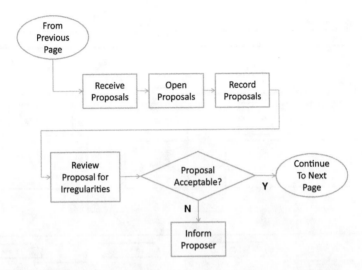

Fig. 9.13 Flow of activities in a process

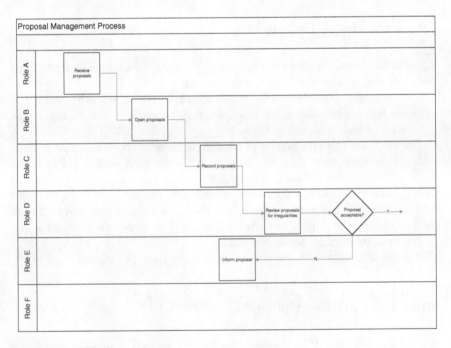

Fig. 9.14 Activities in swimlanes

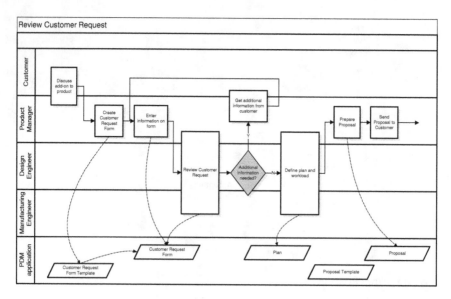

Fig. 9.15 Activities and data in swimlanes

9.9   Data Model, Process Map

Whereas data is usually described in a "model", processes are often described in a "map". However, a process map is also a model.

The swimlane example in Fig. 9.14 showed roles and activities. Adding a swimlane for an application makes it possible to link an activity with the data it uses and the application it uses (Fig. 9.15).

The swimlane models could be of the current situation, or of the desired future situation. In either case, they will help people understand what happens in the process, who does what, what data is used, and so on. The models can be annotated to show who participated in their development, and who validated them. This will help get everyone on the same page. It will also show that everyone involved is on the same page. The models can also be annotated to show strengths and weaknesses, and opportunities for improvement.

Another type of activity documentation is a Use Case Description.

9.10   Use Case Description

A Use Case describes, from the user viewpoint, an interaction between a user of a system and the system. It can be used, during system development, to show the expected behaviour and to clarify requirements.

As an example, a Use Case could describe the login to a system. The first lines of a first draft might look like Fig. 9.16.

The system user starts the application
The application requests a user name and a password
The user enters a user name and a password
The system validates the user name and password, and presents the initial screen

Fig. 9.16 First lines of a first draft of a use case

A standard format is often used for a Use Case Description (Fig. 9.17). This helps to make sure the description is complete. And it makes it easier to write, communicate and agree about Use Cases. The required information often includes information such as Use Case Name, Use Case Purpose, Actors (such as document author, document approver); pre-conditions/Initial State/Start Conditions of the Use Case; Use Case Steps; the End State/Post-Conditions of the Use Case; and exceptions or variants.

The Use Case (Fig. 9.16) shows the system validating the user name and password, and presenting the initial screen. Hopefully that would be the Normal Flow of Events. But it's also possible that the system can't find the user name and/or password, or considers the password to be invalid. In that case, the Alternative Flow needs to be documented and followed.

Many Use Cases are needed to define the scope of a complete system. A Use Case Diagram (Fig. 9.18) brings together several Use Cases. A Use Case Diagram is one of the three UML behaviour diagrams.

9.11   Use Case Diagram

A Use Case describes graphically the interaction between a user of a system and the system. The users, the "actors", are represented by matchstick people. A Use Case is shown in an oval.

Fig. 9.17 Use case description

```
Use Case Number:
Use Case Name:
Version:
Description:
Purpose:
Actors:
            <Actor 1>
            <Actor 2>
Assumptions:
Pre-conditions:
Normal Flow of Events:
            <Step 1>
            <Step 2>
            <Step 3>
            <Step 4>
Post-conditions:
Alternative Flow:
            <Step 1>
            <Step 2>
            <Step 3>
Special conditions:
Notes:
```

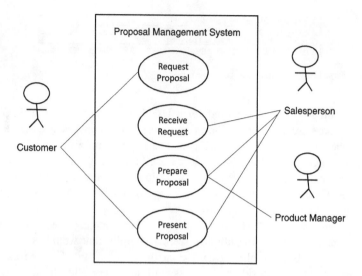

Fig. 9.18 A use case diagram for a proposal

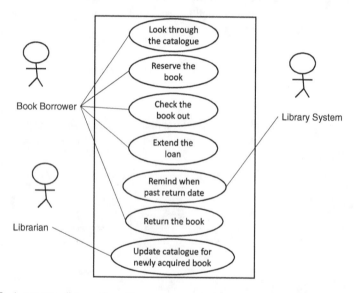

Fig. 9.19 A use case diagram for a library

Sometimes, as in Fig. 9.18, a Use Case Diagram shows the actors outside the system. Sometimes, as in Fig. 9.19, a Use Case Diagram shows the system as an actor.

product	product data	applications
people	**Process**	organisation
metrics	methods	equipment

Fig. 9.20 No process is an island

Product Development process too slow
Exact status of projects not known
Engineering Change process takes too long
Time lost in processes waiting for information
Superfluous tasks, adding no value, are carried out
Process leads to unnecessary travel and meetings with partners
Steps in the Ordering process create rework

Fig. 9.21 Potential process-related challenges for a company

9.12   No Process Is an Island

No process is an island (Fig. 9.20), isolated from the rest of the company. All processes are closely related with other PLM components. They are also influenced by other forces within the company, and outside the company.

For example, to improve performance, two processes may be merged into one. In the future process, only one document may be needed instead of two. The process documentation will need to be changed, and people will need to be trained on the new process and documents. The fields in the documents will be changed, and a new template developed. There will be changes to the application creating the document. There will be changes to the application managing the document. A performance indicator may need to be changed if it was previously based on an activity or entity in one of the two merged processes. The company's Quality Manual will be updated to show the new process.

9.13   The Challenges

The specific process-related challenges that a particular company faces could come from several sources (Fig. 9.21).

9.14   The Way Forward

The issues described in this chapter all relate to processes, but they can't be solved in isolation from the other components of PLM. Their solution, within the PLM environment, is addressed from Chap. 21 onwards.

Chapter 10
PLM Applications

10.1 Introduction

Like most things in the PLM environment, "applications" are referred to by different names by different people (Fig. 10.1).

And like most things in the PLM environment, there are many of them. More than fifty different classes (groups) of applications are described briefly in the next section. Many of these are commonly referred to by a two-, three- or four-letter acronym. Figure 10.2 shows some examples.

Within some of the groups of applications, there are more than one hundred different applications provided by different vendors. In total, there are thousands of different applications in the PLM environment.

In theory, applications bring many benefits (Fig. 10.3). There may be some problems as well. For example, if their underlying logic is wrong, they will always give the wrong result. And they may hide the most important information in a mass of data.

One of the challenges of PLM for a particular company is to identify the applications that are most relevant to the activities on which the company wants to focus its efforts. The role, and the potential benefits and disadvantages, of each application should be clearly understood so that it's possible to see if, and how, it could best fit into a PLM solution.

10.2 Applications Overview

This section describes, in alphabetical order, some of the most frequently encountered groups of applications. Issues related to many of these applications are addressed in the next section.

© Springer International Publishing Switzerland 2016   199
J. Stark, *Product Lifecycle Management (Volume 2)*,
Decision Engineering, DOI 10.1007/978-3-319-24436-5_10

computer programs	Information Technology (IT) systems	software packages	application programs
IT tools	Information Systems (IS)	IS applications	application software
systems	programmes	tools	computer systems

Fig. 10.1 Some different names for applications

CAD	CAE	CAID	CAM	CAPE
CAPP	CASE	CSM	DMU	EDI
ECM	EDA	EDM	FEA	IPM
KBS	LCA	MRP	NC	PDM
PM	RP	SCM	TDM	VR

Fig. 10.2 Examples of application acronyms in the PLM environment

calculate faster	work longer hours
calculate more reliably	do repetitive work better
calculate more precisely	work at lower cost
are always present	work with large quantities of data

Fig. 10.3 Some potential benefits of PLM applications

10.2.1   BOM Applications

BOM applications manage Bills of Materials (BOMs). A BOM is a structured description (Bill) of the "things" (Materials) that make up a product or assembly. Depending on the level and the product, the things could be the ingredients of a food product, or the intermediate sub-assemblies that make up an assembly of a mechanical product.

There can be several BOM structures in the product lifecycle. Users at different times in the product lifecycle want to work with the structure that best suits the work they have to do. Common structures are the eBOM (Engineering BOM) which describes the things in a product from the viewpoint of a design engineer, mBOMs (Manufacturing BOM) which describes the things in the product from a manufacturing viewpoint, and sBOMs (Service BOM) which describe them from the viewpoint of a Service Engineer. Managing different structures for many products, yet keeping them coherent, is almost impossible manually. BOM applications were one of the earliest applications of computers in manufacturing industry.

BOM applications support the creation and modification of BOMs, maintain coherence between the different structures, and provide different views on the product depending on its position in the product lifecycle.

10.2.2   Compliance Management

Compliance Management applications help ensure that product-related activities are carried out in accordance with standards, rules and regulations. These may be company-internal standards, best practices and guidelines. They may be external standards of international and industry organisations, governments and business partners.

10.2.3 CSM Applications

Component and Supplier Management (CSM) applications enable selection of approved components. They provide access to an Approved Supplier List and a common database of components, parts or ingredients.

10.2.4 CAD Applications

Computer Aided Design (CAD) is an umbrella term for applications using interactive graphics techniques that are used in translating a requirement or concept into a design.

CAD can help companies that want to improve quality and reduce development cycles and costs. A design built with CAD should be of much higher quality than one made by traditional manual means. The model in the computer is accurate and unambiguous. Many things can be done with it that are just not possible, or would take far too long, to carry out manually. A 3D CAD model of a part can be displayed graphically so that the designer sees what it really looks like. The model can be rotated, viewed from different angles, and magnified so that the designer sees the details on the screen. Any errors can be corrected immediately.

It costs less to design with CAD than with traditional means. Although it may cost as much to develop the initial design, everything after should cost less. Above all, CAD is much faster than manual techniques. As with the cost, there may not be much difference on the initial design, but after that everything goes much faster. Once a part model has been built, it can be used throughout the development process. People don't have to re-develop it for each task. Once a part model has been developed and released, it's available for use on other products. It may be possible to use it on another product in exactly the same form or perhaps with a slight modification. In either case, much less work is needed than when developing a completely new design, so a lot of time will be saved. Companies that use CAD can find it easier to get closer to customers. For example, the customer can be brought in and shown the design on the screen and asked if it satisfies requirements. If it doesn't, changes can be made much quicker than they would have been if it was necessary to wait for a physical prototype to be built before the customer could be involved again. Once the design is agreed, it can be communicated immediately to the team responsible for the next phase of the lifecycle. They can start working right away. And as soon as they are finished, the following phase can start.

10.2.5   CAE Applications

Computer Aided Engineering (CAE) is an umbrella name for applications used in the design engineering and manufacturing engineering functions. Sometimes, though, it's given other meanings. Sometimes, it's used only to mean all computer-based tools used in design engineering. On other occasions, it's used to mean only those applications that are used at the front end of the design engineering process, prior to detailed design.

10.2.6   CAID Applications

Computer Aided Industrial Design (CAID) allows an industrial designer to model a design in three dimensions and see the result immediately on the screen. Through the use of shading, colour, movement and rotation, CAID lets designers create photorealistic images and animations from a core design. The model can then be used to communicate information to other groups involved in product development (such as Marketing and Manufacturing).

10.2.7   CAM Applications

Computer Aided Manufacturing (CAM) is an umbrella term for all applications used in manufacturing engineering activities. These include Computer Aided Production Engineering, Computer Aided Process Planning, Computer Aided tool and fixture design, NC programming and Programmable Logic Controller (PLC) programming. CAM is used in preparing for a wide range of manufacturing processes, including the cutting of metals, fabrics, leather and composites, and the forming of metals, plastics, rubber, leather, composites and glass. It's used in preparing for paint spraying, composite laying, deburring and parts assembly.

10.2.8   CAPE Applications

Computer Aided Production Engineering (CAPE) applications are used to digitally model a manufacturing plant, production line or work cell. They enable simulation of production processes for particular products in a "Virtual Factory".

10.2.9   CAPP Applications

Computer Aided Process Planning (CAPP) applications are used in the generation of process plans. Process plans describe the operations which a part must undergo.

They define the sequence of production operations, specify the tooling, detail the speeds, feeds and coolants, and define set-up and run times. They work with either a "variant" or a "generative" approach. In the variant approach, a new plan is created by modification and adaptation of an existing plan to meet the specific requirements of a new part. In the generative approach, predefined algorithms are used to generate a plan on the basis of the characteristics of the new part.

10.2.10 CASE Applications

Computer Aided Software Engineering (CASE) applications are used to support some or all of the phases of the software lifecycle. There are basically three types of CASE applications: those that are used in planning, those that are used in analysis and design, and those that are used in code-related activities.

10.2.11 CIM

Computer Integrated Manufacturing (CIM) is an umbrella term, dating from the 1970s, for all applications used in engineering and manufacturing activities. The benefits of CIM are achieved through automation and integration. Repetitive activities are fully automated, and creative activities are computer-assisted, with the computer doing the routine part of the activity, leaving people free to do the creative part. And the time wasted in expediting information and errors due to obsolete, inaccurate data can be eliminated with integration.

10.2.12 Data Exchange Applications

Data Exchange applications, also known as data translation applications or data conversion applications, transform data from one data basis to another data basis. For example, a circle may be represented by three points on its circumference in one CAD application, and by its centre and radius in a second CAD application. A data exchange application would translate the circle from the representation of the first application to the representation of the second application.

10.2.13 DECM Applications

Digital Engineering Content Management (DECM) applications manage and enable the creation, generation and use of engineering content, for example, for CAD parts catalogue applications and for product and sales configurators.

10.2.14   Digital Manufacturing Applications

Digital Manufacturing applications help with activities such as manufacturing cost estimating, factory layout and simulation, and process planning. They are also used to program robots, machine tools and inspection equipment.

10.2.15   DMU Applications

A Digital Mock-Up (DMU) is a computer-based model of a real product. The model can be used to carry out many activities such as display and analysis.

10.2.16   EDI Applications

Electronic Data Interchange (EDI) is the exchange across a telecommunications network of documents and other information such as engineering drawings, receipt advices, purchase orders and advance shipping notices between the computer in one company, and the computer in another company. The information is transmitted in a standard format in which the various contents of the documents are arranged in a pre-described way.

10.2.17   EDA Applications

Electronic Design Automation (EDA) is an umbrella term for applications used in designing, testing and producing electronic products. An alternative name is ECAD (Electronic Computer-Aided Design). ECAD applications include Schematic Capture, Printed Circuit Board (PCB) Layout, and Integrated Circuit Simulation.

10.2.18   ECM Applications

Enterprise Content Management (ECM) applications capture, manage and deliver a company's electronic content and documents.

10.2.19   EDM Applications

Engineering Data Management (EDM) applications are similar to Technical Document Management applications and Product Data Management applications.

They provide an electronic vault for drawings and documents, as well as mechanisms to index and access them.

10.2.20   FEA Applications

Finite Element Analysis (FEA) applications, also known as Finite Element Method (FEM) applications, are used to compute the deformation, or other response, of an object in response to a particular action or event.

10.2.21   Geometric Modelling Applications

Geometric Modelling applications are used to design, or model, the shape of a part or product.

10.2.22   Haptic Applications

Haptic technology provides the sense of touch. Haptic-supported applications provide information to a user through the sense of touch. Without haptic input, the user only receives information through the sense of sight. Haptic technology can help, for example, to create and control objects in a virtual environment.

10.2.23   IM Applications

Innovation Management (IM) applications manage Product Development and Research & Development (R&D) processes and projects.

10.2.24   IPM Applications

Intellectual Property Management (IPM) applications are used to manage intellectual property rights.

10.2.25   KBS

Knowledge Based Systems (KBS) are applications that aim to allow the experience and knowledge of humans to be represented and used on a computer so as to increase people's decision-making ability. Knowledge Management includes many knowledge-related activities (Fig. 10.4).

The Knowledge part of Knowledge Management includes knowledge of anything about a company, such as its customers, its products, its competitors and it partners. It could be knowledge from the past, current knowledge or, perhaps most valuable, future knowledge (foresight). Figure 10.5 shows some examples.

There are many application areas of Knowledge Management (Fig. 10.6).

10.2.26 LCA Applications

Life Cycle Analysis (LCA) applications are used to evaluate the environmental impact of a product.

10.2.27 MRP 2 Applications

Manufacturing Resource Planning (MRP 2, or MRP II) applications are used to plan the resources (such as equipment and materials) of a manufacturing company.

creating knowledge	capturing knowledge	analysing knowledge	storing knowledge
indexing knowledge	classifying knowledge	validating knowledge	synthesising knowledge
searching for knowledge	finding relevant knowledge	making knowledge available	making use of knowledge

Fig. 10.4 Some activities of knowledge management

Type of Knowledge	Example
current internal knowledge	the BOM of a company's product
current competitive knowledge	user profiles of a competitor's best market segment
future internal knowledge	a company's next product design
future competitive knowledge	user profiles of a competitor's target market segments
breakthrough knowledge	how to make a product for half the cost
breakthrough knowledge	user profiles of the 10,000 early adopters of a new product

Fig. 10.5 Examples of types of knowledge

Area	Use
customer knowledge management	using knowledge about customers to provide customised product information or customised service information
knowledge databases	containing experience of best practices across a wide range of industries
knowledge retention systems	conserving knowledge of how people work in different activities
virtual educational organisations	enabling rapid education and training at the knowledge consumer's workplace

Fig. 10.6 Examples of application areas of knowledge management

10.2.28 NC Applications

Numerical Control (NC) programming applications are used to develop programs that control NC machine tools.

10.2.29 Parts Catalogue Applications

Parts Catalogue applications allow a company to enter its products into an electronic catalogue, where they will be available for search, selection and purchase.

10.2.30 Parts Libraries

Parts Libraries make available a range of preferred parts from a library included in an application.

10.2.31 Phase-Gate Applications

Phase-Gate applications manage the progress of a project (such as a product development project) through a set of phases separated by gates.

10.2.32 Portfolio Management Applications

Portfolio Management applications provide an overview of a company's pipeline of development projects. They allow managers to take trade-off decisions based on the risks/rewards of the product portfolio against a company's strategic objectives. Portfolio Management enables assessment of resource allocation against top-level strategic goals and risk/reward expectations. They show the interdependencies of resources, intermediate deliverables, and other information. To facilitate analysis, Portfolio Management applications provide a range of display options such as pie charts and graphs. They offer many possibilities for rolling up, filtering, and grouping projects to focus on specific issues. Many parameters can be plotted, such as revenue versus cost, impact versus probability, and market share versus Net Present Value.

10.2.33   PDM Systems

Product Data Management (PDM) systems are similar to Technical Document Management applications and Engineering Data Management applications. They provide an electronic vault for product data, as well as mechanisms to protect, index and access it.

10.2.34   PM Applications

Project Management (PM) applications assist in the planning, organisation and tracking of resources to achieve project objectives.

10.2.35   RP Applications

Rapid Prototyping (RP) is the application of 3D printing to prototyping. It's the production of a physical prototype directly from a computer-based model of a part or product. Rapid Prototyping systems are used to rapidly produce an accurate prototype from a CAD model. The prototype can be in a variety of materials including investment casting wax, PVC, and polycarbonates. An RP system can produce a prototype in a few hours compared to the days or weeks of conventional prototyping techniques. RP technologies include selective laser sintering, ballistic particle manufacturing, stereolithography, instant slice curing, and direct shell production casting. A physical prototype is a good visualisation and communication tool for people unaccustomed, unable, or unwilling to work with an image on the screen. It provides a common language for people from different functions and eliminates misunderstanding. The parts produced can be used as fit and function models. They can be used as design verification tools, and patterns for other manufacturing processes. They can be used to check interference, and to test the ease of assembly and maintenance.

The activity prior to rapid prototyping is the development of a computer-based model of a part or product. This is the normal CAD design activity. It has to be carried out whether a prototype is going to be produced by rapid prototyping or traditional means. Once the CAD model exists, a physical model can be produced directly by one of the rapid prototyping technologies, whereas with traditional means, drawings of the CAD model would be produced, manufacturing engineers would decide how to produce it, and then it would be manufactured. Rapid prototyping cuts out these steps. It saves the time associated with them. It also saves their cost, and eliminates the possibility of transcription errors and misinterpretation.

10.2.36   Requirements Management Applications

Requirements Management applications gather and manage user, business, regulatory, technical, functional, process and other requirements for a product. Requirements Management applications are used to identify, document, analyse, prioritise, track, communicate and agree on requirements, and control changes to the requirements.

10.2.37   Reliability Management Applications

Reliability Management applications address the ability of a part or product to perform under given conditions for a specific time period.

10.2.38   Simulation Applications

Simulation applications are used to study the performance of a system before it's been physically built or implemented. They can be used at many stages of the product lifecycle. Simulation can be used to study the likely performance of a strategy without actually implementing it. It can be used to study the performance of a product or a process without actually building or implementing it. Simulation involves the development of a computer-based model of a part or product, the development of a computer-based model of the environment in which the part or product will be used, and the testing of the part under different conditions of the environment. This is followed by analysis of the behaviour of the part. Often the result of simulation will be modification of the model of the part to improve its behaviour.

The models of the part and the environment may be built graphically using a CAD system or they may be input in the form of equations. Computer-based simulation is cheap and effective. It makes it easier to evaluate before implementing. It allows errors to be identified and corrected before they are implemented. Models can be built, tested and compared for different concepts. "What-if" analysis can be carried out. Recommendations for improvement can be made. Simulation helps meet the objective of developing products faster because it doesn't require the time-consuming activities of building physical models of the part and the environment. Instead it uses the models designed in the computer which would normally be the basis for building the physical models. Time is saved because it's not necessary to build the physical model. In addition, even more time is saved as modifications are made to the computer-based model and the simulation is repeated. Simulation is cheaper than the traditional methods of building and testing a physical model. There are savings in reduced material costs. There are savings because all the activities of defining the process for making the prototype, and then building it, and testing it, are no longer needed. Quality is improved because it's possible to define and test many more potential designs using a computer-based model of the part than when using physical prototypes.

10.2.39   SCM Applications

Software Configuration Management (SCM) applications manage the development and modification of software.

10.2.40   TDM Applications

Technical Document Management (TDM) applications are similar to EDM and PDM systems. They provide an electronic vault for all drawings and documents, as well as mechanisms to index and access them.

10.2.41   Technical Publication Applications

Technical Publication applications allow a user to author, manage, publish and deliver a technical publication such as a User Manual.

10.2.42   Translation Management Applications

Translation Management applications translate text from one language (such as English) to another (such as Polish).

10.2.43   VR Applications

Virtual Reality (VR) is the application of computer simulations, based on 3D graphics and special devices, of an environment, that allow a user to interact with that environment as if it was real.

10.2.44   VE Applications

Virtual Engineering (VE) brings together Virtual Reality, engineering computation, geometric modelling and CAE/CAD/CAM technologies.

10.2.45 Virtual Prototyping Applications

Virtual Prototyping is the construction and testing of a virtual prototype, or Digital Mock-Up (DMU). A digital mock-up is a computer-based model of a physical product that can be viewed, analysed, and tested as if it were a real physical model. Virtual prototyping uses 3D models created in CAD systems for activities such as assembly/disassembly verification, design reviews and visibility verification.

10.2.46 Visualisation and Viewing Applications

Visualisation and Viewing applications are used for visualising, viewing and printing product and process definition data.

10.2.47 3D Printing Applications

3D Printing applications produce parts or products directly from a computer-based model of the part or product.

10.2.48 3D Scanning Applications

3D scanners are used to create a cloud of points corresponding to the surface of a part. A 3D scanning application can be used to check that a manufactured part corresponds to its design specifications. For example, an OEM outsources a part, and sends the CAD file to the supplier. The supplier makes the part, scans it, and e-mails the point cloud to the OEM who checks it against the CAD model.

10.3 Issues of Applications

Although there are thousands of applications, and they are used for many different things in very different situations, there are issues that are common to many applications (Fig. 10.7).

Naming	Change, Version Management	Island of Automation
Definition, Scope, Functionality	PLM APPLICATION	Ownership, Training, Support
Architecture, Overlap, Duplication	Data Management, Exchange	Integration, Interfaces, Customisation

Fig. 10.7 Issues common to many PLM applications

10.3.1   Ambiguous Name and Unclear Scope

The functionality of an application is often unclear from the name of the application. Many umbrella terms such as Computer Aided Design, Computer Aided Engineering and Computer Aided Manufacturing are used with a wide range of different meanings. Different vendors include different functionality within apparently similar applications.

It's often unclear, from the name, exactly what an application does. Some groups of applications, for example Computer Aided Manufacturing, Computer Aided Production Engineering and Digital Manufacturing applications, are similar, and have overlapping functionality.

10.3.2   Islands of Automation

There are thousands of applications in the PLM environment. Any one of these may, in some circumstance, be required to work independently of other applications. As a result, it needs all the functionality that makes it usable. This could include, for example, user interface, mathematics and data management functionality. To be able to work independently of other applications, it must be able to work as an "Island of Automation". It's an advantage to be able to work independently of other applications. However, that also brings problems. For example, it results in duplication of functionality. And it slows the flow of data.

For various reasons, sets of files tend to belong to the application that wrote them. One reason is that many application developers are unwilling to inform application users about the structure and meaning of the data that their applications use and produce. As a result, the data often remains understandable only to the application that wrote it.

Just as each application can be an Island of Automation, unconnected to the mainland of other applications, for each Island of Automation, there can be a corresponding Island of Data, again unconnected to the mainland.

Each new application program has the potential to become a new Island of Automation and to create a new Island of Data.

One part of a company, such as a department, can overcome the problem of having many Islands of Automation, in part, by using only an "integrated set of applications" from a single vendor. In this case, some of the physical data transfer problems will be reduced, and the vendor may provide some means for improving the flow of information between the individual applications. In all but the smallest departments, however, it won't be feasible to buy only an "integrated set of applications" from a single vendor.

Since most companies will have needs that can't be met effectively by a single set of applications from one vendor, other applications will have to be acquired to handle these needs. As a result, new Islands of Data will arise. In the future, as current technologies evolve, and new technologies are introduced, it can be expected that new Islands of Automation will appear. They may be of great importance to individual companies, who will acquire them even if they aren't integrated. For the foreseeable future, companies will have to cope with incomplete integration, application-related files and the resulting problems of working with information that's connected in real life but unconnected in the applications they use. These problems, through redundant data, redundant data entry, redundant conversion of data, and redundant application functionality lead to increased operating costs.

10.3.3 Departmental Islands, Supplier Islands

Even if one part of a company, such as the design engineering function, could overcome all these problems and consolidate all its computing and communications activities into one Island of Automation, it would still face the problems of working with the Island of Automation in the Manufacturing function, and with the design engineering function in partner companies. It's unlikely that these companies would have chosen exactly the same Island of Automation solution. They could have chosen a different solution, or they might have decided to work in an environment that's not integrated.

10.3.4 Interface and Integration Need

Many PLM applications (Fig. 10.8) need to share and exchange product data (such as part numbers, version numbers, product costs) with other applications in the company.

10.3.5 Overlapping Data Management Functionality

Since most PLM applications have to be able to work in a stand-alone mode, they need to be able to store the data that they create and use. For example, they may

ERP applications (which manage all sorts of company assets, inventories, capacity, schedules, forecasts, orders, costs and revenues),
SCM applications (which manage all sorts of material and financial across the supply chain)
Maintenance, Repair and Operations (MRO) applications which track the status of products, their configurations, repair processes, and upgrade status
CRM applications (which manage all sorts of customer information including customer requests, requirements, experience and problems)
Marketing and Sales applications (which help implement high-impact Marketing strategies and effectively empower sales associates to see sales trends and identify customer needs earlier)
NC controllers (which drive motors on tools to produce components and products)
Human Resource Management applications (which realise the potential of employees, with up-to-date information about performance, payroll, benefits, and career path)

Fig. 10.8 Applications that may need to be integrated with PLM

need to store product names and engineering drawings. The application developer develops specific data management functionality to do this. However, the developers of all other applications will also develop specific, but different, data management functionality for their applications. The result is overlapping, duplicate functionality.

10.3.6   Different User Interfaces

As well as storing data in their own specific ways, many PLM applications have their own specific user interfaces. And their approach to other common functions, such as maths functions, may also be specific to each application. Each time that someone uses one of these programs they waste time in first learning, and then remembering, the specifics of the program.

10.3.7   Organisational Match

Application programs often reflect the organisational environment for which they were created. In the past, organisations have tended to be departmental, and application programs matched the functionality and data needs of a particular department. This limits their usefulness for use across the lifecycle.

10.3.8   Limited Operating Environment

To manage products across the lifecycle, most companies use computers of various types from different vendors. Some of the computers may be stand-alone, others linked together over various types of networks and connections. Some may be in a Cloud. The computers may run on a variety of operating systems. Some of these will be proprietary, not standardised. Others will, in principle, conform to a standard. However, even those that are, in principle, standardised, may have minor differences, particularly between different versions and releases. Many PLM applications only run on one operating system, so aren't usable on all computers in the company.

10.3.9   Versions

The different versions of applications, such as CAD applications, are a potential source of problems. The capabilities of successive versions of a CAD application can be incompatible. The application vendor may upgrade the application with the intention of providing better functionality and richer information content. However, by doing so, it may create the situation where an earlier version can't make use of all data created under the new version, and the new version can be limited in its ability to use data created under the earlier version.

10.3.10 Legacy Applications

The computer hardware and operating system at the heart of an application, such as a CAD application, are also a potential source of problems. If past trends continue, hardware currently in use won't be in use in 20 years' time, yet some companies will need, in 20 years' time, to access data currently being created. It could be difficult for users to re-create exactly the present environment, unless the company intends to archive its computers, operating systems and application programs, as well as its data.

10.3.11 Neglected Functionality

An application will primarily address the activity that it's supposed to support, such as geometry definition, or technical publication, or packaging graphics creation, with data management being a secondary (and ineffectively implemented) function. Applications typically focus on activity-specific functions. Although they create product data, they have limited product data management functionality for data definition, structuring, organisation, storage, retrieval, archival, communication, exchange, protection, distribution and tracking.

10.4 Grouping the Applications

There are thousands of applications in the PLM environment. To make it easier to understand which applications are most suitable for a particular company it's useful to structure them in a manageable way.

The application functionality that will be needed in a PLM solution can be grouped and described in different ways for different purposes. Different companies will look for different groups of functions. Not all functionality will be needed by a particular company.

10.4.1 First Grouping

Applications can be grouped in many different ways. For example, they could be put in the following three groups, with the first two groups being focused on Design and Manufacturing (Fig. 10.9).

Design Focus	Manufacturing Focus	General	
Requirements Management	Rapid Prototyping	Viewers	Content Management
CAD	Factory Simulation	QFD	Image Management
Idea Management	Robot Path Analysis	Process Mapping	Knowledge Management
Discovery	NC Programming	Process Definition	Visualisation
Concept Management	BOM Management	Project Management	Collaboration
EDA	Routing Definition	EDM	Data Exchange
Recipe Development	Calibration	PDM	Data Translation
Plastic Behaviour Analysis	Tool Management	Document Management	Compliance Management

Fig. 10.9 Three groups of PLM applications

10.5   Generic and Specific PLM Applications

PLM applications can also be put into two groups, generic applications and specific applications.

"Generic" applications (Fig. 10.10) are applicable to all kinds of companies, all types of products, and all types of user within those companies.

It will be seen that, as an example, the first application in the list, "data management", is a generic application which is needed by a design engineer in the automotive industry, but also by a project manager in the pharmaceutical industry. They both have enormous amounts of data to manage. Similarly, the fifth item in the list, "collaboration management", is applicable in any situation where people in different locations are working together.

Often, all the above applications are needed for most people working in product-related activities. That isn't the case for the applications in the second group (Fig. 10.11). These are much more specific to a particular context. This second group contains more specialised applications that are needed by particular people, departments, functions or industries.

Data Management / Document Management
Part Management / Product Management / Configuration Management
Process Management / Workflow Management
Program Management / Project Management
Collaboration Management
Visualisation
Integration
Infrastructure Management
Product Idea Management
Product Feedback Management

Fig. 10.10 Generic PLM applications

Fig. 10.11 Task-specific PLM applications

Product Portfolio Management
Idea Generation Management
Requirements and Specifications Management
Collaborative Product Definition Management
Supplier and Sourcing Management
Manufacturing Management
Maintenance Management
Environment, Health and Safety Management
Intellectual Property Management

The first item in the list, "Product Portfolio Management", has very specific functionality that's only needed by a few people in a company. Similarly "Collaborative Product Definition Management" will have functionality specific to the needs of people who define the product. The Generic and Specific PLM applications are detailed in the following sections.

10.6   Generic PLM Applications

Generic PLM applications (Fig. 10.10) are those that are applicable to all kinds of companies, all types of products, and all types of user within those companies.

10.6.1   Data Management/Document Management

These applications enable a company to store and make available data (documents/drawings/files) throughout the entire product lifecycle in a controlled-access secure distributed environment. They enable activities such as version management, revision control, classification, search, analysis and reporting.

10.6.2   Part Management/Product Management/Configuration Management

These applications enable a company to manage products, product structures and product attributes throughout the entire product lifecycle in a controlled-access secure distributed environment. They enable activities such as version management, revision control, classification, search, analysis and reporting. They enable improved design, part and module reuse.

10.6.3   Process Management/Workflow Management

These applications enable a company to map business processes, to define and automate simple workflows (such as change approval and release workflows, and the change management workflow) and ensure compliance with regulatory requirements from organisations such as the FDA and the ISO. Templates enable common, repeatable processes. Workflow management includes routing templates, paths, lists, logic and rules. It can include notification management.

10.6.4 Program Management/Project Management

These applications enable a company to plan, manage and control projects and programs. They enable stage, gate, milestone, and deliverable control. They provide visibility into a project's status in terms of progress and costs. They shows interdependencies such as those among project resources and intermediate deliverables. These applications provide a range of display options such as dashboards, cockpit charts, pie charts and graphs.

10.6.5 Collaboration Management

These applications enable geographically-dispersed teams and individuals to work together in a secure, structured, virtual working environment using up-to-date product information. They offer a wide range of functionality (Fig. 10.12).

10.6.6 Visualisation

These applications provide viewing, visualisation and virtual mock-up capabilities.

10.6.7 Integration

These applications enable exchange of product information between PLM applications (such as between CAD applications). They enable exchange of product information between PLM and enterprise applications such as ERP and CRM.

10.6.8 Infrastructure Management

These applications manage services of infrastructure such as networks, databases, and servers.

calendars	schedules	e-mail	messaging
tweeting	electronic whiteboards	discussion groups	virtual meeting sites
web conferencing	videoconferencing	audio conferencing	collaborative blogging
collaborative content co-authoring	chat rooms	intranets	shared project spaces
portals	vortals	project directories	social networks

Fig. 10.12 Examples of collaboration management tools

10.6.9   Idea Management

These applications enable product ideas to be captured and analysed, appropriate actions to be initiated, and progress to be tracked.

10.6.10   Product Feedback Management

These applications enable feedback about the product to be captured, analysed and made available where needed.

10.7   Task-Specific PLM Applications

Compared to the generic PLM applications, the task-specific applications (Fig. 10.11) are much more specific to a particular context. This task-specific group contains more specialised applications that are needed by particular people, departments, functions or industries.

10.7.1   Product Portfolio Management

These applications enable review, analysis, simulation, and valuation of a company's Product Portfolio of existing products integrated with the pipeline of development projects, showing estimates of sales and reuse, and showing the effects of decisions such as introducing new technologies, making acquisitions and launching joint ventures. They support the analysis of risks/rewards for different scenarios. They enable tracking and analysis of product costs against target costs and profit. These applications provide a range of display options, dashboards, cockpit charts, pie charts and graphs, with possibilities for rolling up, filtering, and grouping projects to meet various objectives.

10.7.2   Idea Generation Management

These applications enable systematic management of the generation of ideas for new and improved products.

10.7.3   Requirements and Specifications Management

These applications enable a company systematically to gather, analyse, communicate and manage product requirements describing market and customer needs. They enable a company to systematically manage and standardise product specifications.

10.7.4   Collaborative Product Definition Management

These applications enable the definition of products by people and teams from different companies working at different locations.

10.7.5   Supplier and Sourcing Management

These applications enable purchasing teams to collaborate with other team members and external suppliers for various activities such as reviewing, selecting, and purchasing custom and/or standard parts. They support qualification of new suppliers and tracking of supplier performance. They enable early involvement of suppliers, giving them real-time access to relevant product information. They enable product quality planning and use of Quality Templates. They enable the purchasing process to be streamlined, and prevent over-limit purchases.

10.7.6   Manufacturing Management

These applications enable realisation teams to simulate, optimise and define the realisation process and understand the relationships between product, plant, and manufacturing processes.

10.7.7   Maintenance Management

These applications enable customer support and maintenance teams to optimise processes, get better customer feedback, carry out activities more effectively, and better manage part and equipment inventories.

10.7.8   Environment, Health and Safety Management

These applications enable deployment and management of business processes complying with regulations of organisations such as the ISO and the FDA for the development, production, use and end-of-life of a product.

10.7.9   Intellectual Property Management

These applications enable the valuation and management of the intellectual property represented by a company's products and related services.

10.8   Applications and Data Management

Some of the applications used in the product lifecycle are file-based, some are database-based. File-based applications store data directly in files under the control of the computer's operating system. Another program on a different computer can't easily access the data in these files.

One reason for this is related to the operating system and the communications network. It results from the difficulty of physically transferring data from a file created in an application running under one operating system on one computer, to another application on another computer running under another operating system, or another version of the same operating system.

Another reason is that information about an object such as a part or product isn't independent of the application that created it. The knowledge about the structure and meaning of the data in the file is often only available in the application that wrote the file, and not available to the other application. As a result, even if the latter application were able to access the data physically, it wouldn't be able to understand it.

10.9   File-Based Data Management

File-based applications use operating system functions to store data in files. In a very small organisation with few contacts to the outside world, it may be possible to manage the data by using an essentially directory/file-based data management system. Different parts of the file name can be given specific meanings. One part of the file name can represent a project name, another can represent a version number, and a third can represent the release status. A library of frequently-used parts can be built up using similar techniques. The operating system's password structure and privilege rights can be tailored to restrict access to authorised users. Archiving and backups can be handled automatically. For some time, it will be possible to work like this.

However, users will eventually grow tired of file names like truck7_roof43_front2_right5_try7.ds8. They will be handicapped by their lack of ability to add extra information to the file name. They won't want to share data with other users, as they will be worried about data corruption. Even in a small organisation, it doesn't take long to reach the limits of file-based data management techniques.

10.9.1   Problems with Files

In a large organisation, it's almost impossible to manage file-based data efficiently. Users lose track of their data in a sea of files. It's difficult to share data between applications. There's a lot of data redundancy. Duplicate items of information are held in several applications. It's difficult to control access and to maintain security of data across several independent file-based applications. It's difficult to apply uniform rules and standards to data that are held in several file-based applications, especially as a lot of the data logic may actually be in the applications and not with the data in the files. As the data aren't completely independent of the application, any changes to the application are likely to affect the physical and logical structure of the data. The result of this is that an enormous amount of time and money is spent in maintaining existing applications, thus holding up new developments.

10.9.2   Files and Fields

Each user in the file-based environment would use applications that have their own associated files of data. Each of these applications, for example, for sales, mailing and delivery, could have a different structure of records and fields (Fig. 10.13).

One application might have one record per customer. Another might have one record per product. A third might have one record per purchase. Each application would store the data it requires in fields. As some of the information, such as customer names, would be needed in several applications, there would be redundant data. Information such as customer addresses would also be needed in several applications (such as invoicing, mailing list, delivery routing) and would be repeated in several files. When data needs to be updated, there would always be the risk that some references wouldn't be updated. A change of customer address might be entered into the mailing list file, but not on the delivery routing file. Purchase orders, payments received, and credit ratings might be in files controlled by different applications. As a result, if these were not all updated properly, an order might be turned down because the most up-to-date payment information wasn't available to the credit control application.

Application		Fields						
Sales		Product#	SalesAss.	Commision	SalesMgr.	Prod.Gp.	Date	Cus.Name
Mailing		Mr.Ms.Dr.	FstName	ScdName	Address1	Address2	Town	PostCode
Delivery		Cus.Name	Address	Town	PostCode	Parcel#	Truck#	Date

Fig. 10.13 Field structures of different applications

10.10   Databases and Data Management

10.10.1   Database Management Systems

To overcome the problems with data in files, another approach is to store data in a database.

A database is a collection of interrelated information stored with minimum redundancy, usable by several applications, but independent of these applications. A database management system (DBMS) is software that manages the database. Users are only allowed to access the database through the DBMS.

The DBMS approach is a top-down, global approach to data. It's oriented towards managing data, rather than towards providing specific functionality to help an individual user carry out their particular tasks in the product lifecycle (Fig. 10.14).

The DBMS overcomes the problems of file-based data. It provides a common repository for data, elimination of redundant data, improved security, and access control. Applications using the DBMS don't have their own data files. All users of the data, from whatever department, make use of applications that go through the DBMS to add, view, or use data. As soon as information, such as a change of customer address or the receipt of a payment, is in the database, it's available to all users.

10.10.2   DBMS in Commercial Environments

A simple example from the commercial environment will illustrate some of these points about the DBMS approach. Many of the data models in the commercial environment can be thought of, like spreadsheets, as having horizontal rows (records) and vertical columns (fields). All information on a given entity, such as a customer, is stored in a record. Each record includes many fields. For a customer record, the fields could hold information on items such as customer number, customer name, address, telephone number, entry date, purchases-to-date, and credit rating.

provide multiple users, throughout an environment, with efficient, secure, and convenient access to data
provide for physical and logical data independence
provide for explicit data validation rules and standards
provide for controlled access, security, and recovery
reduce data redundancy
allow data to be shared by many different users with different requirements

Fig. 10.14 Aims of the DBMS approach

10.10.3   *Commercial and Product Database Differences*

There are major differences between databases in the PLM environment, and
commercial databases of the type commonly used for bank customer data, payroll
data and airline reservation data. Commercial databases typically contain only
alphanumeric information, not the mixed data types of the PLM environment.
Often, they don't handle graphic data.

In commercial databases, all records relating to a particular item (such as a
customer) have similar lengths. In the PLM environment, data on two products of
the same type (such as two gear wheels) could be in the form of a different number
of records of different record lengths. The data structure of the two products could
also be different, reflecting different product structures.

In the commercial database environment, a transaction, such as a seat reservation, typically lasts a few seconds or minutes. In the PLM environment, a transaction, such as the design of a new part, will generally last several hours or days. In
the commercial data environment, the applications are built around the DBMS.
New data is then added to the DBMS. The majority of data should therefore be in
the DBMS. In the PLM environment, a lot of the data already exists elsewhere and
isn't in a DBMS. Many of the major creators of data, such as CAD applications,
don't feed their data into a DBMS, but into separate files. Some of the records in the
PLM environment are so long that they can't be handled by commercial DBMS.

In the commercial data environment, the relationships between data are generally
static, well-specified, and simple. In the PLM environment, they are frequently
changing, unclear, and complex. The PLM environment is characterised by versions and alternatives of the same basic part or product. These occur less frequently
in the commercial environment. In the commercial environment, modification to
one record won't normally require modification to a large number of other records.
In the PLM environment, modification to one record, say a part design, could lead
to a large number of modifications to related records.

In view of the above differences, few PLM organisations would benefit from use
of a DBMS as used in the commercial environment for more than a small fraction of
their product data.

10.10.4   *A Metadata DBMS*

However, in view of the obvious disadvantages of file-based applications, and the
potential advantages of database management systems, product companies recognise that a DBMS approach could be helpful for managing product data.

For various reasons (in particular, the great differences between the different
types of data to be stored), the solution of putting all product data in a single DBMS
is generally rejected. However, there's an alternative approach, lying between the
two extremes of a purely file-based, non-DBMS solution and a full DBMS acting as

file location	file name	creating application	creator name	owner name
object status	creation date	release procedure	release date	reviewer name
releaser name	part name	data type	data structure	data format
access rights	version number	promotion level	project name	references

Fig. 10.15 Examples of metadata

a repository for all product data. It's to leave the data in the files in which they were created, and only put in the DBMS a limited amount of data about the data in each file. This "data about data" is metadata.

It's usually the metadata (Fig. 10.15), rather than the underlying data, that's stored in the DBMS in the PLM environment.

The files containing the underlying data can be maintained in their native format in their usual locations. Or they can be stored in a special location, with an added level of security to prevent unauthorised access. The DBMS can automatically track what's happening to the file, automatically writing and managing the metadata. The metadata can be extended to include everything that needs to be known about the information in the file, and the way that it fits into the PLM environment.

Users can request searches of the metadata records to find specific information. A user may want to find where a given file is stored. It may be that there's no corresponding file, but a drawing on paper, in which case the metadata could point to the physical location of the drawing. Once a file is found, the DBMS can retrieve it and modify its access status (for example, to view-only).

10.10.5 Database Vocabulary

Among the major elements of a DBMS are the data model, the data dictionary, the data definition language, the data manipulation language, data management functions, and query programs. These are described in more detail below.

A few users of the database may need to have a detailed understanding of the way that product data is physically stored. However, these details are of no interest to most users. They will only want to know about, for example, the current value of a particular field.

To obviate the need for all users to work at the most detailed level, a multi-level approach is used. The lowest level is the physical level. Here, low-level data structures are described in detail. This level describes how data are actually accessed and stored (for example, on a particular disk and with a particular layout on the disk). The typical user doesn't want to work at this level, but at a higher, logical level, at which it's possible to interact with individual attributes of parts, or with complete models of objects.

Data models are used to describe the structure of a database. The data model describes the organisation of relationships between items of data, as well as the constraints on data. There are several different data models in use.

Physical data models describe data at the lowest level.

At the next level up, the logical level, there's a description of the data actually stored in the database, as well as the relationships between them. This level is sometimes referred to as the conceptual level. A logical data model describes the overall, global view of the organisation's data. This description is independent of the way that data is physically stored.

Record-based logical models and object-based logical models describe data at the conceptual level. The most commonly used record-based logical models are the hierarchical model, the network model and the relational model. They are described in more detail below.

As most users don't want to access all of the information in the database, they don't need to be aware of the entire logical level description. They only need to work with a certain well-defined part of the database called a "view". They are provided, at the view level, with a conceptual description of that part of the database that's of interest to them. This is sometimes called a logical data sub-model.

The overall design of the database is called the database schema. Corresponding to the levels described above, there'll be a physical schema, a conceptual (logical) schema, and sub-schemas corresponding to views. The logical schema contains the names of record types and fields, and their relationships. Constraints can be applied in the schema to control access and to maintain security and integrity.

A schema is defined using the data definition language. This is used to define data elements and the relationships between them. The data definitions are compiled to give a set of tables that will be stored in the data dictionary (DD). The data dictionary maintains data definitions and the rules governing data access. The DD contains the schemas and sub-schemas (logical structure) of the database, describing the view that a particular application or user has of the data. The DD maintains information on the location of a particular data element, as well as corresponding access and retrieval methods.

The data manipulation language (DML) allows access by an application program to the underlying data. The DML allows users to access, insert, and manipulate data. With a procedural DML, a user specifies what data is needed and how to get it. With a nonprocedural DML, a user just specifies the data that is needed.

Most DBMSs offer query languages that are easy to learn and easy to use, and can be used independently of applications to allow casual users direct access to data. Generally, they also include report generators that allow users to describe the format and contents of screen layouts and reports.

10.11 Data Models

The three most common data models are the hierarchical, the network, and the relational data model.

10.11.1 Hierarchical Data Model

In database terminology, an entity is something, such as a product, a part, a machine, or a customer, about which the company will maintain data. An entity is a basic object that's distinguishable from other objects and about which data can be stored. A set of descriptive attributes is associated with each object.

In the hierarchical model, entities are organised in a tree, or hierarchical structure. The uppermost level of the hierarchy has no "parent" above it. At all other levels, entities have one parent. Each entity can have one or more "children" at the level below it. An example of such a structure is a world composed of continents, each made up of countries, each with cities, each with streets (Fig. 10.16). In the hierarchical model, entities are implemented as records. Each record is made up of a collection of fields. Records are connected together through links, or pointers.

Only one hierarchical structure can exist in the model. It's defined by the database designer. Since data can be organised in the way that it will be used, the hierarchical model can provide very high performance.

With only downward steps allowed in the model in Fig. 10.18, starting from the top level, any city street anywhere in the world can be reached in four steps, through choice of continent, country, city and street.

In a hierarchical model, if an attempt is made to access data other than by the chosen path, performance can be very poor. If the access sequence is from parent to child, the model is efficient. If, however, it's from child to parent, or from child to child, it can be very inefficient.

In the model in Fig. 10.16, there's no direct path from Birmingham to London, since only downward steps are allowed. The only paths from the city of Birmingham are to the streets of Birmingham. The only way to London from

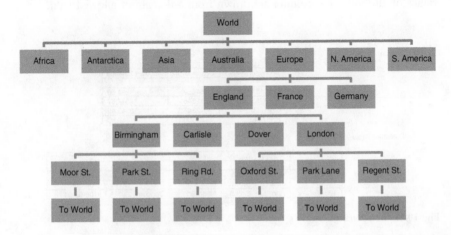

Fig. 10.16 Part of a hierarchical model

Birmingham involves going down to Birmingham street level, going back to the
World level, and then going down to London. It requires five steps.

Rearrangement of the structure of a hierarchical data model isn't trivial, and may
involve changing each individual record.

10.11.2   Network Data Model

The network data model is similar to the hierarchical data model, except that it
allows entities to have many parents at the level above them, while still having one
or more children at the level below. Entities are represented by records. Each record
is made up of fields. Records are connected by links.

Network and hierarchical databases can be tuned to become very efficient for
specific applications. However, this requires expertise and the introduction of
redundant data and data access paths. Perhaps most importantly though, their major
disadvantage is that the applications that use them need to be aware of the database
structure. If the structure changes, all the applications may need to be changed.

10.11.3   Relational Data Model

In the relational data model, data are represented in two-dimensional tables known
as relations. Each table (relation) is made up of horizontal rows (tuples) and vertical
columns (attributes). There's no hierarchy in or between tables. The tables are
"flat".

Each table has a distinct name (Fig. 10.17). Each row is unique, and each
column has a distinct name. Rows are associated by common column entries. The
values of attributes in a column are drawn from a domain of allowed values.

Account Table

Account Number	Customer ID	Account Balance
FR-245784	64859	234.50
AG-3765	12346	4347.67
CH-1206	12346	10543.45
PR87543	28560	10.50

Customer Table

Customer ID	Name1	Name2	Street	City	State	Date of Birth
12345	Anna	Smith	Main Street	Columbus	OH	06-10-94
12346	Jack	Brown	High Street	Derry	NH	04-05-87
12347	Ann	Green	North Bridge	Concord	MA	06-05-94

Fig. 10.17 Relational tables

Rows in the Account table are uniquely identified by the Account Number, the primary key in that table. In the Account table, the Customer ID isn't a unique identifier as a customer can have two or more accounts.

Rows in the Customer table are uniquely identified by the Customer ID, the primary key in that table. In this case, the Customer ID links information in the two tables.

The data dictionary stores information about the relations (such as the names of relations and attributes, and the names and definitions of views).

The relational data structure allows applications to become independent of data. The application tells the DBMS what data it needs, and the DBMS finds the data. As a result of the structure of the relational model, data can be added or deleted from the database without the need to change applications. New tables can be added, and new rows and columns added to existing tables, without reorganising the existing database. New tables can be produced by combining existing tables. The data independence offered by the relational data model cuts the workload and increases the flexibility of programmers. Another major advantage is that little or no knowledge of detailed data structure is required by casual users.

Relational databases offer flexibility, but can suffer from poor performance when large volumes of data are stored. Individual operations are generally slower than if programmed in a standard programming language. Query languages may be limited to one-dimensional indexes. This can make searches in large databases very slow. Searches go faster with the type of order-preserving structures associated with hierarchical databases. In the complex environment of PLM, relational databases also suffer from only having simple data types and structures. Although relational databases aren't currently the solution for all the product data, they do represent an appropriate technology for management of product metadata.

Some of the problems of relational databases can be overcome through enhancements. For example, to overcome problems associated with handling very long data items, a file system can be embedded in the DBMS. Special data types can be included to cater for hierarchical relationships. More use can be made of object-based methods. In the relational model, the data is stored in flat tables, but the relationship between data has to be maintained by the application. In the entity-relationship model, an object-based logical model, the relationships are considered to be "data", rather than part of the application.

10.11.4 Object-Oriented Data Model

An object-oriented database contains objects defined by an object-oriented data model that includes objects, attributes, constraints, and relationships. Each object is an instance of a class of similar objects with common characteristics. Associated with each class is a description of common methods and procedures for building and handling objects within that class. The data and the methods are stored together. This technique, of storing both information describing objects and the methods or

Fig. 10.18 No PLM
application is an island

product	product data	process
people	**PLM applications**	organisation
metrics	methods	equipment

operations that work on the objects, is known as encapsulation. Users can only work with an object through its methods. A message is sent to the object capsule, requesting that a particular encapsulated procedure be applied to the object.

New objects can be created from existing objects, inheriting all the character-istics of the existing object. Compared to relational databases, object-oriented databases offer new features of encapsulation, inheritance, and polymorphism, the ability to uniformly apply procedures to a set of objects.

10.12   No PLM Application Is an Island

No PLM application is an island (Fig. 10.18), isolated from the rest of the company. All PLM applications are closely related with other PLM components. They are also influenced by other forces within the company, and outside the company.

For example, if a new application is implemented, documentation will have to be developed to show how it should be used and supported. People will have to be trained to use and support it. The application will change some steps of a process, so the process description will need to be changed. The new application may replace existing applications, so these will need to be retired. The data in the existing applications may be archived, or may need to be migrated to the new application. The new application may enable new working methods. These will need to be documented. People will be trained to use them.

10.13   The Challenges

The specific PLM application-related challenges that a particular company faces could come from several sources (Fig. 10.19).

Fig. 10.19 Potential PLM
application-related challenges
for a company

poorly used applications
under-used applications
high costs of applications
insufficient user support
wildcat applications
duplicate applications
too many applications
too much customisation
too many interfaces
too many changes
slow response to new technology
not integrated with the business

10.14   The Way Forward

The issues described in this chapter all relate to PLM applications, but they can't be solved in isolation from the other components of PLM. Their solution, within the PLM environment, is addressed from Chap. 21 onwards.

Chapter 11
The PDM System

11.1 Introduction

Like most things in the PLM environment, PDM systems are given different names by different people (Fig. 11.1).

And like most things in the PLM environment, PDM systems are often referred to by a 2-, 3- or 4-letter acronym (Fig. 11.2).

Apart from PDM systems, there are, of course, many other applications that manage product data. For example, ERP applications, product testing applications and technical publishing systems all work with, and manage, some product data. However, the primary purpose of these applications isn't to manage product data.

PDM systems can manage product data across the lifecycle. However, in the early 21st Century, they often only manage a part of a company's product data. Usually this is product data created in product development applications. Other product data is managed by other applications (Fig. 11.3).

11.2 PDM System Overview

There are eight basic components of a PDM system (Fig. 11.4). In different PDM systems, the functionality corresponding to these components may be distributed differently between different modules. And, in different PDM systems, these components may have different names. As a result, these components aren't always easy to see when investigating a PDM system.

The first component is the Information Warehouse. Product data is stored in the Information Warehouse.

The Information Warehouse Manager controls and manages the information in the Information Warehouse. It's responsible for such issues as data access, storage

© Springer International Publishing Switzerland 2016
J. Stark, *Product Lifecycle Management (Volume 2)*,
Decision Engineering, DOI 10.1007/978-3-319-24436-5_11

Document Management System	Engineering Data Management System
Technical Data Management System	Engineering Document Management System
Engineering Drawing Management System	Product Data Management System
Technical Documentation System	Product Information Management System
Master Data Management System	Product Knowledge Management System
Engineering Knowledge Management System	Technical Document Management System
Engineering Information Management System	Technical Information Management System

Fig. 11.1   Some different names for PDM systems

EDMS	PDM	PKM	PDMS	EKMS	PDM/EDM	DMS
PIM	TDM	EDM/PDM	MDM	EIM	TIM	XDMS

Fig. 11.2   Some acronyms for PDM systems

Product data						
.... examples	Ideas Proposals Requirements Cost estimates Profit estimates Prototypes	Detailed product and process specifications	Test data, analysis data	Manufacturing data, process plans, MBOMS	Field data, complaints
.... created by ..	Word processing, spreadsheet	CAD, recipe development applications	CAE, Test	MRP2	Word processing, spreadsheet
.... managed ..	in files	by PDM	in files	by ERP	in files

Fig. 11.3   Examples of product data managed by PDM systems and by other applications

Information Warehouse
Information Warehouse Manager
Infrastructure
System Administration Manager
Interface Module
Product And Workflow Structure Definition Module
Workflow Control Module
Information Management Module

Fig. 11.4   Eight components of a PDM system

and recall, information security and integrity, concurrent use of data, and archival and recovery. It provides traceability of all actions taken on data.

The PDM system requires a basic infrastructure of a networked IT environment. The infrastructure usually includes computer and communications hardware and software, a range of graphics terminals, printers, plotters, and other devices. It may include other data management systems. There will probably be a communications network that will be used for both local and wide area communications, and for both short transfers (such as messages) and long transfers (such as files).

The fourth component of the system is the System Administration Manager. This is used to set up and maintain the configuration of the PDM system, and to assign and modify access rights.

Users and other applications access the PDM system through the Interface Module. It provides a standard, but tailorable, interface for users. The Interface

Module supports user queries, menu-driven and forms-driven input, and report generation. It provides interfaces for applications such as CAD, document scanning, electronic publishing and ERP.

The structure of the information and processes to be managed by the PDM system is defined by the Product and Workflow Structure Definition Module. The workflow is made up of a set of activities, to which information such as roles, resources, events, procedures, standards and responsibilities can be associated.

Once initiated, workflow needs to be kept under control. This is the task of the Workflow Control Module. It controls and coordinates activities. It manages, for example, the engineering change process.

The exact structure of all products and information in the system is maintained by the Information Management Module.

11.3 Importance of the PDM System

A PDM system is one of the most important elements of a PLM solution. It can manage all the product data created and used in the PLM environment.

Whatever the PLM Strategy that's chosen, it's probable that the PDM system will be a major constituent. PDM gets product data under control. And, unless the product data in the product lifecycle is under control, it will be difficult to get the product under control.

The PDM system is used in the management of product data throughout the product lifecycle. PDM systems provide support, in the complex environment of PLM, to the many activities of the lifecycle such as design, sign-off, the sharing of data between multiple users, the tracking of engineering change orders, the management of design alternatives, and the control of product configurations.

Any system that's put in place to manage product data must have functionality to limit actions on data to what's allowed, and be sufficiently powerful to maintain control. Yet it must also be sufficiently flexible to support the changes that typify the product development and support environment. The system needs to be able, for example to allow partial, or early, release of data. It needs to allow changes to be made as work progresses. It needs to be able to manage activities that are still primarily paper-based as well as those that are computer-based. It needs to be able to handle scanned paper documents as well as the data created by other applications. It needs to be able to work with different levels of data definition, and with different representations of data.

At first sight, it may appear that the need for PDM results from the need to manage the large volumes of product data generated by computer-based applications. However, it's actually the business reasons, the needs to improve productivity and to respond better to customers, that have become the driving force to achieve better management of product data. PDM oversees the creation and use of product information throughout a product's life. It's by improving the use, quality and flow of product data that PDM makes it possible to reduce lead times and product costs,

Fig. 11.5 Potential
performance improvements
with PDM systems

better use of resources
better access to information
better reuse of design information
better control of engineering changes
reduction in development cost
better support of customer use of the product
reduction in lead times
improved security of product information

and improve competitivity, market share and revenues. PDM systems offer the potential for many improvements (Fig. 11.5).

A PDM system can provide exactly the right information at exactly the right time. Having digital product data under PDM control helps attain the objectives of improved product development and support in several ways (Fig. 11.6).

For most of the product lifecycle, information is all-important. It's all people can work with when the product doesn't physically exist. Product data is a strategic resource, and its management a key issue.

11.4   The Eight Components

11.4.1   Information Warehouse

The Information Warehouse stores information regardless of its medium (such as paper, tape, disk) or physical location (which could be, for example, in engineering, in manufacturing, on-site, or off-site). It can handle all the information in the company, or the Extended Enterprise, thus permitting centralised control of distributed data. Information only has to be entered once into the Information Warehouse. All information is indexed and traceable, and can be searched for. The Information Warehouse acts as a single source for all product information.

The Information Warehouse stores all types of product data. Information can be of varying sizes and formats. Information may be text, numeric, or graphic. It may have been created internally or externally. Information will be in various states

With a PDM system ..
.. it's much quicker to access and retrieve product data than it is to access paper documents.
.. costs are reduced. Once the data is in the system it can be displayed on a screen. There's no need to pay someone to get the document. There's no need to make a physical copy.
.. quality is much better. The information shown on the screen is the information in the computer.
.. data is secure. It doesn't get torn, it won't be mislaid, it won't be the wrong version.
.. the information is available almost immediately. There's no need to wait for a document until someone gets back from lunch, or recovers from being sick.
.. time is saved. There's no need to wait while someone who recently joined the company asks his or her boss where a particular document can be found.
.. many problems are avoided. There are no longer issues such as someone else having the document, or someone modifying the document but not telling anyone, or someone modifying the entity described in the document but not modifying the document.

Fig. 11.6 PDM systems help improve product development and support

(such as in-process, in-review, released) depending on its position in the lifecycle. Information will have various structures which can be stored in the Information Warehouse. The Information Warehouse can also store different alternatives and versions.

Apart from product definition data, the Information Warehouse will also store information such as relationships, workflow models, and product configurations. It will store computer applications and technical manuals. It will store procedures and standards. For a given assembly, for example, the Information Warehouse may store information on workflow models, parts in the assembly, relations between the parts, specifications, CAD models, drawings, process plans, tooling drawings, test results, and field information. Similarly, information on hierarchical structures, such as a car and the corresponding powertrain and transmission, can be stored in the Information Warehouse.

11.4.2   Information Warehouse Manager

The role of the Information Warehouse Manager is to store incoming information securely and with integrity, to provide controlled access, and to protect product information. The Information Warehouse Manager manages the Information Warehouse. It controls and guards the data in the Information Warehouse. The Information Warehouse Manager allows data to be entered into the Information Warehouse and retrieved. Once retrieved, it can be transferred, modified, and copied.

The Information Warehouse is sometimes referred to as a repository or a library. The Information Warehouse Manager is sometimes referred to as a librarian module.

The Information Warehouse Manager works in a distributed computing environment and a multi-organisation, multi-company environment. It must be able to control, for example, a text file on one vendor's computer in one company that will be required by a CAD user on another vendor's PC in a supplier company.

The Information Warehouse Manager must be able to keep track of information outside the company, for example, information that's with suppliers. To support a variety of users and tasks, the Information Warehouse Manager allows for multiple views of data and multiple levels of data. The Information Warehouse Manager provides check-in/check-out facilities for individual files and sets of files. It's used to set up and maintain parameters describing data characteristics. It stores and makes available information under allowed access conditions. It provides access to information through a range of permission levels. These allow access to be controlled by a variety of criteria such as user, product, project, group, device, state of information, and type of information.

The Information Warehouse Manager limits access to authorised users. Data can be given a range of classifications from user-private to public. Some users will have view-only rights. Some will be able to copy data. Others to read and write data. At various well-defined times in a project, or in a product's life, these rights may change. For example, once design information has been released to manufacturing, designers will no longer be able to modify it, but manufacturing engineers will be able to modify it. The access conditions may change as the data moves through the product lifecycle. Depending on its status, data may be read-only. To maintain integrity, multiple simultaneous update will be prevented.

The Information Warehouse Manager supports private data bases (single-user), project data bases (multi-user, linked to project lifetime), and product data bases (multi-user, existing parts). It can manage information stored in simple files or in hierarchical or relational databases. It can manage information on paper. The Information Warehouse Manager provides security information on all unauthorised attempts to access data. It can provide an audit trail of all action taken on data. The Information Warehouse Manager automatically backs up the information it receives and is able to recover all information lost as a result of computer or human problems. It has responsibility for systematic and on-demand archival. This may be to electronic or traditional media.

11.4.3   Infrastructure

The basic infrastructure of the environment includes computers and a communications network. The PDM system runs in a multi-vendor computer environment. Some computers may be in the company. Some may be in other companies in the Extended Enterprise. Some may be in the Cloud. A variety of data-creation applications such as recipe formulation, CAD/CAM, document scanning, electronic publishing, structural analysis, and process planning will run on these computers. The infrastructure will probably include workstations, personal computers, tablets and smartphones as well as other devices. These may range from smart 3-D devices to less sophisticated 2D view-only terminals. There may be some other data management applications, such as relational data base management systems, in the environment. The infrastructure also includes other input devices, such as scanners, and output devices, such as printers and plotters.

The communications network will include both local area networks (LANs) and wide area networks (WANs), so that information can be communicated both on one site and between sites. Both short messages and long data files will have to be transmitted on the network. A message may be only a few words long. On the other hand, the volume of a data transfer, such as a product description file, will be much longer.

The infrastructure must be able to handle electronic messages, some generated automatically as a result of events occurring, that need to be distributed to system users who may be based locally or may be far away. They may even be on supplier or customer sites. Electronic messages could inform users and managers that an event has occurred and that work on the following task should now proceed. A message could, for example, inform a supervisor that a design has been completed and should now be reviewed.

11.4.4   System Administration Manager

The System Administration Manager is the component of the PDM system that allows the initial configuration and environment of the system to be described. It will also be used to handle the changes that will occur in the environment.

The specifications of the initial configuration will address, for example, the computers, data bases, data storage devices, networks, applications, workstations, plotters, printers, and other terminals within the environment for which the System Administration Manager will be responsible.

The System Administration Manager will be used to define users and applications in the environment, and to define and modify the access rights of individual users.

11.4.5   Interface Module

People and applications will want to communicate with the PDM system. Suitable interfaces will be needed for both classes of user. In some cases, people will want to access directly the data management system from a terminal without going through another application. They may want to query data held by the PDM system. They may want to look at a part attribute or a workflow description. Users may want to know what work they should do next. They may want to check on existing parts or look at test results. The user interface needs to support queries of many different types as efficiently as possible. The interface should be common to all the graphics devices in the environment.

In other cases, the users of the system will be applications that store, create, modify, process, or otherwise make use of data managed by the PDM system. An efficient and secure interface is needed for access by these applications. The applications could include recipe development, CAD/CAM, document scanners, software development applications, technical publication applications, business applications, and other data management applications. They could also include word-processing and spreadsheet applications. They may exchange large or small volumes of data with the PDM system. The data could be on the same computer as the PDM system, or elsewhere.

The user interface should be easy to understand and use without lengthy training. People from many functions will make use of it. It should be suitable for casual users, yet also offer efficient facilities for frequent users. The interface should include an online help facility.

The user interface should include both menu-driven and forms-driven approaches. Users should be able to tailor both menus and forms to their own requirements. The interface should offer report generation facilities. Again, it should be possible to tailor these to user requirements. Some users may want to view, or to print, information on products and parts. Others may want to generate reports on project status, change history, or attempts to gain unauthorised access to data managed by the PDM system.

11.4.6   Product and Workflow Structure Definition Module

The Product and Workflow Structure Definition Module is used to define the initial structure of a product. It's also used to define workflows. And it allows these structures to be detailed as work progresses. The product structure defines the information requirements of a product throughout its lifecycle. There'll be various generic classes of information, such as ingredient descriptions, assembly drawings, part drawings, NC programs, and user manuals. Each class of information will have characteristic attributes and tasks. The product structure describes the information that's needed, or is produced, at each phase of the lifecycle. The workflow is defined as a set of tasks, characterised by resources, events, associated information, responsibilities, decision criteria, procedures to be used, and standards to be applied.

The product structure and the workflow structure are closely linked. For each group of products, there will be a specific product structure and a corresponding workflow structure. The product structure defines all the information describing a product or part. Some of this information will be created or used at each step of the workflow. New product structures can make use of parts of existing structures. In some cases it will be easiest to start at the beginning of a process and define the information that's to be created at each step of the process. In other cases it may be easier to start at the end of the process (for example, with a Bill of Materials) and work backwards, identifying at each step the information that would be necessary to generate each information item. Initially, this task is very difficult, and a variable degree of detail is essential for the structure so that work isn't held up by the unavailability of detailed information on the structure. The product structure has to be sufficiently flexible so as to be able to handle changes to information structures and items. The Product and Workflow Structure Definition Module should offer the possibility to add or associate information when enough information isn't available to define a product, workflow, or relationship.

The workflow may be described for the entire product lifecycle or for individual processes. The level of detail needed to describe the workflow is variable. The workflow may be described either by starting with the end product and working backwards to the beginning, or by starting with a blank sheet and working forward to the end product. The workflow can be split up into projects or processes. In turn, these can be broken down into phases and steps. A workflow structure may be built up from existing steps, or a new structure may be built. In the same way that the product structure has to be sufficiently flexible to handle changes to the product structure, the workflow structure has to be able to handle changes to workflow.

The workflow structure can be used to control either individual steps in the product workflow (such as document creation) or the entire workflow. It has to include all of the information required to make this possible. It will become the source of work statements, defining the activities to be carried out and the resources to be used. For example, it should specify which users should be involved in the design of a particular part, the applications they should use, the information they will need, the information they should produce, the procedures they should follow, and the approval process. The approval process work statements should specify the roles and authorities of the individuals involved, the rules and requirements for sign-off, and the process to be followed if sign-off is refused.

During a process, "events" occur. An event marks the end of one activity and the beginning of another activity. Events need to be identified and included in the workflow definition. Completion of a design, and initiation of a periodic maintenance activity, are typical events. The activities that lead to and follow each event should be specified. The messages that should be communicated when the event occurs, or if it doesn't occur within a specified time, have to be defined in the workflow definition. The review, approval, and release workflows have to be defined. The engineering change process has to be modelled. The various steps, reviewers, approvers, and sign-off procedures have to be defined. The definition should include hierarchies so that, if an activity doesn't occur, it will be passed automatically to the next highest authority. Since many parts of the process are similar or repetitive, some workflow structure elements will be repeated throughout the overall workflow. The creation of change requests and orders is a typical example. Automatic process sequences can be set up to handle tasks such as the provision of copies to individuals named on a distribution list. Once the product and workflow structures have been defined, the PDM system can manage data and workflow. Formal description of the workflow will make it possible to identify unnecessary activities, as well as those that could be run in parallel.

11.4.7 Workflow Control Module

For a particular task, the product structure and the workflow structure will have been defined by the Product and Workflow Definition Structure Module.

The Workflow Control Module then manages the workflow of the various activities in progress, and monitors progress. It controls the progress of projects in an event-driven mode. It maintains status information on ongoing projects.

Once initiated, the Workflow Control Module is in control of the activity. It assigns tasks to individuals, informs them of the resources to be used and the procedures to be followed, initiates the associated actions, and maintains status information. If necessary, the Workflow Control Module can remind users of standard operating procedures, and can check that standards information is accessed. It distributes data and documents to the individuals as needed. When the task is finished it can, for example, request a review, or promote the design and initiate the next step of the process. It can enforce promotion rules. If the person responsible for the next step is absent, it can automatically pass the work to the most suitable replacement or the next highest authority. It manages the review, approval, communication, and archival of information.

Automatic process sequences handle such tasks as providing copies to individuals on a distribution list. Following the rules specified during the definition of the workflow and product definition structures, the Workflow Control Module controls versions and manages the engineering change process.

The Workflow Control Module, on the basis of the workflow and product structures, ensures that all necessary information is available before releasing parts to manufacturing.

The Workflow Control Module monitors the occurrence of events. When an event occurs, it initiates a previously defined set of activities. If an event doesn't occur at the expected time, the appropriate level of management will be alerted.

When an event occurs, the Workflow Control Module sends, in accordance with the defined structures, the appropriate messages. Engineering change requests and change orders can be rapidly transmitted to inform all interested parties of impending and actual changes. Interested parties can be informed of upcoming events. The Workflow Control Module can notify other people that a change has been requested. It can initiate messages based on parameters captured at each step. It can notify downstream users that modifications have been made to upstream information.

The Workflow Control Module will keep status information up-to-date, and ensure that information is handled as planned. At any time, the Workflow Control Module will be able to display the exact status of each process that it's managing. It can track and report the status of tasks in process. It can produce progress reports at specified times, showing, for example, how much lead-time has been consumed. The Workflow Control Module maintains an audit trail of activities relating to the process.

Performance analysis can be carried out on activities and on information access. The impact of proposed changes can be analysed and assessed. Resource-level loading can be coordinated, and schedule visibility maintained, for all related tasks.

Once set up for all activities in the environment, the Workflow Control Module should be able to promote parallel, rather than serial, workflow, thus reducing lead times. Increasingly, the Workflow Control Module should play an expert role, checking that input is legal and consistent, and that correct procedures are being followed, and automatically organising activities so as to minimise lead time.

11.4.8   Information Management Module

The information items include all the product data needed to specify, build, test, install, operate, and maintain the product, and to support its end of life. This will include information such as specifications, drawings, lists, programs, reports, and installation manuals. The Information Management Module is used to describe the exact configuration of a particular product throughout its life. It relates components, subassemblies, and assemblies. It supports multiple assembly levels, multiple hierarchies, and multiple membership. The Information Management Module maintains a complete history of the product through design, manufacture, delivery and field use to end-of-life. The status of all information (such as in-process, in-review, released) is maintained by the Information Management Module. It maintains the configuration of a given end product, managing all the required information. For a given product, for example, it may maintain information such as bill of materials, goes-into lists, assemblies, relations between the assemblies, CAD models, drawings, analysis results, recipes, parts lists, process plans, NC programs, tooling drawings, test results, and field information. The Information Management Module can distinguish between the as-designed, as-planned, as- built, as-installed, and as-maintained configurations of the product.

The Information Management Module maintains exact configuration information on each individual product. It supports multiple versions and alternatives of data. It takes account of engineering changes. It maintains information about the relationships between information, such as the creation of a document from a particular template.

The Information Management Module offers the possibility to navigate product structure by paging down and traversing the workflow and information structures. It allows information to be accessed in many ways, such as by model number and by part number.

11.5   Benefits of PDM

Different companies will have different drivers for implementing a PDM system. Some examples are shown in Fig. 11.7.

The reasons for implementing a PDM system can be divided into two classes. In one of these, the PDM system appears to alleviate some of the problems that occur in the product development and support environment. In the other, it appears to pro-actively and positively impact operations across the product lifecycle. Although these two classes of reasons can be treated separately, in practice, they're closely related. Reasons in the first class address the resolution of existing problems. Reasons in the second class go one step further, and address the potential for further

Fig. 11.7 Some reasons for implementing a PDM system

centralise the control of data
eliminate redundant data
enable better product development
improve access to data
improve data management
improve supplier management
reduce costs
reduce the number of IS applications
reduce time to market
replace a legacy PDM system

Benefit Category	Example of Benefit
Information Management	provide a single, controlled vault for product information
	maintain different views of information structure
	provide faster access to data
	manage configurations
Re-use of Information	make available existing designs for use in new products
	reduce duplicate data entry
Workflow Management	make sure the most appropriate process is followed
	improve distribution of work
	ensure procedures are followed
Engineering Change Management	speed up Engineering Change distribution, review and approval
	provide status information on engineering changes
Business Performance Improvement	improve product quality
	reduce overhead costs
Business Problem Resolution	reduce scrap
	reduce product liability costs
Functional Performance Improvement	increase engineering productivity
	reduce inventory
	develop better cost estimates
Product Development Management	improve project co-ordination
	increase product development schedule reliability
	provide high-quality management information
Product Development Automation	automate the sign-off process
	automate the transfer of data between applications
IS Effectiveness Improvement	integrate Islands of Automation
	link data bases together
	remove unnecessary systems
Product Development Infrastructure	support product development practices and applications
	distribute data and documents electronically

Fig. 11.8 Eleven categories of benefits of PDM systems

improvement. The reasons can be grouped into eleven categories. In each category, most of the reasons can be related both to the resolution of current problems and to pro-active improvement of activities across the product lifecycle. The eleven categories, and examples of benefits in each category, are shown in Fig. 11.8.

11.6   Common Issues

There are many PDM systems on the market. The detailed data management of individual companies are different. Implementations of PDM systems tend to be company-specific. Nevertheless, there are several issues common to most PDM systems (Fig. 11.9).

Naming, Functionality, Scope	Change, Version Management	Interfaces
Data Model, Workflow	PDM SYSTEM	Ownership, Funding, Support
Fit in IS Architecture	Customisation, Installation	Everyday Use

Fig. 11.9 Issues common to most PDM systems

11.6.1 Naming, Functionality, Scope

The functionality and scope of a PDM system are often unclear from the name of the system.

The scope of PDM systems can be very different. Some may have a lot of functionality and have been designed to support product data across the lifecycle. Others may only have limited functionality, and be focused on specific parts of the lifecycle. Some may be focused on the data management needs of a particular industry.

Some systems suffer from problems such as incomplete functionality, malfunction of the system, poor response time, and unavailability of the system on a wide range of platforms.

Among the key factors in determining the functionality that a company needs from a PDM system will be the quality, quantity, and coverage of the applications that are already in place. What do these applications do? What functionality do they have? How will they fit with PDM? Is PDM seen as a replacement for these applications? Is it an add-on? To what extent should it be integrated with them?

PDM systems can offer a wide range of functionality including information management, change management, process management and product structure management. Some, or all, of these functions may already be present in a company's existing applications. Product structure may be managed in parts master, BOM and MRP applications. Process management may be addressed in project management applications. Some information management functionality may be built into other applications such as a CAD application, or may be in an application developed in-house.

The required functionality will also depend on the way the company is currently organised, and the way it will be organised in the future. If everybody is on one site, then multi-site functionality may not be needed. On the other hand, if users are spread over several locations, multi-site functionality will probably be needed. If product development is carried out in teams, or the company has taken a Concurrent Engineering approach, then corresponding functionality would be looked for in the PDM system.

PDM systems range from simple, off-the-shelf packages with basic functionality to complex tailorable systems with wide-ranging functionality that can be further developed to exactly fit a company's requirements.

11.6.2 Change, Version Management

One of the driving forces for PLM is the high level of change in the product environment. However, change can be an issue for PDM systems.

A new version of a PDM system can raise problems. The previous version may have met a company's requirements perfectly. The change, while being of great value to most companies, may raise problems for others.

Changes to other applications can also be an issue. If the data structures in an application interfaced to the PDM system change, then the interface may need to be changed. If many applications are interfaced to the PDM system, and each one changes a few times each year, the situation can get complicated.

11.6.3   Interfaces

PDM systems need to share and exchange product data (such as part numbers, version numbers, product costs) with other applications in the company. Some interfaces may be provided by the vendor as part of the PDM system. Others may be developed with the Application Programming Interface (API) provided with most PDM systems. The API enables the team supporting the PDM system to create, using the system's objects and routines, any additional functionality that's needed. For example, an interface may be needed between the PDM system and a costing application that has been developed in the company.

Many PDM systems have their own specific user interfaces. It can take a long time for users to learn the details.

11.6.4   Data Model, Workflow

There may be issues related to the flow, use and quality of product data. There may be problems with the cost of entering information in the system. The system may not be able to handle all data types. It may not be able to store data where it's needed. There may be incompatibility between data structures. Classification mechanisms may be inappropriate. There may be no way of encouraging re-use of information.

Problems may arise with particular types of documents. The system may only be set up to handle certain types or formats of documents, and not be able to handle others. It may only be able to handle a limited number of variants of a particular document. It may only be able to apply the same release process to all documents of a particular type, even though the company has different ways of releasing them.

There may be problems with storage and communication of data. The PDM system may only be able to store all data in one physical location, yet use of data may be required at two or more locations. In some cases, it will be the cost of communicating information between different sites that's the problem, in other cases it may be security or confidentiality.

There may be problems with the structure of information. Different departments may structure the same information in different ways and unless the system is capable of accepting different structures (or views) for the same information, there may be issues as people try to ensure "their" structure is chosen as the standard.

Problems may arise if a company has several naming, numbering and classification conventions, but the system is limited to one convention.

11.6.5 Ownership, Funding, Support

A PDM system is cross-functional. It doesn't belong to any one of the functional departments. As a result, it may be unclear who is responsible for it. It may not be obvious how it will be financed. It may not be obvious which practices should be followed in addressing it, which jargon should be used to describe it, or which rules should be followed when managing information in a PDM system. There may be problems cost-justifying the PDM system. Which department or departments should pay the costs of a system that's used by several departments? How should costs be distributed so that the department that gets the most benefit pays the most? How can the running costs of the system be shared equitably? This is especially difficult to achieve if the system is installed in one department, supported by people from another department, and used by people from many other departments.

Insufficient investment is a common issue with PDM systems, as is the use of inappropriate project cost-justification calculations. These may generate over-optimistic expectations. The targets put in place to drive the implementation and use of PDM may be inappropriate or even unattainable.

11.6.6 Fit in IS Architecture

With a primary purpose of managing product data, a PDM system needs to be closely linked to the other PLM applications that create and use product data. The PDM system needs to fit seamlessly into the company's IS architecture.

11.6.7 Customisation, Installation

At installation time, all sorts of issues can arise with a PDM system. Its implementation may take much longer than expected. The people who planned for PDM, and selected the PDM system, may pull out before the system is installed, leaving implementation in the hands of people who neither understand the objectives nor are motivated to succeed. Insufficient training may be given to users and the system support team. There may be no guidelines describing how the system should be used. There may be problems with the system itself. It may not work the way the vendor claimed it would, or it may have bugs, or it may not be documented, or there may be no procedures showing how it should be used.

As another example of the type of problem that could occur, a PDM system vendor may propose to a company that a system integration partner will implement the system. The system integrator may not have enough people with the right skills available, so asks one of its partners if it has someone suitable. The partner doesn't, but knows some people who have the skills. These people work hard at doing what they're told to do. The only problem is that they've never seen either the company or the vendor. Not surprisingly their deliverables aren't exactly what the company wanted. Then, of course, there are lots of meetings at which the company blames the vendor, and the vendor blames the integrator. None of which adds any value for the users.

Another potential problem, with its roots in the past, is the unorganised state of most information that's under the control of traditional manual information management systems. Provided that it's been possible for a person who's been in the company for the last thirty years to lay their hands on a particular document within a few hours, most people have been happy. PDM systems don't work like that.

Other implementation problems can be due to poor understanding and definition of the processes. Problems may arise because the system doesn't address the parts of the overall process that the company is interested in. Problems can also arise at the level of individual activities if the system doesn't work the way the company wants to carry out specific activities such as release and engineering change control. It's important that the process be understood and clearly defined. Otherwise it's going to be difficult to use PDM to support it. PDM can't be used effectively if key issues are unclear (Fig. 11.10).

Another problem that can arise is that the people who are supposed to look at the process issues can't agree among themselves as to what the process should be.

11.6.8   Everyday Use

Some issues may arise when the system starts to be used on an everyday basis (Fig. 11.11).

One source of problems can be the interfaces between the PDM system and other applications. Unless all the interfaces exist, some users will work entirely outside the PDM system rather than sometimes inside and sometimes outside. Problems may arise if new developments promised by the vendor don't appear. In-house

| it's not clear how information flows in the process |
| it's not clear what the information is being used for |
| it's not clear who has access rights at different times |
| the individual steps of the process aren't clear |
| it's not clear what happens at each step of the process |
| it's not clear what conditions must be met before moving to the next step |

Fig. 11.10 Lack of clarity concerning information in the process

| errors and inconsistencies in the system |
| missing functionality |
| lack of training and support |
| lack of funding |
| lack of interfaces to other applications |
| failure to make the necessary organisational changes |

Fig. 11.11 Potential issues when everyday use starts

system developments can also be a source of problems. Sometimes, developments won't be made because funding is cut or because they have low priority on the waiting list. Another problem that may arise after installation is that the project budget, in particular the training budget, is slashed. The funding of the PDM system support team, the group that should make sure the system works on an everyday basis and should provide everyday support to users, may be cut. As PDM gradually takes hold, some departments may feel they're losing control or power. As a result they may start to block its use and hinder further development.

There may be issues related to top management such as lack of commitment, lack of leadership, lack of support and lack of patience. Problems at the middle management level may be due to conflicts with personal goals, empire-building, and fear of loss of power. Users may fear that the PDM system may play a Big Brother role, or may lead to job losses. Problems can also arise if the members of the PDM project team don't work together effectively.

Other issues may affect everyday use of PDM (Fig. 11.12).

Users may run into problems. For example, the system may crash or malfunction frequently. Users may complain that it's not user-friendly and takes too long to learn to use. They may suffer from poor response time as the amount of product data in the system increases. System upgrades may be necessary and the result may be that system use becomes too expensive. In some areas, the system may not behave as expected, and time-wasting workarounds may be necessary. As users get to know the system, they may find that functionality has been oversold. Functions they need may not exist, or may only be partially implemented. The system may only handle a limited number of document types. Documentation and on-line help may not exist for some key functions. There may be no guidelines describing how the system should be used. Necessary customisation may be too difficult or too time-consuming for users.

The people involved in PDM system administration and support will hear all about the problems that users are having. They may also have their own problems (Fig. 11.13).

| lack of agreement and co-operation between departments |
| difficulties in getting cross-functional activities to occur |
| departmental barriers preventing information flow |
| departments using different definitions |
| departments using different standards |
| issues with customers and suppliers |

Fig. 11.12 Issues affecting everyday use

| the system may be difficult to set up for more than a prototype |
| system administration may be inflexible, time-consuming and error-prone |
| previously hidden limitations may appear in the definition of roles and processes |
| previously hidden limitations may appear in the creation of reports |
| the system may not work on all the platforms where it's needed |
| the system may be difficult to integrate with other applications |
| the vendor may be unable to provide good, well-trained support staff |

Fig. 11.13 Possible concerns of the PDM support team

As time goes on, it may become clear that the wrong vendor was chosen. New versions may be delivered late, lack promised functionality, and have quality problems. Maintenance costs may become unacceptably high. There may be no upgrade path between successive versions. Key individuals may leave the vendor. Eventually the vendor may go out of business.

11.7 Little Data Management Excitement

Most business managers find "let's spend a lot of money to manage data" about as interesting as "let's watch paint drying". If you are a technical manager trying to implement a PDM system, make sure you have plenty of reasons why it's needed, and plenty of answers to the negative replies you're likely to receive.

The typical business manager will respond to the technical manager's request for PDM with, "You mean you want to spend $1 million (or $100,000 or $10,000) on a system that's just going to manage data? But you're already managing the data, aren't you? So why do you want to spend all that money? You can't do that, that'll just increase our costs. And at the end of the day there's no real benefit. We'll just be managing the same data a different way. Our costs will go up and our productivity will go down."

"No", says the technical manager, "there'll be many benefits. Everyone will have faster access to product data and developers will no longer waste 30 % of their time looking for it. We'll be able to develop products faster, we'll have fewer quality problems and, because we'll develop faster, we'll need fewer development hours, so our costs will go down."

"Listen", the business manager says, "the only way I can move this business forward is by increasing revenues or cutting costs. This PDM system of yours isn't going to increase revenues. I'll give you three good reasons why.

One, you say it'll get our products to market quicker, but when I look at our Engineering organisation I can't see that changing the way we manage data is going to cut the development time by even 1 %. Two, it takes so long to get a product out in the market, and make money, there'll be no benefit on the bottom line for at least five years. And no-one's going to throw money at a project that doesn't show a return for five years. Three, there'll be such chaos when you try and take the old system out, and put the new one in, that there'll be all sorts of problems for the customers. We can't risk that again.

So, there's no increase in revenues, and as I keep telling you, I've got to reduce costs. Knowing our cost structure, I can't see how buying a new system will do that. It will do the opposite. It will push up costs. But let's try to be positive. The easiest way for us to reduce costs is to reduce headcount. Where will this data management system help us lay off people? Who's doing the data management work today? You said the engineers will save 30 % of their time, but we can't lay off 30 % of each engineer."

"You make it sound so difficult", says the technical manager. "Just cutting the product data management workload will reduce the development hours, so we'll cut costs. And because we'll be faster to market, we can increase market share, and charge premium prices".

"Listen" says the business manager, "We just said that if we want to reduce costs we've got to lay off people. So first you tell me how many people we can lay off in your department in the next three years, and then you write down their names. As for speed to market, it sounds great, but it doesn't mean anything. If we decide to go ahead with this data system of yours, it will take at least two years to select. Remember when we got the CAD system? It will distract everyone's attention from the real issue of developing products faster. It will take three years to bring into operation. Look what's happening with the ERP implementation. And after that it will probably be another four or five years before we'd actually see any product developed with the new system hitting the market. So you're talking about benefits nine or ten years down the road. By that time, I'll have retired, and you'll be in a new job. Just forget it and get those drawings to the plant."

11.8 No PDM System Is an Island

No PDM system is an island (Fig. 11.14), isolated from the rest of the company. All PDM systems are closely related with other PLM components. They are also influenced by other forces within the company, and outside the company.

If a type of document that's managed by the PDM system is changed, and the metadata is changed, then it may be necessary to change the way that the PDM system manages the document. Relationships with other objects may need to be changed. If an application that's interfaced to the PDM system is modified, it may be necessary to modify the interface. The information about users in the PDM system will need to be updated when new people join the company, or when people change roles or positions in the company. When processes are changed, workflows managed by the PDM system may need to be changed. If a new type of product is

Fig. 11.14 No PDM system is an island

product	product data	process
people	**PDM system**	organisation
metrics	methods	equipment

missing PDM functionality	insufficient user support
multiple PDM systems	inappropriate data model
duplicate PDM functionality	limited workflow functionality
deficiencies in the PDM system	too much product data managed outside PDM
poorly designed implementation	high costs for support
too much bureaucracy in system use	slow response time
too much customisation required	long delays in error resolution

Fig. 11.15 Potential PDM system-related challenges for a company

developed, a new product information structure may need to be created in the PDM system.

11.9   The Challenges

The specific PDM-related challenges that a particular company faces could come from several sources (Fig. 11.15).

11.10   The Way Forward

The issues described in this chapter all relate to PDM systems, but they can't be solved in isolation from the other components of PLM. Their solution, within the PLM environment, is addressed from Chap. 21 onwards.

Chapter 12
People

12.1 Introduction

Like most things in the PLM environment, the people who work in it are referred to by different names (Fig. 12.1).

And what they do in the organisation also has many names (Fig. 12.2).

More specifically, they may have a job title. This may say something about the type of work they do, or their level in the hierarchy, or their part of the organisation. Figure 12.3 shows some of the hundreds of names used to describe their jobs. These names may be used in different ways in different companies. What a person with a particular job title does in a particular company is likely to be different from what people with the same job title do in other companies.

There are various names for the units in which people work (Fig. 12.4). These names may be used in different ways in different companies.

12.2 It's a Jungle

From the above, it can be seen that it may not always be crystal clear, from a business card showing a job title and the name of an organisation, exactly what somebody does in a company.

As well as issues with naming, there are many other people-related issues in the PLM environment (Fig. 12.5).

© Springer International Publishing Switzerland 2016
J. Stark, *Product Lifecycle Management (Volume 2)*,
Decision Engineering, DOI 10.1007/978-3-319-24436-5_12

workforce	associates	human capital	workers
human resources	staff	personnel	employees

Fig. 12.1 Some different names for people in the PLM environment

job	work	role	task
position	situation	occupation	mission
assignment	employment	function	trade

Fig. 12.2 Some different names for what people do in the PLM environment

account manager	business analyst	cost accountant	course developer
Crater	CXO	designer	director
database administrator	executive	drafter	environmental staff
documentation clerk	field engineer	financial analyst	HR manager
knowledge worker	IT manager	lease representative	manager
manual worker	marketer	network specialist	PC technician
product developer	product manager	programmer	project manager
regional finance manager	quality manager	recycling director	resident engineer
sales associate	support staff	sales representative	service engineer
software developer	technician	system consultant	system developer
technical support analyst	test engineer	validation engineer	VP

Fig. 12.3 Job titles in the PLM environment

department	function	project	program	crew	section
team	workgroup	division	office	group	gang

Fig. 12.4 Some different names for work organisations in the PLM environment

Industry and product specifics	Experience	Culture
Skills	PEOPLE	Team dynamics
Knowledge	Adaptability	Job and Department Naming

Fig. 12.5 Common people-related issues

12.2.1   Different Products

The skills needed by a company depend on the product it makes. With a particular product, a company will need people with the corresponding particular skills. Another company, with a different product, may not need people with those particular skills, but will need people with other skills.

12.2.2   Different Companies

In some cases, people doing a particular job will be working within a company. In other cases, that particular job may have been outsourced, so the equivalent person is working in another company. Outside the company, many other organisations may be involved with the product (Fig. 12.6).

suppliers	partners	distributors	clients	auditors
governments	regulators	lawyers	bankers	product users
industry bodies	leasers	owners	customers	contractors

Fig. 12.6 Organisations where people may work outside the company

12.2.3 Different Departments

Different companies give their departments different names. In one company, there may be an Engineering Department. In another company, perhaps there won't be a department called the Engineering Department, but there will be a department called the R&D Department that does exactly the same type of work. And a third company may have both an Engineering Department and an R&D Department.

As well as the name, the scope of a department's activities may differ from one company to another. A person doing a particular job in one company may be working within a particular department, such as the Engineering Department. But the person doing the same job in another company may be working in the Manufacturing Department.

12.2.4 Same Job, Different Title

Different companies give different names to the same job. In one company, someone doing a particular job may be called a Product Engineer. In other companies, people doing the same job may be called Product Developers, or Engineers, or Designers.

12.2.5 Same Title, Different Job

People with a particular job title may do very different things in different companies. For example, a company's expectations of a Product Manager may be very different in different companies (Fig. 12.7).

| be a business leader for a product line |
| be a team member among many developing a new product |
| plan for new products, build product roadmaps, gather market requirements, identify new product candidates, develop business cases, define new products at a high level, plan and lead development projects, explain the product to Marketing and Sales |
| define the marketing message, promote the product at trade events, work with the press, monitor the competition |

Fig. 12.7 Different expectations of a product manager in different companies

12.2.6   Different Locations

Across the lifecycle, people may be at many different locations. They may be working on the company's premises. They may be working for a supplier, or a partner, or they may even be the final customer of the company's product.

12.2.7   Different Background

Some of the people in the lifecycle will have an engineering background, others will have other backgrounds such as accountancy, human resource management, marketing and sales. Some of the users of the product will be customers with completely different backgrounds to those of people in the company.

12.2.8   Different Computer Literacy

Many of the users will have different levels of computer literacy. Some will be writing state-of-the-art software. Others will need their secretary or assistant to interface with their computer.

12.2.9   Different Data Need

Depending on what they're doing, different people in the lifecycle will have different product data usage and product data management needs. Some will create data, some will modify it, some will delete it. Others will only want to reference data, perhaps for management purposes.

12.2.10   Different HR Policies

Some of the people in the lifecycle will be working for a company with a Human Resource Department that has developed skills matrices, training plans and career paths. Other people in the lifecycle will be working in companies in which the Human Resource Department is only involved with people when they join the company and when they leave the company.

12.2.11 Different Metrics

Some of the people in the lifecycle will be working in organisations that are measuring performance on one set of parameters (such as Time To Market achievement). Other people in the lifecycle will be working in organisations that are measuring performance according to other metrics (such as scrap rate).

12.2.12 Different Bonus Systems

Some of the people in the lifecycle will be working in an organisation in which their bonus is mainly linked to the performance of the company. In another organisation, there may be no bonuses. In a third organisation, the bonuses may be mainly linked to the performance of each individual.

12.2.13 Different Languages

Some of the people in the lifecycle will be working in organisations with one working language (such as Spanish or German), others in organisations with another working language (such as English).

12.2.14 Different Culture

Some of the people in the lifecycle will be working in organisations in which everyone is expected to say what they think. Others will be in organisations where everyone is expected to keep quiet and just do what their boss tells them to do.

Some of the people in the lifecycle will be working in organisations in which a meeting is held to give formal agreement to an issue that has already been agreed informally. Others will be in organisations in which a meeting is held to discuss an issue in detail from all possible viewpoints, and then take a collaborative decision. Other people will be in organisations in which a meeting is held to discuss an issue in detail from all possible viewpoints, after which a manager will take the decision alone.

12.2.15 Changing Population

Some of the people in the lifecycle will have been working with the company for decades, have wide experience and be able to carry out many activities at the same time. In another organisation, most people will be new to the company, have little experience, and only be able to carry out one or perhaps two activities at the same time.

12.2.16   Different Roles

Independently of their job title, or their job description, people may play several roles. One person may, for example, have the roles of Process Owner, Product Developer, Document Owner and Document Creator for some documents, and Document Approver for other documents. Another person may have the roles of Process Owner and Change Reviewer.

12.2.17   Different Sins

People in the lifecycle are humans, not machines. They aren't always 100 % rational and cooperative. Sometimes their individual needs and objectives play a larger part in their decisions and behaviour than those of the company. In some companies, if they want promotion, it may be important to do something simple that shows short-term success. In others, it may be important to do something complex that will only show success after many years. People in the lifecycle, like other humans, may suffer from envy, gluttony, greed, lust, pride, sloth, and wrath.

12.2.18   Response to PLM

Many people in a company will understand why PLM is important. They'll be able to see where it can help, and what benefits it can bring. Unfortunately for those wanting to implement or improve PLM, there are many people who won't understand the need for PLM. Among those who won't understand will be many whose support is essential to the success of a long-term, cross-functional initiative that will have significant effects on company performance and organisation. From their position in the company, PLM may seem to have a low priority, or even be unnecessary.

PLM is necessary, but many people will initially fail to understand this simple message. They won't react to it in the right way. They'll slow down the speed at which the company can obtain the potential benefits. Many of the people whose support is necessary, and who control the resources needed for success, just won't understand. Their level of understanding will be low and it will take a long, long time to get it to a level where they'll become supportive. Yet it's not possible to implement PLM without support throughout the company. It's not possible to go it

product	product data	process
PLM applications	**People**	organisation
metrics	methods	equipment

Fig. 12.8 No man or woman is an island

alone. PLM is holistic and cross-functional. It's as much an organisational approach as a technological approach, and it needs positive involvement from people at many levels in many functions.

12.3   Nobody Is an Island

No man, or woman, is an island (Fig. 12.8), isolated from the rest of the company. People are closely related with other PLM components. They're also influenced by other forces within the company, and outside the company.

For example, a new manager may be hired to improve performance in a particular department. To change performance they're likely to change working methods. They may hire people who they've previously worked with successfully in other companies. They may bring in equipment used in their previous companies. Old-timers in the company may not like all these changes, and may decide to leave. The new methods may call for new applications. The new methods will probably use new documents. The new methods will be reflected in changes to processes. Process documentation will be changed to reflect the changed processes. There'll be training sessions to make sure everyone understands the new methods. New procedures will be written for the new equipment. New people will be hired to operate it. Job descriptions will be changed to correspond to the new situation. The new manager will implement new performance measures to show that the new methods are leading to better performance. The company may then acquire a company on another continent to help improve business there. The organisational structure will be changed accordingly. In the resulting structure, the new manager is one level further from the top than before. Unhappy about this, and discussing it with a colleague from a previous company, the new manager is offered a more important job in their previous company. And takes the job, leaving the new company to cope with all the changes.

12.4   The Challenges

The specific challenges related to human resources that a particular company faces could come from several sources (Fig. 12.9).

| the company doesn't have the right skill mix |
| people don't have the right experience |
| people aren't sufficiently innovative |
| people don't show enough commitment |
| people aren't adaptive enough |
| personnel costs are too high |
| people don't learn new skills |
| a slow response to new technology |
| people aren't empowered |
| people aren't disciplined enough |

Fig. 12.9 Potential people-related challenges for a company

12.5 The Way Forward

The issues described in this chapter all relate to people, but they can't be solved in isolation from the other components of PLM. Their solution, within the PLM environment, is addressed from Chap. 21 onwards.

Chapter 13
Methods

13.1 Introduction

Previous chapters have addressed products, product data, the processes that create and use product data, the applications that create, use and manage product data, and the people who create and work with product data.

This chapter looks at methods, another component of the PLM environment. Like most things in the PLM environment, "methods" are referred to by different names by different people (Fig. 13.1).

They range from very technical methods to broadbrush management approaches. The constituents of this component of PLM are intended to improve performance (Fig. 13.2).

13.1.1 The Need

Most companies that develop and support products are under pressure from the competitive, global environment to improve performance. They know that, to be successful, a company must be able to supply and support the products that customers require, at the time required by the customer. Product costs must be trimmed so that they correspond exactly to customer requirements. Product functionality must be improved to match these requirements. Customer service must be improved with on-time documentation delivery, reliable delivery times, prompt complaint handling, and easy product repairability.

Products must be brought to market faster. As product lifetimes get shorter, significant market share is lost if a product isn't brought to market at the earliest possible moment. A company that gets to market first can capitalise on late market entry by other companies. Companies see that many key parameters of their performance are heavily dependent on the performance of the activities in the product

© Springer International Publishing Switzerland 2016
J. Stark, *Product Lifecycle Management (Volume 2)*,
Decision Engineering, DOI 10.1007/978-3-319-24436-5_13

techniques	approaches	practices	programs
how to's	methodologies	work methods	best ways

Fig. 13.1 Some different names for methods

reduced time to market	quality improvement	reduced manufacturing costs
improved service	cycle time reduction	reduced Cost of Quality

Fig. 13.2 Typical performance improvements expected with methods

lifecycle, for example, product development time, product cost, service cost, product development cost, product quality and disassembly costs. They want to improve performance in these areas.

13.1.2 Improvement Initiatives

In response to the need for change, many improvement initiatives have been proposed as ways to improve performance across the product lifecycle. More than thirty are mentioned in the next section. Some of the initiatives address product development and support practices (Fig. 13.3). Some address standards and standardisation (Fig. 13.4). There are also initiatives addressing quality, processes, finance and management (Fig. 13.5).

These techniques and approaches have all met with success in one or more companies, and should be understood by companies embarking on PLM initiatives. However, all companies are in slightly different situations, so what is needed for one may not be needed for another. A particular company may only need a few of these techniques and approaches. Another company may be able to benefit from many. Selecting and prioritising such techniques and approaches is part of the activity of defining a strategy for PLM. One of the challenges of PLM for a particular company is to identify the techniques and approaches that are most relevant to the activities on which the company wants to focus its efforts.

DFA	DFE	DFM	DFSS	EMI	FMEA	FTA
JIT	PDCA	PY	QFD	TRIZ	VA	VE

Fig. 13.3 Examples of product development-related methods in the lifecycle

Group Technology	ISO 9000	ISO 14000	STEP	XML

Fig. 13.4 Examples of standards-oriented methods in the lifecycle

ABC		Benchmarking	BPR	CWQC
Continuous Improvement		Process Mapping	Teamwork	TQM

Fig. 13.5 Examples of other methods in the lifecycle

13.2   Overview of Methods

13.2.1   ABC

Activity Based Costing (ABC) is a costing technique used to overcome deficiencies of traditional product costing systems. These, under some conditions, such as high volume and product diversity, can lead to inaccurate product costs. Inaccurate product costs create problems when taking pricing decisions for new product introductions, retiring obsolete products, and responding to competitive products.

13.2.2   Alliance Management

Alliance Management is needed in the extended enterprise environment to ensure that everything works well among the many participants, including partners, suppliers and vendors. Alliance Management usually includes more than just contract compliance. It can also include governance activities, building and maintaining alignment, and ensuring that relationships move forward.

13.2.3   Benchmarking

Benchmarking is a good technique for helping to understand the performance of other organisations that are believed to have more effective operations. The organisations that are benchmarked may be direct competitors, suppliers, or even companies in other sectors. If the other organisations are found to have more effective operations, then the benchmarking organisation can try to understand how they work, and why they are better. It can then start to improve its own operations, and will be able to set itself realistic performance targets. Benchmarking is often carried out in two steps. Usually the first step is investigation of best practices. The second step is identification and application of metrics.

Benefits of benchmarking are shown in Fig. 13.6.

Figure 13.7 shows the six phases of benchmarking activity.

| making the organisation's relative performance very clear |
| providing clear quantitative targets to management |
| providing targets that are not just theoretical visions of the future, but are already practical reality in other companies |
| getting immediately useful ideas for better products and processes |
| providing impetus for management to start behaving proactively, and to look for ways of working which will bring significant improvements |

Fig. 13.6 Benefits of benchmarking

| setting up a team. Training the team to carry out the activity |
| understanding the organisation, and assessing its strengths and weaknesses |
| selecting industry leaders and   competitors to benchmark. Understanding their strengths and weaknesses. Researching these companies to improve understanding. (A lot of preparation is required before visits to these companies. Without preparation, benchmarking teams only get a fraction of the possible benefit from visits.) |
| visiting the companies to improve understanding |
| reporting on what has been seen and understood (without reports, companies only get a fraction of the possible benefit from visits.) |
| incorporating what has been learned, and improving on it |

Fig. 13.7 Six phases of benchmarking

About 25 % of the workload for a typical benchmarking exercise is to get the team up and running, and trained in useful methods for analysing operations. About 25 % of the effort goes on understanding the company's own performance in a chosen area. About 25 % of the effort is to prepare for, and conduct, the external benchmarking studies. The remaining 25 % captures the lessons learned, and prepares to put them to use.

13.2.4 BPR

Business Process Re-engineering (BPR) is a technique used to significantly reorganise a company's business processes with the objective of making very large improvements in performance (such as 80 % reduction in product development time). BPR generally involves redesign of a company's processes and the application of Information Technology to the new processes.

13.2.5 CWQC

Company-Wide Quality Control (CWQC) is a technique in which all parts of the company co-operate to improve all aspects of company operations.

13.2.6 Concurrent Engineering

Concurrent Engineering is a technique to bring together multidisciplinary teams that work from the start of a product development project with the aim of getting things right as quickly as possible, and as early as possible. Input is obtained from as many functional areas as possible before the product and process specifications are finalised. Getting the development correct at the start reduces downstream difficulties in the product workflow.

Name	Activity
configuration identification	determining the product structure, selecting configuration items, documenting items, interfaces and changes, and allocating identification characters or numbers
configuration control	addresses the control of changes to a configuration item after formal establishment of its configuration documents
configuration status accounting	is for formal recording and reporting of the established configuration documents, the status of proposed changes and the status of the implementation of approved changes
configuration audits	are carried out to determine whether a configuration item conforms to its configuration documents

Fig. 13.8 Four activities of configuration management

13.2.7 CM

Configuration Management (CM) is the activity of documenting initial product specifications, and controlling and documenting changes to these specifications. CM is a formal discipline to help assure the quality and long-term support of complex products through consistent identification, and effective monitoring and control, of all of this information. ISO 10007:2003 provides guidance on the use of configuration management within an organisation. Applicable across the product lifecycle, it describes the configuration management responsibilities and authorities, the process and the planning, as well as the four activities (Fig. 13.8) of configuration identification, change control, configuration status accounting and configuration audit.

13.2.8 Continuous Improvement

Continuous Improvement is an approach of incremental change aimed at making many small-scale improvements to current business processes.

13.2.9 COQM

Cost of Quality Management (COQM) is an approach to reducing the Cost of Quality (COQ). The COQ is the sum of all the costs incurred throughout the product lifecycle due to poor quality. These costs occur when the product doesn't have perfect quality first time and every subsequent time. The Cost of Quality is made up of four types of quality cost (Fig. 13.9).

Type of Cost	Examples
internal failure costs	due to failures such as rework, scrap and poor design that the customer doesn't see
external failure costs	due to failures that occur after the product has been delivered to the customer. They include warranty claims, product liability claims and field returns
appraisal costs	the costs of measuring quality and maintaining conformance by such activities as inspection, testing, process monitoring and equipment calibration
prevention costs	the costs of activities to reduce failure and appraisal costs, and to achieve first-time quality. These activities include education, training and supplier certification

Fig. 13.9 Quality cost components

13.2.10 Customer Involvement

During the product lifecycle, there are usually times when the customer would like to be in direct contact with the manufacturer, for example, from their tablet or smartphone. The customer might like to see detailed operating instructions for a new product, or a video of how to assemble some furniture, or some information about where to pour the engine oil. Having real-time, detailed knowledge about the product would enable the customer to be informed about use of the product (such as a smartphone) and/or its services more cost-effectively. And the customer could be informed that some components need to be replaced otherwise the product will fail. Information about product use and status could be used to inform the customer of the most cost-effective way to return the product. These are moments of real Customer Involvement, and it's at moments like these that the best information can be collected from the customer. The direct involvement of the customer provides them access to information they want, and in return they provide useful information to the manufacturer.

Such customer involvement is much better than just "listening to customers". How can a Focus Group really know, better than a real customer, what works best in a particular situation with the product? It's only by involving customers that companies can understand their experience, behaviour, needs and expectations, and apply that understanding for future products.

Technologies such as RFID enable detailed information about the use of products to be automatically collected. This "Voice of the Product" provides a lot of data about the way that products are really used. Together RFID and Customer Involvement provide a good overview of the customer.

13.2.11 DFA

Design for Assembly (DFA) techniques aim to reduce the cost and time of assembly by simplifying the product and process (Fig. 13.10). For example, tabs and notches in mating parts make assembly easier, and also reduce the need for assembly and testing documentation. Simple z-axis assembly can minimise handling and insertion time.

| reducing the number of parts |
| combining two or more parts into one |
| reducing or eliminating adjustments |
| simplifying assembly operations |
| selecting fasteners for ease of assembly |
| designing for parts handling and presentation |
| minimising parts tangling |
| ensuring that products are easy to test |

Fig. 13.10   Examples of DFA techniques

13.2.12   DFE

Design for Environment (DfE) is an approach that integrates all environmental considerations into product and process design.

13.2.13   DFM

Design for Manufacture (DFM) techniques are closely linked to Design for Assembly techniques. However, they are oriented primarily to individual parts and components rather than to DFA's sub-assemblies, assemblies, and products. DFM aims to eliminate the often expensive and unnecessary features of a part that make it difficult to manufacture. It helps prevent the unnecessarily smooth surface, the radius that is unnecessarily small, and the tolerances that are unnecessarily high. DFA and DFM need to be carried out at the conceptual design stage before major decisions have been taken about product and process characteristics.

13.2.14   DFR

Design for Recycling (DFR) aims to increase the level of recyclability and to ensure that recycled material keeps as much of its value as possible.

13.2.15   DFSS

Design For Six Sigma (DFSS) includes various methodologies to achieve six sigma performance. (At Six Sigma quality, a process must produce no more than 3.4 defects per million.) DMAIC (define, measure, analyse, improve and control) is generally used for an existing product or process. Other DFSS methodologies are used to design a new product for Six Sigma quality.

| using energy as efficiently as possible |
| minimising use of toxic chemicals |
| taking care of air and water quality |
| efficiently using and recycling materials |

Fig. 13.11 Taking account of sustainability needs

13.2.16   DFS

Design for Sustainability (DfS) is an approach which recognises that products and processes must take account of environmental, economic, and social requirements. It integrates these into product and process design (Fig. 13.11).

13.2.17   Design Rules

Design Rules, Design Guidelines and Design Axioms bring together principles of successful design.

13.2.18   DTC

Design to Cost (DTC) techniques address cost-effectiveness from the viewpoints of system effectiveness and economic cost. The economic costs may be limited to design engineering costs and/or production costs. In other cases, operations and support costs may be included. Cost is addressed by establishing an initial design, comparing estimated costs with an allocated budget at the system or subsystem level, and addressing any cost inconsistencies through subsequent redesign or re-evaluation of requirements.

13.2.19   EMI

Early Manufacturing Involvement (EMI) brings manufacturing engineers into design activities that take place early in the product workflow, rather than only bringing them in once the design engineers have finalised a product that will be difficult or impossible to manufacture.

13.2.20   ESI

Early Supplier Involvement (ESI) brings suppliers into development activities early in the product workflow, rather than only bringing them in to manufacture some of the parts. ESI is particularly important for companies that want to focus on upstream customer, specification and product design activities, where they can best use their resources. They will want to outsource downstream activities where they are not cost-effective (such as detailed drafting) or are less competent (such as parts manufacture), than organisations more specialised in these areas. Suppliers for these activities have an important role to play. Companies want to make the best possible use of suppliers with the aim of getting a customer-satisfying product to market as early as possible. This means involving the supplier right at the beginning of the process, when the major modules of the product are being defined. The supplier can then be given the job of designing and manufacturing a complete sub-assembly. Suppliers are expected to provide fast response, to be responsible, to be reliable and to have excellent skills, knowledge and experience concerning particular parts or activities. The company will want to have long-term relationships with a small group of excellent, knowledgeable and trusted suppliers.

13.2.21 FMECA

Failure Modes Effects and Criticality Analysis (FMECA) is a technique for identifying the possible ways in which a product or part can fail, the corresponding causes of failure, and the corresponding effects.

13.2.22 FTA

Fault Tree Analysis (FTA) is a technique that uses a hierarchical decomposition technique for analysing faults. A fault tree is a diagram showing the interrelationships between failures and combinations of failures.

13.2.23 GT

Group Technology (GT) is a technique to exploit similarities in products and processes so as to improve the overall efficiency of operations.

13.2.24 Hoshin Kanri

Hoshin Kanri (Policy Deployment) is a technique of step-by-step planning, implementation, and review for managed change. It's a systems approach to management of change in business processes. A system, in this sense, is a set of co-ordinated processes that accomplish the core objectives of the business. Policy Deployment cascades, or deploys, top management policies and targets down the management hierarchy. At each level, the policy is translated into policies, targets and actions for the next level down.

13.2.25   JIT

Just In Time (JIT) is a waste-reduction management technique for improving business processes. Originally thought of mainly as a technique for reducing material inventory, it has grown to become an enterprise-wide operating philosophy with the basic objective of eliminating all non-value-added activities and other waste.

JIT can be applied to all activities in the product workflow, including those that traditionally have taken place in product development. JIT aims to eliminate non-value-added activities. On the shop floor, there are many non-value-adding activities with material. They include unpacking, inspecting, returning, storing, material handling, and rework. The results of applying JIT on the shop floor include reductions in stocks, delays, overruns, errors, and breakdowns. These vices can be found, in one form or another, throughout the product workflow. JIT techniques can be applied to remove this waste from the product workflow before automation is introduced.

JIT focuses on continuous flow. On the shop floor, in a non-JIT environment, manufacturing is carried out in a set of discrete steps with WIP (work in process) being moved to and from stores at each step. This leads to all sorts of problems, such as long manufacturing cycles, damaged goods, unnecessary transportation of goods, and unnecessary storage areas. A parallel can be drawn with the traditional product workflow, which is also unnecessarily split up into a set of discrete steps, with much paper and many unfinished projects in various states of progress at several stages of the process. Many of the related actions (such as collecting, controlling, communicating, and copying data) do not add value.

13.2.26   Kome Hyappyo

Kome Hyappyo is a philosophy that refers to the spirit shown by Kobayashi Torasaburo. Receiving a hundred bales of rice to distribute to the impoverished people for whom he was responsible, he sold it to pay for a school, thus greatly increasing the value of the rice. In product lifecycle terms, it's the willingness to sacrifice a small short-term gain, in an individual function, for important long-term benefits across the lifecycle.

13.2.27 Lean Production

Lean Production appeared in the West in the early 1990s. Key concepts include "only add value in production", "eliminate waste", and "flow value from the customer". Some interpretations of Lean Production also include a focus on people and their knowledge. Six Sigma and JIT are seen as tools to support a Lean Production approach.

13.2.28 LCA

Life Cycle Assessment is a methodology used to understand the main impacts arising in each phase of a product's life. Life Cycle Assessment involves the calculation and evaluation of the environmentally relevant inputs (such as minerals and energy) and outputs (such as energy savings) and the potential environmental impacts (such as emissions and solid waste) across the lifecycle of a product, material or service. It aims to increase the efficiency of the use of resources, and reduce waste and liabilities.

13.2.29 LCD

Life Cycle Design incorporates disposal and recycling issues at the early stages of product development. All the issues related to a product's useful life are considered at the outset. So are those involving the product once its useful life is over. Life Cycle design includes evaluation of environmental protection, working conditions, resource optimisation, company policies, lifecycle costs, product properties and ease of manufacture. The goals of lifecycle design include ease of disassembly, ease of assembly, fast and safe decomposition, lowest cost to find/recover, and lowest cost to recycle.

Lifecycle techniques such as lifecycle design, lifecycle analysis and lifecycle assessment will play an increasingly important role in product development. Trends in the environmental area will significantly change design, production and distribution methods. Because of regulations and consumer pressure, manufacturers are becoming responsible for the environmentally safe disposal and recycling of their products. Insurance companies may require documentation of a lifecycle approach before accepting a company as an insurance risk. Environmental agencies require lifecycle documentation. Companies want to respond to customer demands for environmentally safe products with labels that show their environmental correctness. Being environmentally correct is a valuable marketing asset as many customers are taking the issues of environmental protection, occupational health, and resource use seriously. Auditors address environmental issues in company annual reports.

Current lifecycle practices often concentrate on the activities within a company, but they will be extended to cover the complete cradle-to-grave cycle including use of raw materials, production methods and usage/disposal patterns.

A lifecycle approach offers the possibility to change the business model. When one customer has no further use for a product, another customer may want to use it. Or parts of it may be re-usable in other products. Or the aggregated primary materials may be reusable. Without a lifecycle approach, issues like these aren't addressed at the development stage, with the result that much of the product is disposed of at the end of life, or only a few of the raw materials are recycled. Without a lifecycle approach, OEMs typically don't focus on how products are recycled, or how recycled products, parts and material can re-enter the lifecycle. However, when they do come back into the lifecycle, they change the needs for new parts. The more parts that come back, the fewer the new parts that will have to be produced. The product has to be developed to ensure optimum use and re-use, both during its life, and after it gets to the end of its life.

13.2.30   Open Innovation

From about 2000 onwards, the scope of Innovation Management changed. Previously it had mainly addressed product innovation in R&D. Since then, the scope has been extended to include innovation in processes, delivery and business models. "Open innovation" has emerged. It's a way for a company to search for and acquire innovation outside its borders. Innovation methods include Edward DeBono's "Lateral Thinking" and "Six Hats", Eric von Hippel's "Lead User Analysis" and Genrich Altshuller's "TRIZ".

13.2.31   Phase/Gate Methodology

Phase/gate and stage/gate methodologies split the product development activity into separate phases (usually between four and six phases). When product development projects are carried out, people from different functions work together in each phase, carrying out the tasks defined for that phase and producing the required deliverables. At the end of the phase is a "gate", at which a cross-functional team, including managers, reviews results for that phase. The cross-functional team only allows the project to proceed through the gate, to the next phase, if it meets pre-defined targets.

Without a well-defined development and support methodology, it's unlikely that people are going to work in harmony across the lifecycle. A well-defined methodology lets everybody know exactly what should be happening at all times, and tells them what they should be doing. It defines the major phases and explains what has to be done in each phase. It shows how the phases fit with the company organisation and structure. It shows the objectives and deliverables at the end of each phase, and the way phases connect together. It shows which processes, systems, methods, techniques, practices and methodologies should be used at which time in each phase. It shows the human resources that are needed, the people, skills and knowledge, and their organisation. It shows the role and responsibilities of each individual and the role of teams. It shows the role of management, project managers, functional reviewers and approvers. It describes the major management milestones and commitments. It describes the metrics used in the process.

13.2.32 PDCA

The Plan-Do-Check-Act (PDCA) cycle is a technique for continuous improvement of any activity or process. In the "plan" step, a plan of action is generated to address a problem. Corresponding control points and control parameters are generated. The plan is reviewed and agreed. In the "do" step, the plan is implemented. In the "check" step, information is collected on the control parameters. The actual results are compared to the expected results. In the "act" step, the results are analysed. Causes of any discrepancies are identified, discussed and agreed. Corrective action is identified.

13.2.33 *Platform Strategy*

Platform strategies and module strategies aim to reduce costs and improve quality by maximising re-use of parts in different products. Modular products that make use of common parts allow the variety required by marketing, while limiting the workload on the manufacturing function.

13.2.34 *Poka-Yoke*

Poka-yoke existed as a concept for a long time before the Japanese manufacturing engineer Shigeo Shingo developed the idea into a technique for achieving zero defects and eventually eliminating quality control inspections. The methods are sometimes called "fool-proofing", but recognising that this term could offend some managers and workers, he came up with the term poka-yoke, generally translated as "mistake-proofing" or "fail-safing" (to avoid (yokeru) inadvertent errors (poka)). The use of an additional locator pin to prevent misalignment of the workpiece is an example of poka-yoke. Appropriate use of poka-yoke by a washing-machine manufacturer could ensure that the machine doesn't start until the fixtures used for transport have been removed.

13.2.35   Process Mapping

Process Mapping is carried out to understand, analyse and design business processes.

13.2.36   Project Management

Project Management approaches help manage a company's product-related projects in an effective way.

13.2.37   QFD

Quality Function Deployment (QFD) is a step-by-step technique for ensuring that the "voice of the customer" is heard throughout the product development process so that the final product fully meets customer requirements. The first step of QFD is to identify and capture customer requirements, wishes, expectations and demands. In the following steps, these are translated by cross-functional product teams into the corresponding technical specifications. QFD uses a series of related simple matrices and tables as the tool for translating the voice of the customer first to design specifications, then to more detailed part characteristics, then to show the necessary process and technology characteristics, and finally to show the specific operational conditions for the production phase. The interrelated tables and matrices develop to form what is often called, due to the roof-like shape of some of the tables, a "Quality House".

13.2.38   Roadmapping

Roadmapping helps portfolio managers and product managers to identify new opportunities, and to identify the product and technology activities required to get a product to market.

13.2.39   Reliability Engineering

Reliability Engineering is a technique to improve the reliability of a product or process. Reliability engineering encompasses the activities of planning, measuring, analysing, and recommending changes with the aim of improving the reliability of a product or process. Reliability is the ability of the product or process to perform its functions in a defined environment over a given period of time. The failure rate over the product lifecycle, or during operation of the process, measures the probability of not meeting these requirements. In order to increase capability while saving resources, reliability and maintainability must be formalised concurrently during the design process, and must be valued and measured as much as performance and other design criteria.

13.2.40 Robust Engineering

Robust Engineering is the engineering of products and processes such that they will work satisfactorily throughout their lifetime in spite of the disturbances that are bound to occur. Robust engineering takes, as a starting point, the fact that it's impossible to prevent or control all variations throughout a product or process lifetime. In consequence, they should be designed to be as immune as possible to the variations that will occur due to a variety of sources in the product or process environment. Typical sources of variation for a product are the manufacturing process and in-use deterioration.

13.2.41 Simultaneous Engineering

Simultaneous engineering is similar to concurrent engineering. It brings together multidisciplinary teams that work together from the start of a product development project to get things right as quickly and as early as possible in the project. The overall intention is to get a quality product to market as soon as possible. Sometimes, only design engineers and manufacturing engineers are involved together in concurrent product and process development. In other cases, the cross-functional teams include representatives from purchasing, marketing, accounting, the field, and other functional groups. Sometimes, customers and suppliers are also included in the team. Multidisciplinary groups acting together early in the workflow can take informed and agreed decisions relating to product, process, cost, and quality issues. They can make trade-offs between design features, part manufacturability, assembly requirements, material needs, reliability issues, and cost and time constraints. Getting the design correct at the start will reduce downstream difficulties in the workflow.

13.2.42 Software Development Methodologies

Software development methodologies were introduced to manage software development. In 1970, W.W. Royce published an article called "Managing the development of large software systems: concepts and techniques." This referred to the Waterfall system development methodology (requirements, analysis, design, coding, testing and maintenance). Such methodologies divide projects into phases, and use deliverables and approvals to maintain control.

The Agile approach takes an iterative approach to software development. In the first step, a high-level project plan and a high-level view of the targeted system are defined by the project team working closely with user representatives. Then, working from the high-level plan, the next steps are defined by the project team. The team then works on these steps, again in close collaboration with users, to detail them and carry out the activities. A more detailed view of the system is created. This is reviewed and validated (or not) by the users. The next steps are agreed and then executed. At each step, a prototype is built to meet the user's apparent requirements. Experience of its use provides input for the next step. Each step builds on the results of previous steps. Compared to the Waterfall approach, the Agile approach involves users throughout the project, and repeatedly tests the most up-to-date proposal for the system. This offers the possibility to identify any needs for change and to make corresponding adjustments as early as possible.

13.2.43 Standards

Many standards can be applied in the product environment. Examples include ISO 9000 (Quality System), ISO 14000 (Environmental Management), ISO 10303, the SEI Capability Maturity Model (CMM) developed at the Software Engineering Institute in Pittsburgh, and the AQAP (Allied Quality Assurance Publication) 2000 series, the NATO requirements for an Integrated Systems Approach to Quality through the Life Cycle. ISO 10303 is the standard for the computer-interpretable representation and exchange of product data (STEP).

13.2.44 SPC

Statistical Process Control (SPC) is used during the production phase of the product lifecycle to reduce variation and to help correct whatever is wrong. All production processes fluctuate over time, but, provided they are stable, they will stay within certain well-defined limits known as control limits. If a process gets out of control, the fluctuations go beyond the control limits. At the heart of SPC is the statistical analysis of engineering and manufacturing information. Facts, data and analysis support the planning, review and tracking of products throughout the lifecycle. SPC is based on the use of objective data, and provides a rational, rather than an emotional, basis for decision-making. This approach recognises that most problems are system-related, and aren't caused by particular people. It ensures that data is collected and placed in the hands of the people who are in the best position to analyse it, and then take the appropriate action to reduce costs and prevent non-conformance. If the right information isn't available, then the analysis, whether it be of shop floor data or engineering test results, can't take place, errors can't be identified, and consequently, errors can't be corrected.

13.2.45 STEP

STEP, the ISO 10303 standard for the computer-interpretable representation and exchange of product data, includes many Application Protocols. The Product Life Cycle Support (PLCS) standard is designated Application Protocol 239 (AP 239).

13.2.46 System Engineering

System (or Systems) Engineering is an interdisciplinary approach aiming to enable the realisation and deployment of successful systems. The Systems Engineering effort spans the whole system lifecycle. It focuses on defining customer needs and required functionality early in the development cycle, documenting requirements, then proceeding with design synthesis and system validation while considering the complete problem. System integration issues are dealt with early on, rather than later in the development cycle.

13.2.47 Taguchi Techniques

Taguchi's experimental design techniques allow designers to experiment with a large number of variables with relatively few experiments. Genuchi Taguchi started developing them in the 1950s. The Taguchi approach is particularly relevant to the parameter design phase in which the designer sets the value of design parameters, assigning specific values for product and process parameters to get a stable reliable product.

13.2.48 Teamwork

Teamwork involves a group of individuals, often from several functions, working together ("simultaneously" in time, often "co-located" in space), sharing information and knowledge, and producing better and faster results than they would have done if operating independently and in serial mode. In the traditional product lifecycle, people in different departments worked one after the other on successive phases of development. Compared to that approach, teamwork is a new technique. It requires people to behave differently, to take decisions differently, and to be measured differently. Team objectives have to be set and controlled. The team has to be managed, and may need to be coached.

13.2.49   TCO

Total Cost of Ownership (TCO) approaches help companies understand how much it will cost, not just to purchase a product, but also to use, maintain and retire it.

13.2.50   TQ

Total Quality (TQ) is a description of the culture, attitude and organisation of a company that aims to provide, and continue to provide, its customers with products and services that satisfy their needs. The culture requires quality in all aspects of the company's operations, with things being done right first time, and defects and waste eradicated from operations.

13.2.51   TQM

Total Quality Management (TQM) is an approach to the art of management that became popular in the West in the early 1980s. Figure 13.12 shows the key points of TQM. Continuous improvement of all operations and activities is at the heart of TQM. Because customer satisfaction can only be achieved by providing a high-quality product, continuous improvement of the quality of the product is seen as the only way to maintain a high level of customer satisfaction. As well as recognising the link between product quality and customer satisfaction, TQM also recognises that product quality is the result of process quality. As a result, there is a

| customer-driven quality |
| TQM leadership from top management |
| continuous improvement |
| fast response to customer requirements |
| actions based on data and analysis |
| participation by all employees |
| a TQM culture |

Fig. 13.12 Key points of TQM

focus on continuous improvement of the company's processes. This will lead to an improvement in process quality. In turn this will lead to an improvement in product quality, and to an increase in customer satisfaction.

13.2.52   TRIZ

TRIZ is a way of systematically solving problems and creating suitable solutions. It was invented by a Russian engineer and scientist, Genrich Altshuller. TRIZ is known in English as the Theory of Inventive Problem Solving. The TRIZ acronym comes from the Russian original.

13.2.53   VA and VE

Value Analysis (VA) and Value Engineering (VE) are techniques in which a multifunctional team measures the current value of a product or its components in terms of functions that fulfil user needs. "Value Analysis" is applied to existing products, whereas "Value Engineering" is carried out during initial product development. However, the principles are very similar, and the term value analysis is sometimes used where value engineering would be more appropriate. Value analysis should be carried out by a multidisciplinary team (including design, marketing, production, finance, service) with the aim of finding the most cost-effective solution for a particular product that is consistent with customer satisfaction. The team develops and evaluates alternatives that might eliminate or improve component areas of low value, and matches these new alternatives with the best means to accomplish them.

13.3   Some Characteristics of Methods

Although there are many methods, and they are used for many different things in very different situations, there are issues that are common to many methods (Fig. 13.13).

Naming	Evolution, Version	Relationship with IS
Definition, Purpose, Scope	METHODS	Relationship with processes
Duplication, Overlap	Implementation	Commitment

Fig. 13.13 Issues common to many methods

13.3.1   Unclear Name

The functionality of a method is often unclear from the name of the method. Many umbrella terms such as Lean Production and Total Quality Management are used with a wide range of different meanings. Different proponents include different functionality within apparently similar methods.

It's often unclear, from the name, exactly how one method differs from others. Some groups of methods, for example DfE, DFR and DfS, may appear to be similar, and have overlapping functionality.

13.3.2   Overlap Between Methods

There's often overlap between the objective, scope and activities of the methods mentioned above. It can be different, for example, to distinguish between Simultaneous Engineering and Concurrent Engineering. And the terms Value Analysis and Value Engineering are often used to apply to very similar activities.

13.3.3   Overlap Between Methods and Applications

Many methods are supported by related applications. Often these aren't integrated with other applications, and create new Islands of Automation.

13.3.4   Confusion Between Methods and Processes

People sometimes find it difficult to distinguish between processes and methods. Processes are company-specific. Methods are more general in nature. They aim to improve performance in a particular area in any company.

13.3.5   Duplication of Existing Activities

Many methods duplicate, to some extent, existing activities. A technique such as Life Cycle Design is, of course important, but most companies will already have been doing product design for many years before embarking on Life Cycle Design.

They need to make sure that, if they apply Life Cycle Design, they only add new value-adding activities, or selectively improve existing activities, and remove any resulting duplication.

13.3.6   Unclear Definition

Many methods are developed with input from many sources, such as companies, industry organisations, academic organisations and consultants. Each of these has its own viewpoint and focus. Unless an independent organisation of some sort is set up, there's rarely a unique definition of the technique.

13.3.7   Unclear Metrics

Without a clear definition or scope, it's difficult to know how to measure the impact of a particular technique. Unless a lot of care is taken, any improvement that's claimed to result from one improvement method, could actually be due to another technique or to external factors. For example, an improvement in Time to Market might be attributed to use of a phase-gate approach, but could actually result from outsourcing the development of product modules to suppliers.

13.3.8   Difficult to Implement

Without a clear definition and scope, and with overlap with other techniques and activities, it can be difficult to implement a technique. And unless it's implemented, it's not going to provide any benefits.

13.3.9   Method Evolution and Confusion

After a method appears to achieve good results, its practitioner will want to maintain the momentum. Conferences will be organised to explain the method. The scope of the method will be expanded so that those who benefited from the initial version can achieve further improvements. Newcomers will be surprised to see there are now two versions of the method. They will wonder if they should start with the initial version and achieve the benefits it offers, or start immediately with the second version.

13.3.10   Market Push

New methods are regularly described in the industry press. They're discussed by industry analysts. Their intended advantages are highlighted. Generally these initiatives are referred to by a positive-sounding acronym. It's said they're easy to introduce and to use. They're expected to result in major productivity gains. It's said they're being implemented by "all world-class companies" or "more than half the companies in the Fortune 500", or "more than 60 % of companies in a particular industry sector". Those who don't invest in them are claimed to be dinosaurs and to live in the Stone Age. (The logic of this isn't clear as the Stone Age didn't start until more than 60 million years after dinosaurs died out.)

13.4   No Method Is an Island

No method is an island (Fig. 13.14), isolated from the rest of the company. All methods are closely related with other PLM components. They are also influenced by other forces within the company, and outside the company.

For example, to improve performance, a company may decide to implement a new technique. People are sent to training courses to learn how to apply the technique. They receive voluminous documentation, and details about various applications supporting use of the technique. They visit other companies to see how the technique can be applied, and to find out which application is best. Before the technique can be used, some business process descriptions are modified to take account of the technique. Potentially affected people are trained about the new processes. An application is purchased to support use of the technique. It's interfaced with the company's PDM system, as this will manage the data produced by the new application. A new performance measure is introduced to measure the benefits of using the new technique. When the technique is used for the first time, it's found that it overlaps with another technique.

13.5   The Challenges

The specific method-related challenges that a particular company faces could come from several sources (Fig. 13.15).

product	product data	process
PLM applications	**Methods**	organisation
metrics	people	equipment

Fig. 13.14 No method is an island

| lack of support for methods |
| lack of knowledge about methods |
| duplication of methods |
| overlap with other methods |
| integration of methods |
| conflicting methods |

Fig. 13.15 Potential method-related challenges for a company

13.6 The Way Forward

The issues described in this chapter all relate to methods, but they can't be solved in isolation from the other components of PLM. Their solution, within the PLM environment, is addressed from Chap. 21 onwards.

Chapter 14
Facilities and Equipment

14.1 Introduction

Like most things in the PLM environment, facilities are referred to with different names by different people (Fig. 14.1).

And there are many special names for facilities used for specific activities. Some examples are shown in Fig. 14.2.

Similarly, as with most things in the PLM environment, equipment is referred to with different names by different people (Fig. 14.3).

And there are many special names for equipment used for specific activities. Some examples are shown in Fig. 14.4.

And, as for most things in the PLM environment, there are many acronyms for equipment. Figure 14.4 included the Point of Sales (PoS) kiosk, Quartz Crystal Microbalance (QCM) monitor and Ultraviolet (UV) light system. Figure 14.5 shows some other acronyms.

Figure 14.5 includes acronyms for analogue-to-digital converter; automated storage and retrieval system; bed-of-nails; electromagnetic compatibility tester; electrostatic discharge; high efficiency particulate air; manufacturing execution system; supervisory control and data acquisition; Universal Serial Bus; Universal Product Code.

In total, there are thousands of different machines and tools in the PLM environment. One of the challenges of PLM for a particular company is to identify the facilities and equipment that are most relevant to the activities on which the company wants to focus its efforts. The role, and the potential benefits and disadvantages, of each facility and type of equipment should be clearly understood so that it's possible to see if, and how, it could best fit into a PLM solution.

© Springer International Publishing Switzerland 2016
J. Stark, *Product Lifecycle Management (Volume 2)*,
Decision Engineering, DOI 10.1007/978-3-319-24436-5_14

facility	factory	installation	mill	plant
shop	site	works	workshop	yard

Fig. 14.1 Some generic names for facilities

brewery	cannery	clean room	disposal site	distillery
distribution centre	forge	foundry	laboratory	loading dock
mint	packing plant	prototype line	refinery	repair shop
service centre	store	test tower	wind tunnel	wrecker's yard

Fig. 14.2 Some specific names for facilities

apparatus	appliance	device	equipment	instrument
machine	machinery	mechanism	plant	tool

Fig. 14.3 Some generic names for equipment

aerator	bagger	crusher	digester	extruder
fixture	granulator	hopper	incinerator	Jig
label applicator	kiln	mandrel	nut inserter	oven
PoS kiosk	QCM monitor	robot	shredder	tension meter
UV light system	vision system	wet scrubber	3D printer	3D scanner

Fig. 14.4 Examples of names of specific equipment in the PLM environment

ADC	AIE	ASRS	ATE	ATM
BoN	CMM	CNC	DNC	EMC
ESD	HEPA	SFDC	MES	PLC
NC	SCADA	PLC	USB	UPC

Fig. 14.5 Examples of equipment and tool acronyms in the PLM environment

14.2   Characteristics

Even though there are so many different types of equipment, many issues are common to many or all of them (Fig. 14.6).

14.2.1   Range, Specialities

There's a very wide range of equipment and facilities. Many are so specialised that unless you have worked in a particular industry you may not even have heard of

Range, Specialities	Lean, High throughput	Automation
Simulation	EQUIPMENT	Standards
High Cost	Know-how, Training	Tuning, Feedback

Fig. 14.6 Issues common to many types of equipment

them. Unless you've worked with calendars, ellipsometers, fluffers, flying probe systems, light curtains, machine guards, magnetic sweepers and sheeting systems you might wonder what they are and what they do.

14.2.2 Lean, High Throughput

Whatever the equipment, companies are always looking for high quality equipment and output. They want flexibility, fast response and short set-up time.

14.2.3 Automation

Automation is widely used in facilities. Advantages of automation include increased throughput and better quality than human operators. However there are potential pitfalls. Special industrial computers and controllers may be needed, such as Programmable Logic Controllers (PLC), Numerical Control (NC), Computer Numerical Control (CNC) and Direct Numerical Control (DNC). Manufacturing Execution Systems (MES), Material Handling Systems (MHS), Shop Floor Data Collection systems (SFDC), and Supervisory Control And Data Acquisition (SCADA) systems may be needed.

As with any use of computers in the PLM environment, the result can be islands of automation that aren't integrated with the rest of the company's information systems.

14.2.4 Standards

The wide range of specialised equipment, and the related range of specialised numerical controllers lead to an environment in which only a few specialists in a company really know what's going on with a given piece of equipment. Companies would like a standardised environment, but there are few standards in the equipment environment.

14.2.5 Simulation

Companies would also like to be able to simulate the behaviour of equipment before making real-life tests. Simulation can save time and money.

14.2.6   High Capital Cost

A lot of equipment has a high capital cost. Its purchase has to be rigorously justified. Once installed it has to be used as much as possible to ensure it meets financial targets and pays back the investment.

14.2.7   Know-How, Training

With so much specialised equipment, the know-how about a particular piece of equipment is often restricted to one or two individuals in a company. And as the manufacturer of the equipment will bring out frequent new versions to stay ahead of competitors, the equipment operators will probably need to follow frequent training courses.

14.2.8   Tuning, Feedback

Often, equipment will be almost human, with two apparently identical pieces of equipment behaving differently, as if they had their own character. Tuning of the equipment may take a long time. Although people working frequently with the equipment may get to understand its behaviour, it can be difficult to feed this information back to other people who should know how the equipment behaves.

14.3   No Facility Is an Island

Facilities and equipment aren't islands (Fig. 14.7), isolated from the rest of the company. Facilities and equipment are closely related with other PLM components. They're also influenced by other forces within the company, and outside the company.

For example, if a new machine is implemented, it may be necessary to hire a new operator, or to train another operator about the new machine. Operator schedules may be need to be changed. In addition to the operator, it may be necessary to hire or train a process developer, hire or train maintenance staff, and hire or train for simulation. Product developers may need to be trained about the new machine.

product	product data	process
PLM applications	**Facilities and Equipment**	organisation
metrics	people	methods

Fig. 14.7 Facilities and equipment aren't islands

Fig. 14.8 Potential
challenges related to facilities
and equipment

lack of repeatability
under-used
unreliable
lack of appropriate skills
problems with cost-justification
under-achievement
lack of support from the supplier
lack of standards
difficult to integrate with other equipment
difficult to upgrade

The new machine may lead to changes in some steps of processes, so the process descriptions will need to be changed. Procedures may need to be rewritten. If the machine has a controller, and software, it may be necessary to get the IT department involved. The machine may need to be connected in a network. Documentation will have to be developed to show how the machine should be used and supported. A new application may be needed to create the control software. The new application may replace existing applications, so these will need to be retired. The data in the existing applications may be archived, or may need to be migrated to the new application. The new application may enable new working methods. These will need to be documented. New metrics may be mentioned in the cost justification for the machine. New measuring activities will be needed after implementation. Changes may be needed to organisation structures to create a new team to work with the machine.

14.4   The Challenges

The specific challenges that a particular company faces with facilities and equipment could come from several sources (Fig. 14.8).

14.5   The Way Forward

The issues described in this chapter all relate to facilities and equipment, but they can't be solved in isolation from the other components of PLM. Their solution, within the PLM environment, is addressed from Chap. 21 onwards.

Chapter 15
Metrics

15.1 Introduction

A metric is a quantifiable attribute of an entity or activity that helps describe its performance. It's something that can be measured to help manage and improve the entity or activity. Examples include Cost of Quality, Net Present Worth and Time To Market.

Like most things in the PLM environment, "metrics" are referred to by different names by different people (Fig. 15.1).

And like most things in the PLM environment, there are many metrics. Even in one small area of PLM, such as Engineering Change, it's easy to find more than a dozen metrics (Fig. 15.2).

A few of the many other metrics in the PLM environment are shown in Fig. 15.3.

Many metrics are commonly referred to by an acronym. Figure 15.4 shows some examples.

In total, there are thousands of potential metrics in the PLM environment. Many of them are related to quality, resources, and time. Many, such as Discounted Cash Flow, Internal Rate of Return, Return on Assets, Return on Capital, and Return on Investment are related to costs and revenues.

One of the challenges of PLM for a particular company is to identify the set of metrics that's most relevant to the activities on which the company wants to focus its efforts. The role, and the potential benefits and disadvantages, of each metric should be clearly understood so that it's possible to see if, and how, it could best fit into a PLM solution.

© Springer International Publishing Switzerland 2016
J. Stark, *Product Lifecycle Management (Volume 2)*,
Decision Engineering, DOI 10.1007/978-3-319-24436-5_15

key performance indicator (KPI)	key performance parameter	measure
measurement statistic	metric	performance measure
performance metric	performance rating	performance standard

Fig. 15.1 Some different names for metrics

monthly/annual numbers of approved, pending, processed, and rejected changes
number of changes resulting from errors made implementing previous changes
number of changes not related to products
percentages of Full Track and Fast Track changes
percentages of changes submitted by different groups
average time (days) taken to prepare an ECR
average time (days) taken to approve an ECR
average time (days) planned to implement a change
average actual time (days) taken to implement a change
average time (days) taken for a Fast Track change
average time (days) taken for a Full Track change
average cost ($) for a Fast Track change
average cost ($) for a Full Track change
percentage of approved changes not implemented

Fig. 15.2 Some metrics related to engineering change

number of new products per year	% of information on electronic media	cost of rework
number of times data is recreated	cost of IS as % of company sales	new product revenue
number of projects started per year	number of patents	parts count
number of projects completed per year	revenue per engineer	R&D spend
number of projects abandoned per year	spend on IS as % of earnings	Time To Market
number of defects per product family	number of product families	response time to RFQ
spend on product development	% of business processes defined	new product ramp-up time
ratio of PLM support staff to value-adding	% of products annually obsoleted	level of part reuse
difference between planned and actual	span of control	number of customers

Fig. 15.3 Examples of metrics in the PLM environment

ARR	BAPM	COQ	DCF
IRR	KPI	NPV	NPW
ROA	ROC	ROI	TTM

Fig. 15.4 Examples of metric acronyms

15.2   Characteristics

Each of the many metrics has its individual characteristics. Although some characteristics are specific to just one metric, many are common to many or all (Fig. 15.5).

15.2.1   Naming, Definition

Metrics are a system of measurement that characterises an entity. The entity could be an army, a company, a person or a product. It's often said that you can't manage what you can't measure, and without metrics it's difficult to describe an entity, set

Naming, Definition	Business Relevance	Current Value
Target Value	METRICS	Number, Priority
Balance, Consistency	Level	Reporting, Action

Fig. 15.5   Common characteristics of many metrics

targets, monitor progress, track results or fix problems. There are hundreds of possible metrics. They include the number of ships in a navy, the batting average of a baseball player, the revenues of a company, headcount, productivity and quantity of drawings.

Metrics need to be clearly and correctly named and defined. If they're incorrectly named, or if there's more than one definition of a metric, then people will be confused.

Metrics help an organisation to set targets for its annual improvement plans and to measure the progress that it is making. Metrics help a company to understand its performance, the performance of its competitors, and the behaviour of its customers.

15.2.2   Business Relevance

Many of the metrics, such as Return on Assets (ROA) and Earnings, that are a useful guide to company performance for the CEO, the business press and stock-holders, are less useful for PLM. For example, the company's assets may include the headquarters skyscraper, the value of which will probably be much more closely related to the state of the local market for office space than to the effectiveness of the PLM activity. Changes in Earnings may be more closely linked to fluctuations of the dollar on the foreign exchange market than to changes in the level of useful product functionality. Product-related metrics such as the product's revenues, and the costs associated with a product, are needed to manage various aspects of the product and the product environment.

The metric set is often biased towards what's easy to measure, even if this information is of little value. However, it's more useful to have just a few high-value metrics, even if they're hard to measure.

15.2.3   Current Value

For each metric there's a current value and there can be target values for the future. For example, the current value of the headcount metric could be 50 people and the target value in the year 2025 could be 100 people.

15.2.4   Metrics and Targets for PLM

PLM helps achieve improvements in many areas, such as Financial Performance, Time Reduction, Quality Improvement and Business Improvement.

In the area of Financial Performance, possible metrics and targets could be to increase the value of the product portfolio by 20 %, or to reduce costs due to recalls, failures and liabilities by 75 %.

In the area of Time Reduction, possible metrics and targets could be to reduce time to market by 50 %, or to reduce engineering change time by 80 %.

In the area of Quality Improvement, possible metrics and targets could be to reduce defects in the manufacturing process by 25 %, or to reduce customer complaints by 50 %.

In the area of Business Improvement, possible metrics and targets could be to increase the rate of introduction of new products by 100 %, or to increase product traceability to 100 %.

15.2.5   Metrics and Targets Example

The following text is from the 1993 Annual Report of Varian Associates, Inc. It illustrates some of the metrics being used. It shows the past, current and target values.

> Our goal remains to make Varian a world-class organisation that delivers world-class results. Although we are by no means at that stage - yet, noteworthy progress is being made. The following examples illustrate how we are moving ahead with each of the five key concepts that drive our strategy to achieve Operational Excellence.

> Customer Focus. On-time delivery has moved up from around 50 % in 1989 to nearly 90 %, with several of our operations now routinely delivering everything on time.
> Commitment to Quality. Our cost of quality (scrap, rework, warranties, etc.) continues to fall and is now approaching 10 %, a quality cost saving of well over $20 million annually over the past two years.

> Fast, Flexible Factories. Cycle time in our factories has been slashed by two-thirds, from around 150 days to an average of 50 days.
> Fast Time to Market. Product development cycles have been compressed on the order of 50-60 %, reducing a process that often took years down to a matter of months in many cases.

> Organisational Excellence. Our organisations are flattening out, replacing slow-moving bureaucracies with more flexible structures where only one or two layers separate a business unit's general manager from its manufacturing employees on the line.

In the example above, Varian Associates selected five key metrics: on-time delivery; Cost of Quality; cycle time in the factory; product development cycle; and organisational structure. Another organisation could work with the set of metrics and targets shown in Fig. 15.6.

Another organisation might work with the metrics and targets shown in Fig. 15.7.

Metric for area to be improved	Target Value
hierarchical layers of management	4 to 6
span of management	10 to 20
amount of work to be cut out	25%
size of operating unit	less than 500 employees
size of corporate staff	cut by 70%

Fig. 15.6 A set of five metrics

Metric	Current value	Target value
product families	2 families	3 families
product generations	1 generations	3 generations
cross-functionality	2 functions involved	5 functions involved
development time	18 months	6 months
juggled projects	4-6 projects at a time	1-2 projects at a time
cost of quality	25%	5%
project start-up time	2 weeks	2 days

Fig. 15.7 A set of eight metrics

15.2.6 Number and Priority

Different organisations use different numbers of metrics. Some only use one or two metrics, others many more. If few metrics are used they generally include headcount and budget. The number of metrics used by the organisation should be kept as small as possible. A set of between eight and ten metrics is often appropriate. This may mean eliminating some metrics that look interesting. However, it will focus attention on those that remain.

Metrics should be prioritised so that it's clear which are the most important. A metric that has a priority as low as 20, even if it does look interesting, is going to be much less useful than one with priority 1 or 2. Maintaining it in the set of metrics will result in wasting time and money on measuring and targeting it. And it will take the focus off the most important metrics.

15.2.7 Balance and Consistency

It's relatively easy to identify metrics. Usually, the problem is not so much how to identify them, but how to ensure that they're correctly measured, consistent, generally agreed, and correspond to company requirements and objectives.

The targets must be consistent. They mustn't conflict. This means going into a lot of detail to understand the objectives and potential implications of each target. Although this will take time and effort, it will be cheaper in the long run than rushing unthinkingly into unwise implementation.

15.2.8   Level

Once targets start to be proposed, it will become apparent that they're at different levels and address different time-scales. There'll be high-level business targets, functional targets, product and process targets, and information system targets. There'll be a mixture of long-term, medium-term and short-term targets.

Metrics will be needed to address the various levels and types of activity in the product lifecycle, and the different phases of the lifecycle. For example, they'll be needed at the level of the product development organisation and the product support organisation. They'll also be needed for individual projects and people. The budget is an example of a metric that can be applied at the level of a development or support organisation. The budget (in dollars or as a percentage) can be compared with that in other companies.

15.2.9   Reporting and Action

It's relatively easy to identify metrics. It's usually more difficult to set up the approach for capturing actual values.

Once values have been captured, it's relatively easy to report them. It's usually much more difficult to take the appropriate action.

15.3   Improvement Projects

Every year, companies have the opportunity to invest in a variety of new and on-going short-term and long-term projects. These could include introducing new products, improving manufacturing productivity, developing the corporate image, improving working conditions, improving processes and implementing new applications. Someone, somewhere, has to select the most suitable projects. Different projects will require approval from different levels of management. In general, the less money and time the project needs, the lower the level at which approval is given.

An Engineering Manager may be able to start a small-scale project in the Engineering Department with funds from the departmental budget. On the other hand, it's unlikely that a project that's intended from the outset to be enterprise-wide will be funded this way. Instead, its sponsors will probably need to apply for business development funding, or for the right to use departmental funds cross-functionally. This means that the project will effectively be in competition for funds with projects from all parts of the company, and that the funding decision will be taken by top management.

Top management has a difficult task in choosing which projects to fund. Most of the projects will appear very important. They will often involve a large initial investment, have a major effect on the company in the long-term, and have the potential for creating major upheavals. Top management is unlikely to understand the projects in detail, so will be heavily influenced by the people proposing projects, and the written proposals. Even if a proposal appears to be very profitable, it may lose out to a proposed project that appears even more attractive. Even if it's presented very well, it may lose out to a proposal that's presented even better. The project may not be accepted because top management has a poor opinion of some of the project team members.

In some companies, because it's so difficult to get a project accepted, people exaggerate the benefits of their proposal. This can be dangerous in the PLM environment. If expectations are raised too high when the project proposal is made, and can't be achieved in practice, top management is likely to punish the proposal team and/or the people who try to implement the project. Although this is true for all projects, it seems particularly relevant for large-scale PLM projects. These tend to have cross-functional aspects, involve changing the way people work, and address numerous skeletons in cupboards that haven't been opened for years (such as product structures, classification systems, product development work practices, and inter-departmental interfaces).

When a PLM project proposal is presented to management, it should contain a financial justification that shows the required investment and running costs, the expected benefits, the expected return, the risks associated with the investment, and the effect of the investment on other areas of the company. Without such a justification, top management will be unable to decide either if the project is worthwhile, or if it's a better choice for investment funds than other projects.

15.4 Project Justification Vocabulary

Some basic project justification vocabulary is needed. It includes cash flow, book value, net income, discounting, tax, income, payback, net present value, return on investment. A company's annual report is a good place to start. Its vocabulary of revenues, earnings and income is familiar.

Somewhere at the back of the annual report are the financial statements. Usually there are three important sets of figures. These are the Income Statement (or profit and loss account), the Balance Sheet, and the Statement of Source and Application of Funds (or Statement of Changes in Financial Position, or Cash Flow Statement).

The Income Statement doesn't give a cash view. It's prepared on the basis of matching the annual sales and expenditures of the company. Matching is a standard accounting practice used to try to give a fair view of what's really happened in the year under review, and to remove effects that really belong to other years. Generally it's fairly easy to identify sales made during a year. It's more difficult to identify the costs incurred during the year. If an expensive machine or system has been bought during the year, and is expected to be used over five years, then it wouldn't be correct accounting practice to include the total cost of the machine in the year's profit and loss figures, with nothing included in the following years. Correct accounting practice would be to match some fraction of the machine's cost (known as the depreciation) against sales in each of the five years. Depreciation is an annual charge representing the gradual inclusion of a one-off investment in a project over the project's lifetime.

As a result of the matching process, there's a difference between the situation (known as the book view) as it appears in the Income Statement, and the reality of the company's finances (the cash view). If the equipment cost $10,000 in 2014, then the cash view would identify a cash outflow of $10,000 in 2014. On the assumption that the company is making an outright purchase of the equipment (the calculations are different for leasing and rental), it pays 100 % of the price in the first year, and will pay nothing in the following years. The book view, though, would see a depreciation charge of $2000 in each year from 2014 to 2020. In different ways, each view is correct in particular circumstances. You just need to remember whether you're dealing with a book view or a cash view.

The financial statements of a company show the cost of sales. For a manufacturing company, this is primarily made up of material expenses and production-related expenses. The financial statements also show the net income. This is (very roughly speaking) the result of subtracting the cost of sales, administrative expenses, depreciation, tax, interest on loans, and other expenses, from the annual revenues. As the calculation of net income involves subtracting depreciation from net sales, it's a book value. In the example above, $2000 a year would be subtracted from the annual sales for five years to represent the depreciation of the machine.

Since depreciation is subtracted from net sales before the tax charge is calculated, it affects the tax paid. The more that can be subtracted in the way of depreciation, the lower the tax charge. To prevent abuse of this deduction, the amount of an investment that can be depreciated, and the period of depreciation, are governed by law. Various methods of depreciation are allowed, but once a company has chosen a particular method, it's expected to retain it for consistency in the presentation of its results. Computer software is often depreciated over 3 years. Under the straight-line depreciation method, the amount of depreciation is constant each year. For software, it would be one-third of the price each year for 3 years.

As the net income shown on the Income Statement is calculated after making allowance for depreciation, and different companies allow for depreciation in different ways, the real meaning of the net income is not immediately obvious to a layperson. To make things a little more transparent, the Statement of Source and Application of Funds was introduced. This looks at the change in the company's cash position over the year. It shows the Source of Funds (cash inflows such as sales and interest income), and the Application of Funds (cash outflows such as taxes paid and the purchase of fixed assets).

Depreciation doesn't appear on the Statement of Changes in Financial Position. Going back to the example above, all $10,000 of the cost of the machine would be included in the Application of Funds of that year, and nothing in the other years.

For financial evaluation of individual projects, it's better to work with a cash view than a book view, and to be able to see the exact movements of cash into and out of the project.

Cash inflows resulting from the project are primarily increased sales. There may also be some tax credits to consider. Cash outflows include payments for equipment, labour, sundries, operating expenses, and taxes.

The difference between the total cash inflows resulting from a project and the total cash outflows is called the cash flow of the project. A positive cash flow for a project indicates that, over its lifetime, cash inflows resulting from the project will exceed cash outflows.

15.5   Time Value of Money

As the effects of the PLM project will be spread over several years, it's necessary to take account of the "time value of money". This term describes the fact that $100 received this year doesn't have the same value as $100 received in previous or future years. This is best explained by an example. Assume that $100 received this year can be invested at a fixed 10 % annual interest rate. Then, in one year, it will be worth $110, and in five years it will be worth $161.051. It will be worth $61.051 more than $100 received in five years' time. The time value of money has to be considered when calculating the costs and benefits that occur in different years of a PLM project.

With a 9 % interest rate, would $147 received in 5 years' time be worth more than $124 received in three years' time? The answer isn't immediately apparent because there are so many variables. To make the values in project calculations easier to compare, all future values are converted to their value in the present (Year 0 of the project). This technique of working backwards from a future time at which a value is known, to find its present value, is called discounting. Often, the cost-justification of a project will look at performance over five years. The initial investment in the project is assumed to take place in Year 0. The resulting cash flows take place in Year 1, Year 2, Year 3, Year 4 and Year 5.

In the above example, it was seen that $100, invested at a fixed 10 % annual interest rate, will be worth $161.051 in five years. When discounting, the calculations are carried out the other way round, starting from a known future value, and calculating the equivalent present value. For example, assuming the same constant 10 % annual interest rate, $100 received next year has a present value of $90.90, and $100 received in two years has a present value of $82.64. Similarly, with this discount rate, $165.01 received in five years' time has a present value of $100.

To calculate the present value of cash inflows of a five-year project, an expected interest rate has to be chosen. This rate will be used to discount the cash flows. It's often called the discount rate. For example, a company might decide that a project should generate a rate of return of 20 % per year.

If the expected annual cash flows received at each year end are $110,000 (Year 1), $130,000 (Year 2), $160,000 (Year 3), $220,000 (Year 4), and $280,000 (Year 5), their present values are calculated by discounting them by 20 % to give $91,666 (Year 1), $90,278 (Year 2), $92,593 (Year 3), $106,096 (Year 4), and $112,526 (Year 5). At this 20 % discount rate, there's a big difference between the sum of the present values of the cash inflows ($493,159) and the sum of the undiscounted flows ($900,000).

A similar calculation could be carried out for the cash outflows of the project. If these are expected to be $700,000 (Year 0), $302,000 (Year 1), $340,500 (Year 2), $377,000 (Year 3), $416,500 (Year 4), and $354,000 (Year 5), their present values, also calculated by discounting them by 20 % would be $700,000 (Year 1), $251,667 (Year 1), $236,458 (Year 2), $218,171 (Year 3), $200,858 (Year 4), and $142,265 (Year 5).

Once the discounting calculations have been carried out for cash inflows and outflows, it's easier to see whether the project is going to meet management requirements. The calculations are simple. It's much harder to develop reasonable values for the cash inflows and outflows. Before coming to that subject though, there are a few other standard calculations to understand.

Several methods are used to express, in understandable and comparable, terms the profitability of different projects. Four of them will be outlined: Accounting Rate of Return (ARR); Payback time; Net Present Value (NPV); Discounted Cash Flow Return On Investment (ROI).

The ARR is obtained by expressing, as a percentage, the ratio of the accounting income (including depreciation) generated by the project to the total investment. This is a quick calculation, but its inclusion of depreciation means that it's not all that useful from the project cash-flow point of view.

Payback time is the time required for a project's revenues to equal the cash outlay. If the investment in the project is $1,000,000 and the annual revenue is $400,000, then the payback time will be 2.5 years. Payback is a quick and dirty calculation. It doesn't take account of the time value of money, or of revenues occurring after the payback period. It gives a quick, approximate feeling for a project's viability.

15.6   NPV and ROI

The Net Present Value (NPV) of a project at any given time is calculated by subtracting, from the investment, the sum of the discounted cash flows up to that time.

$$NPV = -I + \sum_{t=1,n} \frac{(Rt - Ct)}{(1 + DR)^{**}t}$$

where I = investment in Year 0; n = project lifetime in years; Rt = revenue in Year t; Ct = costs in Year t; DR = discount rate

For example, if I = \$1000, DR = 20 %, n = 2 years, R1 = \$850, R2 = \$1050, C1 = \$250, C2 = \$300 then

$$NPV = -1000 + 500 + 520.83 = \$20.83$$

When the NPV is positive, the discounted cash flows are greater than the initial investment, so the project is earning more than the discount rate in use (in this case more than 20 %). If the NPV turns out to be negative, then the project is making less than the discount rate in use.

In some cases, particularly when making comparisons between several competing projects, it may be enough to know the NPV. In other cases, it may be more useful to know the exact return of the project (also known as the Internal Rate of Return, or the ROI). This is the discount rate that corresponds to the Net Present Value of the project being equal to zero, i.e. the investment is exactly equal to the sum of the discounted cash flows. Once again, a simple calculation is all that's needed. Putting NPV = 0 in the above equation, and solving for DR,

$$I = \sum_{t=1,n} \frac{(Rt - Ct)}{(1 + DR)^{**}t}$$

so DR = 21.6, i.e. the rate of return of the project is 21.6 %.

Even when all the cash flows of a project have been identified, and the ROI calculated, questions of the type "but what if…?" still remain. Sensitivity analysis and risk analysis try to answer them. Sensitivity analysis identifies the items that critically affect the project calculations. Risk analysis provides a range of possible values for the outcome of the project, rather than a single value.

Sensitivity analysis is used to look at each cash flow item individually, and answer the question "what would be the effect on the project's ROI if all other items have been estimated correctly, but this particular one over-estimated or under-estimated by x percent?" Each item can be checked in this way, and usually it's found that there are a few items that have much more influence on the ROI than the others. For example, a 10 % variation in one item may lead to a 10 % change in the ROI, whereas the same variation in another item may only lead to a 1 % change. When the analysis has been carried out, the items that have the most influence (are the most sensitive) should be re-examined to make sure that they're based on as reliable and accurate information as possible.

Risk analysis is carried out to estimate the probability that the ROI will be met. One way of doing this is to assign probabilities to expected values for each cash flow item. Thus, instead of assuming that the value for a particular cash flow item will be $7000, it could be estimated that there's a 5 % probability that it will be $5000, 10 % that it will be $6000, 70 % that it will be $7000, 10 % that it will be $8000, and 5 % that it will be $9000. Similar probabilities could be calculated for the other items. The ROI would then be calculated as a function of these probabilities. The result would show the range of values for the ROI, and the probability associated with each value.

15.7   Cost Justification

There are three major areas of cost-justification to describe. These are costs, benefits, and the overall approach. They will be addressed below in the specific context of a PDM system.

Identification of the costs associated with a project is usually not too difficult. Costs are generally divided into initial investment costs (Year 0) and on-going costs (Year 1, and following years). Typical initial investment costs are shown in Fig. 15.8. Typical costs in following years are shown in Fig. 15.9.

Apart from the costs of PDM directly related to the PDM system, there may be other, more indirect costs (Fig. 15.10).

| initial investment in the PDM system (software) |
| initial investment in complementary software (if required) |
| initial investment in hardware for the PDM system (if necessary) |
| initial investment in communications hardware (if necessary) |
| initial investment in communications software (if necessary) |
| awareness training and education |
| consultancy costs |
| system selection costs |

Fig. 15.8 Typical initial investment costs

further investment in the PDM system (software)	PDM system (software) maintenance costs
further investment in complementary software	complementary software maintenance costs
further investment in PDM system hardware	further investment in communications hardware
further investment in communications software	communication charges
costs for customising the PDM system	loading data in the PDM system
PDM system management and operations	cleaning product data
development of new working procedures	development of interfaces
on-going training and education	modification of existing procedures
participation in conferences and user groups	on-going consultancy

Fig. 15.9 Typical sources of costs after the initial investment

| cost of retraining personnel |
| cost of headcount reduction |
| cost of restructuring product information |
| cost of feasibility studies |
| prototyping |
| planning |

Fig. 15.10 Costs not directly related to the PDM system

| labour costs |
| quality costs |
| costs of introducing new products |
| costs of modifying existing products |
| field support costs |

Fig. 15.11 Some costs that could be reduced with PDM

| reduce the number of product developers in the process |
| reduce the number of product development support staff |
| reduce the cost of materials used in products |
| reduce the cost of quality |
| reduce energy consumption in the process |
| reduce finished stocks and work in progress |
| reduce the cost of holding products |
| reduce the cost of storing information |
| reduce warranty costs |
| reduce penalty costs |
| reduce rework costs |
| reduce documentation costs |

Fig. 15.12 Benefits that result in a reduction in costs

In addition to the above costs, there's another set of costs that needs to be understood. These are the costs of doing business without PDM. Reduction of these costs is a source of potential benefits. Some of these costs are shown in Fig. 15.11.

The figures for these costs, in the absence of PDM, should either be known, or can be estimated fairly easily. However, the situation is different for the benefits expected from the PDM project.

There are only two types of benefits possible, those that result in a reduction in costs (Fig. 15.12) and those that result in increased revenues. The five costs mentioned above correspond to the first category.

15.8   Identification of Benefits

The major difficulty in PDM cost-justification is the identification of the benefits. These can be thought of as having a multi-level structure with the overall business benefits at the top level, and the more detailed benefits at the lower levels. The overall business benefits result from the addition of all the detailed benefits. At the highest level of the structure is the "overall business benefit resulting from PDM". It may be expressed in broad terms such as "increased gross margin", or "improved competitive position". For a particular company, it might be expressed more quantitatively as "gross margin increased by 1 %", or "we will become the most competitive player in the make-to-order market". To calculate the "overall business benefit resulting from PDM", the "costs of PDM" have to be subtracted from the "benefits of PDM".

One of the major objectives of PDM cost-justification is to convince top management that it makes sense to invest in PDM. Most top managers will be interested

| increase the number of customers. For example, by reducing the overhead activity of product developers, PDM allows them to develop products for new customers |
| increase the product price paid by customers. For example, by increasing product quality, PDM enables customers to be charged increased prices |
| increase the range of products that customers can buy. For example, by improving product structure management, PDM enables more customer-specific variants |
| increase the number of products of a particular type that a customer buys. For example, by increasing product quality, PDM allows customers to dispense with second sourcing |
| increase the percentage of customers re-ordering. For example, by increasing product and service quality |
| increase the frequency with which customers buy. For example, by getting products to market faster and more frequently |
| increase the service price paid by customers. For example, by using PDM to improve the quality of existing services |
| increase the range of services that customers buy. For example, by using PDM to support additional services |
| get customers to pay sooner. For example, by developing and delivering products faster |
| sell surplus capacity. For example, with much of the overhead burden removed, product developers will have more time available for value-adding tasks. This may be sold to other companies |

Fig. 15.13 Some ways in which PDM can increase revenues

primarily in the "overall business benefit resulting from PDM", not in the low-level benefits. If they see a clear overall benefit, then they will probably at least consider the investment. Of course, in a particular case, there may be reasons (such as lack of funds) why they don't make the investment. Most top managers will rely on lower-level managers to see that the calculations leading up to the "overall business benefit resulting from PDM" have been carried out correctly. They'll expect finance managers to have checked the mathematics of the calculations, and, to a certain extent, they'll assume that the finance managers agree with whatever figures have been used in the calculation. The "overall business benefit resulting from PDM" results from simple calculations carried out on "costs of PDM" and "benefits of PDM". These calculations may be carried out in spreadsheets.

In addition to the "costs of PDM" and the "benefits of PDM", the "time value" of costs and benefits has to be considered in the calculation. The "time value of money" plays an important role in PDM cost-justification. Cost-justification of a PDM project takes account of costs and benefits over several years. As a particular cost (or benefit) may appear as $100, or $110, or $121, or $133.10, depending on the timing assigned to it, there's plenty of scope for massaging the figures. The "overall business benefit resulting from PDM" resulting from the calculations can be improved by delaying "costs of PDM", and bringing forward "benefits of PDM".

The "benefits of PDM" can be divided into two parts. One of these is the "increase in revenues resulting from the introduction of PDM". The other is the "decrease in costs due to the introduction of PDM".

At the top level, there are about ten ways in which PDM can increase revenues. These are shown in Fig. 15.13.

15.9 Project Calculations

The benefits and costs occur over several years, and the results of the calculations will probably be reported in several tables. In Fig. 15.14, the annual "benefits of PDM" and "costs of PDM" are not discounted (for example, the $1 million benefit

All figures in $M (undiscounted)						
	Year 0	Year 1	Year 2	Year 3	Year 4	Year 5
benefits of PDM		1.0	1.0	1.0	1.0	1.0
costs of PDM	2.0					
annual cash flow	-2.0	1.0	1.0	1.0	1.0	1.0
cumulative cash flow (undiscounted)	-2.0	-1.0	0.0	1.0	2.0	3.0

Fig. 15.14 Undiscounted costs and benefits

All figures in $k (10% discount rate)						
	Year 0	Year 1	Year 2	Year 3	Year 4	Year 5
benefits of PDM		909.0	826.4	751.3	683.0	620.9
costs of PDM	2000.0					
annual cash flow	-2000.0	909.0	826.4	751.3	683.0	620.9
cumulative cash flow (discounted)	-2000.0	-1091.0	-264.6	486.7	1169.7	1790.6

Fig. 15.15 Discounted costs and benefits

entered in Year 1 is the actual benefit expected in that year.) The Net Present Value is $M 3.0. The payback time is 2 years.

In Fig. 15.15, the annual costs and benefits are discounted to take account of the decreasing value of money. A discount rate of 10 % has been used. The same $1 million benefit is still expected in Year 1, but its discounted value ($909 thousand) is now used in the calculation. The Net Present Value is $k 1790.6. The IRR is 41.05 %.

In these calculations, the initial investment is made in Year 0. The initial benefits occur in Year 1. All benefits are assumed to occur at the end of the year in which they appear. The benefits that appear in Year 1 are assumed to occur at the end of Year 1, which is why they're discounted.

The Cumulative cash flow row shows the cumulative cash flow at the end of each year.

The Net Present Value shows the value (in Year 0) of all the discounted costs and benefits of the project in the time period examined. When the Net Present Value is negative, the rate of return of the project is less than the discount rate chosen for the calculation. When the Net Present Value is positive, the rate of return of the project is more than the chosen discount rate.

The Internal Rate of Return (IRR) is the discount rate that corresponds to the Net Present Value being zero. When the NPV is zero, the discounted benefits are equal to the discounted costs. For the project shown in the above table, it can be calculated that for the Net Present Value to be zero, the discount rate has to be about 41.05 %.

This is illustrated in Fig. 15.16, where the discount rate used was 41.05 %, and the resulting Net Present Value is $k −0.5, which is close to zero.

In Fig. 15.14, the payback time is visible. The payback time is the time it takes for a project's benefits to equal the initial investment. With undiscounted values, the payback time for this project can be seen to be two years. In Fig. 11.5, where discounted values are used, the payback time is not visible. But it can be estimated to be about 2.3 years. The payback time increases as the discount rate is increased. This is because the benefits are increasingly devalued, so take longer to equal the costs. Usually, the payback time is only quoted for the simplest case (illustrated in Fig. 15.14), that of undiscounted costs and benefits.

The IRR is often quoted. Sometimes it's also referred to as the ROI (Return on Investment).

To make the calculation showing the discounted cash flow calculation more complete and understandable, the row of undiscounted benefits of PDM can be included (Fig. 15.17).

The benefits of PDM can be of two types. The first type is an increase in revenues resulting from the introduction of PDM. The second type is a decrease in costs due to the introduction of PDM. To make the calculation more complete, these two types of benefit can be introduced as separate rows. Figure 15.18 results from assuming that half the benefit of PDM comes from an increase in revenues, and the other half comes from a decrease in costs.

The calculation can be further detailed by including the individual components of the "Costs of PDM" and the "Benefits of PDM".

All figures in $k (41.05% discount rate)						
	Year 0	Year 1	Year 2	Year 3	Year 4	Year 5
benefits of PDM		708.9	502.6	356.3	252.6	179.1
costs of PDM	2000.0					
annual cash flow	-2000.0	708.9	502.6	356.3	252.6	179.1
cumulative cash flow (discounted)	-2000.0	-1291.1	-788.5	-432.2	-179.6	-0.5

Fig. 15.16 Discounted costs and benefits, NPV close to zero

All figures in $k (10% discount rate)						
	Year 0	Year 1	Year 2	Year 3	Year 4	Year 5
benefits of PDM (undiscounted)		1000.0	1000.0	1000.0	1000.0	1000.0
benefits of PDM		909.0	826.4	751.3	683.0	620.9
costs of PDM	2000.0					
annual cash flow	-2000.0	909.0	826.4	751.3	683.0	620.9
cumulative cash flow (discounted)	-2000.0	-1091.0	-264.6	486.7	1169.7	1790.6

Fig. 15.17 Inclusion of undiscounted benefits

All figures in $k (10% discount rate)						
	Year 0	Year 1	Year 2	Year 3	Year 4	Year 5
increase in revenues due to PDM (undiscounted)		500.0	500.0	500.0	500.0	500.0
reduction in costs due to PDM (undiscounted)		500.0	500.0	500.0	500.0	500.0
benefits of PDM (undiscounted)		1000.0	1000.0	1000.0	1000.0	1000.0
benefits of PDM (discounted)		909.0	826.4	751.3	683.0	620.9
costs of PDM	2000.0					
annual cash flow	-2000.0	909.0	826.4	751.3	683.0	620.9
cumulative cash flow (discounted)	-2000.0	-1091.0	-264.6	486.7	1169.7	1790.6

Fig. 15.18 Further extension of the calculation

15.10 Benefit Asset Pricing Model

Benefits fall into different classes such as directly quantifiable benefits, indirectly quantifiable benefits, and synergy effects. In the Benefit Asset Pricing Model (BAPM), published by Schabacker in 2001, benefits of these different classes are built into a BAPM Portfolio. The portfolio of benefits can then be treated similarly to a portfolio of capital market investments. For example, directly quantifiable benefits behave similarly to cash investments, while indirectly quantifiable benefits behave more like stocks. The BAPM method uses similar methods and procedures to those used for the risk limitation of capital market investments to evaluate future performance. Expected performance levels are calculated from previous implementations, and appropriate levels of risk applied. BAPM was originally used for the evaluation of benefits from CAD/CAM systems and PDM systems. It has since been used, more generally, for the evaluation of the benefit yield of other PLM project investments.

15.11 No Metric Is an Island

No metric is an island (Fig. 15.19), isolated from the rest of the company. All metrics are closely related with other PLM components. They're also influenced by other forces within the company, and outside the company.

For example, in one company, at the beginning of the year, the scoring scheme for the bonus of the Manufacturing VP was changed to depend on the monthly output and the production scrap rate. As a result, he changed the scoring scheme for the bonuses of his managers and team leaders to depend on the monthly output and the production scrap rate. And he told them that, to achieve the maximum bonus, they had to avoid scrap and minimise down-time.

They analysed performance in previous years, and found that one of the two main sources of scrap was the start-up of new products. The other main source was a group of troublesome products that, for unknown reasons, always ran into problems.

As they continued the analysis in more detail, they noticed that, prior to market release of a new product, a sign-off was needed from both the Manufacturing VP and the Manufacturing Line Manager for the new product. This hurdle had been put in place a few years previously to ensure that Engineering didn't rush new products to customers before everything was ready.

product	product data	process
PLM applications	**Metrics**	organisation
equipment	people	methods

Fig. 15.19 No metric is an island

unbalanced	ineffective
Inconsistent	not clearly defined
lack of documentation	not enough
not relevant	too many
lack of training	multiple interpretations possible
poor reporting	lack of follow-up
difficult to relate to performance	difficult to measure

Fig. 15.20 Potential metric-related challenges for a company

As they continued analysing they found, in a management report, that a new plant, set up to serve emerging markets, was running at low output.

Over the following months, sometimes the Manufacturing VP had reasons not to give approval for a new product. At other times, a Manufacturing Line Manager had reasons not to give approval. As a result, the scrap from new products dropped to zero.

As for the troublesome products that often led to scrap, they were transferred to the new plant. The new plant had no experience of these products. They did their best, but the result was that scrap rates were even higher.

No metric is an island. In this case, a small change to the way that personal performance was measured led to no new products getting to market and to increased scrap.

15.12 The Challenges

The specific metric-related challenges that a particular company faces could come from several sources (Fig. 15.20).

15.13 The Way Forward

The issues described in this chapter all relate to metrics, but they can't be solved in isolation from the other components of PLM. Their solution, within the PLM environment, is addressed from Chap. 21 onwards.

Chapter 16
Organisation

16.1 Introduction

An organisation is an arrangement of resources, within a particular environment, to achieve a specific objective.

Like most things in the PLM environment, "organisation" can be referred to by different names by different people (Fig. 16.1).

And, as with most things in the PLM environment, there is a wide variety of organisational types (Fig. 16.2).

And, as with most things in the PLM environment, there are many different resources to organise in the PLM environment (Fig. 16.3). And many related activities.

Previous chapters have described components of the environment for developing, manufacturing and supporting products across the lifecycle. They include applications, metrics, people, products, product data and techniques. To make all the resources and activities manageable, organisational structures are defined.

For each of the resources there are several alternative approaches. There are many ways the resources can be organised in response to the questions "what do we want to be?", "how should we organise our resources to best develop and support products?" and "how do we implement the corresponding change?"

16.2 Characteristics

Each of the resources that has to be organised has its individual characteristics. Although some characteristics are specific to just one resource, many are common to several (Fig. 16.4).

© Springer International Publishing Switzerland 2016
J. Stark, *Product Lifecycle Management (Volume 2)*,
Decision Engineering, DOI 10.1007/978-3-319-24436-5_16

arrangement	composition	configuration
form	formation	interrelation
makeup	order	ordering
organisation	structure	system

Fig. 16.1 Some different names for organisation

cell	functional	hierarchical	matrix	pyramid
departmental	geographical	hybrid	project	team
flat	group	line	product-focused	virtual

Fig. 16.2 Different types of organisation

applications	equipment	facilities
methods	metrics	people
products	product data	processes

Fig. 16.3 Some of the resources to be organised in the PLM environment

Scope	Complexity, inter-relations	Identification
Detailing	ORGANISATION	Documentation
Communication	Implementation	Responsibility

Fig. 16.4 Common characteristics of organisation of different resources

16.2.1   Scope

The boundaries of the resources being organised are important. They need to be defined clearly. It's important to make sure that nothing within the scope is ignored. It's important to avoid trying to organise resources outside the boundaries. That's likely to cause problems with the owner of such resources.

16.2.2   Complexity, Inter-relations

All of the components of PLM, including business processes, people, product data, working methods, information systems, interfaces and standards, are necessary to transform ideas into products which meet a company's objectives, meet customer requirements, comply with regulations, and meet environmental objectives. PLM is holistic. Taken singly, none of those components can completely manage a product, but taken together, they achieve the objectives. The whole is greater than the sum of the parts.

There are many resources to manage, and high volumes of many of these resources. And, as if the wide scope and high volumes didn't make it difficult enough, there are complex and changing relationships to manage between products, components, customers and suppliers.

The processes for defining, making, using and supporting a product can be complex, expensive and time-consuming. They may contain several thousand tasks. These may be carried out by many different companies working in different places round the world. All these tasks can be linked in many ways. There's an infinite number of ways of organising and carrying them out in the product lifecycle.

Hundreds, or even thousands, of people may be directly involved in these tasks. In processes that involve hundreds or thousands of people and tasks, it's not uncommon to find that no one person understands the overall process. Each person has a detailed understanding of their own role and tasks, but no-one is able to describe the overall process in any detail.

Just as there are many tasks and relationships to organise for a process, there are also many entities and relationships to manage for each of the other components of PLM. In addition, there are relationships between the components. For example, if the product structure is changed, the data structure may change, and the layout of machines may change.

16.2.3   Identification

There are many ways to organise the components, and many alternatives are possible. It's important to identify the right organisation for each resource, both individually and as part of the overall set of resources.

There are many ways that each of the resources can be organised. For example, the components of the product have to be organised and described in a product structure. Many structures are possible. Product data has to be organised, and relations between data elements maintained. The machines that produce products can be organised in different layouts.

Information systems may be in-house or outsourced. They may be in a Cloud. They may be distributed or centralised, discrete or integrated. Tasks and activities can be organised in different ways into processes. Activities and processes can be organised in different ways on different sites. Processes may be sequential or concurrent. Many will be carried out in-house, some may be outsourced. They may be lean and taut, or poorly-organised and tangled. There may, or may not, be well-defined project phases with strict milestones, gates and phase reviews. Product development may be restricted to assemblies, with components being outsourced. Assemblies may also be outsourced.

Activities such as sales, installation, support and recycling may be in-house or outsourced. Sales and support may be web-based or traditional. The approach to product development may be the glamorous mega-project, or the less exciting but more reliable process of continuous incremental improvement. Services may be released with the initial product, or with future versions. Development may be primarily by individual functions, or it may be multi-functional. The practices to be used in product development may be the traditional ones of literature survey and physical prototype, or they may embrace all the most modern technologies and practices.

Human resources can be structured in different ways. People may be structured by task, skill, project, product, client, geography, distribution channel, or type of equipment used. People may be organised in hierarchical or flat organisations. They may be organised and work in teams, as individuals, or as departments. They may be geographically distributed near clients, or work at a centralised location. Everyone may be employed by the company, or some may be employed by contractors.

16.2.4 Detailing

Once a general approach to the organisation has been decided, there's a lot of work to define the details.

16.2.5 Documentation

Once the details have been worked out, the organisation has to be documented. The documented organisation can have different names, depending on the resource being organised (Fig. 16.5).

Among the organisations for products are product structures and bills. For product data, the organisation can be documented in a data model. For processes, there are process maps and flow diagrams.

The organisation of people can be documented in an organisation chart (Fig. 16.6), which is a control-structure model. A company organisation chart can

architecture	diagram	list	picture	report
bill	form	map	plan	structure
chart	graph	model	register	table

Fig. 16.5 Different ways of documenting the organisation

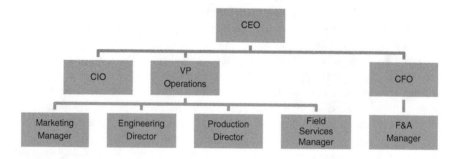

Fig. 16.6 Organisational chart

show the various departments of a company, the groups within the departments, the teams within the groups, and the individuals within the teams. An organisational chart is a model. It shows the structure of an organisation, its main components, and the relationships between them.

16.2.6 Communication

Once the organisation has been detailed and documented, the next step is to communicate it.

16.2.7 Implementation

And once the organisation has been detailed, documented and communicated, the next step is to implement it.

16.2.8 Responsibility

In some cases, the responsibility for organising the resources is clear. In other cases, it's much less clear. For example, few people in a company can decide about the type of organisation chart shown in Fig. 16.6, and everybody knows who they are. However, when it comes to organising some other resources, such as product data, the responsibility is usually much less clear. In some cases, there may be no overall organisation of the resource and no responsible person.

16.3 Changing Environment and Organisation

Creating the organisational structure is one of the most challenging tasks in PLM. However, the environment for products at the beginning of the 21st century is different from that in previous centuries. PLM is a new approach and differs from previous approaches. There aren't yet proven PLM organisations that can be replicated.

Each epoch has its own requirements, structures and figureheads. Some lessons can be learned from the past.

The first part of the 19th century saw great engineers such as Isambard Brunel (1806–1859), apparently able to engineer any individual "product" single-handed, whether it be a bridge, ship, railway, tunnel or hospital.

The second part of the 19th century saw great industrialists such as Andrew Carnegie (1835–1919) and John D. Rockefeller (1839–1937). When he was 13, Carnegie's family emigrated from Scotland to Allegheny, PA where he started work in a cotton mill. After opening his first steel plant in 1875, he went on to control 25 % of the US steel industry. In 1901 he sold his vertically integrated business to J. P. Morgan for $480 million and became the richest man in the world. One of the world's first great philanthropists, he established thousands of free public libraries to enable self-education. Rockefeller went into the oil refining business in the early 1860s. He recognised that an efficient vertically-integrated oil company would beat the multitude of poorly-organised small and medium-sized competitors. By 1868 he headed the world's largest oil refinery. From 1870 when he created Standard Oil, he bought up competitors and consolidated oil-refining. By 1878, Standard Oil controlled about 90 % of US refining capacity. When Standard Oil was broken up in 1911, his fortune stood at nearly a billion dollars.

Carnegie and Rockefeller worked in process manufacturing industries. In the early 20th century came the great industrialists, such as Henry Ford (1863–1947), in discrete mechanical manufacturing industries. In 1927, Ford's Rouge River complex was the most fully integrated car manufacturing facility in the world. It had its own power plant, blast furnaces, steel mill, and glass plant. The efforts of tens of thousands of manual workers were organised minute by minute. This was the time of Frederick Winslow Taylor (1865–1915), the father of scientific management.

The second half of the 20th century saw great industrialists in electronic and software manufacturing. Maybe the future will choose Bill Gates of Microsoft and Larry Ellison of Oracle as the figureheads for the software industry. And perhaps Bill Hewlett and Dave Packard of Hewlett-Packard for electronics manufacturing, and Tim Berners-Lee for the Web.

The second half of the 20th century was also the time of W. Edwards Deming (1900–1993) and a focus on statistics and quality. In 1986 he published "Out of the Crisis", which described his 14 Points for Management, and led to Total Quality Management. The 14 points are as applicable now as they were then, but like Taylor, Deming worked at a time before computing and communications technology were available, omnipresent and understood to have such wide-ranging applications for manufacturing industry.

The second half of the 20th century was also the time of C. Northcote Parkinson and Laurence Peter. In his 1958 book, "Parkinson's Law", Parkinson described how time is wasted, non-value adding work expands, and organisations add non-value adding employees and unnecessary levels of hierarchy. In his 1969 book, "The Peter Principle; why things always go wrong", Peter showed how organisations advance employees to their highest level of competence, and then promote them to a level at which they are incompetent.

Figure 16.7 shows characteristics of early 20th century organisational structures that were successful for industrialists such as Ford.

Figure 16.8 shows characteristics of early 21st century organisational structures. This is the environment for PLM.

vertical integration
huge corporations with tens of thousands of manual workers
many layers of management
a focus on "metal-bashing"
functional specialisation

Fig. 16.7   Characteristics of early 20th century organisational structures

horizontal integration (across the Extended Enterprise)
many small and medium companies
few layers of management
fast evolution
a focus on electronics, software and information
small numbers of knowledge workers from different functions working together in teams

Fig. 16.8   Characteristics of early 21st century organisational structures

16.4   No Organisation Is an Island

No organisation is an island (Fig. 16.9), isolated from the rest of the company. Organisations are closely related with other PLM components. They're also influenced by other forces within the company, and outside the company.

Changing the organisation of one of the components of PLM can affect many other components. Changing the organisation of processes, for example, will lead to changes in documents describing the processes. This may lead to a need to train people working in the processes. Changing the organisation of the processes may lead to some activities being supported by a different application, which may have to be modified to support additional tasks. A change in the organisation of processes may lead to changes in the way some performance indicators are measured.

16.5   The Challenges

The specific organisation-related challenges that a particular company faces could come from several sources (Fig. 16.10).

product	product data	process
PLM applications	**Organisation**	metrics
equipment	people	methods

Fig. 16.9   No organisation is an island

each resource organised independently
some resources not organised
some resources without owners
objectives for some resources are not clear
responsibilities for some resources are not clear
lines of authority for some resources are not clear

Fig. 16.10   Potential organisation-related challenges for a company

16.6   The Way Forward

The issues described in this chapter all relate to organisation, but they can't be solved in isolation from the other components of PLM. Their solution, within the PLM environment, is addressed from Chap. 21 onwards.

Chapter 17
Reasons for Implementing a PDM System

17.1 The PDM System in the PLM Environment

In today's high-tech companies, product development and support wouldn't work without computers and a wide range of applications. In practice, though, the applications are often discrete and poorly integrated. In many ways, they create as many problems as they solve. PDM systems are the tool that can link them together, and make sure that they fit together well in support of the overall process.

By improving the control of product information and activities, PDM systems help reduce lead times, reduce costs and improve quality.

Reduced lead-times open up new market opportunities and improve profits. They also reduce market risk by reducing the time between product specification and product delivery. The sooner that customers use a product, the sooner their feedback can be incorporated in a new, improved version. If quality is improved, not only will customers be pleased, but there will be a reduction in scrap, recall and rework. Corresponding administrative activities, and their costs, will be reduced.

PDM will improve product development productivity. Product development managers will know the exact design status. They'll be able to assign resources better, and release designs faster and with more confidence. Design engineers will know which parts are available and which procedures should be followed when designing new parts. Manufacturing engineers will be able to see how similar parts have been made previously. Everyone will be able to rapidly identify approval mechanisms.

Concurrent Engineering and Simultaneous Engineering, with their emphasis on product teams made up of individuals from different departments and even different companies, and parallel working on processes that were previously carried out in series are only feasible in a well-controlled product development environment, with a high availability of product information and rapid communication of information between individuals. PDM offers a solution. Design For Manufacture techniques require up-front involvement of Manufacturing personnel in the design process.

© Springer International Publishing Switzerland 2016
J. Stark, *Product Lifecycle Management (Volume 2)*,
Decision Engineering, DOI 10.1007/978-3-319-24436-5_17

PDM helps different groups share information and communicate effectively. Many manufacturing companies are trying to focus on the aspects of their products or services where they excel or where their value-added is highest. This implies partnerships or contractor relationships with other companies that will work on the other parts of the product lifecycle. These relationships in turn require efficient information transfer and management between the partners, and close control over projects involving several partners. Again, PDM systems help meet these requirements.

17.2   Two Classes of Reasons

Anyone implementing a PDM system is likely to be faced by people who don't want to change the status quo. There are lots of people in most companies who can provide all sorts of justifications for not doing whatever is proposed. To prepare for such behaviour, anyone who wants to implement PDM needs to be well-armed with reasons. Fortunately, there are plenty of them.

The reasons for implementing a PDM system can be divided into two classes. In one of these, the PDM system appears to alleviate some of the problems that occur in the product development environment. In the other, it appears to pro-actively and positively impact operations across the product lifecycle. Although these two classes of reasons can be treated separately, in practice, they're closely related.

The first class of reasons includes those in which the PDM system is used to overcome currently existing problems. Many of these reasons are related to the rapidly increasing amount of data in the product development environment. There is a general trend towards product development operations becoming more infor-mation—intensive, and a corresponding push to use information—driven processes to support overall business activities. The increased computerisation of product development activities leads to an increase in the volume and availability of product data. Without PDM systems this information can't be managed effectively.

Many companies scan all their engineering drawings into a digital archive. They can use a PDM system to manage the large volumes of data that result, and to control access to, and use of, scanned data. Similarly, to support the growing number and type of CAD activities, a PDM system can be used to help solve the problems arising from the vast number of part files and versions that can so easily be produced. If a company uses several different CAD applications and has to transfer data to suppliers with other systems, the product data exchange task can easily get out of hand. A PDM system can help regain control.

In many companies, there is a low level of integration between the different applications in the product lifecycle. Applications often appear as separate "Islands of Automation", and a lot of time and effort is wasted in managing information in such an environment. It's difficult to control the exchange of product data and the synchronisation of activities between different applications in the absence of well-defined and properly enforced procedures.

A related problem that arises is that the same product data is defined and modified in different parts of the organisation, with the result that there is no agreement as to the correct version. As a result of poor product data management, product developers waste time finding out which version of a part they should work with, or discovering which is the latest version of an existing part. Designs will take longer than expected because so much time is spent looking for product data, and checking that the right version is being used. Sometimes, the wrong information will be passed on to other people in the organisation with the result that their time is wasted, and in some cases, parts are made that will eventually have to be scrapped. A key area here is the interface between the engineering organisation and the manufacturing organisation. Unless the product data release process is under control, time and effort will be wasted. Poor product data management results in wasted resources. PDM systems provide the opportunity to use these resources better.

17.3 Information Management

As there are many reasons in the "information management" category it's best to divide them into three sub-categories: basic data control and management; data access and advanced information control and management.

This categorisation is based on user behaviour and requirements. The first category of reasons, "basic data control and management", contains the reasons likely to appeal to the person responsible for the data resource. This person may have a different title in different companies, such as "data controller" or "document control manager". Assume for a moment that the activity of this person is mainly a fairly bureaucratic control function of making sure that the data is in its place, that it's secure, that it's archived, that transactions are logged, etc. The major responsibility and interest of this person is to ensure control of the data. The data controller is not expected to be interested in the content and use of the data.

The second category of reasons, "data access", contains the reasons likely to appeal to a user of the data, for example, a product developer. The requirements of this user will differ from those of the data controller. The engineer is not interested in issues such as the archiving of other people's data, or the storage of different types of data by people in other functions. The engineer's prime concerns are to be able to find required data very quickly, and to be sure it will be there when next required.

In this simplistic categorisation, it's assumed that "basic data control and management" only addresses the basic requirements of allowing data to be stored, retrieved, released, and archived in a secure environment. These requirements are virtually content-independent. Similar requirements exist for storage of data in other functions such as marketing and sales.

Other control and management requirements fall into the "advanced information control and management" category. This category includes the need to handle relationships and linkages between data, to standardise data formats, and to ensure the unique definition of data elements. Meeting these requirements demands knowledge of the content of the data, which is why they're put in a separate category from basic data management and control, which is content-independent.

In each of the three categories there are many reasons for doing PDM. In each category it appears that for some of the reasons, PDM is a way to alleviate a problem that occurs in the product development environment, whereas for other reasons, PDM appears to be a way to proactively and positively impact activities across the product lifecycle.

17.3.1   Basic Data Control and Management

The reasons in this category address many issues (Fig. 17.1).

In the absence of PDM, people will find many faults with the situation (Fig. 17.2).

the many versions of product data	the sheer volume of data describing products
the large number of users of product data	prevention of access to unauthorised users
the need for back-up and archival of data	maintenance of data required for legal reasons
integrity of data	the different and changing states of product data
security and confidentiality of data	maintenance of data on products with long lives
access for authorised users of data	the different sizes, types, formats and media of data
release of product data	traceability of data, and traceability of access to data

Fig. 17.1 Reasons related to basic data control and management

we can't manage the rapidly growing mass of product data with our traditional manual system. We need a more rigorous and powerful system to do it for us. Often data is misplaced, lost or unintentionally destroyed. Sometimes people can't find the data they need.
when everything was on paper it was relatively easy to manage. Now we have all sorts of data on all sorts of media. Electronic data, such as CAD files and the programs we put in our products, can't be managed with a paper-based system. It must be managed electronically.
there are so many versions of data in existence, and it takes so long to keep track of them manually, that often people get the wrong version. As a result, all sorts of problems occur. Sometimes the result may be that the wrong part or product goes to a customer.
there are so many users of data, and they come from so many different departmental organisations, that sometimes we get mixed up and send them the wrong data.
our manual product data management system is so slow that we don't know the up-to-date status of information. We don't know the current status of projects. We generally know what the status was about 3 weeks ago, but don't know anything more recent because the information is still working its way through the system. Project managers aren't aware of slippage and other problems as soon as they occur, so they can't address them promptly.
we can't be sure that confidential data isn't leaking out to competitors. Sometimes we don't know where data is, so we can't control it. Sometimes we transport it in ways that we can't control.
sometimes someone will unintentionally overwrite someone else's computer files. With manual practices, it's difficult to implement an effective, yet user-friendly, system that will allow us to set up the correct access rights and privileges to product data.
there's so much data and so many people creating, using and changing it, that we can't control it. Even if we only had one problem in every thousand data transactions we'd still have far too many problems each week.

Fig. 17.2 Issues in the absence of PDM

There are many reasons in the "basic data control and management" category why PDM can proactively and positively impact activities across the product lifecycle (Fig. 17.3).

17.3.2 Data Access

The reasons in the "data access" category address issues such as availability of data and access time to data. There are several reasons why PDM can alleviate problems with data access and availability (Fig. 17.4).

There are several reasons in the "data access" category why PDM can positively and proactively impact activities across the product lifecycle (Fig. 17.5).

17.3.3 Advanced Information Control and Management

Reasons in this category address issues such as: the definition of data elements; standardisation of data; and relationships and linkages between data elements.

a PDM system will allow us to get control of our voluminous and disparate product data. It will allow us to save our engineering know-how, and treat product data as a valuable company resource.
we'll be able to control all the different versions of data, and even if, as expected, customisation of new product lines results in more variants and versions, we'll be able to handle them as well.
under certain very well-defined conditions we'll be able to manage the use of unreleased data, and support concurrent engineering activities.
we'll be able to provide a quality service to users throughout the company. We'll be able to serve the data requirements of customers and suppliers.
with a PDM system, we'll have up-to-date status information on product data, so the product development managers will be able to take better decisions. We'll be able to give them better reports on data usage.
with the PDM system, we'll be able to control access to the data. We'll make sure someone doesn't accidentally destroy other people's information. We'll be able to prevent unauthorised access, and keep confidential data away from people who are not supposed to see it.
we'll be able to reduce the amount of paper in circulation.
we'll be able to make sure we meet legal and regulatory requirements.
if anything goes wrong, we'll be able to track back to see what happened.
once the PDM system is up and running, it will do a lot of the work automatically, so we'll be able to reduce the number of people involved in data management and control

Fig. 17.3 Reasons why PDM impacts activities across the lifecycle

with a manual system it can take a long time to get the required data. Perhaps the person who knows where to find it is unavailable, or the data has been misplaced, or lost. Sometimes it takes hours to track information through different files and listings. Because it can take so long to access data, people keep their own unofficial copies. They may not be informed when the official copy is modified, and continue working on a version that is no longer up-to-date.
data access is so slow that people may prefer to use unofficial communication channels. To overcome an urgent problem, a change may be sent directly from one person to another, bypassing the official system. Due to the pressure of work, the change may never be entered into the official system.
with a manual system it takes a long time to get data. Data is sent through the internal mail, or by the postal service. There is a limited number of deliveries each day, and if there's a mistake in the address, it may be several days before the data gets to its destination.
with a manual system, it can take as long to get an answer to a simple question as it does to receive a large volume of data. The system was set up to manage a particular type of data record (such as a drawing), and takes just as long to handle requests for more simple information structures.

Fig. 17.4 Reasons why PDM alleviates problems with data access

with the PDM system, we won't have to waste time waiting to receive data, so we'll be able to reduce lead times.
with the PDM system, the data will always be available. We won't be limited by the unavailability of particular individuals. We'll be able to get on with our work without having to wait, so we'll be able to reduce lead times.
with the PDM system, if we have a simple query we won't have to wait days to get the answer. We'll make queries on-line, and then take the corresponding action. We'll save time and reduce lead times.
data access with the system will be so easy that we'll be able to cut out a lot of administrative paper-shuffling. This will help save time and also reduce overhead costs.
we'll be able to get the right information to the right people in the right place at the right time.

Fig. 17.5 Reasons why PDM positively impacts activities across the product lifecycle

There are several reasons in this category that are related to problems arising in the absence of PDM (Fig. 17.6).

And there are reasons why PDM can positively and proactively impact activities across the product lifecycle (Fig. 17.7).

in the current situation, with manual data management, a given data element is often defined in different ways by people in different departments. When the data is transferred, there is a need to reconcile the different definitions and to re-enter data. This process wastes time and is a source of error.
data in different computer applications is often in different formats. When it's transferred from one application to another, it has to be converted. When this is done manually it takes a long time. Time is wasted and errors may be introduced. During the transfer, some of the information content may be lost. Errors slip through, and may not be discovered for weeks, during which time many incorrect actions and decisions may be taken on the basis of the erroneous information.
in the current situation, individuals and groups define data relationships independently. When data is transferred, the relationships may be lost, or may be misunderstood, or may have to be rebuilt. As a result, errors are introduced, and time is wasted. Examples of the relationships range from complex product structures that are defined in different and uncoordinated ways by different departments, to simple relationships between sets of data files used by particular individuals.
configuration documentation may no longer correspond to the actual product. Unexplained differences arise between as-designed, as-planned, and as-built structures. Incomplete products are delivered. Field problems are difficult to resolve as the product does not correspond to the documentation. New product introduction is delayed by a large number of small problems. Product quality varies, lead times get longer and longer, and product development costs rise.

Fig. 17.6 Problems arising in the absence of PDM

the PDM system will help track simple relationships such as those between specifications and design data.
the PDM system will help relate the view of the product structure that is used by one department to that used by other departments.
the PDM system will offer classification functions and intelligent search functions which will speed up development cycles.
the PDM system will manage the transfer of product data from one application to another, reducing the time this takes, and eliminating conversion errors.
the PDM system will provide a unique definition of each data element. People will not need to waste time and money recreating the definition.
the PDM system will provide information about the product data it manages. People will not need to waste time and money re-inventing data that already exists.
by automatically managing and providing information on linkages and relationships, the PDM system will reduce the time and effort to access and use this type of information.
by automatically providing information on product and process structures, the PDM system will reduce lead times, errors and costs. It will improve the quality of products and services.

Fig. 17.7 Ways in which PDM can positively impact activities across the lifecycle

17.4   Re-use of Information

We'll save time and money by re-using information.

One reason for re-using information is to avoid the cost and the time of repeating a task that has already been carried out. When information is created, it goes through a process of specification, development, production, test, modification, use, storage, etc. The process usually includes several quality checks to ensure that the information is correct. It usually includes decision points where choices are made as to the best way to proceed. The overall process takes time, and costs money. At the end of the process, the information exists and is correct. If exactly the same information is created again from scratch, the process just described will be repeated. It will take about the same time to create the information as it did the first time, and the process will also cost about the same. However, instead of recreating the information from scratch, the information created the first time can be recalled and re-used. This information is known to be correct and usable. The process of recalling it can take very little time and money. It's much quicker and cheaper to re-use information than to re-create it.

If an existing part can be re-used there is no need to go through the wasteful process of re-designing the part, re-designing the process to make and support it, simulating performance, re-developing tools, waiting for prototypes, etc. By re-using an existing part, time and money is saved, and quality is guaranteed. Also, since a new part number is not needed, overhead is not increased, there is no need to extend part files, there is no need to find additional storage space, there is no need to create additional documents, and there is no need for additional working capital for holding costs. Because there are fewer parts, it's easier to locate existing parts, so access times can be kept acceptable. If creation of a new part can be avoided, time and money are saved.

When we re-use information we'll have fewer quality problems.
The first time that information is created it goes through a long process involving a lot of quality checks and validation. It is then used, and any remaining errors are uncovered and corrected. By the time someone wants to re-use information, it can be expected to be more or less bug-free. Of course, this implies that the information that is to be re-used should have gone through a demanding test cycle so that its re-users will not be disappointed. It should also be kept in secure conditions so that it's not inadvertently modified. Any intentional modifications should be clearly documented.

We'll be able to speed up our development cycle.
As re-use of existing information doesn't require any development time, since everything has already been developed, it can reduce development cycles.

We'll be able to improve our engineering change statistics.
Whether the engineering change metric in use is the number of engineering changes, or engineering change cycle time, re-use of information will have a positive effect. As the information has already been used, and can be expected to be correct, there should be no need for engineering change.

We'll be able to reduce scrap. We'll be able to reduce the number of prototypes.
By re-using information that has already been tested and used, the need to carry out
further tests is reduced. As the information exists and is well-defined, the corresponding parts should be perfect and there should be no scrap.

We'll be able to reduce our engineering costs.
By re-using some existing designs, less investment will be needed for the development of new designs. As there will be fewer designs, and fewer changes to
designs, it will be possible to reduce the overhead costs associated with managing,
storing, copying and communicating designs.

We'll be able to make improvements in many areas.
Costs and time cycles will be reduced in all areas where information can be re-used.
All sorts of information can be re-used. Some of this information will define parts,
some will define the processes to make these parts, and some will define the
relationships that link parts to products. Sometimes these parts will be mechanical
or electronic components, sometimes they will be software.

We'll be able to get the productivity gains promised from CAD.
Use of CAD was expected to improve productivity. Many of the benefits were
expected to result from the re-use of data, but it has been difficult to achieve these
benefits. Once CAD information has been created and used, it's often stored and
forgotten. It's difficult to re-use it. Even if the basic geometry file can be found, it's
difficult to find all the related information. Rather than waste time on what is nearly
always a fruitless search, most people just don't bother to try to re-use CAD data.

We'll be able to reduce data re-entry cost and time.
Currently a lot of time is wasted at those points in the product lifecycle where data
is exchanged between different applications. For example, some data from a program on one computer may be processed in a PC spreadsheet before being used in a
program on another computer. Often the data transfer will be done with a print-out
from the first computer, keyboard entry to the PC, a print-out from the spreadsheet,
and keyboard entry to the second computer. Along the way, time is wasted and
errors are introduced. More time is wasted as the data is checked to make sure no
transcription errors get through. With a PDM system, and electronic interchange,
the transfer will be instantaneous and error-free.

We'll be able to schedule more accurately.
A lot of information about the time and cost of design and manufacturing processes
is never re-used, yet this information should be the basis for planning the costs and
schedules of the processes for similar parts and products.

We'll be able to reduce the overall cost of the product.
The development cycle will be shorter because existing information will be re-used.
Design and verification time will be reduced. Less engineer and overhead cost will
be incurred. There will be less need for simulation and less scrap. Re-use of existing
information with known characteristics should allow for a more optimised design.

These individual benefits will all contribute to a reduction in the overall cost of the product.

We'll be able to support our business strategy.
As companies move to product focused businesses, they will re-use more and more parts in their products. The ability to re-use information underlies the ability to re-use parts.

Those companies seeking to differentiate through customisation will need to re-use information supporting the common parts used in customer-specific products.

We'll be able to support concurrent engineering.
Some companies find that their information management capabilities can't support attempts to carry out concurrent engineering. Each department has its own way of defining the same data element so departments can't work together in parallel. Instead, each department has to wait until the preceding departments have finished their work and handed over the information they have created. This information is then converted, translated or reformatted so that it fits the way the department is used to using it. Then the department does its work and, if it finds a problem, sends information back upstream where it's translated back to the original form. It's impossible to work concurrently in this environment. With the PDM system it will be possible for people from different departments to work together in a concurrent engineering team. They can use, share and re-use data from the applications and databases of the various phases of the product lifecycle.

We'll be able to work better with suppliers and customers.
Currently, information transfer with suppliers and customers is often very slow. To maintain security and confidentiality, each drawing and document is produced on paper, signed-out, and mailed. This process wastes time, inhibits close relationships with suppliers, and annoys customers. A PDM system will manage security and confidentiality aspects, and send information directly by EDI (Electronic Data Interchange).

We'll be able to re-use data to trace the source of a problem.
Without a PDM system, some data may get lost or mislaid somewhere in the product lifecycle. Often, there's so much data, that some of it isn't properly recorded. If related faults are ever found in the product, it can be extremely difficult and expensive to create the missing data. For example, one automobile manufacturer didn't fully record all the information about a key component. Later, the supplier found a problem with a small batch of the component, and told the manufacturer which components to recall. The manufacturer hadn't maintained the link between component and car, so had to unnecessarily recall thousands of cars. A PDM system will make sure that information can always be traced and related to particular parts and products.

17.5   Workflow Management

Make sure the most appropriate process is followed
The workflow management module of a PDM system automatically routes product development tasks and information round the organisation in a predetermined way. Provided that the predetermined way is the most appropriate way, the workflow management system will make sure that the most appropriate process is followed.

Clean up product development
To identify the most appropriate way to carry out product development, the organisation will need to analyse what is currently being done. When it does so, it will probably find a lot of bottlenecks and duplicate, unnecessary and non-value-adding activities. These can be removed, saving time and money.

Get control of product development
Once the process has been analysed, and the most appropriate layout of tasks has been identified, the organisation will, at last, have the opportunity to take control of the process, and make sure that things are done properly.

Improve resource utilisation and project co-ordination
With the process clearly defined, and under control, the workflow module of the PDM system can select which people should receive tasks at a particular time. For example if there are three people capable of doing a four-day low-priority task and two of them are busy on higher priority tasks for the next week, while the third person will complete a medium-priority task tomorrow, the system can assign the four-day task to this person for the day after tomorrow. The workflow module ensures the tasks get done in the right order by the most suitably qualified people available without continually rescheduling the entire work plan for the rest of the project. It can also ensure that those who do not need to be involved in a task are allowed to focus on their work without unnecessary disturbance. The productivity of individuals will increase, as will overall product development performance.

Improve project visibility and status information
The workflow management module controls the flow of work in the project, so can provide information as to the status of individual tasks and the overall project. This will allow developers and managers to take better decisions.

Provide an audit trail
The workflow module controls the flow of work in a project, so it can provide information as to the tasks carried out on the project. This information could include, but not be limited to, the name of the person who carried out the task, when they started working on it, who they received it from, when they stopped working on it, etc.

Provide security
By only sending work and information to people who have the right to receive it, the workflow module can restrict access to selected individuals.

Provide the basis for an automated product development environment
By automatically controlling the flow of work, the workflow control module provides part of the necessary infrastructure for an automated environment. This could be the first step towards a very fast, low-cost and effective automatic product development activity.

Reduce the cost of product development
By automating the overhead activities of project scheduling and task distribution, the workflow module enables a reduction in overhead product development personnel.

Reduce product development cycle time
Workflow management will save time in many ways. Product developers will not sit around with nothing to do while they wait for work to get to them. The system will make sure that all relevant information is sent to an individual when a task is assigned. As soon as one task is finished, the module will inform the person responsible for the next task that they can start work. Apparently small savings of this type add up to significant benefits. These time savings translate into cost reduction, faster cycle times and improved productivity.

Reduce rework
By making sure the right things happen at the right time, workflow control leads to a reduction in scrap and rework.

Provide the benefits of electronic information flow
One of the functions associated with workflow management is electronic transmission of information. In its absence, traditional methods of distribution would be used. The transfer of information, on paper, to a distant site might take several days. The electronic workflow system only requires a few minutes. This fast transfer time allows for interactive and responsive working modes between geographically distant sites. It saves time, cuts costs and provides efficient and error-free communication.

Reduce the volume of paper in circulation
Information can be sent across the network in electronic folders, reducing the need to put information on paper and then produce and distribute multiple copies.

Make sure that information is available when needed
The workflow management module will know when a task is going to start, so it will be able to assemble the information associated with the task well in advance of task initiation. This will help reduce the time that would otherwise be lost by only finding out after task initiation that all necessary information is not available.

Bring geographically distant engineers together
The workflow management module will be able to work across national barriers. It will also be able to work across departmental and corporate barriers. Once the most appropriate work routing has been determined, tasks and information can be sent anywhere on the network. Distant team members can work together. Concurrent

Engineering team members working in different locations and different countries can be supported.

Help reduce the number of face-to-face project meetings
With better control of the product lifecycle, and electronic transfer of information across the network, there will be less need for face-to-face status meetings. A one-hour meeting of eight people uses up one day of project time. Reducing the number of administrative meetings will make more time available for value-adding activities.

Ensure required procedures are followed
Predetermined streamlined sequences for sign-off, release and change processes can be set up and used. Special sequences can be developed to permit other activities such as partial release under certain conditions.

Help manage the engineering change process
In some companies, engineering changes account for about half of the time and cost of product development. Anything that can help increase the visibility, or reduce the cost, of engineering change will save money.

Automate uninteresting procedures
The workflow module can ensure that standard repetitive procedures such as archival and back-up are carried out automatically.

Improve the reliability of product development schedules
Based on the results and statistics of previous projects and tasks, the workflow module will be able to develop optimised schedules for forthcoming projects.

Improve communication
The messaging capabilities of the workflow module can be used to keep all project team members informed of what is happening, and what is about to happen, on the project.

Make sure that things are done the right way
In addition to ensuring that the predetermined flow of work is carried out, the workflow module can also ensure that the right practices, applications and design principles are used.

17.6   Engineering Change Management

We'll get better status information on our changes
Our manual engineering change management system is so cumbersome we don't know the up-to-date status of changes. They may number several hundred for a project. Usually we know what the status of a change was about a month ago, but don't know anything more recent because the change is still working its way through the system. With the PDM system, if we have a simple query we won't

have to wait days to get the answer. We'll cut it back to less than a day. We'll be able to make queries on-line, and then take the corresponding action, thus saving time and money.

We'll be able to reduce the costs of change management
Once the PDM system is up and running, it will do a lot of the overhead activities associated with change management automatically, so we'll be able to reduce the number of people involved in change management and control. We'll be able to cut out a lot of administrative paper-shuffling. That will help save time and reduce overhead costs.

We'll be able to make sure the most appropriate change process is followed
Once the change process has been analysed, and the most appropriate layout of tasks has been identified, the organisation will, at last, have the opportunity to take control of the change process, and make sure that things are done properly. The change management module of a PDM system can automatically route change information round the organisation.

We'll improve change resource utilisation and co-ordination
With the process clearly defined and under control, the change module of the PDM system can select which people should receive tasks at a particular time. The change management module ensures the tasks get done in the right order by the most suitably qualified people available without continually rescheduling the entire plan. It can also ensure that those who do not need to be involved in a change can focus on their work without unnecessary disturbance. The productivity of individuals will increase, as will overall product development performance.

We'll have an audit trail of change activities
The change module of a PDM system controls the flow of work, so it can easily provide information as to the tasks carried out on the project. This information could include, but not be limited to, the name of the person who carried out the task, when they started working on it, who they received it from, when they stopped working on it, etc. If anything goes wrong, we'll be able to track back and see what happened.

We'll reduce cycle time for engineering changes
PDM will improve the change process in many ways. The system will make sure that all relevant information is sent to each individual as soon as possible. As soon as one task is finished, the module will inform the person responsible for the next task that they can start work. If someone doesn't respond within a certain time, the system will pass the task to the most appropriate person. Apparently small savings of this type will add up to give significant benefits. These time savings should translate into cost reduction, faster cycle times and improved productivity.

We'll get the benefits of electronic information flow
One of the features of the change management module of a PDM system is electronic transmission of information. In its absence, traditional manual methods of

information distribution would be used. The transfer of information, on paper, to a distant site, might take several days. The electronic change system only requires a few minutes. This fast transfer time allows for interactive and responsive working modes between geographically distant sites. It saves time, cuts costs and provides efficient and error-free communication.

Use of PDM will bring geographically distant people together to work on changes
The change management module will be able to work across national barriers. It will also be able to work across departmental and corporate barriers. Once the most appropriate change routing has been determined, tasks and information can be sent anywhere on the network.

We'll be able to make better change management plans
The PDM system will provide information about past changes, such as how long they were planned to take, how long they took, and where problems occurred. It will also provide information about the future availability of resources. Good information about the past, and about the future, are what we need to make good plans.

We'll be able to make more reliable estimates of change costs and times
For a particular change, we'll be able to refer back to similar changes to see what kind of resources were needed, how much time they took, how much they cost, and the extent to which they followed the plan. We'll use this information to provide better estimates for a change.

Use of PDM will make changes easier to manage
Automated reporting on the progress of a change is faster, cheaper and less liable to suffer from transmission and transcription errors than manual reporting. With the right information available, changes will be easier to manage.

We'll be able to run changes in parallel and get them done quicker
With PDM, we'll know what's happening on a change, and we'll be in control. We'll do things that were too risky without PDM because we weren't sure of the status of different changes, we didn't know how much progress had been made, and weren't sure of the value of information until a change was released. With PDM, status will be clear, even down to the level of individual sub-tasks and components of the change.

We'll be more responsive to change
The PDM system will keep us up-to-date with change progress, so if a change is needed to a change it will be easy to respond. We'll know the exact status of the change. We'll know what needs to be done. We'll decide what to do and then send the new plan out over the network.

17.7   Overall Business Performance Improvement

We'll be able to increase revenues in all sorts of ways
PDM will allow us to make better proposals which will lead to more sales. Product developers won't waste their time on data retrieval and management activities, so they will have more time available for value-adding, revenue-generating activities. By improving product structure management, PDM will enable more customer-specific variants, so we will be able to increase the range of products that customers can buy. PDM will help us increase product quality. As a result, we'll be able to increase the price paid by customers for products and services. By increasing product quality, PDM will allow customers to dispense with second sourcing. This will allow us to increase the number of products of a particular type that we can sell to a customer. Many companies lose about 10–15 % of their customers each year through poor quality service. By increasing product and service quality, PDM will help us increase the percentage of customers re-ordering. By helping us get products to market faster and more frequently, PDM will increase the frequency with which customers buy, and so increase revenues. By allowing us to develop and deliver products faster, it will help us get customers to pay sooner. This will improve our cash flow.

Use of PDM will help reduce costs
PDM will help us reduce costs all the way along the product lifecycle. The PDM system will help us reduce the number of errors we make, so we'll be able to reduce unwanted engineering changes. We'll reduce scrap, rework, erroneous order and manufacture of parts, obsolete parts, penalty costs, warranty costs, recall parts, and product liability costs.

We'll increase profits
By helping us increase revenues and reduce costs, PDM will help us increase profits.

We'll have more customers
PDM will allow us to increase the range of products we can develop and support. The extended range will attract new customers.

PDM will help us to offer better customer service
PDM will support all phases of product development and support from the identification of customer requirements to the provision of excellent field service. PDM will allow us to get the right information to people when they need it, enabling them to provide better, on-time customer service. The PDM system will allow us to increase the range of services available to customers. With much of the overhead burden removed by the PDM system, service engineers will have more time available for responding to customers. At the front end, PDM will help us manage customer requirements, and product and process specifications. It will help us make better use of existing knowledge, and re-use existing parts. Out in the field, PDM

will provide maintenance engineers with the information they need to support customers' products, so they'll be able to respond better.

We'll be able to communicate better with suppliers
The PDM system will allow us to transfer information to suppliers quickly and cheaply. We'll use electronic information transfer instead of the slow, expensive and error-ridden communication of paper-based information.

We'll increase productivity
With PDM, we'll reduce overhead costs and non-value-adding activities. We'll increase output, yet our costs will decrease. The result will be a gain in productivity.

PDM will stimulate re-engineering of our processes
Implementation of PDM will force people to address product workflow and the use of product data in the product lifecycle. Once they start looking at the product lifecycle in detail they'll see it's extremely ineffective. They'll joke about the silly wasteful things that go on. There'll soon be pressure from management to re-engineer the entire product lifecycle with the aim of achieving major performance improvement.

17.8   Resolution of Business Problems

We'll be able to satisfy customers
The PDM system will allow us to get our product lifecycle under control. From the moment we have the customer specifications, all the way through to the moment we retire the product, the PDM system will have all our product data under control. It will make sure the Voice of the Customer is heard all the way through the process.

PDM will make sure the product doesn't break down when it gets to the customer
We have had all sorts of configuration problems and quality problems because our manual document management systems don't work properly. For example, people have worked with the wrong version of a replacement part and then sent it off to the customer. The customer has tried to use it, and of course it doesn't fit. The customer gets really annoyed and threatens to sue. With PDM we'll be able to control our product data and our configurations, so we won't have this type of problem.

PDM will allow us to reduce product prices
With PDM, product development and support won't take so long, so won't burn so many hours of effort. PDM will allow us to cut out all the time usually wasted while people wait to get the information they need. With the PDM system in place, we'll no longer have to pay a large group of people to manage information manually and to transport it from one person to another. All these reductions in our costs will be reflected in lower prices for our customers.

PDM will allow us to get products to market faster

Once we've cut all the waste out of the process, and the PDM system is getting the right information to people when they need it, the development cycle will be much shorter. We'll take less time to get products to market. From the moment a design is agreed, much less time will pass before the product is in the customer's hands. For our consumer products this means the products will be much more fashionable, their specifications taking in all the latest trends that our market researchers have sniffed in the wind. It means our industrial products will be more hi-tech and modern, benefiting from all the most recent technologies.

PDM will allow us to increase market share and increase profits

PDM will allow us to reduce costs and prices, and get really up-to-date products to the market. Customers will love them. They'll buy our products rather than those of our competitors. Our market share will increase. We won't reduce prices quite as much as we reduce costs, so everyone will be happy. The customer will pay less, and we'll increase our profit margin. We'll increase our market share in what looks like a fairly stable market, so we'll be selling more products. With more profit per product and more products being sold, our profits will leap ahead.

PDM will allow us to customise products profitably

In the past, some customers have asked for customised products. Of course we answered "Yes, we can customise, but it may cost you more, maybe 5 or 10 %". They are happy to pay a little more for something that is exactly what they want, so go ahead on that basis. The problem is that we can't control our costs. Even if the customer doesn't horse around changing the specs we overrun our cost limits. Our design engineers change things without thinking of all the related parts and inter-faces that will change and the downstream activities that will result. We often finish up over-spending by 50 %. There's no way the customer will pay that. With PDM, things will change. People will be able to foresee the consequences of their actions, and we'll make money on customisation.

PDM will allow Field Service to provide customers with better service

Our field personnel don't get on too well with headquarters. They claim the doc-umentation from Engineering is useless as information is missing, and drawings and lists don't correspond to the products on customer sites. Due to the poor service, Field Service built a library of all drawings, and developed a set of as-built, as-installed, and as-maintained lists. Because they don't have a maintenance budget for this activity, the end result is chaotic. However, once we have the PDM system it will take over management of all data, drawings, and other documents, and will make available the information that people need when they need it. Our Field Service people have portable PCs, and they'll get up-to-date information as and when they need it.

PDM will help us deliver products on time

Our product development and support applications aren't integrated. Design work gets done on CAD Unix workstations, but the project planning application runs on a

PC, so doesn't know what progress has been made. Our internal mail service isn't computerised, so once a document has been mailed, no-one knows where it is, or when it will arrive. There are lots of unknowns in the system, so people have to do a lot of guesswork. Quite often the project manager will have the feeling the project is behind time so will reorganise the remaining tasks. The resulting pandemonium has to be seen to be believed and the end result is always that we overrun our schedules. With PDM this won't happen. The system will know the status of each activity and the location of each document. If any slippage is noticed, it will carry out the necessary corrective action to bring the project back on time.

We'll be able to meet regulatory and standards requirements with PDM
A lot of OSHA and ISO regulations and standards address document management. With a manual system, there's the danger that documents get lost, mislaid, sent to the wrong place. There's the danger that someone makes a change without documenting it, or documents it incorrectly. A PDM system can carry out document management tasks automatically and can do them much better than people. It will keep us up to standard. It won't need a vacation, and it won't get sick.

17.9 Functional Performance Improvement

We'll be able to respond better to Requests For Proposal
With PDM, we won't have to write each new proposal from scratch. We'll be able to look for similar proposals we've made in the past, take out the best bits and include them in the new proposal. We'll also be able to look at the cost estimates in previous proposals and compare them to the real costs. This should allow us to come up with realistic, competitive prices for our proposals. As we won't have to create everything for a new proposal, there will be less work to do, so it will get done quicker and cost less, and there should be fewer errors. The end result will be that the customer will get a high-quality reply very quickly and we'll feel happy with the content. Customers will appreciate our rapid response, especially on rush jobs, and so we'll get a higher proposal acceptance rate.

We'll be able to improve our Marketing and Sales performance
With all the past and current product information and customer specification information under control, it will be much easier for us to evaluate future opportunities. We'll have a much better feel of the market, and will be able to launch much better Marketing activities. We'll also be more flexible when it comes to responding to market demand. When we do get customers, it will be harder for them to change their requirements without paying more, as all the information on discussions we've had with them, and the agreements we've made with them, will be available on the PDM system. Hopefully, the end result will be that customers think more about their requirements upfront, so once a project is started, it will go through much quicker and stay within our estimates. If we can stay on-time and

on-budget, everyone will be happy, and the customer will come back with a repeat order.

We'll increase productivity
There will be many gains once we start using PDM. We'll spend less time looking for data, we'll re-use existing designs, we'll be able to manage our CAD data better. We'll save time by sending product data electronically round the network. We'll automate the release cycle so it will be faster. We'll get our engineering changes under control. We'll have better status information available on our development projects. We'll improve the reliability of schedules. We'll reduce paper manipulation. The list goes on and on, but the end result is that we'll reduce our overhead costs and increase our productivity.

Manufacturing will be able to operate much better
We in Manufacturing will be involved earlier in product development activities, so will be able to help get Design Engineering under control. As a result, we won't have to try to produce unmanufacturable designs, and there will be less scrap and rework. There'll be a similar effect when we get the change process under control. With the information available on the PDM system we won't have all the problems we get now with paper-based documents from Engineering. We won't have to re-enter data by hand, so won't have that source of error to contend with. Once we can get Engineering under control, we'll only produce what is needed, and will reduce inventory.

Our Installation activities will benefit
Often, when our Installation teams go out to the installation site, they find that a lot of the parts don't fit. Between Marketing, Engineering, Manufacturing and Logistics there are always all sorts of misunderstandings and the end result is that we can't install as quickly as promised. Of course, we always get the blame because we're at the end of the chain. With the PDM system in place, all the upstream functions should be able to follow through on a particular order, so what gets delivered to the customer site should be exactly what we have to install.

Field Service performance will improve
Without a PDM system in place, we in Field Service have a really tough job. Often we have no way of knowing what's been configured in a particular product. Engineering might send us an "as-designed" configuration, but all sorts of things can change after this is produced. Often, Engineering will have changed components but not told anyone. Some of the sub-assemblies might have been purchased, and have different components to the initial in-house design. Manufacturing might also have made changes. At the Assembly and Test stages, where there's always a lot of pressure to get the product out the door, additional changes might be made and not get properly documented. With the PDM system in place, everything will be properly documented, so we in Field Service will be able to do a great job for the customer.

The IS Department will profit from PDM
Once the PDM system is up and running, we in the IS Department won't have to waste so much time building interfaces for new databases, and between engineering and business applications. We'll be able to cut costs by removing unnecessary intermediate applications. PDM will lead to a reduction in the work we have to do, so we'll be able to address important IS issues better.

The performance of the Quality Department will improve
One of the major problems of the Quality Department is that a lot of the company's documentation isn't under control and doesn't correspond to reality. A PDM system is ideal for getting documentation under control. We'll be able to get all the documents under control and meet all the OSHA, EPA, NRC and ISO requirements. We'll have far fewer quality problems, so the customers will be happier and we won't be wasting our time and money.

17.10   Better Management of Product Development Activities

We'll be able to make better plans
The PDM system will provide information about past plans and activities, such as how long they should have taken, how long they took, and where problems occurred. It will also provide information about the future availability of development resources. These two types of information, about the past and about the future, are what we need to make good plans for product development.

We'll be able to make more reliable estimates of project costs and times
For a particular project, we'll be able to refer back to similar projects to see what kind of resources were needed, how much time projects took, how much they cost, and the extent to which they followed the plan. We'll use this information to provide better estimates for future projects.

Use of PDM will make projects easier to manage
Automated reporting on project progress is faster, cheaper and less liable to suffer from transmission and transcription errors than manual reporting. With the right information available, projects will be easier to manage. We won't need to go hunting through printouts every time we need some data.

We'll be able to run tasks in parallel and get work done quicker
With PDM, we'll know what's happening on a project, and we'll be in control. We'll do things that were too risky without PDM because we weren't sure of the status of different tasks, we didn't know how much progress had been made, and weren't sure of the value of information until a task was finished and a design released. With PDM, status will be clear, even down to the level of individual sub-tasks and information elements. We'll be able to run tasks in parallel and allow for partial release.

We'll be able to make sure the right process is followed
Once we've identified the best processes, we'll want people to use them. Without PDM, it's hard to know what people are doing. We can't look over everybody's shoulder. The PDM system will tell us what's going on—and signal any divergence from the planned process.

We'll improve communication between managers and developers
It will be much easier to get information to people in the process. Press a button, and the information will go out over the network to everyone who needs it. They will be able to respond immediately over the network. This will be much more effective than the current situation where we can only communicate to everybody at the weekly Monday morning meeting.

We'll be able to improve the distribution of tasks to product developers
The system will be aware of the tasks that need to be done, and the current assignments and skills of our developers. It will make the best assignment of tasks, then send the tasks and related information over the network.

Use of PDM on our projects will result in benefits for the whole business
PDM will bring our projects under control. We'll improve performance on release management, change management, version management and configuration management. People will be able to access information faster and re-use existing information. As a result, the project will cost less and take less time, and the product will be more likely to meet customer specifications. Everyone will be a winner.

We'll be more aware of what's happening on a project. We'll be able to make better decisions
Without PDM, it's often difficult to know what's happening on a project. Some people just don't want to tell you what's happening. Others would tell you, but they don't know what you want to know. The system will automatically provide status information, and periodic and on-demand reports. With all the information that the PDM system provides, we'll have the facts to take the right decisions.

We'll be more responsive to change
The PDM system will keep us up-to-date with project progress. If a client wants to change the specs, we'll find it easier to respond. We'll know the exact situation of the project. We'll know what needs to be done. We'll decide what to do and then send out, over the network, the new plan. Without PDM, this would take weeks. It would create chaos. With PDM it will happen quickly and smoothly.

We'll be able to co-ordinate projects better
Without PDM, it's difficult to co-ordinate a project. You don't know who's doing what, or when they will finish. When you have to change the plan, you're working in the dark. With PDM, these problems go away. You have the information. You're in contact with the team. You know what you're doing. You lead the project. Without PDM, it's very difficult to manage several projects together. If you change something on one project to help another, you often finish up slowing them both

down. The PDM system, with detailed knowledge of all projects, will make it easy to manage several projects.

We'll have a way of saving company know-how for future use
The PDM system will classify and save information created at great expense during each project. On future projects we'll be able to refer back to this, and understand what was done and why it was done. We'll be able to make the best use of our company know-how, perhaps even re-using information.

We'll make better use of resources. We'll improve quality, cut costs and cut product development cycles
There's a real business benefit to using PDM. We'll do things at the right time in the right order. We'll overlap them when we can. We won't waste time because some information or some person isn't ready. We'll save time, and cut out the waste. We'll finish ahead of schedule and under budget.

We'll work better with other departments, suppliers and customers
Without PDM, we may not know what's happening in another department, so it can be difficult to work with other people. We can't tell them that we don't know what we're doing, so have to slow them down until we find out. With PDM, information will be at our fingertips, so we'll work better with other groups.

We'll be able to implement change to business activities
Without PDM we don't master our business activities. They just run themselves, and it would be dangerous to try to change them. We don't understand exactly what's happening, so any change could be risky. We'll need to clarify the processes for PDM. Once they're clear, we can start to improve them.

17.11   Automation of Product Development Activities

We'll improve the automation of direct product development activities
In the past we've put in CAD, CAM and CAE applications to automate direct product development tasks such as analysis, design, drafting and part programming. Later we realised that we didn't get the productivity benefits we expected because these applications only address parts of these activities. For example, if you are doing design work with CAD, you often want to look back at previous designs and you often want to look through old files, but you can't do that if they're not on the CAD application. CAD applications have a lot of functionality for geometric modelling but they don't have much functionality to support tasks like design management and file management, so people waste time looking for information, or even worse, they re-create existing designs. The PDM system will provide the missing support functionality, such as information management, for direct product development activities that are already partially automated.

We'll automate all our product development activities, and get all the benefits of
automation
In the past we mainly automated direct product development activities. Now we'll
use PDM to automate both direct and indirect product development activities. As a
result, all these activities will get done quicker, there will be fewer quality problems, and we'll save money. Machines work much faster than people, and are more
reliable, so the work will get done faster and with fewer mistakes. Computers cost
very little, so there will also be a cost reduction as computers take over from people.

We'll automate indirect product development activities
With PDM, we'll be able to automate a lot of indirect activities such as information
management and communication, information storage and back-up, archival,
sign-off, design release, configuration management and project management. They
will get done better, faster and at a lower cost.

PDM will automate the information management and transfer processes
In the past, we had to transfer information manually on paper. It was a slow process
in which mistakes often arose. People entered data in the wrong field. Drawings
went missing or got mixed up. People picked the wrong version of a design or
document. Documents sent by internal mail went to the wrong person or arrived
late. Once we've automated with PDM, we'll be able to get the right information to
the right person at the right time for less cost. Comparing PDM to the manual
process is a bit like comparing fax and telephone. When you send a fax you send a
brief, concise message. It's quick, cheap and clear. With the telephone you have to
add all sorts of politeness, discussion and small chat about other things. There's a
danger of misinterpretation. It takes longer and is more expensive.

We'll be able to automate better than in the past
When we automated in the past, we put in computers and applications and tried to
do CIM. One of the reasons it didn't work very well was that we didn't automate
the information. We automated individual tasks without automating the environment around them. For example, the information input, output and management
stayed manual. PDM will help us automate better. For example, it will automate
information management and information communication.

We'll be able to integrate already-automated activities
We've got a lot of Islands of Automation. These are unconnected applications that
we bought to automate individual product development activities. We've CAE,
CAD and CAM Islands. They are automated but the links between them aren't.
When we transfer data from the CAE application to the CAD application, we do it
on disk or on paper, and we waste time making sure we've got all the right files and
information. Data transfers involve manual intervention and lead to loss of time and
quality. The PDM system will do this kind of work. It will automate the connections between Islands.

We'll automate product development management activities
In the past, automation of the product development environment hasn't really addressed management activities. It has been limited to tasks such as design and analysis. PDM can take automation to the next level. It has the potential to manage the product development process and to manage product development projects. The PDM system knows about all the information in the system, and it knows about the process and the people. Once we set the system up properly, it will be able to take over many management activities. It will help product development managers to develop strategies and plans. At the project level, it will tell people what to do and when they should do it.

We'll automate workflow between product development activities
As soon as one product development activity is finished, the next one should start. Often, in the past, one activity would be finished, but we'd lose time before the next one started. Someone would forget to tell the next person to start work, or the next person wouldn't find all the information they needed. We wasted a lot of time like that. PDM will automate the workflow, and cut out the waste. It will tell the next person to start work, and make sure they have all the information.

Automating information is the first step to getting product development work done by expert systems
Once the PDM system has got the information automated, an expert system will be able to apply rules and start doing the work itself. We'll need less people in the process, and it won't matter so much if people leave or are sick.

PDM will help us automate activities such as QFD and DFM
Practices such as QFD and DFM create, and make use of, a lot of product information. PDM can help us with the input, output and management activities associated with this information.

We'll have push-button product development
To start with, PDM will just automate the information and the workflow. Then it will start managing projects and practices. Eventually it will manage all our product development activities. We'll just have to enter customer specifications, and push a button. The new product will come out at the other end of the process.

17.12   IS Effectiveness Improvement

We'll replace our inefficient current applications with a more powerful tool
Currently we have many applications that all do some part of the work that the PDM system will do. They weren't designed with PDM in mind. The result is that the overall system is not optimised, and performance is not as good as it could be. Once we've got the PDM system in place, we'll tune it to fit the product lifecycle, and performance will improve dramatically.

We'll gain by linking applications together
The PDM system will allow us to link applications together. It will help us build the bridges between Islands of Automation. We'll be able to get rid of unnecessary interfaces and artificial barriers. The result will be that information will flow much better. Because there will be much less manipulation and conversion of data, there will be fewer errors, and quality will improve. We won't lose time with unnecessary activities.

We'll reduce the overall cost of IS
The PDM system will allow us to link applications together easily. We'll be able to remove unnecessary interfaces and applications. When we install the PDM system, we'll remove any old applications that do work that will be done by the PLM system. We'll remove applications that aren't adding value. That means we'll need fewer applications, and we won't have to develop, purchase and maintain so many applications. The end result will be a reduction in overall IS costs.

We'll provide better service to our customers
Our customers in product development and support will benefit from the introduction of the PDM system. They'll find it easier to concentrate on their value-adding activities. Once the IS organisation has eliminated all the unnecessary interfaces, applications and databases, it will have much more time to spend on the important activity of supporting users in product development and support. They won't have to wait so long for the developments they're asking for.

We'll be able to reduce staff levels
We'll need fewer IS staff because there will be less in-house development work to do and fewer interfaces to maintain. We'll need fewer people in the Engineering Computer Systems organisation. With one underlying PDM system, and market-place interfaces between the PDM system and other applications, there won't be so much IS development work to do.

We'll be able to reduce the paper flow
Paper is an expensive medium for information communication and storage. We'll save money by transmitting and storing data electronically. We'll be able to sell off the storage space that we'll no longer need. We'll use the PDM system's electronic mail functionality to communicate internally, and EDI to communicate with suppliers and customers.

We'll benefit from standardisation
We'll have the same PDM system in use across all the departments. We'll be able to standardise the database management systems we use, standardise the operating systems, and standardise the development languages. We'll have common PDM training sessions for people from different departments. The end result will be that we'll save time and money.

We'll manage the information resource better
With the PDM system in place, there'll be a single controlled source for all product information. We'll provide automatic back-up and archiving. We'll be able to

standardise data definitions throughout the company. We'll be able to avoid the data duplication and re-definition that we've had in the past. We'll be able to ensure that information is secure, but available to users who have appropriate access rights. We'll make sure users get the correct version of information. We'll be able to save company know-how.

We'll provide simultaneous access to business data and product data
In the past, we had one application that gave users access to business programs and information, and another application that gave them access to engineering programs and data. People couldn't get at the data in both applications at the same time. With PDM, this problem will go away.

We'll use PDM to manage our computer programs
Applications have to be managed just like other product data. There are all sorts of versions and releases of both the software we purchase and the software we develop in-house. We'll use PDM to manage the software just like it manages other product data.

We'll have an effective overall IS strategy
The product lifecycle runs from product conception to disposal, and we should have an IS strategy that supports this cycle. In the absence of PDM we couldn't have one. There were so many little bits of application all over the place, and so many discontinuities in the cycle. With PDM, we'll benefit from a clear overall strategy. We'll develop a technology road map, creating the infrastructure to support cross-application integration. We'll be able to get properly organised, reduce costs and improve performance. We'll be able to develop synergies between applications that weren't possible without PDM.

We'll demonstrate that we've changed our paradigms
With the introduction of PDM, we'll show our customers that we've changed from a twentieth century paradigm to a twenty-first century paradigm, and from a hierarchical to a horizontal mindset. We'll establish product data and document standards, and design a corporate integration architecture so that formerly fragmented and disparate information is provided to users in the format they want.

We'll be able to normalise organisational structures
The introduction of a system like PDM that runs across organisational boundaries provides new opportunities for the IS organisation. Individual departments can have difficulty in taking responsibility for a cross-functional system because they don't have company-wide responsibility. Organisations, such as the IS organisation, that do have such responsibility, can play an important new role.

We'll be able to take advantage of advanced IS technology
We'll be able to build up a database of company know-how. We'll implement Knowledge Management applications. We'll use Artificial Intelligence techniques to exploit our knowledge to the full.

17.13   Infrastructure for Effective Product Development

PDM will provide the infrastructure for effective product development
At the moment, we have a lot of Islands of Automation and unconnected networks and databases. All the applications, networks and databases belong to people in different groups and departments, so we don't really have an overall computing infrastructure for product development. Instead, we have bits of infrastructure owned by different groups. PDM will be the part of the infrastructure that unites all the other bits of infrastructure and creates the overall infrastructure.

PDM will support the product development process
At the moment, people don't think in terms of the overall product development process. Instead, they think in terms of the departments or groups on the organisation chart. This type of thinking has led to the situation where we have so many Islands of Automation. To improve product development performance, we have to start thinking and working in terms of the overall product development process. PDM is the ideal system for this. It manages product data throughout the process, and doesn't belong to any individual development department. It supports people throughout the process providing them the product data they need when they need it.

PDM will allow data to flow throughout the product lifecycle
At the moment, people are only interested in the way that data is used in their own group or department. No-one thinks about the best way to use information throughout the product lifecycle. Within each group and department, people use the data structures, rules and formats that suit them best. This creates problems when they communicate data with other groups. A lot of data conversion and translation has to be done, and invariably leads to errors and misunderstandings. With PDM, we'll overcome these problems. PDM will be the layer of the infrastructure that bridges all these gaps and discontinuities, and allows product data to flow freely from one end of the product lifecycle to the other.

PDM will help clarify and improve the product development process, then support its use
PDM systems have workflow management functionality as well as information management functionality. When people use PDM to manage their data, they will be forced to think about the way it flows through their part of the process. When they start using the workflow management functions, they will have to understand the process in detail. When they do this, they will find that the current process is inefficient, and they will take steps to improve it. PDM will support the activities of the improved workflow.

PDM will support lifecycle practices
At the moment, product-related practices are focused on individual departments. In the past, it didn't make sense to have lifecycle practices because people in different

parts of the process didn't work together and didn't have common data formats to work with. With the introduction of PDM things will change.

PDM will support teamwork
In the past, it's been difficult for people in different departments to work together. All the applications, languages, practices and databases they used were department-specific. When people from different departments have tried to work in a team there has been no common infrastructure to support them. They've all been supported by infrastructure supporting their own departments, but there hasn't been a common base for cross-functional teamwork. PDM will provide that common base, and enable cross-functional teamwork.

PDM will allow people to access whatever data they need
In the past, people have generally been able to find product data that belonged to their group or department fairly quickly. It's been much more difficult for them to access needed data that belongs to other departments. They've had to go through lengthy, time-wasting, bureaucratic procedures to access it. With PDM, provided people have suitable access rights, the infrastructure will allow them fast access to product data anywhere in the company or in the supply chain.

PDM will integrate applications throughout the company
In the past, applications have belonged to individual groups and departments who've worked with their own programs and product data. These applications have created Islands of Automation and Islands of Data. They correspond closely to the needs of the people in the group or department, and usually work efficiently for them. However, there is always a problem when data has to be communicated to other groups. Often the other group needs the data in a different format, so there is a conversion process to go through. This takes time and may result in errors. A physical data transfer is needed to get the data from one computer to another. Sometimes this is done on a disk, sometimes by printing out data from one application and entering it at the keyboard of the other application. In both cases, time is wasted and errors may occur. PDM will integrate applications and eliminate errors.

Use of PDM will save time
Once the PDM system is in place as the backbone of the product lifecycle, it will allow data to flow smoothly through the lifecycle. Time will be saved because there will no longer be the discontinuities, barriers and gaps in the lifecycle where time has been lost in the past. People won't lose time waiting for access to data from other parts of the company.

Use of PDM will save money and improve quality
Once the PDM system is in place, it will allow data to flow through the product lifecycle. We'll be able to remove duplicate and redundant applications and databases from the infrastructure. We won't have to pay for their maintenance and

upgrade. We won't have to pay for the conversion of data between previously unconnected applications. We'll avoid all the errors that result from manually transferring data between previously unconnected applications.

17.14   Questions About the Future Role of PDM

There are many generic reasons for implementing a PDM system. However, the specific reasons will be different for each company. It's important that companies understand what is specific about their use of PDM. To help get this understanding, it's useful to answer the following questions about the future role of PDM.

In the long term, how will PDM create better conditions for people developing products?
There are so many theoretical reasons for introducing PDM that it's useful to try to understand exactly what its introduction will result in. For example it's great to say that PDM will lead to better engineering change control, but what does this really mean? How many engineering changes does the company want in the future?

How will PDM help the company be more flexible in responding to changing business cycles?
If business gets better, PDM needs, for example, to be able to support the rapid introduction of new products at a reasonable cost. If it gets worse, PDM needs to offer tools to help reduce cost. Where will the initial emphasis be?

How will PDM help the company respond more flexibly to changing customer demand?
Apart from shorter development cycles and better configuration management, how can PDM help the company be more flexible?

How will PDM, a cross-functional technology, affect the functional organisations in the company?
In most companies, people, applications and information are organised on a functional basis. PDM is a cross-functional technology. So how will it be used, and managed, in and between functional organisations?

Is there sufficient flexibility in the company to deal with the changes and problems that will probably occur when PDM is introduced?
PDM is cross-functional, so its introduction is likely to disrupt the existing way of working. Which problems are likely? How will they be solved? What changes will be necessary? Is the company flexible enough to implement these changes successfully?

Are relationships and working practices with customers and suppliers compatible with PDM?
A lot of product data is used outside the company, for example by customers and partners. It's not enough to decide how PDM will work within the company. It's also necessary to understand how it relates to customers and suppliers.

How is PDM expected to help the company to grow?
Unless companies grow, they're likely to disappear. In which ways will PDM help the company to grow? If they aren't known, then is PDM really necessary?

How will PDM affect business results?
This is an obvious question, but it needs to be answered. Until it's answered, top management is unlikely to show much interest in PDM.

How will PDM affect investment needs?
This is a similar question, but addresses a different issue. Until it's shown how much investment PDM requires, and how this will affect other investments, top management can't take the decision to invest in PDM.

How will PDM help to promote sales, stability and lasting improvements?
This question is on the critical path. PDM may be the greatest technology in the world, and a particular system may have the best functionality, but unless the long-term benefits can be demonstrated in business terms, then top management will not support its introduction.

Answering questions like these not only helps to relate PDM to the business context of the company, but may also highlight areas that would otherwise be overlooked.

Chapter 18
Forewarned Is Forearmed

18.1 Reasons and Replies

We were told "simplify, don't automate"

The rule wasn't "simplify—don't automate". It was "first simplify, then automate". Simplification only gives some of the gains. The rest come when technology and applications are applied to simplified processes.

Years ago we tried to do Computer Integrated Manufacturing (CIM). It didn't work. PDM is a reincarnation of CIM. It won't work.

CIM worked for some companies, but not for others. The companies it worked for knew what they were doing, and they understood and simplified their processes before applying computer systems. PDM isn't a reincarnation of CIM. It doesn't focus on integrating computer systems, but on improving the quality, flow and use of product data in a company. It's linked to concurrent engineering, an activity that aims to simplify and improve engineering processes much as JIT simplified and improved manufacturing processes. Some companies will understand they have to clean up their processes before implementing PDM. Others won't.

These days we're into culture changes. We're doing Total Quality Management (TQM), not technology. We tried technology and systems, but they never gave good results.

Culture changes such as TQM are only part of the answer. A company with TQM and good technology will be more competitive than a company with TQM and no technology. Companies that neglect technology will be losers. Computer systems and technology may well be difficult to manage, but this doesn't mean that they can't, or shouldn't, be managed. Since, in addition to being difficult to manage, they're also extremely important, a lot of management time and effort needs to be invested in ensuring that they're used as effectively as possible.

PDM? Another acronym? We learnt our lesson long ago. No acronym soup here.

There are a lot of acronyms in use. TQM, JIT, CAD, PDM and PLM are all acronyms. People who are instinctively opposed to all acronyms may not last long in management positions. They should overcome this type of automatic emotional reaction, and concentrate on thinking about the possible benefits of applying the technology behind the acronym.

We're re-engineering the corporation, so we don't need PDM.

Information Technology is the enabling component of corporate re-engineering. Any serious attempt at re-engineering will address functions that use product data. PDM will be a key technology in many re-engineering activities.

As a top manager, I can't be involved in low-level technical issues like PDM. As a top manager, I'm here to take decisions, not to be involved in the business. Give me the figures, and I'll tell you what to do.

PDM helps reduce the time to introduce new products, and the cost of developing them. It helps reduce the cost of new products, and improve the quality of products and services. Companies pioneering PDM found it can help reduce engineering costs by 15 %, product development cycles by 25 %, engineering change time by 30 %, and the number of engineering changes by 40 %. If it can do all this, it can really affect the bottom line, so it should be of interest to top management.

Introduction of PDM implies understanding the processes that use product information. It only takes a few months to understand them. During these few months it becomes apparent that these processes have not been overhauled for years and can be made much more efficient, even without a PDM system. Improving the processes will reduce cycle times and reduce errors. This should translate, with very little investment, into reduced costs and improved quality. Once the processes have been improved, the introduction of a PDM system leads to a further reduction in cycle times, and an additional improvement in quality. The system can be used to make sure that the process doesn't slip back into the old, inefficient way of working.

PDM isn't a low-level technical issue. It's an issue that impacts the company at all levels, from management strategy to data transfer. However, unless management is involved with PDM and accepts responsibility for its success, it will become a low-level technical issue having little effect on company performance. If on the other hand, management does take responsibility, understands what PDM can do, and organises resources to achieve the potential goals, then PDM will be an important high-level business issue. For PDM to achieve its goals, it has to be used cross-functionally. Only top management has the authority to ensure that PDM is used cross-functionally, i.e. in departments reporting to different managers. The successful use of PDM is only possible if top management has the vision and ability to influence the company to implement it.

Top managers will benefit in the long term from the introduction of PDM. Their companies will be more competitive, with reduced time cycles, reduced costs and improved quality. Customers will be more satisfied with the company's products and services. Shareholders will be satisfied with management performance.

Top managers need to learn more about new approaches such as PDM and concurrent engineering. Unless someone proposes, and involves them in, some educational or training sessions, it's unlikely that they'll learn enough to become involved. Consequently the promoters of PDM should either prepare a management-oriented PDM training session for top managers, or arrange for them to attend a management-oriented PDM seminar or conference.

The economy is in a recession/downturn, so we're in a cost reduction phase. We can't spend any money on new, virtually unknown technologies.

We know that the economy is in a recession, and we have to reduce costs. At the beginning of the PDM project we propose using our own staff to improve our engineering process. Over 6 months we believe we can remove $x,000 costs. When we've done this, and demonstrated it to you, we'd like to invest half this sum in further improvements. This would allow us to remove another $y,000 costs. When we've done this and demonstrated it to you, we'd like to invest half this sum in further improvements. This step by step bootstrapping approach will minimise risk and finance our progress.

These days we're looking for fast payback projects, 18 months is the maximum. PDM's payback would be much longer than that.

We understand the need for fast payback projects. We also believe it's best to work on a very limited number of projects. We've split our projects into two groups —continuous improvement projects and re-engineering projects. We've selected four continuous improvement projects. They are all short-term, low-cost, limited-benefit, low-risk, fast-payback projects. They're not revolutionary, but they'll keep us moving forward. We don't expect the CFO to understand all the details of these projects, but as they're all short-term projects, it will be easy to judge their success or failure. Their success will pay for the re-engineering project.

We have one re-engineering project. This is a major project, and it's expected to have a major effect on the business. Not surprisingly, it will take some time to carry out the project, and we believe it's not realistic to talk of an 18-month payback. We believe the benefits will be massive, but it will take several years of hard work from people in many departments to achieve them. Through all these years top management will be heavily involved.

Before we start these projects, we need to discuss them thoroughly, and make sure everybody, including top managers and finance managers, understands what we're doing. We'd like to have consensus on the project's objectives and approach. In view of the importance of the project, and the importance of succeeding, we'd like management and finance representatives in the project team. We're sure their involvement will help us achieve a breakthrough.

We have other priorities. PDM hurts everyone. It's not a win-win. We have so much work, we don't have time to handle anything else. PDM is too difficult. There's no support from top management for PDM, so it won't work. Our reward system isn't geared to cross-functional activities like PDM. PDM wouldn't work here, there's too much inertia in this company. PDM works when users trust management, nobody here trusts management. We could do more, but management won't let us. Make one mistake in this company, and you're finished, maybe even fired. We don't have the skills and capabilities to handle PDM. Big projects like PDM need really good project managers, we only have a few really good people you could trust with PDM, and they're all overloaded and fully booked.

One way to get some progress in this type of organisation is to exploit top managers' desire to be with, and to behave like, other top people. A world-renowned management guru can be invited to present the vision of PDM to top management in a high-level presentation. However, there's always the danger that another management guru will manage to sell them another acronym during a similar presentation. PDM will be taken off the list, and the new acronym will become the flavour of the month. Fortunately, in this type of company there are always a few extremely hard-working middle managers who devote their entire lives to the company. It's often only with the backing of this type of person that PDM can be introduced. Not surprisingly, top management accepts the overload and sacrifice of such people, who they see as a little mad, and in return, allows them to have an occasional idea of their own, "a bee in their bonnet". It may not be easy to convince such a hard-working, devoted lifelong servant of the company of a concept like PDM, but once this type of person becomes a believer, they'll support the idea, and work for it, for many years. They'll probably succeed because they know how to overcome all the obstacles that will be put in their way, and the general lethargy that surrounds them. They know which people can be counted on to do good work. They know how to organise the workload so that they do everything top management has told them to do, but still have resources to do what they believe to be right and important.

We're product developers. We have to develop new products, not play around with computers.

We understand that the main priority is developing products. A PDM system would help you manage your projects better and reduce unnecessary overhead. As a result you'll be able to spend more time developing products, and develop them more effectively. In turn this will mean fewer design faults in new products, so less rework, and more time for developing new products. PDM is a win-win.

Our product developers won't accept changes to the way they work.

Product developers will accept something that helps them do what they like doing. Why not give it a try?

Users don't want to be spied on by Big Brother systems like PDM.

PDM isn't Big Brother. It's a tool to help people get the right data when they want it. Instead of wasting their time and then getting the wrong data.

We have CAD, so we don't need PDM. We don't need PDM because our CAD has a data management module.

Few CAD applications manage data well. They were developed to help people define product geometry, and most of the effort went into developing functions that directly aid the design process. Most CAD applications only have data management functionality to manage the data they produce. PDM systems have been developed specifically to manage product data produced and used in many applications.

No system can handle all our data.

In principle, a PDM system (or systems) could manage all the company's data. However, to be useful, a PDM system doesn't have to manage all the data. It just has to manage some of the product data more effectively than existing manual data management systems.

They want to replace us with a PDM system.

A PDM system carries out uninteresting and repetitive work. It doesn't replace creative product developers. It allows them to do the development work they want to do.

PDM response times are too long. Product developers can't wait all day for information.

Experience shows that response times of well-implemented PDM systems are reasonable. They're much shorter than response times from manual data management systems.

We were given the responsibility two years ago of creating a computer integrated business. PDM isn't addressed in our master plan, so it can't be implemented yet.

PDM is a technology that will help the company significantly improve its overall performance. Until it's included in the overall IS master plan, we'll run it in prototype mode on our "other expenses" budget.

PDM? It's too early, come back in five years when it's matured.

PDM technology has been under development since the mid-1980s. It has passed through the phase in which it could only be used in research mode. It can now be used in everyday operations.

It's all vendor hype. Forget it. Even the PDM sales guys we've talked to don't understand it.

During the growth phase of any technology, there will always be some vendor salespeople who haven't fully understood the technology. This doesn't mean that there's anything wrong with the technology.

Why do we need a PDM system? We already have a relational database management system.

Many PDM systems run on relational database systems. The development work that PDM vendors carry out builds additional functionality, specifically to handle product data, on top of existing relational data management functionality. Several thousand person-years of development have gone into some PDM systems. As a result they contain a lot more functionality for handling product data than a basic relational database management system.

We weren't asked to help when the Engineering Department started to use CAD, so we're not going to help with PDM. The last time we spoke to our computer vendor, we were told that companies like ours don't need PDM yet.

The IS Department must work together with users in other departments if it's to succeed. It's ideally qualified to look after some parts of PDM. Other departments are better qualified for other parts. The IS Department has to understand the business needs of users and of the company, and then look for corresponding solutions. In previous decades, the IS Department and its computer vendor could identify new technologies, and then look for potential applications within the company, but in today's world, it's more important to solve existing needs than to create new systems for which there's no real demand.

We in Marketing have been saying for years that the Engineering Department is unmanageable. Every time we tell them what kind of a product the customers want, they go and develop something completely different. If we tell them they need PDM, they'll deliberately go and do something else.

The Engineering Department welcomes the opportunity to work closer together with Marketing. We expect the impact of our use of PDM on Marketing to be minor. We know we need to improve our Customer Focus, so we'll support the introduction of Quality Function Deployment (QFD). We look forward to working closely with Marketing on that.

From a Manufacturing viewpoint, the Engineering Department is chaotic, but it's not our job to sort out their mess. We've asked them several times to clean up their Bills of Materials, but they're just not interested. It's the same with changes. They're always wanting to change parts they've only just released to us. And when they send us documentation, it's never complete. It's taken a long time, but now we've worked out how to be as independent as possible from their mess. We certainly don't want to be involved with them on PDM.

One of the main reasons for the introduction of PDM is to get control of the information released to Engineering's customers. In the past, there have been some problems in this area, and these were due to the lack of a tool to fully control the interface. We now have this tool—PDM. To ensure that it does not create extra problems for Manufacturing, we'll install PDM in several phases. Initially we'll run a prototype of PDM within the Engineering Department. Only when this is successful will we address the link to Manufacturing. We'd like people from Manufacturing to participate with this prototype so that they can learn about the system, and then help us to develop the interface with Manufacturing.

We're glad to be in the Field, well away from the Engineering Department. Those guys have no idea how to run a business, or look after a customer. They never seem to know the configuration on a given site, and they never seem to know which versions they released. They often develop replacement parts which can't be used in existing products. To overcome these problems, we developed our own drawing management system, our own parts classification system, and our own reference manuals. Apart from the spare parts inventory, we also have a buffer inventory in case things really get out of hand. There's no way we'll have anything to do with their PDM. We have to serve our customers, not spend all day fooling around with systems that don't work properly, and people who don't know how to work professionally.

PDM will help improve the quality of the information available in the field. In the past, manual systems have been used to control product information. These systems have been overwhelmed by the increasing amount of data needed to support extensive customisation of products. The introduction of PDM in the Engineering department will help to regain control of product data. Once PDM has been successfully implemented in the Engineering department, it'll be possible to extend its use to other departments.

Departmental empires just want to be separate.

To improve service to customers, both external and internal, the barriers between different departments have to be removed. Many of these barriers were built to make it possible for each department to be relatively independent, and thus organise itself to work as efficiently as possible. In the future, the company has to act as a whole, not as a group of independent departments, and the aim will be to improve the effectiveness of the overall company, not the efficiency of each individual department.

We saw PDM once, and it didn't look useful. PDM is so simple we are going to build our own system. PDM is very good, but we can do it better.

The concepts of PDM are very easy to understand, and many people believe PDM systems are correspondingly easy to develop and implement. However, there's no need to redevelop the wheel. There are many PDM systems available on the market. It's extremely unlikely that even the most brilliant product development IS support group is going to develop, within reasonable time and cost limits, a PDM system that offers functionality anywhere near that of systems currently available on the market.

PDM isn't the type of product development computing system we need. Our engineers really need C++ support.

Many people in product development organisations find it difficult to get enthusiastic about PDM systems. They consider them to be unattractive, technologically backward, and irrelevant to the real needs of product developers. Some people feel that PDM systems have little to contribute to improving new product development cycles, and that rather than investing in PDM, a company should invest in systems that can be directly linked to improving the productivity of product developers. However, in practice, PDM systems do have a major impact on improving product development productivity. They free product developers from a lot of overhead tasks, leaving them free to concentrate on more interesting and value-adding activities.

We can't do PDM because it wasn't included in the 5-year plan. We can't do PDM because we've used up this year's budget. Our management only lets us buy systems from 3 vendors, and none of them has a good PDM solution.

Past mistakes shouldn't be allowed to hinder progress. Ways can always be found to get the PDM message across to management without pointing a finger at the "guilty".

We'd like a PDM system, but the potential users say they don't want one.

If people are unaware of the benefits of PDM, they should be made aware of them. They should be shown examples of the use of PDM by other companies, in particular, use by competitors.

We'd like to do PDM, but the others don't want to.

If you would like to do PDM, but other people don't want to, you should explain to them why it's so important. Tell them how benefits such as reduced product development cycles, increased quality, and fewer and faster engineering changes will improve their working life.

We'd like to do PDM, but we don't have the power.

If you would like to do PDM, but don't have the power to impose it, you need to identify those around you who do have the power, and convince them of the need for PDM.

We'd like to do PDM, but we don't have any influence.

If you would like to do PDM, but think you don't have any influence, think again. Talk to your colleagues. Help them understand the advantages of PDM. Once they've understood, you'll no longer feel alone and without influence. Together you'll be able to spread the PDM message. Before long, PDM will be supported by so many people that implementation will occur.

We can't do PDM, because it isn't in the plan we have to follow.

If you can't do PDM because it's not in the plan you have to follow, then you need to explain the importance of PDM to the person who set the plan. It's unlikely that the plan specifically excludes PDM. Most likely you can bring about some change. Perhaps a small PDM project can be slipped into the "other activities" section of the plan. This could be a good start.

We'd like to do PDM, but we don't have the resources.

You would like to do PDM, but don't have the resources? Be serious, how can you possibly not have the resources for one of the most important technologies of the decade. Perhaps you need to adjust your priorities. Start by cutting out some of the waste, and you'll soon have the people, time and money for PDM.

We'd like to do PDM, but it's a bit too soon. We'd have liked to do PDM, but it's a bit late now.

Unless your company is out of business, it's never too late to start PDM activities. It may be too early to implement PDM, but it's never too early to start making people aware of PDM's benefits.

Your ideas on PDM are good, but we'd need to adapt them to our company.

Some people would rather reinvent their own PDM system than find one that's available on the market. Others would rather reinvent the reasons for using PDM than use other people's reasons. Fortunately, this is less expensive than reinventing an PDM system. Usually it strengthens beliefs in PDM, and leads people to become fervent supporters of their own PDM cause. Make sure that the reasons are reinvented by a team, not by one person unable to make the dream come true.

I don't have time to talk to people who want to talk about PDM. PDM doesn't apply to me. I'm not interested in PDM. PDM isn't my job. PDM isn't in my area.

These remarks imply that the person isn't interested in PDM, and doesn't even want to continue the discussion. Sometimes the answer can be taken at face value, but it may be that the brief rejection covers hidden reasons. Perhaps the manager doesn't have time to talk to you about PDM. Alternatively, the manager may have a little time available, but be afraid that your PDM presentation is so badly prepared that you are going to waste valuable management time. Make sure your presentation is concise, well-organised, and addresses the business factors that are of concern to the manager. If appropriate, send the manager a note asking for a meeting. The note should outline the purpose of the meeting, its content, the expected result and the benefit for the manager.

Some managers will claim that PDM doesn't apply to them. This may well be true, in which case you should ask which other managers they recommend you should talk to about PDM. However, it may be that PDM does apply to the manager but he/she is either not aware of it, or is trying to avoid it. If it's a question of lack of awareness, prepare a brief presentation (two or three pages) for the manager and either discuss it in a short meeting, focusing on the advantages for the manager, or send it to them. If the manager is trying to avoid PDM, try to get a meeting that will focus on the business advantages of PDM for the manager, and identify potential PDM users.

I decided to do something else. Other things are more important than PDM.

Apparently the manager has taken a rational decision that other topics are more important than PDM. What you need to do is to get PDM included in the "other topics". Find out which topics are seen as more important. Understand the problems that are being addressed by these topics. Show how PDM can help solve the problem, or increase the benefits of other solutions that are proposed.

I know what I'm doing, I don't need PDM.

This is a clear rejection of PDM and is the most difficult to deal with. Find out why the person is so sure they don't need PDM. Understand how PDM could help them. Prepare the case to present PDM, but hold back for several months. In the meantime, look to see who else might be a candidate for PDM. Roles change, and, given time, recognition of needs may change.

I don't understand PDM. I don't see why we should change. Our advisor says PDM isn't important.

Although the manager is giving a reason for not doing PDM, a lot of room is being left open for further discussion. An admission of not understanding PDM leaves the way open to a proposal to increase PDM awareness. A reply indicating an inability to see the need for change should be met with a reasoned outline of the need for change, and a proposal to demonstrate the way PDM supports such change. Assigning the responsibility for the decision to an advisor may just be a way of avoiding the blame, or it may be intended to open up a discussion with the advisor. This could lead on to increasing the advisor's PDM awareness, and demonstration of the benefits of PDM.

It will never work. You'll never get all the data on-line, and, if you do, there'll be too much data and not enough information.

Trying to put all a company's product data into a PDM system in one bite is a sure way of creating problems. It's much better to start by putting in the data for a particular project, and then gradually expanding the amount and type of data managed by the system. It will probably be many years before all the data is in the system, but at all times the system will be of use.

Things will be out of control, no-one will know what's happening. You won't know who signed off. It will be anarchy if there's not someone there to organise things.

Today's PDM systems follow the rules they get from people in the product lifecycle. They can be run with an audit trail to show who did what, and why the system acted in a certain way.

It's too risky. When the system crashes, or there's a virus, either we'll lose all the information, or we won't be able to access it.

Techniques have been developed to recover from system crashes. These can be applied to a PDM system to make sure that it's possible to recover from a system crash. Virus detection programs can be used to make sure viruses are discovered before they corrupt the system.

We won't be able to find all the information.

Once information has been notified to the PDM system, it's more difficult to lose track of it than it is to lose track of information in a manual product data management system.

PDM systems aren't intelligent enough to get the right information. PDM systems aren't able to associate different bits of information together, or to assemble packets of information. PDM systems are not intelligent and they have a very limited memory. They don't have any knowledge of the way the company works. They don't know how information was used in the past. Users can't call in and say to the system "Oh, you remember that part Jack did for Bill about ten years ago......".

Current PDM systems aren't as intelligent as human beings. They're only aware of the data about products that's specifically fed to them. For many years to come, people who've spent a long time in a company will be able to outperform PDM systems in many ways. On the other hand, for simple, repetitious, boring activities that relate to data known to the system, the system will outperform human specialists.

It's all electronic. If you don't have a terminal, you won't be able to access the information. The guys on the shop floor need drawings to look at, not disks. PDM systems are too slow, skilled document management staff are much faster. PDM systems are just an added cost, we already have the people to do the job. So what's the point in having a system as well?

When a PDM system is implemented, there's a very long transition phase from the old "all-manual" data management system to the future "all-electronic" environment. For many years, there'll be occasions when the system doesn't come up to the expected performance level. There'll be many occasions when data isn't available in electronic form when it's needed. As a result it will have to be made available on paper. There'll be occasions when the system will fail to find data because someone has forgotten to enter that data. As a result, the system may seem slow and not very useful. During this period, human specialists will need to work with the system, understanding why it doesn't work in a particular case, providing data and making up for its deficiencies. During this period, the company will be paying for both the system and the humans. However the service should be much better and, as time goes on, the human component should gradually diminish. There'll be many opportunities for those of today's data management specialists who have the required skills to adapt to the new electronic environment.

Chapter 19
FAQs About PDM Implementation and Use

19.1 What PDM Functionality Do We Need?

Among the key factors in determining the functionality that a company needs from a PDM system will be the quality, quantity, and coverage of the systems that are already in place. What do these systems do? What functionality do they have? How will they fit with PDM? Is PDM seen as a replacement for these systems, is it an add-on, to what extent should it be integrated with them?

PDM systems offer a wide range of functions, such as information management, change management, process management and product structure management. Some or all of these functions may already be present in a company's existing applications. Product structure may be managed in parts master, BOM and MRP applications. Process management may be addressed in project management applications. Some information management functionality may be built into other applications such as CAD, or may be in an application developed in-house.

The functionality needed is influenced by the type of product the company makes, and its position in the value chain. A company that makes complex high value-added, customer-specified products with lots of mechanical and electronic components will need product structure management functionality, whereas a company that repetitively manufactures simple, commodity, low value-added products, will have much less need for product structure management.

For power plant and refinery operations, the main functionality required may be that which makes data readily available to operators and maintenance staff, so that downtime can be minimised.

The PDM functional needs of prime contractors and suppliers are often different. A supplier may only have responsibility for managing its design data, and not be responsible for the management of data relating to the purchase, manufacture, installation and use of the parts it supplies. This would also mean that it wouldn't be

© Springer International Publishing Switzerland 2016
J. Stark, *Product Lifecycle Management (Volume 2)*,
Decision Engineering, DOI 10.1007/978-3-319-24436-5_19

looking for company-wide process and project functionality in a PDM system. On the other hand, an engineering contractor would be much more interested in process and project management, and in the overall management of product data throughout the product life.

PDM systems can provide, along with other functionality, management and control of product data and product workflow. Usually, managers realise that such functions can be very helpful in the battle to reduce cycle times, reduce development costs, and improve product quality. However, some of the users may not be able to understand the need for these functions. They may see PDM as a way to impose a Big Brother culture, in which they lose their freedom, are under continual surveillance from the system, and will be fired by the system if it finds they don't meet work targets. Often, users who feel like this will be in companies where managers and users apparently live in different worlds, and there's no feeling of working towards a common goal. The functionality a company requires from an PDM system should take account of cultural issues like this. There would be no point in buying functionality that none of the users are going to use. It would be better to start with functionality in other areas, such as data management, where the users feel more comfortable, and only introduce other functionality once the basic system has been accepted.

The required functionality will also depend on the way the users are currently organised, and the way they'll be organised in the future. If the users are all on one site, then multi-site functionality isn't needed. On the other hand, if users are spread over several locations, multi-site functionality probably will be needed. If product development is carried out in teams, or the company has taken a Concurrent Engineering approach, then corresponding functionality would be looked for in the PDM system.

In some companies, there's very little direct communication between departments. Engineering drawings may be "thrown over the wall" to Manufacturing. In such an environment, the Engineering Department might decide to invest in a PDM system just to be able to manage and retrieve product data, but not to communicate it to other functions, or to handle cross-functional engineering change management. If this is the case, functionality that addresses inter-departmental requirements would not be needed (at least initially). Attempts to use PDM for cross-functional data management before walls between functions have been broken down generally fail. Similarly, attempts to use PDM functionality for process management will fail if the company hasn't really understood, in detail, the processes it uses to design, manufacture and support its products.

PDM systems range from simple, off-the-shelf packages to complex tailorable systems that can be further developed to exactly fit a company's requirements. A company shouldn't buy a system that needs a lot of tailoring if it doesn't have the resources to tailor it. The resources required are either money (to pay a vendor or system integrator to tailor the system) or system development staff (to do the tailoring in-house).

Two other key factors that affect the functionality that's needed are the objectives that management has set for PDM, and the money that's available for PDM. These will obviously differ from one company to another, underlining the fact that it's impossible for an outsider without any knowledge of a particular company to say what PDM functionality is needed. Instead of asking for outside help before trying to understand how the company works, companies should first try as hard as possible to understand the way they carry out activities in the product lifecycle, and the way product information is used, stored and communicated. If they can get this understanding, they'll find they don't need much outside help. If they can't get it, they're probably suffering from major structural problems. More than likely, these will prevent them implementing PDM successfully.

19.2 Who Should We Involve in PDM?

The company's culture is one of the most important factors in deciding who should be involved with PDM. Without a lot of proactive management involvement it's unlikely that the company is going to behave any differently from the way it has behaved in the past. As a result, it's likely that the people who have been involved with systems like PDM in the past will be involved with PDM.

For example, there are companies where the engineering computer systems organisation reports to IS. If the IS manager wants to run PDM in this company he/she has the power to do so. There are other companies where the engineering computer systems organisation reports to the Engineering VP, and if the Engineering VP wants to be responsible for PDM, will be.

There are also a lot of companies where the situation isn't so clear. In some cases, the CAD support team will see PDM as an extension of their responsibilities, in some cases the field service systems support team will see it as their job, and in others the Manufacturing systems support team will claim responsibility.

Another issue that affects PDM responsibilities is the extent to which top management maintains long-term commitment. There are companies in which top management will initiate PDM activities and keep them moving for many years. In this environment, PDM should succeed wherever it's located. In other companies though, top management interest and support will disappear if there isn't immediate success. In this case, the PDM team leader needs to have a solid power base to help overcome any slippage in implementation. If PDM is only being implemented in one department this is less of a problem, but if the intention is for it to be cross-functional, the project leader needs to be very powerful and have a lot of support.

In some companies, top management will not actively support PDM, assuming that if it's really important, then middle managers will make sure that it's implemented. In these companies, the danger is that departmental warlords will fight over the right to introduce PDM, and in doing so, prevent its successful implementation.

True PDM is a very cross-functional activity, and it won't succeed without the co-ordinated support of all the major departments involved.

The vision of PDM, and its relationship to the company's vision is another key factor. If the company has a very functional view of its future, then a vision of PDM as a cross-functional system may be difficult to implement. If the Engineering VP is only measured on the success of the Engineering Department (e.g. measured on the number of engineering drawings produced per year), while the Manufacturing VP is only measured on the success of the Manufacturing Department (e.g. measured on shipments per month), neither may see any reason to get involved with cross-functional PDM.

Apart from the relative strengths and objectives of the IS and the Engineering Departments, there are other organisational factors that will affect the extent to which different people are involved in PDM. In some highly centralised companies, there's an Organisation Department or a Projects Department that's responsible for the structure and organisation of operations. This Department could be very involved in defining how PDM should be used, but would play no role in PDM use or operation. Some companies have distributed Divisions organised along product lines, but a centralised core engineering function. Here the central engineering group could play a leading role in introducing PDM, while the Divisions could take responsibility for implementation and use.

Part of the answer to the involvement in PDM is directly related to the industry sector and the type of product. In aerospace and automotive companies there will often be a group with special responsibility for PDM-like functionality, and in many cases this group will already have implemented a partial PDM solution. In the mechanical engineering sector, involvement in PDM is more likely to come from the IS Department than in the electronics sector where the Engineering function tends to be more powerful. Of course, these are only generalities. Individual companies, even making similar products, have very different organisations, and their particular structures may even result from the relative influence of individual VPs.

The availability of individuals who can lead the PDM effort, take responsibility for its implementation, be responsible for its use, manage its day-to-day operations, and ensure its further development, will depend on the cultural and organisational issues addressed above. PDM implementation and use requires a good blend of skills, and if one part of the company is dominant, certain skills may be missing. If the IS organisation is very strong, it will be easy to find people who know about data modelling and data bases, but it may be difficult to find those who combine a good knowledge of IS with knowledge of the details of product development.

A key factor in deciding future PDM involvement is past performance with PDM. If the people who have been involved in the past have been successful, they'll probably be allowed to continue. If they haven't been successful, a change of responsibility may be needed. Implementing PDM is difficult and, often, a lack of success is due to a lack of power, influence and general PDM awareness.

19.3   How Do We Cost-Justify PDM?

For questions involving cost-justification, the two key factors appear to be the industry sector, and the extent to which the company's culture allows intangible benefits to be used in cost-benefit calculations.

The industry sector is important because there a few sectors (such as leading-edge aerospace, automotive and electronics sectors) where companies feel they must get involved with PDM and long-term product data management requirements. The reasons for this may be either that they know their competitors are doing it, or because their customers are requesting it. In these cases, cost-justification plays a minor role, and PDM system introduction and selection is based primarily on the technical capability of particular PDM products to meet short-term requirements.

Although this is the way things are in practice, it could be argued that it doesn't lead to the best results for the users, since they're not forced to understand in detail how they'll use PDM, or how they'll measure its success. "You get what you measure", so users who aren't forced to define business-related measures of PDM success aren't likely to develop solutions that solve business problems. They may judge success by measures such as the number of PDM seats installed, and this may have no relationship to business requirements such as reducing product development times, or increasing product quality. Even in sectors where cost-justification may appear unnecessary, it can therefore be helpful to address it.

In most sectors, the question of cost-justification has to be addressed and answered in detail. First of all, it has to be shown that the money to be spent on PDM could not be more beneficially invested elsewhere in the company. This shouldn't be too difficult. In theory, PDM can lead to major benefits, e.g. reducing product development times by 25 %, reducing engineering cost by 15 %, and improving product quality. There are not many other investments that a company can make which offer this type of benefit.

Once this general comparison has been made, a cost/benefit analysis has to be carried out to highlight the differences between individual PDM systems. Whatever basis of calculation is used for this analysis, whether payback, Net Present Value, or Discounted Cash Flow Return on Investment, the problem is basically the same. The costs of PDM are easy to identify and measure, but the benefits are very difficult to measure. As the costs are easy to measure, they don't cause a problem for cost-justification. PDM vendors can tell you how much their systems cost, what kind of support they need, and how much system maintenance will cost in the future.

It's the measurement of the benefits of PDM that causes the problems. Benefits can only occur in one of two ways. Either they come from a reduction in the company's costs or from an increase in sales. At the level of a general comparison between PDM and another technology, or capital expenditure, it's not necessary to go into a lot of detail about exactly where these benefits will occur. The calculation takes place at a macro level, at which it may be acceptable to assume a 25 % reduction in lead time, resulting in a reduction from 12 to 9 months. However, when comparisons are made between different PDM systems, there has to be a deeper understanding of exactly how, where, and why, improvements will occur.

The relationship between the use of a particular PDM system and these factors isn't always clear. Where it's clear, the benefits are referred to as tangible, or direct, benefits. Where there's not a direct relationship, the benefits are referred to as intangible benefits. Tangible benefits are relatively easy to identify, but there are not many of them, and they're often difficult to achieve in practice.

One of the most tangible benefits is to shut down a department or group, and fire everyone who was connected with it. In theory, the company knows how much the department cost to run, so it will know how much it will save by closing it. In practice, of course, it's extremely difficult to shut down an entire department, or group, and fire everyone involved with it. Some of the people generally have useful know-how, so they're kept on, some of the functions are transferred to other groups, so these will have less time to do whatever they were doing before. Then there's the issue of overhead absorption. Overhead is often shared out between departments, so if one department is closed, the overhead charge on other departments will increase. Extending this logic, it soon becomes apparent that, under most circumstances, the actual benefit will be nowhere near what it might appear to be. Another issue that's difficult to handle is that some direct benefits achievable with PDM could also be achieved by redefining work practices, and may not be fully attributable to the introduction of PDM.

For most companies, the major benefits to be expected from PDM are intangible benefits related to increased sales. This is where the cultural problem starts, since there's no generally agreed principle for estimating intangible benefits. The financial formulae are not the problem, since they're well-known. The problem is to know what values to use in the formulae. The types of benefit that can occur are also well-known. The problem is to assign agreed values to them.

For example, one person in a company may estimate that PDM will reduce product costs by $10 million, and because of reduced product development cycles, increase sales by $50 million. Another person may estimate a sales increase of $30 million, and a reduction in product costs of $30 million. A first step to finding a reasonable and widely accepted figure is to take a cross-functional approach, in which people from all departments, including finance, are involved in the cost-justification process from the beginning, so that they all really understand what they're doing and take responsibility for the conclusions. The Finance Department has to be involved from the outset, so that it really understands the values proposed, can explain the calculations, and can help to find consensus among the departments.

One of the results of this activity is generally to show that a lot of the costs of PDM are in Engineering, while the benefits are found in other departments. Looked at from a pure Engineering view, the costs may exceed the benefits, particularly in the short term. Without a cross-functional approach this could mean that the PDM opportunity would be neglected.

19.4   Does PDM Fit with Concurrent Engineering?

People sometimes ask if a company needs both PDM and Concurrent Engineering, and if so, should it introduce PDM before, or after, Concurrent Engineering, or should it do the two together. The answers to these questions depend a lot on the current situation in a company, the company's culture, and the way it's organised.

In practice, many companies claim to be doing Concurrent Engineering if they manage to carry out some parts of the product development process in parallel rather than in series. Others claim they're doing Concurrent Engineering if they have set up teams, or have bought a Design For Manufacture program, or trained people in Quality Function Deployment. The common characteristic among these activities seems to be something to do with working on several activities at once.

Most activities in product development exist to create product data or to communicate it to someone else. A lot of data is created to describe activities. Deciding if the activity needs to be addressed before the information is like trying to decide if the chicken comes before the egg. Most people have an underlying understanding of the concepts of "chicken" and "egg", and the relationship between them, yet still find the chicken and egg question difficult to answer. The corresponding underlying understanding is generally missing in the Concurrent Engineering/PDM debate, hence the general difficulty in answering the question.

Concurrent Engineering isn't a computer system, and it isn't based on a computer system. Concurrent Engineering is a philosophical approach to the way engineering should be carried out. It's easy to see if a system is implemented, but extremely difficult to measure to what extent a philosophy is being followed. Philosophies are "soft" issues. They are difficult to sell, implement and measure. In general, computer systems are "hard", and they're easier to sell, implement and measure. The relationship between PDM and Concurrent Engineering is one of synergy. Concurrent Engineering provides the right "soft" environment, and PDM provides the "hard" backing to make sure it works. The system makes sure that engineering is carried out in the way the Concurrent Engineering philosophy says it should be. In the absence of PDM, there may be nothing to prevent people going back to serial engineering.

If a company has a good, well-balanced understanding of all these issues, it probably won't even try to prioritise Concurrent Engineering and PDM. Implementation will come naturally. A little PDM and a little Concurrent Engineering. Then a pause to see what progress has been made. Then some more PDM and some more Concurrent Engineering. The two will be implemented side-by-side in the effort to achieve a particular goal.

An unbalanced view leaning too far towards Concurrent Engineering will produce a surfeit of activities involving teams, quality circles, ISO 9001, process modelling, QFD, and the DFX acronyms (Design for Manufacture, Assembly, Sustainability, Decommissioning, Environment, etc.). Of course all these are important but, until information use, structure, flow and processing are addressed, only a fraction of the expected results will be obtained.

Similarly, many companies have an unbalanced view leaning too far towards PDM. This will lead to a lot of discussion about distributed databases, entity relationships, Level 2 data modelling, schemas, interoperability, enablers, EAI and STEP. Of course, all of these are important, but until the process issues are addressed, only a fraction of the expected results will be obtained.

It's often difficult to find where the right balance lies, and just how receptive a company's culture and organisation will be. Positive signs include use of applications that manage the flow of work through engineering processes, or that apply rules to the way that product development is carried out. The extent to which different departments work closely together is another key indicator. If they can work closely together, it often implies that they have understood how their processes use data, and how product data has to be structured and organised for shared use. Another key measure is the extent to which people are trained to think beyond the boundaries of their own department.

Companies that don't understand these issues, or can't understand that both Concurrent Engineering and PDM are important, need to find out more about them before starting implementation. Otherwise the implementation will be costly, time-wasting and, eventually, a failure.

Companies that focus only on Concurrent Engineering will probably succeed in the short-term, but then find they can't build on this success, and may even find that it slips away from them. Companies that focus on small-scale PDM, and, for example, only apply it for CAD data, will achieve very limited results. Companies that focus only on PDM, and try to apply it company-wide without addressing the process issues, will fail because they'll run into endless problems caused by inefficient and uncontrollable processes.

Unbalanced champions, of one side or the other, need to understand that Concurrent Engineering and PDM are both partial answers to the problems of uncompetitive organisations. They both aim to improve product quality. They both aim to reduce lead times. They both aim to reduce costs.

Organisations aim to provide customer satisfaction. Companies need to have an overall picture of what they're going to do to achieve this objective. They should recognise that they need to improve the management of both data and processes, and that the resulting action plans should address concurrent improvement of product data management and product workflow management.

19.5   How Should We Introduce PDM?

One of the most important factors in deciding how to introduce PDM is the level of PDM understanding in a company. If the company understands very little about PDM, then a feasibility study will be helpful. If it has a basic, but incomplete, understanding, then a prototype will be helpful. If it really understands how PDM will be used, then it may need neither feasibility study nor prototype.

There are two types of feasibility to address. There's the feasibility of PDM as a technology, and there's the ability of a particular company to apply PDM. It's fairly easy to demonstrate the feasibility of PDM as a technology. There's a lot of published information, and PDM vendors can generally show results from their customers. It's much more difficult to demonstrate the feasibility of using PDM within a particular company. However, it's also much more useful, because it implies understanding how to use PDM, where to use PDM, and how to organise for PDM. It means understanding where the benefits of PDM will occur, and how they'll relate to the corresponding costs. This type of feasibility study really helps a company improve its understanding of PDM.

A well-planned prototype can be a useful tool for increasing the overall awareness of PDM, and learning enough about it to show that wider-scale implementation will be successful. All too often, though, prototypes take place in conditions that don't represent normal use, or they're hijacked by a few individuals with the result that the overall gain in experience is negligible. A prototype carried out by the Central Research Laboratory to test the use of PDM is unlikely to be meaningful, whereas a prototype supporting a small, real-life, market-driven development project should be very useful. It must be remembered, though, that the prototype needs to be well-planned. The objective and deliverables must be clearly defined before the prototype starts. The experience gained must be regularly communicated. Unless the prototype takes place on a live project, it's unlikely to yield maximum benefit. If the project isn't essential, or is seen to be unrealistic, interest will soon flag, people will move, or be moved, to essential projects, and the eventual result may be disillusion with PDM. In such a case, it might have been better not to have started the prototype. A prototype is by no means obligatory. Some successful users of PDM did not prototype. Some companies that did prototype created unnecessary expense and problems.

In a Big Bang approach, a company could switch overnight from only using manual product data management to only using computer-based PDM. Unless a company is extremely well-organised, and has all its processes in order, a Big Bang approach should not be tried. Mega-projects like this rarely work. A step-by-step approach is much more likely to lead to success. Although a Big Bang approach may appear quick and simple, it's generally the opposite. A successful Big Bang only comes after a very long preparation and planning phase, during which many people lose interest and are demotivated by the lack of results. A step-by-step approach, starting with an overall multi-step plan, is more likely to succeed. As each step is implemented, people can see and appreciate the progress. Should any problems occur, these can be resolved before they get out of hand.

The decision as to where the initial implementation of PDM should take place, will vary from one sector to another. In some industry sectors, there's enough pressure from competitors and customers to overcome any doubts about getting started with PDM. In these sectors, Engineering is often a strategically important function, and able to act fairly independently. As a result, there's a tendency for PDM to be implemented bottom-up, starting within the Engineering Department, and spreading out to other areas of the company. The opposite occurs in sectors where Engineering is less important, and the Engineering Department is comparatively weak. In these cases, top management may be the driving force behind PDM, and may view initial use in the Engineering Department as an uninspiring approach.

The choice between implementing PDM in one department or cross-functionally depends a lot on the current organisational structure. Companies that are used to working with cross-functional teams and projects will find it easier to implement PDM cross-functionally than those where each department is surrounded by high walls. For the former, a cross-functional project will be a good place to start a prototype of PDM. In a company where the departments are very separate, cross-functional PDM will be nearly impossible to implement, so PDM will probably start within a single department, even though this may make it more difficult to go cross-functional later.

The choice between top-down and bottom-up implementation depends a lot on the culture of the company. A bottom-up approach is going to be difficult in a company where management is used to running things top-down. More likely, this type of management will favour a top-down, Big Bang approach. In companies where there's a more consensual, less autocratic, culture, a step-by-step, department-by-department approach will probably be preferred.

The computer systems in the company also play a part in the decision. If there are a lot of different, unconnected applications in place, each needing to be interfaced to PDM, then a Big Bang approach, or an immediate company-wide approach, will be difficult. On the other hand, if most of the applications are already integrated, it should be easy to integrate them with PDM. The way that applications were implemented in the past will also affect the approach. Companies often prefer an approach with which they have succeeded in the past, rather than one which has failed.

Another factor that weighs heavily in the decision is the skill level in the company. There's no point in suggesting a company-wide or Big Bang approach if the company doesn't have the people available to support or use it, and can't afford the money to buy in outsiders to do the work. PDM requires a lot of effort for implementation, training, cleaning data, data load, user support, and process improvement. The requirements at each stage of the product lifecycle are different, and different skills are needed at each part of the process.

Perhaps the most important factor in choosing the approach is the PDM experience so far. If the general level of PDM awareness is low, then the approach should be cautious and include a lot of training. If there has been little support from management, it's unlikely that a top-down approach will be appropriate. If different departments can't agree on what they want to do with PDM, then a company-wide approach will be unsuccessful. If a department hasn't shown any interest, it wouldn't be sensible to propose starting with PDM in that department. If no-one has taken the work done so far very seriously, it's probably too early to start thinking about implementing PDM, and the best approach is to try to increase awareness.

19.6 Should We Buy or Make PDM?

Off-the-shelf systems can be purchased and implemented quickly. The systems are mass-produced and aimed at the mass market, so they're relatively cheap, they work, and they can be used immediately. They are used by many companies, so a site visit can be arranged to show the system in action. The disadvantage of an off-the-shelf system is that it may not correspond exactly to a particular company's needs. It may not be powerful enough to handle all the company's product data, it may not have all the functionality that's required, or it may not allow the company to carry out its processes exactly as it did in the past.

On the other hand, a system can be built, or tailored, so that it fits perfectly to a particular company's needs. It can handle the types and volumes of product data the company uses. It has precisely the functionality the company needs. Its workflow functions correspond exactly to the way the company wants to work. The disadvantage of a tailored and unique system is that it's built expressly to the company's requirements. This takes time, so the system isn't available immediately. It also requires development effort, so the system isn't as cheap as an off-the-shelf system. Since the system is unique there's no way, prior to purchase, to visit a company and see it being used. Since the system is unique, it will require a lot of maintenance effort.

In some industry sectors, PDM needs are already well-known and relatively standard, so companies in these sectors will find that many PDM systems either meet or come close to meeting their needs. PDM vendors have had time to develop basic systems that fit the "average" requirements of these industries. In other industry sectors, a company, particularly if it's the first company in the sector to get into PDM, may find that none of the PDM systems on the market offers appropriate functionality.

Whether, or not, a company finds an appropriate off-the-shelf PDM system also depends greatly on its processes, applications, organisation and current approach to product data management.

A company is more likely to find a suitable off-the-shelf system if its product workflow is simple, and its processes are not too complicated. Many PDM systems only have basic workflow management functionality, but this is sufficient for the many organisations that have simple, straightforward processes.

A company is more likely to find a suitable off-the-shelf system if the other applications it uses are widely used by many companies. For example, many PDM systems have tight interfaces to one or more CAD applications, but most of these interfaces have been built to one of a small number of popular CAD applications. Companies using one of these CAD applications are likely to find that the vendor of an off-the-shelf system has developed an interface to their system. Off-the-shelf PDM system vendors are not interested in investing heavily in building interfaces to CAD applications with few users.

A company is more likely to find a suitable off-the-shelf system if its organisation is well-structured and clearly defined. It can be difficult enough for a well-organised company to define its requirements clearly, and select a corresponding system. If there's an overlap of warring departments, each with its own uncoordinated requirements, it will be almost impossible to find a ready-to-use system.

A company is more likely to find a suitable off-the-shelf system if it has little exchange of data with customers and vendors. Exchange of data is often very company-specific so, if data is exchanged with several vendors, customisation will probably be needed.

A company is more likely to find a suitable off-the-shelf system if it has already gained control of its product data. If, however, it's not clear who owns which data elements, and for example, there's multiple and inconsistent definition of each data element, then it's unlikely that an off-the-shelf system will give good results.

19.7 Should We Outsource PDM?

Outsourcing of PDM occurs for a variety of reasons. Some companies find that it's cheaper for them to outsource PDM than to do it themselves. They outsource and save money. Other companies don't have the human resources to run PDM in-house, so outsource it.

In other cases, a company will outsource because it wants to focus on the activities that it does best, the activities where it gains competitive advantage. Many companies compete on the quality of their design work, and they want to focus all their resources on producing the best designs. Such companies may feel that there's little to be gained by building up a large in-house team to manage their product data, so they outsource this activity.

On the other hand, there are, of course, many cases where companies believe that the management of product data is a potential source of competitive advantage. They may believe, for example, that a skilled in-house team can provide just the right information service that's needed to win competitive advantage, and that by outsourcing they would lose this opportunity. They may believe that an in-house team will provide better access to information, enable the incorporation of more existing designs in new products, and reduce new product development cycles.

Most companies' PDM needs lie between the two extremes of companies that see no benefit from in-house product information management, and those that see it as a major source of competitive advantage, and they'll have divided feelings about the real value of in-house PDM. There will be people in these companies who will see PDM as an overhead, an added cost, another Information Systems activity that wastes valuable management time. There will be other people who believe that their product data is, or would be, of great value when managed by people with the appropriate skills.

In such companies, the debate is often emotional and quickly becomes polarised. People on both sides of the argument become irrational believers in their cause. In such a situation, it's often useful to get a neutral external specialist to look at the issues, identify possible solutions, and carry out a cost/benefit analysis. Often, some possible solutions may have been overlooked, since there's a wide range of outsourcing options. In some cases, PDM is outsourced to an organisation that comes in-house to manage the system. In other cases, the information itself may go outside to the outsourcing organisation.

It will often be found that many secondary factors have to be taken into consideration when examining the issues related to outsourcing, for example the abilities of the current PDM team, the likely future availability of PDM specialists, the current status of the organisation's product lifecycle processes and applications, and the views and structure of the IS organisation.

Another set of factors to be addressed includes those in the external environment. The reputation of the company that will take on the outsourcing activity will have to be examined. Other companies receiving a similar service should be asked about the level of service it provides.

Questions need to be asked about the PDM packages currently supported by the service company, the interfaces that are available, and the availability of staff to carry out development work. Security issues should be addressed. Increased information security is one of the reasons many companies give for using a PDM system. Outsourcing the PDM activity could lead to security problems. The service company must be able to guarantee data integrity and confidentiality.

Other issues to be addressed include those concerning the ownership of the PDM systems used, and the rights to use the systems if the outsourcing agreement is terminated. A similar issue arises with custom-built software. What rights will the organisation have to use special software developed by the service company if the outsourcing agreement is terminated? Another issue is the cost of custom-built software. How can the organisation be sure it's not over-charged? How can it be sure that any additional software will be developed in a reasonable time?

The costs and responsibilities of setting up the outsourced service have to be defined. Each party must be clear as to the tasks it will carry out. Who will input data into the new system? Who will be responsible for data and system conversion?

The costs and responsibilities of operating the outsourced service have to be defined. Each party must be clear as to the tasks it will carry out, and the boundaries of its sphere of action. Rules have to be defined about the location of data, the on-site and off-site storage of data, and the transport of data. Access rights to the data have to be defined for both parties. Data ownership has to be considered. Although all the data may initially belong to the organisation that has decided to outsource its PDM activities, later the service provider may develop new information, and claim rights to it. Data corruption is another key issue. Who will be held responsible for corrupt data? Will the responsibility for correcting corrupt data depend on the source of corruption? Who has the responsibility for producing user documentation? Who has the responsibility for producing system documentation?

Operating conditions and performance criteria must be defined. Backup and recovery procedures and schedules must be agreed. Availability of service, and access times to data have to be defined.

Cost quotations from the service provider need to be examined in detail. How much is fixed cost? How much is time-and-materials? How much has only been estimated? What is potentially the maximum cost? What additional costs might arise?

Future requirements need to identified. Can they be met by the service provider? Which computers would they run on? Could they be run in-house? Would they be run by a third party?

Cancellation issues need to be addressed. If the organisation cancels the outsourcing agreement, what rights will it have to the programs that have been used? What rights will it have to programs that have been developed or converted? Will it get all its data back? Will it be able to keep any documentation that the service provider has developed?

Although it may appear from the above that an outsourcing agreement may be very difficult to set up, in fact so difficult that outsourcing should not be considered, it should be remembered that most of the above questions should also be answered by an in-house PDM team. In some cases, it may be that the in-house PDM team believes it does not need to answer such questions, or does not have the time to answer them, but it should realise that by answering them it will be providing a useful service to many users.

Chapter 20
Barriers to Successful Implementation of PDM

20.1 The System Barrier

This category contains problems related to the PDM system and the system vendor. It includes problems such as incomplete system functionality, malfunction of the system, poor response time, and unavailability of the system on a wide range of platforms. The system may suffer from limited customisability, lack of up-grading, using too much computer power, and limited interfaces. This category also includes problems such as oversold functionality and the vendor going out of business.

Problems related to the PDM system can surface at different times. They can occur during start-up, during initial use, when new versions appear, when the system vendor has problems, or at any time in everyday use.

Some problems may become apparent as soon as the system is installed. For example, the system may crash or malfunction frequently, users may complain that it's not user-friendly and takes too long to learn to use, and the response times may be too long.

After the start-up period, system problems can come from two sides, the users and the support staff. Users may continue to run into the problems encountered at start-up. In particular, they may suffer from poor response time as the amount of product data in the system increases. System upgrades may be necessary and the result may be that system use becomes too expensive. In some areas, the system may not behave as expected. Time-wasting work-arounds may be necessary. As users get to know the system, they may find that functionality has been oversold. Functions they need may not exist, or may only be partially implemented. The system may only handle a limited number of document types. Documentation and on-line help may not exist for some key functions. There may be no guidelines describing how the system should be used. Necessary customisation may be too difficult or too time-consuming for users.

The people involved in PDM system administration and support will hear all about the problems that users are having. They may also have their own problems.

© Springer International Publishing Switzerland 2016   371
J. Stark, *Product Lifecycle Management (Volume 2)*,
Decision Engineering, DOI 10.1007/978-3-319-24436-5_20

The system may be difficult to set up for more than a prototype implementation. System administration may be inflexible, and too time-consuming and error-prone. Previously hidden limitations may appear in many areas such as definition of roles and processes, creation of reports, and location of data. The system may not actually work on all the platforms where it's needed. It may be difficult to integrate with the design, analysis, manufacturing and project management applications the company is using. The vendor may not be able to provide good, well-trained support staff who really understand the system in detail.

As time goes on, it may become clear that the wrong vendor was chosen. New versions may be delivered late, lack promised functionality, and have quality problems. Maintenance costs may become unacceptably high. There may be no upgrade path between successive versions. Key individuals may leave the vendor. Eventually the vendor may go out of business. It's probable that many of today's PDM users will run into these problems and many of today's PDM vendors will meet this fate. This is probably not as catastrophic as it sounds. There's usually plenty of warning before a vendor disappears. Provided the system is working when the vendor goes out of business, it should continue to work. There will be time to look around for a new system. By this time, PDM systems will have moved on to another generation and have much more functionality. The company will have learned from its experience. It will have made progress in organising its information and processes. When it gets its new system, it will be able to move on and make much better use of PDM.

20.2   The People Barrier

Some of the reasons in this category are related to top management, some to middle management, and some to users. Other reasons are related to the PDM project team, or to the influence of other people such as consultants. Top management may be a source of problems for reasons such as lack of commitment, lack of leadership, lack of support and lack of patience. Problems at the middle management level may be due to conflicts with personal goals, empire-building, and fear of loss of power. Users may fear that the PDM system may play a Big Brother role, or may lead to job losses. Problems can also arise if the members of the PDM project team do not work together effectively.

Top managers should be among the leading supporters of PDM since they'll benefit greatly by its success. However, top managers are likely to have many other things apart from PDM on their minds, and may not provide the expected support. Top managers may be so involved with other issues that they don't become sufficiently aware of the potential benefits of PDM before it's implemented. As a result, if things go wrong, they may not have a sufficiently deep understanding of its importance to keep providing the necessary support. Also, if they aren't sufficiently aware of the benefits, they may not be able to provide the necessary leadership, with the result that the implementation drifts out of control. Again, if they aren't sufficiently aware of the issues and benefits of PDM, they may be disappointed by the lack of rapid payback, and not be willing to be patient and wait for results to come through. Later in the project, top managers may not want to implement the organisational changes that may be necessary for the success of PDM.

To prevent these problems arising, it's important to make top managers aware of all the issues (e.g. likely costs, benefits, problems) surrounding PDM before the project starts. If they don't take any interest and don't have the time to listen, then it's best not to start the project, since it's probably going to fail due to lack of management support. Provided that managers do become aware of the issues, the next important point is to avoid overselling the benefits of PDM. Far too many people do this because they believe strongly in PDM and they expect everything will work out as they hope. In accordance with Murphy's Law though, things won't work out, and those who have oversold the potential benefits are likely to be in trouble.

The next important point is to have a step-by-step plan for implementation, with progress to the next step being dependent on completion of the previous step. This way, the project and the associated risk should be under control. It's important to make it clear right from the start that a PDM system will only provide significant benefits if the appropriate organisational changes are made. It's a good idea to have an interlocking plan in which system usage and organisational improvements run side by side at each step. This way, the project can be held back until top management makes the promised organisational improvements, and the danger of continuing the implementation without any prospect of the expected benefits can be reduced. It's a good idea to have a fall-back plan in reserve in case top management does decide to change its mind and not go ahead with organisational change.

It's often middle managers and project managers who most impede implementation of PDM. They may have the responsibility and power to make it work, but they also face many of the risks accompanying its implementation. The changes and benefits that PDM may result in, may not be in line with a middle manager's personal goals.

The organisational changes accompanying the introduction of PDM quite often reduce the power of functional departments, and consequently, the power of the managers of these departments. Another problem is that the costs of PDM often fall in one department, but the benefits in another. As a result, from the departmental point of view, PDM may appear unattractive. It's important to make it clear before implementation starts what the likely costs and benefits are going to be, and how they're going to be accounted for. That way, everyone can see prior to implementation what the planned results should be, and the plan can be adjusted if necessary before implementation starts.

Project managers on tight schedules may not want to run the risks and dangers of using an unknown system. However, there has to be a guinea-pig project that's the first user of PDM. To reduce the risks, it's best to run a pilot for several months to iron out any obvious problems and get people used to working with the system. It's not only the risks that have to be considered. There's glory for the project manager who accepts the possible risks, but nevertheless uses PDM and gets significant benefits.

The users of the system are another potential barrier to the effective use of PDM. If they decide they don't want to work with the system, then it's clearly not going to meet objectives, so it's important to identify their potential problems and try to avoid them. There may be things they don't like about the system. These could include the user interface, the response time, and the fact it doesn't work the way they used to. It may not be the system they wanted.

To overcome these problems, it's best to involve the users in the work of the PDM project team. Make them fully aware of the reasons for its introduction, getting them involved in system selection, and make sure they receive suitable training. It's good to get them to understand why performance has to be improved, and involve them in performance improvement.

There can also be other reasons why users may not like PDM. They may see it as a Big Brother system that will look over their shoulders, watching to see how they are behaving and telling them what to do. Other users may fear that its introduction will result in job losses. Worries such as these can be eased by keeping users fully aware of the reasons for introducing PDM. However, the underlying reasons are cultural, and imply a lack of trust between users and management.

20.3   The Project Team Barrier

One potential problem with a project team set up to introduce PDM is that it doesn't function as a team. Instead it functions as a set of individuals, each with their own agenda. Other problems can be that the team leader isn't a good leader, that the procedures of running the team take precedence over the content of work, and that the team doesn't have clear objectives. The result of these problems will be that the team doesn't function effectively, and as a result, either doesn't develop a solution or doesn't develop the right solution. To avoid these problems, a lot of effort has to be put into setting up the team and the project. It's important to select a suitable team leader and to make sure the people in the team know how to work together and know what they are working towards. It may be necessary to give some training in team-working before starting the project.

Once the project has started, it may be found that some of the team members think that they already know everything, want to select a system very quickly, don't want to involve users in the project, have already taken their final decisions, and don't intend to implement or use the system.

Selecting a system is just one of the many activities that have to be considered when introducing PDM. Team members need to understand that their role is to introduce PDM, not just select a system. It's unlikely they'll know everything. They'll probably need to learn a lot about the way the organisation manages product data, and the way it could manage product data in the future. They'll need to learn about PDM systems, workflow, cost-justification, overcoming resistance to change, etc. They'll be able to get some of this information from journals, the Web, books and conferences. They'll probably learn a lot by visiting other companies that are introducing or using PDM.

Users should be involved in the PDM project team for several reasons. They know how the organisation works today, and they know the type of functionality they need. Also, if they participate in the project they are far more likely to work with the eventual solution than if it's imposed on them. It's never a good idea to have a lot of team members who are only involved in the project to select a system, and not in implementing and using the chosen solution. Again it's a question of involvement. It's easy to choose a solution if you don't have to use it, you're not putting your own head on the block. People who are going to use the selected system are likely to be much more thorough in their evaluation, and be influenced by practical considerations rather than academic theory or corporate correctness.

Once the PDM system has been implemented, there are often problems among the PDM system support team. After all the excitement of selecting a system, people may not want to face the tedium of everyday support. The support team may find itself being attacked by the users because the system doesn't function as promised. Of course this is probably nothing to do with the support team. More likely it's because the project team didn't do its job properly. If, as often happens, the project team doesn't take the selected system through to use, then the only people at hand to blame are the members of the support team. The support team may also feel demotivated because the project team overspent its budget, with the result that there's less money available for support. This may mean that the support team doesn't receive the training it needs, or the support team is reduced in size, with the result that there's no-one available to develop necessary procedures, develop standards, or keep records of system use. To avoid these problems, it's best to make sure that some members of the support team participate in the project team, and that one of the tasks of the project team is to define the tasks, responsibilities, resources and budget of the support team.

20.4 The Process Barrier

If all goes well, the PDM system, once installed, will support the flow and use of information throughout the processes of the product lifecycle. It will manage the workflow and make sure that people have information when they need it. Unfortunately, things don't always turn out so well. Many implementation problems are due to poor understanding and definition of the processes. Problems may arise because the system doesn't address the parts of the overall process that the company is interested in. Problems can also arise at the level of individual activities if the way the system works, doesn't correspond to the way the company wants to carry out activities such as release and engineering change control.

It's important that the process be understood and clearly defined. Otherwise it's going to be difficult to use PDM to support it. How can you support something you don't understand? PDM can't be used effectively if it's not clear how information flows in the process, what it's being used for, who has access rights at different points in the process, what the individual steps of the process are, what happens at each step, or what conditions have to be met before moving from one step to another.

In many companies, it's difficult to find anybody who understands the overall process in detail. Usually very few people are interested in understanding it, and many don't even understand what is meant by 'the process'.

As many people don't understand the process, a lot of training and analysis will be needed to get them to understand it, and to get it defined. This will take a long time and cost a lot of money. The progress of the PDM project will slow down and its costs will rise. Management sponsors of PDM will be unhappy with the lack of progress and will blame the PDM team members. Some team members may decide that they'll focus on managing product data as they can't see how to manage the process.

Another problem that can arise is that the people who are supposed to look at the process issues can't agree among themselves as to what the process should be. They may be from different groups or departments, and have different views of the process that can't be reconciled quickly. As long as they continue to disagree, and fail to define the process, the PDM project will make slower progress than expected.

When the current process is finally understood, it will probably turn out to be very disorganised, much longer than it should be, and lacking quality assurance. As a result, use of the process will be causing development cycles to be too long and development costs to be too high. Suggestions will be made that the process should be improved. While it's being improved, more time will be lost. Management may not want to make the changes required to bring about the necessary improvements. The process may be so inflexible, and so deeply anchored by historical and departmental barriers, that it's difficult to change.

All these problems will slow down the introduction of PDM, creating many problems for the PDM team. A major problem for the team will be that as a result of their inability to move forward, they'll fail to meet their targets. The promises they made to get a particular activity working with PDM won't be kept. People will lose confidence in the team. As the team waits for the process to be re-organised, costs will rise, and before long all the PDM money will have been spent, but there'll be little to show for it.

Even worse, people in the process will claim that instead of reducing development cycles and costs, and improving quality, the introduction of PDM has done the opposite, lengthening developing cycles, increasing development costs and creating quality problems. They may well be right. It's quite likely that as a result of efforts to change the process there'll be an initial phase, before the new process is understood by everybody, during which some things take longer and many problems occur.

Rather than face all the above problems, the PDM team may decide that it's best not to get involved with process improvement. Unfortunately, there are also many problems with a 'papering over the cracks' solution. In most companies, the current process is so disorganised that unless it's addressed and improved, the PDM project will fail to meet its targets, and as a result the PDM team will suffer. The PDM team may have claimed for example that the introduction of PDM will reduce product development cycles by 20 %. It's quite possible that the development cycle can be reduced by 20 %. But this target can't be met by applying PDM to a disorganised development process. The development cycle time is just the time it takes to go through the development process, so to reduce development cycles the development process has to be reorganised. PDM can then provide the necessary rigour and support for better use of information and better flow of work in the process.

The PDM team must avoid proposing solutions that correspond to one person's or one department's view of the process but aren't accepted by others. If they don't, they'll find that many people will refuse to use the proposed solution. And as a result, there'll be fewer gains from PDM than there should be.

To avoid problems with the process, the PDM team should separate out the process activities from the PDM activities. If possible, process improvement should be addressed before the PDM project starts, and the team should make it clear in the project plan what has to be done on the process side, and what will happen if these activities aren't carried out. A step-by-step approach should be taken in which PDM activities are linked with process improvement activities in each step. Until the process improvement activities of a step have been carried out, the team shouldn't attempt to start the PDM activities of the following step.

20.5   The Organisational Structure Barrier

The fundamental source of problems in the organisation category is the difficulty of working in a departmental organisation. Problems that may affect use of PDM include lack of agreement and co-operation between departments, difficulties in getting cross-functional activities to occur, departmental barriers preventing information flow, and departments using different definitions and standards. Problems can also arise outside the company, for example with customers and suppliers.

In the departmental organisation, each department is completely responsible for its activities. Each department has its own leader, its own budget, its own practices and techniques, its own computer systems, its own information, its own processes, its own offices, its own jargon, etc.

The root of the problems that occur when a company tries to make use of PDM in a departmental organisation is that PDM is cross-functional. It doesn't belong to any one of the functional departments. As a result, it's unclear who is responsible for it. It's not obvious how it will be financed. It's not obvious which practices should be followed in addressing it, which jargon should be used to describe it, or which rules should be followed when managing information in a PDM system.

There will be problems cost-justifying the PDM system. Which department or departments should pay the costs of a system that's used by several departments? How should costs be distributed so that the department that gets the most benefit pays the most? How can the running costs of the system be shared equitably? This is especially difficult to achieve if the system is installed in one department, supported by people from another department, and used by people from many other departments.

In the departmental organisation, in which every department has to be responsible for everything within its own walls, each department will claim ownership of the product data it uses, and define and structure it the way it believes is best for its own particular needs. A lot of product data is used by several departments. Often it's defined differently, and given different structures, by each department that claims to own it. An attempt to introduce a cross-functional approach to product data in this environment is likely to meet strong resistance from anyone who feels that they may lose some power or that they'll have more work to do.

Departments may not be able to agree on the purpose of the PDM system. One department may see it as a parts list/BOM system, another as a system for managing CAD drawings. People in the field may believe it's a fast way to access the correct version of data relating to products on customer sites.

It may be difficult to get the departments to work together to introduce PDM. Some departments may consider themselves superior to others. They may be unwilling to include their particular objectives in the overall requirements for the system.

Even if people can agree in principle about the requirements, they may not be able to agree about the system to be purchased. If the system has a good parts list module it may seem ideal for one department, which will be willing to purchase it, yet its lack of an interface to an application used in another department may render it of little value. Even if the system does have all the interfaces and modules required by the different departments, it may be difficult to agree in which order the modules should be implemented.

When it comes to use of the system, the same types of problem will occur. Each department has its own working procedures and will not want to change them. This may mean customising the system to work differently for each department.

The best solution to these organisational problems is to take, from the start, a professional, business-oriented, phased and cross-functional approach to PDM. The approach should be professional, because like any other project, the PDM project will fail unless it's managed properly. The approach should be business-oriented, because it will fail unless it has clear objectives of business success. The approach should be phased, because implementing PDM is a massive and lengthy task that can't be done in one bite. The approach should be cross-functional, because the creation and use of product data is cross-functional.

Ideally the project should start with a cross-functional awareness exercise. This will help people in all functions to understand what PDM can do for them. It will help people to start thinking about their real needs. It will help management to begin to understand the relationship between company objectives and PDM. In most companies, there's some kind of a cross-functional management group or committee that meets regularly and can sponsor this type of awareness activity. If the awareness exercise shows that there's potential for PDM, then the committee can give the go-ahead for the next phase, and propose a cross-functional feasibility study. The committee can define the objective of the study so that it will take account of the needs of the various departments. The committee can get help from an experienced neutral outsider during the study. By taking a top-down cross-functional approach, right from the start, a lot of inter-departmental problems can be avoided later in the project.

20.6   The Funding Barrier

The 'funding' category of barriers includes issues related to funding and cost-justification. Insufficient investment is a common problem, as is the use of inappropriate project cost-justification calculations. These may generate over-optimistic expectations. Measures and incentives put in place to support implementation and use of PDM may be inappropriate or even unattainable.

During implementation of PDM, all sorts of problems arise with project funding. The underlying problem is that there's never enough money to fund the project properly. Usually enough money is found for system purchase, but not for other activities such as training, installation, development of working procedures, and paying the project team. Even if the project team thinks it's got enough money, there's a good chance that some of it will be withdrawn before it's been spent. Even if the project team thinks it's got enough money for its own activities, it may find that related activities haven't been funded properly. Often the success of a PDM project hinges on the success of these related activities, such as process improvement and the definition of information structures, flow and use. Loss of funding on related activities can cause just as many problems as loss of funding on the PDM project itself.

Another problem that can occur with project funds is that they come at the wrong time, for example, at a time when they can't be used. Often money suddenly appears towards the end of the year. This may be because the annual budget hasn't been spent, and rather than let it appear that the money isn't needed, it may be given to the PDM project to consume. Unfortunately, such a sudden windfall may be of no use because activities and resources will have been arranged well in advance, and there'll be no-one available to do the work the new funds would pay for.

Sometimes a funding problem arises because the departments that are potential users of PDM can't decide who will get the benefits of its use, and as a result can't agree the proportions in which they should fund it. Each department funds it as little as possible, in the secret hope that the others will fund it more, in the hope of getting themselves more benefits for less investment.

Many project teams focus too much on selecting a PDM system. They should remember that avoiding problems with funding is just as important. If they can't get enough money to carry out the PDM project properly, then it's not going to matter which system is selected. Any system will fail if there's not enough money available to support its implementation.

Financial problems can also occur after the system has been selected, installed and brought into everyday use. If the benefits of use aren't apparent immediately, some users will withdraw their support and they'll try to reduce their spending on PDM. There'll probably also be problems because some departments will claim the others are getting more benefit, so should pay a higher share of the running costs. Also, once the PDM project has got to the stage of the PDM system being in everyday use, it will no longer be considered as an exciting new technology. Many people will lose interest and start looking at newer technologies. Instead of spending money on improving PDM, they'll switch their spending to some newer technology they hope will give massive benefits for very little effort.

PDM cost-justification is always a problem because the costs of PDM are very clear, but the benefits aren't. The potential benefits can only be estimated. Different people will make different estimates and, to be on the safe side, an accountant evaluating the project will probably pick the most conservative estimate of a potential benefit. The result will be that the apparent benefits don't appear very significant. They may not even equal the costs, with the result that it will appear that there's no reason to invest in PDM. In an effort to make things look better, the project team may inflate their estimates of the benefits. This is a dangerous step because someone may hold them to the inflated benefits and cause them a lot of trouble if they're not achieved. In the same way that the value assigned by an accountant to a potential benefit may be very different from the value the team feels most appropriate, other measures of performance may also be deformed. If these figures are then used as measures of success, it may be very difficult for the project team to meet them. Even though the project succeeds, it may appear to fail.

To help overcome these problems it's important to put things in the right context and get management to understand the wide-ranging and strategic nature of PDM. PDM will become the backbone of the product lifecycle. Implementation will take a lot of time and money, but it's necessary. Some medium-sized companies calculate PDM project size by multiplying the number of employees by $1000 (an estimate they can check with PDM vendors and companies using PDM). A company with 1000 employees could be looking at a project cost of $1 million. Three to 5 years may elapse between the initial interest in PDM and widespread productive use. Such an effort can't be made lightly. But since it must be made, it should be made as effectively as possible. The project team should make it clear from the start that implementation of PDM isn't a simple system selection project, but involves complex organisational issues and requires significant financial resources. Management needs to understand how much funding will be needed. The project team should show the benefits that will result from this spending. A phased implementation plan linking investment and benefit will allow management to understand why money is being spent, and discourage the withdrawal of funds necessary for achieving benefits.

20.7   The Information Barrier

This category of reasons groups all the problems related to the flow, use and quality of product information. There may be problems with the cost of entering information in the system. The system may not be able to handle all data types. It may not be able to store data where it's needed. There may be incompatibility between data structures. Classification mechanisms may be inappropriate. There may be no way of encouraging re-use of information.

These reasons cover a wide range. At one extreme, product information is such an under-valued asset that, in some cases, information will have been lost, and it won't be possible to find a copy of it for the PDM system. At the other extreme, information is of great value. Information is a vital working resource and people who use it will be unwilling to give up control for fear of not being able to do their job properly. In other cases, information is a source of power, and the people who draw their power from it will be unwilling to give up their power base. The issues of ownership and control of information are at the source of many of the problems that arise when implementing PDM.

Before information management is automated, the ownership of information isn't a big issue for individuals. Everyone has a copy of the information they need and the information is controlled by the Document Management Store/Reprographics. If people want to have their own copy of some information, they simply take a copy and put it in their drawer or on their shelf. With PDM it's not so simple. After all, the system is there to make sure that information is being used better, so it's not going to like someone referring to a three-month old document that, according to the system, they aren't allowed to access. It can be a time-consuming and meticulous process to work out exactly who has the right to access information, and once it's clear, a lot of people will be unhappy to find they don't have the rights they had in the past.

Another problem, with its roots in the past, is the unorganised state of most information that's under the control of traditional manual information management systems. Provided that it's been possible for a person who's been in the company for the last thirty years to lay their hands on a particular document within a few hours, most people have been happy. The system hasn't needed to be very effective. With PDM this changes. One of the reasons for using PDM is to make sure people don't have to wait several hours to access a document they need urgently. The PDM system should be able to make the document available almost immediately. This implies a very well-organised information base because the system can't go through all the weird logic that an old hand uses to find documents in strange places.

Problems may arise with particular types of documents. The system may only be set up to handle certain types or formats of documents, and not be able to handle others. It may only be able to handle a limited number of variants of a particular document. It may only be able to apply the same release process to all documents of a particular type, even though the company has different ways of releasing them.

There may be problems with storage and communication of data. The PDM system may only be able to store all data in one physical location, yet use of data may be divided equally between two locations.

In some cases, it will be the cost of communicating information between different sites that's the problem, in other cases it may be security or confidentiality.

It's often a problem to know what information to put into the PDM system. Some people would claim it's best to start to use PDM on a new project and only put data that's created on that project in the system. At the other extreme, some people claim the main requirement of PDM is to access legacy data, and that all existing data should be referenced in the system. It's usually impossibly expensive to input all existing data, so problems arise as different groups and departments squabble about which documents should be put in the system, and which should be left out.

There are often problems with the structure of information. Different departments may structure the same information in different ways and unless the system is capable of accepting different structures (or views) for the same information, there'll be problems as people try to ensure 'their' structure is chosen as the standard.

Problems may arise if a company has several naming, numbering and classification conventions. Fast access to information usually requires some fairly intelligent mechanisms in the PDM system so that people don't have to know all the attributes of a document to be able to access it. Given some rough search criteria, the system should be able to come up with some likely choices. This can't be done if the conventions that the company uses aren't compatible.

A related problem is that of re-use of information. One of the reasons for using PDM is to promote the re-use of information. If the system can't find information that people can re-use, then it's not going to be much use, and there's not going to be a significant increase in the re-use of existing parts and designs.

Similarly, to re-use parts, users must be able to quickly and easily find the parts that meet their needs. This means the part characteristics must be available in the PDM system as searchable attributes. If useful part attributes aren't available, users won't be able to find a part to re-use.

All the problems that arise with information are close to the heart of product data management. These are problems that need to be brought to the surface during the awareness stage of the PDM project. During the planning phase, the PDM project team must take action, as early as possible, to make sure that they are solved. By the time the system is selected, they must be solved. They are much more difficult to solve if they aren't dealt with until the system is installed. In practice, if they haven't been solved by installation time, then expect the PDM project to flop.

20.8   The Installation Barrier

After many months of hard work, the PDM team has selected a PDM system, and installation is about to start. At last, after all the effort, the system will soon be in use and everyone can relax a little. Well, not quite. At installation time, all sorts of problems can arise and it's very unlikely that anyone is going to be able to relax. Implementation may take much longer than expected. The PDM project may be poorly managed. The people who planned for PDM, and selected a system, may pull out before the system is installed, leaving the project in the hands of people who neither understand the objectives nor are motivated to succeed. Insufficient training may be given to users and the system support team. There may be no guidelines describing how the system should be used.

There may be some problems with the system itself. It may not work the way the vendor claimed it would, or it may have bugs, or it may not be documented, or there may be no procedures showing how it should be used. If problems like these appear, it's really the fault of the PDM project team. During the selection phase, they should have carried out detailed tests of the system to ensure that everything existed and worked according to specification.

Then of course, there can also be many problems that aren't related to the system. There may be problems due to money, time, and above all, people.

Financial problems will occur if the budget is withdrawn or cut, or if the cost of the system is more than expected. It may be that there's not enough time available. Things may take longer than expected, or people may have to work on higher-priority tasks. To avoid these problems, the PDM project team must do everything possible to make sure it estimates correctly the amount of time and money needed for the installation. It should develop an installation plan in which activities, funding and results are closely linked. The reason for this is to be able to show that if funding is withdrawn, then certain activities can't be carried out and the corresponding benefits can't be achieved. This way, the person who cuts the funding, and not the team, takes responsibility for the failure to achieve the benefits.

Often the team will be faced with people who have said that they'll support the project, promise funding, effort and time, and then withdraw their support once others have committed their time, money and effort to the project. To avoid getting into problems because of this type of person, the team has to avoid getting into a situation where it makes promises that can only be kept if someone else keeps their promises. The team must only make promises if it's sure that it has the corresponding resources (such as time, money, and people). Often this means it has to work in a step-by-step mode, only promising to carry out small steps for which it's sure it has the resources. It shouldn't promise big steps if it isn't sure it will have the resources. Of course, if it actually has the resources, it can carry out larger steps. But it should still have a contingency plan to cover the cases where funding is withdrawn or other promises are not kept.

Even if money is available, other problems may arise. People may not have been trained properly, or they may not be available because they are needed on higher-priority activities. Users may not want to use the system. They may not have been consulted in the choice of the system and see no reason to use it. They may spend a lot of time looking for bugs in it, or showing why it doesn't really fit the work they do. In other cases, the introduction of the system may stimulate users to understand what they really need. The introduction of a PDM system with limited workflow management may lead them to claim that what they really need is a workflow management system with limited data management functionality. To avoid this type of problem, some user representatives should be included in the project team from the beginning of the project. They should be aware of all the reasons for introducing PDM. They should be aware of the problems it will help overcome and the benefits it will provide. The user representatives on the project teams should be fully involved in the system selection process, and a particular system shouldn't be selected unless it has their support.

Management may also cause problems at installation time. Project managers may decide that they don't want to use PDM on their 'mission-critical' projects. To avoid this happening, the PDM project team needs to get written confirmation from senior management before installation that PDM will be used on particular projects. If the system isn't used on real-life projects then it's not going to lead to major business benefits.

There are many sources of problems in a PDM project, and to make sure the project succeeds, the PDM project team must foresee the problems and take the necessary steps to prevent them happening. Often, when a problem does occur it's too late to avoid the consequences. Remember that implementing PDM is a high-risk activity. Take out good insurance. Be prepared.

20.9 The Everyday Use Barrier

A good PDM project team will overcome all the problems that can arise during the selection and installation phases of the project. Its members will make sure the most appropriate system for the company is selected and installed. However, their work doesn't stop at installation time. They must be ready for any problems that arise when the system is used on an everyday basis. These problems can be due to errors and inconsistencies in the system, missing functionality or lack of training and support. They can be due to lack of funding, lack of interfaces to other applications, and a failure to make the necessary organisational changes.

Typically, the first users of a new system are patient with its teething problems, and put a lot of effort into learning how to use it and overcome its deficiencies. However, once the system is in everyday use, other users won't be so under-standing. They'll view the system as a tool to be used, not as a new technology to be experimented with. They'll expect it to work perfectly and help them in their daily work. If there are bugs, or functionality is missing from the system, they'll claim that they're not able to work with it, and will revert to manual methods. To avoid this problem, the PDM project team shouldn't bring the system into everyday use until it's been thoroughly tested and found to be capable of doing what has been promised. The team should ensure that users of the system receive sufficient training to understand how the system works, and what it can and cannot do.

The people who are expected to use the system on an everyday basis for production work can also be expected to create problems if the PDM project team hasn't developed procedures for use of the system. A PDM system can have all the functionality promised by the vendor, yet still be unusable unless there are procedures showing exactly how this functionality should be used within the company. Everyday users will expect to see procedures showing how to identify, classify, store and access data. They'll want to know how to decide on access rights, and how to ensure confidentiality and security. They'll want to know how to define product structures and how to define the flow of work. They'll want to know about release procedures, who should review and approve data, what work and files are needed, and who has responsibility for promoting work to the next stage.

Another source of problems can be the interfaces between PDM and other applications. Unless all the interfaces exist, some users will work entirely outside the PDM system rather than sometimes inside and sometimes outside.

Problems may arise if new developments promised by the vendor don't appear. Some people will claim that the only reason they supported the introduction of PDM was to have these new functions, and in their absence they'll not use the system. This is a problem that often arises, so the PDM project team should be prepared to respond to it. System selection has to strike the right balance between a vendor with a system that has a lot of functionality, but doesn't propose a lot of new developments, and a vendor with a system that doesn't yet have all functionality, but promises that new developments will result in superior performance.

In-house system developments can also be a source of problems. Sometimes, developments may not be made because funding is cut or because they are low down on the waiting list. Even if the developments are made they may be error-prone and poorly documented, and so be of little use.

Another problem that may arise after installation is that the project budget, in particular the training budget, is slashed. To make sure this doesn't happen, the project team should make it very clear in its plan that training is essential. The team should propose a phase-by-phase approach to implementation in which a particular phase isn't started until all activities of preceding phases have been completed. If the training activities aren't carried out within a particular phase, the following phase can't start, and the team can show which of the expected benefits won't occur.

The PDM system support team is the group of people that should make sure the system works on an everyday basis. They should provide everyday support to users. For budget reasons, the funding of this team may be cut. Again the project team needs to be able to refer back to the plan to show which benefits won't occur.

As PDM gradually takes hold, some departments may feel that they're losing control or power. As a result they may start to block its use and further development. Necessary organisational changes may not be carried out. The project team needs to be able to identify this problem, quantify its effect on expected benefits and, if necessary, get top management to act.

By the time a system gets into everyday use, a lot of people will have lost interest in it. For many people, new systems are exciting and a challenge, but once they are in everyday use, they are no longer of interest. The result of the drop in interest can be that everybody more or less abandons PDM. This magnificent system, that the PDM project team has worked so hard to implement, is left on the shelf. To overcome this problem, the PDM team has to plan for everyday use at the beginning of the project. It has to make sure that all the necessary resources (people, money, etc.) will be available to support the system in everyday use. The plan has to show clear responsibilities so that people can't later pretend that they didn't know of their involvement. In the early awareness stages of the PDM project, the project team must make it clear that the aim isn't just to select a PDM system, but to bring a PDM system into successful everyday use. Selecting a system is easy. The challenge is to make it work effectively.

Chapter 21
Of PLM Vision and Strategy

21.1 Objectives

According to Webster's New World Dictionary, objectives are "the aim or goal aimed at or striven for."

21.1.1 Strategy

Originally the word "strategy" was used in a military context. The word itself comes from the Greek word for a General. In many dictionaries, "strategy" still has a primarily military definition. According to Webster's New World Dictionary, "Strategy is the science of planning and directing large-scale military operations, specifically (as distinguished from tactics) of manoeuvring forces into the most advantageous position prior to actual engagement with the enemy". This is the original context of military strategy, preparation for a battle. In the 19th century, views of war changed. Carl von Clausewitz, a leading influence on German strategy, saw war as the continuation of diplomacy. War became more than a question of battles. It affected everything a country did. This wider view of military strategy can be seen in other definitions. For example, according to the Encyclopaedia Americana, "Strategy in its general sense is the art and science of developing and employing the political, economic, psychological and military forces of a nation." Another definition of military strategy is "In warfare, strategy is the science or art of employing all the military, economic, political and other resources of a nation to achieve the objectives of war".

The following definition reflects the fact that strategy is no longer confined to the military environment. "A strategy is a general method for achieving specific objectives. It describes the essential resources and their amounts which are to be committed to achieving those objectives. It describes how resources will be

© Springer International Publishing Switzerland 2016
J. Stark, *Product Lifecycle Management (Volume 2)*,
Decision Engineering, DOI 10.1007/978-3-319-24436-5_21

organised, and the policies that will apply for the management and use of those resources".

21.1.2   Mission

Webster's New World Dictionary defines a mission as "the special task or purpose for which a person is apparently destined in life." In the business context it can be defined as "the special task or purpose for which a company is destined". For example, the following text appears in a document "Our Mission and Our Commitment" produced by the Coca-Cola Company:

> In February 1989, Coca-Cola formulated its mission for the 1990s. The mission reads:
>
> We exist to create value for our share owners on a long-term basis by building a business that enhances The Coca-Cola Company's trademarks. This also is our ultimate commitment.
>
> As the world's largest beverage company, we refresh the world. We do this by developing superior soft drinks, both carbonated and non-carbonated, and profitable non-alcoholic beverage systems that create value for our Company, our bottling partners and our customers.

This is a mission statement. It describes the purpose of the company.

There's a big difference between strategy and mission. A mission describes the purpose. Whereas a strategy describes the way to achieve objectives.

21.1.3   Vision

The Oxford English Dictionary defines a Vision as: "A mental concept of a distinct or vivid kind; an object of mental contemplation, especially of an attractive or fantastic character, a highly imaginative scheme or anticipation".

Webster's New World Dictionary gives a similar definition of Vision: "A mental image especially an imaginative contemplation".

In the industrial context, a Vision will be a mental image of something in the future. A Vision describes the future state of something, so it's very different from a strategy which describes the way to achieve objectives. It's also very different from a mission. A mission describes a purpose.

21.1.4   Plan

The Oxford English Dictionary gives two definitions of a Plan. The first one is: "A diagram, table or program indicating the relations of some set of objects or the times, places etc. of some intended proceedings". It then gives an example of use of this meaning in 1807 by J. Nightingale: "A local preacher's plan is a paper properly divided and subdivided into columns and squares on which the names of all preachers are inserted, the respective places of their preaching appointments, and the dates of the month".

This is the definition of a plan as a graphical display in which activities and responsibilities are shown.

The Oxford English Dictionary also gives the following definition of a Plan: "A formulated or organised method according to which something is to be done; a scheme of action, project design; the way in which it's proposed to carry out some proceeding". Webster's New World Dictionary gives a similar definition of a Plan: "A detailed method, formulated beforehand, for doing or making something".

This definition defines a plan as a detailed method of doing something. Clearly this isn't a mission (which is a purpose), or a vision (which is a mental image), or an objective (which is an aim). However both "strategy" and "plan" describe how to do something. The main differences are that a strategy is at a much higher level than a plan, and that the strategy is broad-brush. It's "a way to achieve objectives". Whereas a plan is "a detailed method".

21.1.5 Tactics

The word "tactics" comes from the word "taxis" used by the Greeks to describe a military formation led by a "strategus". It was used in the 17th century to mean "the art of disposing any number of men into a proposed form of battle". Once decided, military strategy directs tactics, the use of weapons in battle. Strategy and tactics impose demands on logistics, the use of resources.

21.1.6 Policy

A policy is a general rule or set of rules laid down to guide people in making their decisions. For example, when faced with a situation of Type A, then do X.

A policy is a general guideline that will help people in the company to make decisions without continually referring back to top management for guidance. Typical subjects of policies include use of technology, supplier relationships, management span, quality, investment in new equipment, recruitment, salaries and benefits, and training. Policy statements could address areas such as those shown in Fig. 21.1.

A strategy addresses the key areas where the organisation will strive to gain competitive advantage. A strategy has to be simple and concise, otherwise it will be

Subject	Questions addressed by policy statements
Technology	
	leader or follower?
	maturity level for use?
	develop proprietary technology or use commercially available technology?
Quality	
	prevention or inspection?
Suppliers	
	long-term relationships or decisions based solely on contract price?
Equipment	
	purchase authorisation levels?
Culture	
	management-driven or worker empowerment?

Fig. 21.1 Some questions addressed by policy statements

impossible to implement successfully. A strategy doesn't aim to describe all the detailed issues that may arise in the product lifecycle environment. The issues which aren't explicitly addressed in the strategy document can be addressed with policy statements.

21.2   From Vision to Plan

The overall hierarchy of objectives, visions, strategies and plans is shown in Fig. 21.2.

The mission is at the highest level. It's the special task or purpose of a company. It describes the purpose of a company. However, it doesn't say what has to be achieved to carry out this task. Or how it will be achieved. The objectives are closely linked to the mission. They express at a high level what must be achieved to carry out the mission. The strategy describes the way to achieve the objectives. It defines how resources will be organised. It defines the policies that will apply for the management and use of resources. After the strategy comes the plan. Once the strategy has been defined, it's possible to start planning detailed activities and resources. After the plan comes the implementation.

mission
↓
objectives
↓
vision
↓
strategy
↓
plan

Fig. 21.2 From mission to plan

21.3 A PLM Vision

A PLM Vision is a high-level conceptual description of a company's product lifecycle activities at some future time. It's difficult to look further into the future than a few years. So it's appropriate to develop a vision of what PLM will look like five years in the future.

A PLM Vision represents the best possible forecast of the desired future situation and activities. A PLM Vision outlines the framework and major characteristics of the future activities. It provides a Big Picture to guide people in the choices they have to make, when strategising and planning, concerning resources, priorities, capabilities, budgets, and the scope of activities. There's a saying, "a ship without a destination doesn't make good speed". Without a PLM Vision, people won't know what they should be working towards, so won't work effectively.

Companies need a clear PLM Vision so they don't drift along, going wherever external forces are pushing them. People in the company need a clear agreed PLM destination that everyone can work towards. A PLM Vision for the company will enable all PLM participants and decision-makers to have a clear, shared understanding of the objectives, scope and components of PLM. It's a good basis for future progress. A PLM Vision is a focal point for everybody in the company that says: "this is where we're going". The Vision is a useful basis for communication about PLM between all those involved with PLM, such as executives, IS managers, Product Managers, product developers, service staff, recycling managers and other stakeholders. It allows everybody to "work from the same book" and "sing from the same page".

21.4 Basic Points About the PLM Vision

A PLM Vision will be company-specific. Without knowing a particular company in detail, it's not possible to say what its PLM Vision should be. For example, the Vision of an organisation that produces millions of identical electronic components would be expected to be different from that of an organisation that makes customised forgings.

A PLM Vision is built on the assumption that the company wants to carry out its product lifecycle activities as effectively as possible. Organisations don't proactively set out to perform badly.

A Vision must make sense to others. It has to be unambiguous and easily understandable. It must be believable and realistic, although it may appear to be at the limits of possibility. It must relate to the world of its readers, so that they can find a place for themselves within it. The Vision has to be as realistic and clear as possible. Otherwise it's not going to be useful.

Although a Vision may sound as if it's ghostly and immaterial, the PLM Vision needs to be concrete, clear, complete, consistent and coherent. It needs to be understandable and meaningful to different types of people in the company. It needs to provide people at different positions in the company with different levels of detail. A Vision that's incomplete isn't going to result in much progress.

The PLM Vision helps communicate to many different people an overview of what PLM is, and what it will be, why it's important, and how it will be achieved. The Vision has to be communicated to everybody likely to be involved in the future product lifecycle activities or impacted by them. It wouldn't make sense to have a Vision that's only accepted or understood by its inventor.

There has to be consensus about the Vision. A shared Vision helps everybody to move forward along the same road towards a successful situation and effective lifecycle activities.

The purpose of the Vision is to document and communicate the focus, requirements, scope and components of PLM. It communicates the fundamental "what's, why's and where's" of PLM, and provides a framework against which decisions can be taken. The Vision will make it easier to carry out all the activities that are needed across the product lifecycle to successfully develop and support products. It will guide people through PLM strategy setting and planning, and help with deployment of PLM.

A PLM Vision is the starting point for developing a PLM Strategy, and for developing and implementing improvement plans. In the absence of a shared Vision, people won't have a common picture of the future to work towards, so plans and improvement initiatives might be unconnected or even in conflict.

A Vision should be built by a team of people working part-time on this task over a period of a few months. It doesn't involve the acquisition of any equipment or the implementation of any software. It's a relatively low-cost activity, much less expensive than real-life implementation. Once the Vision exists, it can be used as a support for simulation of various options, again at much lower cost than real-life implementation.

The Vision is a best estimate for the future. It's the most likely Vision out of an infinite number of possible Visions. It's unlikely that the Vision will be the reality in five years. Most likely, new opportunities will arise over the five years and lead to a different reality. And, during the five-year period, the company will be in intermediate states on the way to the Vision, rather than in the Vision state itself.

21.5 Position of the PLM Vision

A PLM Vision isn't an independent stand-alone entity. It has to fit with the company's overall vision of its future, its mission and its objectives. Upstream of the Vision are the company's objectives, vision, strategies and plans.

The PLM objectives result from the requirements of the company. They express at a high level what's expected from PLM. In some companies, PLM objectives may not be provided by top management, so it's up to PLM management to develop them and get them confirmed by top management. Some examples are shown in

"we want total control of our products across the lifecycle from cradle to grave"
"the environmental footprint of our products must be the lowest in the sector"
"we want to be among the fastest product developers in our industry"
"our PLM activity must rank among the 100 most successful in the world"
"we want to eliminate four low-margin products in the next five years, and then be introducing 6 new and 10 upgraded products each year in the US"

Fig. 21.3 Potential objectives for PLM

Fig. 21.3. Sometimes, at the beginning of the visioning process, there may not be any stated objectives such as these, and they'll only become clear as the vision is developed.

One of the first steps towards the PLM Vision is to understand the scope, range and content of product lifecycle activities. The internal and external influences on PLM need to be clarified. Then the Vision of the future PLM environment can be developed. This will provide a picture of the environment, scope, performance and behaviour of the product lifecycle activities that are expected in the future.

The PLM Vision will include a description of the future PLM Strategy, the way that PLM resources will be organised in the future. It will include policies for the future management and use of PLM resources. It will act as a guide to everyone in the organisation who is involved in taking decisions about the future. It will set the scene for all the improvement initiatives that will follow.

Once the PLM Vision has been agreed, a suitable Implementation Strategy has to be developed to achieve it. Once the Implementation Strategy has been defined, it's possible to start planning detailed activities. These plans will address IT applications, modifications to the lifecycle processes, information, organisational structures, and many other topics. Individual projects will have to be identified, and their objectives, action steps, timing and financial requirements defined. The relative priorities of these projects will have to be understood. The projects will have to be organised in such a way that they result in the Vision being achieved within the allowed overall budget and time scale. When planning is complete, implementation can take place.

The process of developing a PLM Vision, and the related Implementation Strategy and plan, is easier to describe than to execute. In practice, it requires a lot of work with, initially, little to show for all the effort. When all the work has been carried out, there should be a very clear and simple link between Vision and implementation. In fact, it should look so simple that people who didn't participate in the process won't believe that it really was so difficult, and took so much time and effort.

21.6 Metrics and the PLM Vision

As described in Chap. 15, companies need relevant metrics to allow them to understand and quantify their current PLM performance, and that of their competitors. Metrics provide parameters that help an organisation to set targets for its implementation plans, and to measure the progress it's making towards achieving the Vision.

Objectives, Vision, Implementation Strategy, plan and metrics are closely related. Although the Vision only describes the overall desired state in mainly qualitative terms, it should include some important metrics and approximate targets. From this Vision, some clearly expressed quantitative targets will have to be defined. Generally, during the visioning stage, various metrics and targets will be mentioned, so the problem isn't so much how to identify them, but how to ensure that they're consistent, generally agreed, correspond to the Vision, and correspond to the time-scale of the implementation.

The targets need a lot of input from management because the Vision initially only provides a hazy view of the future environment and organisation. This needs to be translated into clear, reasonable and achievable targets that can be expressed in business terms.

Once targets start to be proposed, it will be seen that they're at different levels and address different time-scales. There will be high-level business targets, functional targets, product and process targets, and information system targets. There will be a mixture of long-term, medium-term and short-term targets. The targets must be consistent. They mustn't conflict. This implies going into a lot of detail to understand the objectives and potential effects of each target. Although this will take time and effort, it will be cheaper in the long run than rushing unthinkingly into unwise implementation.

21.7 In the Absence of a PLM Vision

Without a Vision, organisations drift along. They don't have an agreed destination towards which everyone can work. Decisions are taken on an uncoordinated day-to-day basis. There may be an underlying belief that the people are good, they know what they're doing, they've done it all before, they're doing their best, and as a result, they'll perform well. In practice of course, they don't perform well. Their aimless behaviour takes them nowhere. Their projects finish late, with cost overruns and quality problems. Their products increasingly lose out to competitors.

Without a Vision, managers don't have any future goals to aim for, so they concentrate on the short term. They become totally focused on quarterly results and day-to-day project management. Some spend their time in endless "strategy" and "planning" meetings that are doomed to failure in the absence of a Vision. Others continually change organisational parameters, driving the people affected into quiet fury.

Without an agreed Vision, individual employees may have their own, conflicting views of PLM. These are unlikely to lead to the company's required performance levels. Without an agreed Vision, decisions in the PLM area will be taken on an individual, uncoordinated, day-to-day basis. For example, one day, one of the managers may authorise purchase of PDM licenses for the design engineers in the

US, a month later another manager may decide to outsource all design engineering activities to China, with the result that there's no-one left to use the licenses in the US. And with the design engineering budget spent, there's no money to invest in managing the activities of the design partners in China.

Without a Vision, engineers get fully involved in everyday operations, and may start creating unnecessary work to keep themselves busy. Some spend their time developing unwanted functionality for products. They refine the mousetrap. Others waste their time developing IT applications to help them carry out their jobs. However, by ignoring the development of user requirements and quality assurance, they render their work virtually useless. Support staff travel unnecessarily to customers just to avoid being seen in the office. Looking busy becomes more important than achieving progress.

None of these unnecessary activities add value for the customer. On the contrary, the money and time expended on them reduces the productivity of the organisation. It's a debatable point whether it's better for people to be doing wasteful things, or for them not to be doing anything on the grounds that they can't take decisions because they haven't been told what to aim for. Debatable maybe, but both are the wrong way to behave.

When people start behaving like this, they become increasingly unaware of the reality of the outside world. It's almost as if they assume that because they aren't doing much, nothing much is happening in the outside world. They look back on a past in which they've done little, and extrapolate it forward to tell themselves they won't have to do any better in the future. This environment can continue for a long time until something out of the ordinary happens. When it does happen, there are usually tears all round. Murphy's Law ensures that things get worse. According to the Bible, "where there is no vision, the people will perish".

21.8   Reasons for Developing the PLM Vision

Several things can happen to lead an organisation to develop a PLM Vision (Fig. 21.4).

Sometimes a Vision will be developed because it's thought to be a good tool for encouraging management to put up money. An exciting Vision can give the

Fig. 21.4 Potential drivers for development of a PLM vision

repeated failure to achieve an objective such as developing a successful product
an urgent need for funds to maintain the current activity level
a realisation that tomorrow's needs will differ from today's
a need for a common approach across the company
current performance being unlikely to result in achievement of objectives
a change or expected change in objectives, such as reducing cycle time
a change or expected change in resources, such as reducing headcount
a change or expected change in competition
a change or expected change in the environment, such as loss of a key customer
the intention to exploit a new opportunity
a need to solve a particular problem
a desire to control the organisation's destiny
the need to motivate people to achieve a particular objective
the need to prevent people working at counter-purposes

impression that things are going to work very well in the future. If it's combined with a clear Implementation Strategy, and a good plan showing how it will be attained, and highlights the financial benefits that can be achieved, management may even provide the requested funds.

Sometimes a PLM Vision will be developed because an organisation feels the need to get control of its products. Without a Vision, it drifts along at the mercy of the whims of the rest of the market. With a Vision, it will be able to set itself targets, make the best possible use of resources, and turn itself into a high-performer. If it can do this, it will be less at the mercy of the surrounding environment. High-performers are unlikely to be attacked without reason. If necessary, the value and reputation they've built up can be traded on the open market.

Without a PLM Vision, people in the lifecycle don't know where they're going, or what changes may be made, so can't take reasoned decisions about what they should do in the future. Sometimes a Vision will be developed because people get tired of going round in circles, continually meeting to criticise their boss, and to discuss what they should do next, and how they can survive in the future. Invariably, without a Vision, their discussions lead them nowhere, so they give up and go back to working on day-to-day activities. But then it starts again. They look round and see that the world is changing, and it gradually dawns on them that their own little world will also have to change. They realise that the way they're operating now won't be successful in the future. And they know they have to do something about it. One way out is through a Vision.

Often the development of a Vision will be provoked by a particular problem which threatens the existence of the company. Typical examples are very poor financial results or the loss of a key customer. Severe problems like these can force top management to take action. Once it's clear that it won't be enough to take short-term corrective action, management may be forced to look to the future. A Vision points the way.

A new opportunity, due to a change in the market environment, or the acquisition of a company with different products, skills, technologies or approaches, can also lead to the development of a PLM Vision. The organisation is unaware of the best way to address the change. Instead of taking a short-term view and responding with the first solution that comes to mind, a PLM Vision may be developed to take account of the future situation.

Sometimes an organisation senses the need for a common goal towards which everyone can work. This may happen because many new hires join the organisation, and need to learn its beliefs and culture very quickly. There may not be the time or the possibility for them to learn the culture, goals and beliefs of the organisation in the traditional way, by working with the old-timers. The solution is to develop a Vision that can be quickly communicated and understood. There needs to be some kind of a focal point for everybody saying "this is where we're going", "this is how it's gonna be". The need for a common goal may also arise because many of the old-timers are fired, and there's no one left to transmit the culture.

21.9   Thinking About Visions

Experience of working with companies to help them develop a Vision of their PLM activity shows that most people have little idea of what a PLM Vision is, or how it can be developed. To demonstrate that Visioning isn't so difficult or unusual, we get them to think about Visioning in an environment that's familiar to them and usually of interest.

So, for a moment let's think about a Vision of your vacation next year. If you think about it for a few minutes you'll be able to create a mental picture of this vacation. Let's call this the Vacation Vision. Now assume that you've asked me to organise your vacation. Unless you can describe your Vision and communicate it to me, then you're probably going to have a "surprise" vacation.

So, tell me about your Vacation Vision. The best vacation of my life, I hear you say. Well, before we go any further, please cut out the generalities. First of all, I wouldn't expect you to be envisioning the worst vacation of your life. Secondly, you've not communicated useful information, so you're wasting your time. Now let's try again. Sunshine and sea, you say? OK, I'll book you on a cruise ship to Morocco in April. No, I hear you say, my Vision isn't Morocco in April. It's California in September, and I'm going to fly. OK, I'll book you into a five star hotel in Long Beach, and you can fly through Chicago to benefit from bargain basement rates. But, I hear you say, you want to go to Monterey, sleep rough and drink Old Tennis Shoes, and you don't want to fly through Chicago because you live in Centennial, Colorado.

A little exaggerated perhaps, but only to point out what may be less obvious in the context of PLM. A Vision has to provide a very clear and complete description. And it must be possible to communicate it without information loss.

In most cases, flowery statements about being #1 are going to be inappropriate in a PLM Vision. Firstly, nobody would aim to be #7,000,000. Secondly, not everyone can be #1, so if you're not really aiming to be that good, don't devalue the rest of the Vision with a statement that's unrealistic. Thirdly, what does it mean to be #1, since no metric is defined?

A vision has to be created right at the start of the activity. You wouldn't want to tell me your Vacation Vision after I've booked your flights and hotels.

Just as your Vacation Vision is unlikely to correspond to that of many other people, it's unlikely that a company's PLM Vision is going to correspond to that of many other companies. And, just as it's impossible to say that there's a "correct" Vacation Vision that all individuals should share, it's also impossible to say that there's a "correct" PLM Vision for all companies. It's to be expected that different organisations will have different PLM Visions. The only real test of whether a PLM Vision is correct, or not, is the extent to which it will allow a company to perform successfully.

One of the problems you'll have when defining your Vacation Vision is to identify all the parameters that are necessary to describe it. As you try to define the

Vision you'll uncover more and more parameters. In the one paragraph above, location, timing, weather, travel and sleeping quarters were soon identified.

Let's think about it some more. We'll start with your flight. I'll put you on the first flight from Centennial so you can get to California as early as possible and as cheaply as possible. As you haven't mentioned special meal requirements, you'll be getting standard airline food on your flights. I've fixed you up with a rental car at San Francisco airport, and I've got you a special 14-day rate.

What do you mean, I've got it all wrong? You didn't tell me you can't get to the airport for the first flight, or that you're a vegetarian, or that you won't need a car, or that you're only going for 10 days.

The same problems arise with PLM, but they'll be much more serious because the PLM Vision has to be communicated to hundreds or thousands of people. And, if you forget a few "details", everyone will organise things the way they see best. And the end result will be chaos.

Whatever the reason for creating the Vision, the creation process and the effect should be similar. People will react positively to a chance to shape the future and improve their working lives. They'll start working together, working to create a picture of an effective product lifecycle environment in which they'll have a rewarding and exciting role.

Although a few people will immediately understand the meaning of PLM, and the importance of PLM for the company, many won't understand what PLM is, many will misunderstand, some people will disagree, and some will ask for more information. It wouldn't be useful to reply to them: "I'm sorry, PLM doesn't exist in the company today, so I can't tell you about it, but in a few years it should be fully deployed. If you ask me then, I'll be able to tell you more". It would be much more useful to reply: "I can tell you what PLM will look like in the future, for example five years from now". "What it will look like five years from now" is a five-year Vision of PLM. Once this Vision has been created, it can be communicated throughout the company so that people will know what PLM is going to look like. It will be easier to make progress if everyone knows where they're going.

It's not unusual or wrong for people to want to know more about PLM. They could have several reasons for wanting to know more. They could want to know more about the PLM environment in which they'll be working. They could want to know more about what they'll have to do to achieve that environment. They could have questions such as those shown in Fig. 21.5.

| what will PLM look like for us? |
| how will it differ from today's situation? |
| how can we prepare for PLM? |
| what training will be needed? |
| what resources will be needed? |
| what products will be managed? |
| what's the lifecycle for our products? |
| who's managing our products? |
| what actions will be needed to achieve PLM? |
| what's in PLM, what's not in PLM? |
| is PLM the same as ERP? |
| are we talking about an Enterprise Application? |

Fig. 21.5 Questions people will ask about PLM

21.10   The Danger of Underestimating Vision

A PLM Vision provides a high-level conceptual description of the way a company will manage its products across their lifecycles at some time in the future, enabling it to perform successfully and continue to perform successfully. The reason for having, and sharing this Vision, is to make it easier to carry out all the activities needed across the lifecycle to successfully develop and support products.

Some people think it's so easy to forecast what will happen in the medium-term future that it's not necessary to develop a Vision. But it's not easy. If they think back a few years, maybe they'll remember the days when, for example, Microsoft, Internet and the Web didn't exist. If it's so easy to forecast the future, it's a pity they didn't create them. Or they can think back a few years and remember the Cold War. Was it all that long ago that there was a daily menu of news about the Berlin Wall, ICBMs, Star Wars, megaton equivalents, mutual assured destruction, and anti-missile missile systems? Yet, in 2003, a decade after the fall of the Soviet Union, there were twice as many major conflicts around the world as at the height of the Cold War. They didn't involve two super-powers of equal and opposing force. Usually they weren't even between nation states. Instead, they tended to be between different groups in the same country, with irregular forces using tactics that, to a large extent, targeted civilians. In view of all the M&A activity in the defence industry in that decade, it doesn't look as if many defence companies foresaw the changes that were coming.

Some people don't like Visions because they don't believe that a mysterious intangible dream can play a role in the business world. Often, this is because they can't understand that a Vision will eventually get translated into sales and money. They would rather start with the result (expected sales) and work backwards to plan for them. Once their plan has been made, they can tell everyone what to do. Or they may just feel Visions are vague and meaningless academic philosophy. Their solution to understanding the future of the product lifecycle could be for top management to delegate the task to a group of high-cost consultants and a few overworked executives, and set them a 6-month deadline for developing the plan for the next ten years.

What happens next is illustrated by a typical case in which the group was made up of three consultants and three VPs. The group was led by the Quality VP acting as moderator. They plunged straight into the details of the new organisation chart, new products, human resource allocation and the IT applications that would be needed. There was never even any real awareness or consensus about the meaning of what they were doing. There was no attempt to develop a common approach, or to think about the future competitive environment. Six months later a three-volume report was ready in time for the deadline, but it was the worst type of committee compromise and quickly shelved. Soon after, I was called in.

It was clear that none of the members of the group had been deliberately destructive. Each one of them had valid aspirations and concerns, and had behaved in a way that had previously brought success. Taken individually, each was aware of the important issues that had to be addressed, and wanted to find a solution. However there had been no real framework in which they could work. There had been no overall agreement about what the future should, or even could, be. The organisations they represented weren't interested in Visions, but wanted results. They already had their own strategies, and only wanted immediate answers to specific problems. In particular, any easy, short-term gains would be greatly appreciated.

The group leader saw himself as a skilled leader and organiser, responsible for setting up meetings, documenting them, and making sure a comprehensive plan (in the corporate format) and a corresponding final report (in the corporate format) were produced. He was focused on the format of the result, and not the content. He saw no reason to understand detailed product issues. The consultants were unaware of the issues in the company, they claimed to be "Technology Strategy Specialists" and "Best Practice Business Process Architects". They didn't want to "get lost in the details of the product portfolio". The VPs were all independently-minded individuals with their own business and personal goals. They knew what the PLM budget was likely to be, and they wanted to get as much funding and benefit as possible for themselves.

As is often the case, the activity was bureaucratic, budget-driven, short-term and over-focused on the plan. The process of developing, agreeing and communicating a Vision wasn't even considered. Representatives of several activities in the life-cycle weren't involved. Their valuable input about the way PLM could be carried out in the future was ignored. As a result, it was unlikely that the eventual "plan" would correspond to their requirements. No account was taken of the strategic business focus. As a result, it was unlikely that the resulting implementation, even if it had not been shelved, would have produced the type of business performance improvements they were looking for.

21.11   Vision Description and Documentation

Different people will have a different understanding about the deliverable that results from the activity of creating a "PLM Vision". Some will think that the deliverable will be a Vision addressing the future global manufacturing industry environment. Others will think it's a Vision of what will happen within a particular company. Some will think the deliverable is a "mental image". Some will think it's a document describing that mental image. Some will think it's a PowerPoint presentation. Some will think it's a voluminous document. It's important to avoid potential confusion by defining the form of the deliverable before the Visioning activity starts.

A PLM Vision addresses the way in which a company manages products across their lifecycles. The PLM Vision doesn't address the way in which a particular product will perform over its lifecycle.

| more competition |
| more outsourcing to China |
| more offshoring to India |
| more application software to choose from |
| more electronics in products |
| more software in products |
| more effects of globalisation to take into account |
| more global competition |
| more product liability problems |
| more sales of products in China and India |
| more advances in technology |

Fig. 21.6 Potential vision of the surrounding business environment

It's also important to make the distinction between the "Vision of PLM" and the "Vision of the Surroundings in which the company will operate in the future". The Vision of the Surroundings will probably be similar for many companies (Fig. 21.6). Most companies have very little control over many of these "surrounding issues". For example, they don't claim to control the level of trade between China and the rest of the world. Fortunately though, it's not the objective of the Vision of PLM to describe a "Vision of the Surrounding Business Environment". The Vision of PLM should just address issues over which the company does have control. The Vision of PLM may make reference, in a paragraph or two, to the expected business environment in the future. But the focus of the Vision of PLM isn't issues over which the company has little control, but product-related issues over which the company does have control.

Several different types of people with very different needs, and very different know-how, should read a Vision document. One of the aims of the Vision is to get a common understanding of the future role of PLM in the company. This implies that the Vision document has to be readable and understandable for all of those people. It should not go into details that some of the readers won't be able to understand. However, it's likely that some readers of the Vision document will find that it contains too many details, while others will find that it contains too few. High-level business executives will want a concise, complete Vision document that they can read quickly, and from which they can reach a decision regarding the future of PLM. They should be able to do this with 5 pages of text and 20 PowerPoint slides. However, when product developers and IS managers read such a concise document, they may see generalisations of little relevance and a few incorrect details which they would like to correct. They may see little value in such material. However, what's more important is that they understand, and agree with, the overall picture and the main points. The details can be reviewed and dissected later.

A lot of work goes into the creation of a PLM Vision. It may take several weeks or months to prepare, discuss, analyse and agree upon the content of a PLM Vision. However, despite all the time and effort that goes into preparing the Vision, there's a danger that, after the Vision document has been discussed and a decision taken about the future of the PLM Initiative, it may not be read again. High-level business executives won't have the time to go back and read the PLM Vision document. They'll be looking for the results of PLM deployment. Product developers and IS managers may not want to reread the document as they feel it contains little, if anything, of value. However, if the Vision document isn't reread, it's likely that there'll be little similarity between the agreed Vision and the eventual deployment and use of PLM. If it's not reread, the high-level specifications it contains will probably be replaced by lower-level shorter-timeframe needs. To overcome this problem, it's important to include expected targets for performance in the PLM Vision, and to plan to review and update the Vision. This will allow management to track progress towards the agreed Vision.

Different readers require different levels of detail about the Vision. All should be interested in a high-level overview. However, none will be interested in all the low-level details. The Vision document should be structured so that it provides a single top-level overview or summary that everyone can read and understand. It should also contain multiple, more-detailed sections addressing specific subjects of interest to particular people or functions.

There's not a standard style or convention for a Vision document. As a result, the "look and feel" of different Vision documents may be very different. Sometimes, a Vision document will be written primarily in the future tense. Written now, it will describe what will be done, and what the situation will be. Sometimes, a Vision document will be written primarily in the present tense. Although written now, it will project the reader into the future, and describe what the situation is in the future. Sometimes, a Vision document will be written primarily in the past tense. Although written now, it will project the reader into the future, and look back at the intervening period, describing what was done. All of these styles are acceptable. However, it's important to avoid mixing them together as that could confuse the reader.

Different readers may be looking for different information in the Vision. Some executives may be looking for information about the activities that are needed in the next five years to deploy PLM. Other executives may be looking for information about the way that PLM activities will be organised in the future, or the way that the company will work after PLM has been deployed. It's to be expected that a PLM Vision will contain a mixture of descriptions of targets (such as "this is where we'll be" and "this is what we've achieved") and activities ("this is what we did" and "this is what we're going to do").

21.12   The PLM Strategy

A strategy describes: the way to achieve objectives; how resources will be organised, managed and used; policies governing use and management of resources.

This general definition leads to requirements for a PLM Strategy (Fig. 21.7).

| show how to achieve PLM objectives |
| show how PLM resources will be organised, managed and used |
| describe policies governing use and management of PLM resources |

Fig. 21.7 Requirements for a PLM strategy

Among the resources that can come into play in the product lifecycle, and therefore should be addressed in the PLM Strategy, are products, product data, PLM applications, processes, methods, people and equipment.

A PLM Strategy describes how resources will be organised and used. PLM Strategies aren't generic. They're specific to individual organisations because they depend on the particular circumstances and resources of the individual organisation, and on its particular environment.

A PLM Strategy will be company-specific. Without knowing a particular organisation in detail, it's not possible to say what its strategy should be. For example, the PLM Strategy of an organisation that develops and manufactures high-performance aircraft engines can be expected to be different from that of an organisation that develops low-cost plastic toys.

PLM Strategies change. Today's PLM Strategy describes how PLM resources are used in today's environment. A PLM Strategy for the future shows how they'll be used in the future.

The PLM Strategy has to be documented and communicated to everybody likely to be involved in the future environment or impacted by it. It wouldn't make sense to have a Strategy that nobody, apart from its developers, knows about, understands or approves.

The PLM Strategy shouldn't be changed frequently. It can take several years to implement a new PLM Strategy. And it can take several years for the effects of a new PLM Strategy to become apparent.

A good, well-defined and well-communicated PLM Strategy is important because it: shows how PLM objectives will be achieved; makes sure resources and capabilities will be used to their best; makes sure everybody knows what's happening; and makes sure all resources are aligned in the right direction.

21.13 An Implementation Strategy

To achieve the PLM Vision, two strategies need to be developed. These are the PLM Strategy and the Implementation Strategy. The PLM Strategy shows how PLM resources will be organised in the future, envisioned environment. The Implementation Strategy shows how resources will be organised to achieve the change from today's environment to the future environment. The Implementation Strategy is sometimes referred to as a Change Strategy or a Deployment Strategy.

The Implementation Strategy may be referred to as a Change Strategy as it enables the company to change from its current PLM Strategy to its future, envisioned PLM Strategy. It's a strategy for change. So it's called a Change Strategy. The Change Strategy may include statements such as "we'll change everything at once", or "we'll make changes one step after the other", or "these are the changes we'll make".

Typically, the PLM Vision will be 5 years in the future, and the Implementation Strategy will be developed for a similar period. It may then be reviewed annually,

provides the best chance of achieving the PLM Vision and future PLM Strategy
makes sure resources and capabilities are used to their best during implementation
makes sure everybody knows what's happening during the implementation phase
makes sure everybody is working towards the same target
enables implementation planning decisions to be taken in a coherent way

Fig. 21.8 Benefits of the implementation strategy

but if initially defined correctly, shouldn't change significantly during this time period.

A good, well-defined and well-communicated Implementation Strategy is important (Fig. 21.8).

The Implementation Strategy is the starting point for developing and implementing the Implementation Plan. It helps everybody to move forward along the same road towards the new environment. However, the Implementation Strategy is just the starting point. It isn't the end of the road. A detailed Implementation Plan will also be needed. It's a good idea to develop the Implementation Plan for the first year of the Implementation Strategy at the same time as the strategy itself. This will help to make sure the two are synchronised.

21.14   Industrial Experience of Visioning

In industry, there's so much pressure from all sides, that people usually just put their heads down and race forward. But, now and again, some of them look up to see where they're going. Fazed by the continuously changing environment, they ask themselves what to do next. Should they search for a way to do existing business better, or should they just pull out of product development, manufacturing and support, and concentrate on marketing, sales and logistics?

I experienced this situation in an interesting way in 1993. A Corporate Vice-President of a Fortune 50 corporation asked me how much it would cost to develop a Vision of the future Engineering Environment. He told me he was tired of his engineering managers implementing short-term uncoordinated improvement projects that led to no measurable impact. He said he was looking for the Big Picture that companies in his corporation could work towards over a five-year period. My estimate of nearly $500,000 to develop the Vision seemed slightly too high for him, but the subject seemed a promising one. I turned to my friend Chris Horrocks, then of Coopers and Lybrand, Boston, to see if we could carry out a multi-client study on the subject. Eventually we found twenty companies willing to participate.

We proposed a good approach. We would visit each of the twenty companies once to find out what they were looking for, identify the people and organisations with the best thoughts on the future environment, build a questionnaire, get the answers, develop the Vision, and then go back to the participants and explain it to them.

Although the twenty companies were in different industries and made all sorts of products (including soft drinks, rockets, helicopters, cameras, computers, cars, and trucks) our discussions with them were surprisingly similar. Obviously there were differences between different industries, particularly where there were strong regulatory forces, but there was more similarity than difference.

Whether in the electronics, aerospace, automotive or other industries, they were all facing challenges of globalisation and rapidly advancing technology. Competitors were bringing out more and more new products, and getting more and more revenues from new products. The lifetime of many products was dropping. As product lifetimes fell even further, the effect of being late with a product would be disastrous. Similarly, producing a product that did not meet customer requirements could be catastrophic. There was no longer the time for trial-and-error. The product had to be right the first time. In the automotive industry for example, Chrysler's Dodge Viper had been designed in 36 months instead of the traditional 60 months. GM's Corsa, launched in early 1993, had 30 % fewer parts than its predecessor, and cost 25 % less to assemble. 30 % of Honda's 1992 Civic came from the previous model, compared with traditional reuse of less than 10 %. Motorola had reduced order-to-manufactured-product time for one basic consumer electronics product from 4 weeks to 2 h. Apple Computer had reduced product development cycles from about 18 months in 1990 to 9 months in 1993. Intel's 586 had a 4-year product development cycle, whereas it had taken 5 years to get the 486 to market. The 586 had 3 million transistors compared to the 486's 1.2 million.

Usually the participants in the multi-client study had clear ideas of the issues they discussed with us, but wanted a better idea of the relations between them. The added value that we could offer was to bring into form what they could feel but couldn't get into shape. During our initial round of meetings, the participants often asked how we would create the Vision. Will you just make it up? Will you dream it? Is it intuition? Is it inspiration? What do you guys smoke? Can you give us an example?

The example we often gave was Martin Luther King's "I have a dream" speech, delivered on the steps of the Lincoln Memorial in Washington, DC in August 1963. It gave a powerful and easy-to-understand vision of a new environment. So powerful that the third Monday in January is a federal legal holiday commemorating him.

Eventually a participant in upstate New York correctly identified our approach to creating the Vision. The Trash Can Method. Put all the ideas in the trash can, shake or stir, and the vision will emerge. With several hundred experts, and several hundred questions, we needed a large trash can. Eventually the vision emerged. It couldn't be explained with a simple one-line answer. What it showed was that there were forces emerging (in particular we identified lead times being cut by an average factor of five, and cost reduction of 10 % per year) which would drive companies in a certain direction, and to respond, they could build upon a common framework we built for the Big Picture.

With the Vision delivered, some of the participants asked what we would offer them next. They wanted to know how they could achieve the Vision. With Chris, by now with IBM Consulting in Atlanta, Georgia, we set off on the next stage in 1994. This was the Engineering Strategy multi-client study which also had about 20 participants. Some of these had participated in the Engineering Vision study. Others, including a utility, hadn't.

At first sight it may seem strange that a utility company would participate in a study with manufacturers of soft drinks, rockets, helicopters, cameras, computers, cars, and trucks. But it's not. There are many similarities between Engineering in companies in discrete manufacturing, and Engineering in companies in process industries. There are many similarities between the problems faced by companies in the discrete manufacturing sector and by those in process manufacturing.

The Engineering Strategy study followed a similar approach to the Engineering Vision study, but was more difficult to report. Although all the participants could aim for a common vision, they were starting from different positions, and we had to take that into account.

With the Strategy delivered, some of the participants asked what we would offer them next. Could we develop the Engineering Plan? Well, we couldn't develop their Plans as a multi-client effort. There wasn't enough commonality between the participants. However we did develop plans for individual companies, demonstrating how Vision, Strategy and Plan link together.

21.15   Progress Depends on Retentiveness

It was the philosopher George Santayana (1863–1952) who wrote, "Progress, far from consisting in change, depends on retentiveness. Those who cannot remember the past are condemned to repeat it." Often the second sentence is quoted alone, but much of the message is contained in the first sentence.

Companies thinking about the future today have similar questions to those raised by the participants in the Engineering Vision study in 1993. There's an on-going desire to step back from everyday pressure to understand the future, and find a way to keep control. Of course, the pressure on the product development function of companies has changed since then. The pressures that existed then are stronger, and new pressures have appeared. Many of the new pressures are appearing downstream in the product lifecycle, in particular during use and disposal.

As in 1993, the approach of developing a Vision, a Strategy and a Plan can be applied to understand how to respond to the pressure. One notable difference with 1993 is that there's now a very clear Vision to aim for. It's the emerging paradigm of PLM which offers a coherent approach to solving the problems and seizing the opportunities. In the early 1990's, the product lifecycle wasn't given the same importance as in the second decade of the new millennium. Leading companies were still looking for ways to operate in a globalised environment. At that time, there were important questions about how to co-ordinate development and production across continental boundaries. Since then, new issues have appeared, all of which lead to more focus on the lifecycle. For example, if a competitor can manufacture in China for 20 % of the cost in the West, how are companies in the West going to make money? How should companies react if regulators are really going to hold them responsible for what happens when people use their products? How should companies react if regulators are going to hold them responsible for what happens to their products at the end of their life? How can manufacturers adjust to the pressures for sustainable development? How can manufacturers react if they're made to pay for the pollution resulting from the transport and use of their products?

21.16   Similarities and Differences

As a consultant, I've been in contact with hundreds of companies. I can only think of a few that didn't consider themselves as special cases. Once, that belief was so extreme that my contract was terminated when I mentioned the way that other companies had improved performance. It was pointed out to me that this company was so different from others that what happened in other companies was of no interest. If I couldn't understand that, then my services would be of no value.

However, one lesson that I learned in 1993, and again in 1994, when we had close contacts with about 40 companies in a short time, was that companies are very similar. Since companies all exist in the same environment, share the need to get products and services to customers, and are staffed with people of similar behaviour and experience, perhaps it's not surprising that they're so similar. This similarity can be denied. Or it can be exploited, to learn from successful companies.

Tolstoy's book "Anna Karenina" starts, "Happy families are all alike; every unhappy family is unhappy in its own way". In the corporate world, this could be rewritten as "successful companies are all alike; every unsuccessful company is unsuccessful in its own way".

Often the main differences that do exist between companies are due either to the particular influences of the industry they operate in, or to the particular individuals who lead them. For example, the definition of geometry data has high priority in aerospace and automotive companies. But it's not a key factor in the pharmaceutical industry, where the focus is on discovering new compounds, and managing clinical trials. The medical equipment industry has to demonstrate compliance with strict regulations. This isn't such an issue in the consumer electronics industry, which is more concerned with managing the BOM across the extended enterprise. Industries with long lifecycle products such as planes, plants and machine tools focus on configuration management to support future access to data about the products. In food and beverage industries, that's much less of an issue.

The behaviour of the person at the top of the company has a profound effect on the company. Originally I was sceptical about this, reasoning that the behaviour of tens of thousands of people in a company must have more effect than that of one individual. But experience proved me wrong. The lesson is important for an activity with such a wide range as PLM. The success or otherwise of a PLM initiative is likely to depend on that one person.

Once individual PLM initiatives get the go-ahead, they may show wildly different results, presumably because they're influenced much more by the particular individuals who participate in them, than by the overall environment. Due to these influences, there's not a unique approach to implementing PLM. Each company has to work out how it can benefit from PLM. And how it should implement PLM.

Chapter 22
Strategies

22.1 Military Strategy

Histories of modern military strategy often start with Napoleon. Between 1796 and 1815 he dominated most of Europe. For hundreds of years before, no-one had achieved such domination. Napoleon had several strengths. After 1804 he was both Emperor of France and commander of the French army, so he controlled both the national and the military strategies. No other general at the time had such freedom. Napoleon also became very experienced. He fought more battles than other generals. Napoleon fought in the name of Liberty. Many of the generals he defeated were fighting in the name of despots. Until Napoleon's time, battles were often formal events requiring mutual agreement to fight and a long set-up time. Napoleon, however didn't just bring his army to a place opposite the opposing army and wait until his enemy was prepared for battle. He carefully selected battlegrounds advantageous to his forces, rapidly concentrated all his forces for battle at a position where his enemy was weak, and forced his enemy to fight by threatening lines of communication and supply.

Carl von Clausewitz saw many of the Napeolonic battles and wrote about military strategy in "On War". One of his theses was that "war is nothing but a continuation of political intercourse with the admixture of different means", in other words, an extension of diplomacy. This leads to the concept of total war involving not only a country's army, but also civilians and economic resources. It also implies political direction in military matters.

For Von Clausewitz, as for Napoleon, victory in war resulted from the destruction of the enemy's forces on the battlefield rather than the mere occupation of territory. To achieve this he identified three targets. These were the enemy forces, their resources and their will to fight. According to Von Clausewitz, defensive warfare offered a stronger position than offensive warfare.

Antoine-Henri Jomini, a contemporary of von Clausewitz, and a staff officer of Napoleon, put forward his ideas of strategy in "Summary of the art of war". Unlike

© Springer International Publishing Switzerland 2016
J. Stark, *Product Lifecycle Management (Volume 2)*,
Decision Engineering, DOI 10.1007/978-3-319-24436-5_22

von Clausewitz, he favoured a strategy of occupation of territory rather than destruction of the enemy. By the time of the American Civil War (1861–1865) the effects of the Industrial Revolution were becoming apparent. Steam power was widely used. Accurate long-range infantry rifles had been invented. The use of steam power for rail and water transport changed the military equations of space and time. As long-range rifles could wipe out a concentrated attacking force before it could get to grips with a well-entrenched enemy, the tactic of frontal attack with concentrated forces was abandoned. It was however used by Lee at Gettysburg, and the disastrous result ended any hopes of victory for the South.

At the time of the American Civil War, Prussia was growing in strength in Europe. The Prussian commander Von Moltke agreed with von Clausewitz that battles are the primary means of breaking the will of the enemy. But he didn't agree that defensive warfare offered the best position. He favoured the offensive. Speedy decisive action with superior forces. This strategy was used successfully against the Danes, Austrians and French. His successor, von Schlieffen, took this approach one stage further with his strategy of annihilation, a decisive battle from which the enemy couldn't escape. While the Prussian generals were focusing on military strategies that focused on victory by offensive action and decisive battle, other strategists such as Delbrück and Mahan were looking at strategy in wider contexts. Delbrück proposed a strategy of exhaustion in which the enemy was worn down by territorial occupation, blockade, destruction of crops, and destruction of commerce. Mahan proposed a change in US naval strategy away from coastal protection and commerce-raiding to command of the seas.

Some of the results of these strategies were seen in the First World War. Apparently the effect of advances in technology, which had already been clear in the Civil War, hadn't been fully understood by the strategists. The introduction of the machine gun and field artillery had tilted the balance in favour of the defender, yet both the French and German commands favoured offensive strategies. Von Schlieffen's plans for an annihilating attack against France were watered down by his successors. The initial German attack in August 1914 failed to achieve its objectives. By November 1914, the opposing armies were faced with trench warfare in which a well-entrenched defender held the upper hand. A long series of inde-cisive but costly battles followed. None of the commanders appeared to grasp the futility of their offensive strategies. One of the worst examples occurred at the Battle of the Somme on July 1, 1916. After a week's preparatory bombardment (alerting the Germans to a major offensive) the British infantry attacked on a 15-mile front. They moved in formation, and at walking pace, towards the German positions. By the end of the day, the British had 60,000 casualties, 20,000 of them dead.

The trench warfare lasted for four years, becoming a war of attrition in which the naval blockade of Germany eventually played a large part in the Allied victory. By the Second World War, new technologies were available, providing the possibility for a very mobile attack capable of overcoming strong defences. As von Clausewitz had foreseen, war then became as much a test of civilian morale and economic strength as of military prowess.

| American Civil War |
| France |
| English Channel |
| Russia |
| Pacific Ocean |

Fig. 22.1 Different contexts of military strategy

The following sections help to illustrate military strategies in specific contexts (Fig. 22.1), and related success and failure factors, and provide the basis for some "Lessons Learned". These are a useful input when developing strategies in the context of PLM.

22.2   American Civil War

The American Civil War started at Fort Sumter on April 12, 1861. The North wouldn't accept that the Union could be divided. The Southern states believed the Union no longer protected their rights and interests. The objective of the North was to prevent the Confederate States from seceding from the Union. The objective of the South was to attain independence.

The population of the North was 21 million. The population of the South was 9 million, of which 3.5 million were slaves. Over 80 % of factories were in the North. So was 95 % of arms production. About 75 % of railroads were in the North. The South couldn't hope to achieve its objective by conquering the North, but in view of its objective didn't have to. Its strategy aimed to convince the North that forcing the South to remain in the Union wasn't worth the cost, and to bring about foreign intervention in its favour.

Whereas the Confederate president Jefferson Davis suggested a purely defensive strategy to meet the South's objectives, others such as Robert E. Lee initially believed the South had to carry the war to the North and defeat the Federal armies on their own ground. After the defeat at Gettysburg in July 1863, the South didn't have the resources for an offensive strategy, and with no sign of foreign intervention, it went on the defensive.

To achieve its political objective, the North had to conquer the South. It had to invade, capture and control vital areas and cities. It had three major military aims. The first was to isolate the South. The second was to cut the South into two parts, East and West of the Mississippi. The third was to capture Richmond, the South's capital. In spite of the South's long coastline, the first aim was largely achieved by 1863. The second aim was achieved in July 1863 when Ulysses S. Grant captured Vicksburg after a long siege, cutting the Eastern part of the South off from supplies in the West.

After the failure of the attack on Petersburg and Richmond in June 1864, the North changed its aim to striking at the Confederate Army and the remaining sources of supply. Sherman's army of over 100,000 marched south into Georgia. In spite of occasional defeats, such as at Kennesaw Mountain, it was far too strong for its Confederate opponents. Atlanta was captured on September 1, 1864. The resulting March to the Sea in November and December 1864, followed by the march up through the Carolinas, cut Lee's army in Virginia off from supplies in the South. Lee surrendered at Appomattox Court House in Virginia on April 9, 1865.

In many ways the American Civil War was the first modern war. About 2.5 million men served in the two armies. The casualties were horrific. About a quarter of the participants died. And a quarter were wounded. There were over 25,000 casualties at Antietam on September 17, 1862. At Gettsyburg there were nearly 50,000 casualties on July 1–3, 1863.

22.3   France

The battle of Crécy in 1346 is remembered as the end of the medieval age of chivalry and the introduction of the English longbow. At Crécy, the English under Edward III took up position with some 4000 men-at-arms in the centre and 5000 longbowmen on the wings. Between them was a sloping valley. The French force under Philip VI was twice the size of the English army. As it advanced up the valley, heavily armed French knights mounted on their war-horses were cut down by concentrated long-distance arrowfire from both sides.

Six centuries later, the experience of the First World War with its static trench warfare in Northern France and its huge losses seemed to show the superiority of the defensive over the offensive. By the end of the war, a superiority of at least three to one was believed necessary for a successful offensive. After the war, a strategy based on defence underlay France's construction of the Maginot line of fortifications between France and Germany. This system of massive self-contained forts ran from near the Franco-Swiss border in the south to Montmédy, south of the Ardennes and the Franco-Belgian border in the north. The French considered the Ardennes impassable to tanks, so not a potential invasion route.

While the French drew the conclusion of the superiority of the defensive from their experience in the First World War, the Germans developed the blitzkrieg, a dynamic war based on the speed of aircraft and tanks. The strategy developed by Guderian was for tanks, concentrated in armoured divisions, to create gaps in the enemy front lines, sweep past, loop round, and create an isolated pocket in which enemy troops would be surrounded and captured by motorised infantry. The ground attack would be supported by dive bombers attacking supply and communication lines.

In 1939, the German tanks were concentrated in 6 armoured divisions, whereas the French tanks were distributed throughout various infantry and cavalry units. In 1939, France's 800,000 standing army was thought to be the most powerful in Europe. The Allied forces were superior to the German forces in terms of numbers and industrial backing. However, their generals had once again prepared to fight the previous war. Germany attacked Holland, Belgium, Luxembourg and France on

May 10, 1940. Most of the French army was assigned to defending the Maginot line. The main German attacks into France were elsewhere, either through Belgium or through the Ardennes. By June 14, 1940 the Germans were in Paris, and on June 22, 1940 an armistice was signed. The Germans lost 50,000 men in achieving the surrender of about 2 million French soldiers.

22.4   The English Channel

From the early 1500s to the end of the Second World War, the military strategy of England was built on control of the seas. Providing the English navy controlled the seas, no foreign army could land in England, England didn't have to support a large army, and it was free to participate as it pleased in European politics and in developing a global empire. Its navy ensured necessary imports of food and other supplies, and could sever an enemy's access to the markets of the world. Examples of the success of this strategy include the Spanish Armada of 1588, and the Battle of Britain in 1940.

In 1534, Henry VIII broke with the Pope and set up the (Protestant) Church of England. His Catholic daughter Mary, who reigned from 1553 to 1558, was married to King Philip of Spain. After her death, Henry's Protestant daughter Elizabeth reigned from 1558 to 1603. Due to the weakness of France at this time and the wealth taken from the New World, Spain was the strongest power in Europe. For various reasons (such as Elizabeth's support of Protestants in the Netherlands, England's refusal to recognise the monopoly of Spanish trade, and the desire to wipe out heresy), in 1588 King Philip of Spain sent his army and navy in the Spanish Armada to attack England. The Spanish navy was expected to gain and hold supremacy in the English Channel long enough for the Duke of Parma's army stationed at Dunkirk to cross the Channel to England. However, the technologically and numerically superior English fleet defeated the Spanish Navy in the English Channel, making the invasion impossible.

In June 1940, after the fall of France, and the evacuation of some 300,000 troops from Dunkirk, Britain stood alone against Hitler's Germany. Its army was vastly inferior to the German army. To conquer Britain, the German army had only to cross the 22 miles of English Channel between France and England. However, Britain's navy was much stronger than the German navy. On August 2, 1940, the Luftwaffe chief Göring issued the Eagle Day directive with the plan of attack to destroy British air power and gain air supremacy over the Channel, and open the way for the invasion fleet of Operation Sea Lion. Germany's strategy was to render Britain's airfields and support installations unusable so that British planes couldn't fly and the German invasion fleet would be able to cross the Channel escorted by the German air force. Germany's air force was much stronger than Britain's. Initially, the British had some 600 fighters, Germany about 1300 bombers and 1200 fighters.

During the summer of 1940, German planes attacked Britain's airfields until they were nearly useless and there were few British planes and pilots left. At the beginning of September, the British retaliated by bombing Berlin. As a result, Hitler ordered the Luftwaffe to switch its attacks from airfields to London and other cities. This change of strategy allowed Britain to repair its airfields, produce more planes and train more pilots. Although the British lost 900 planes in the Battle of Britain, Germany lost 1700, and by the end of September 1940, the British were shooting German bombers down faster than they could be replaced, with the result that the invasion plan was abandoned.

22.5 Russia

Germany's overall objectives at the beginning of the Second World War were German domination of Europe, a continental empire embracing all Europe including the European part of the Soviet Union, and equal rank for Germany with Britain, Japan and the US.

A non-aggression pact was signed between Germany and the Soviet Union in 1939. This was seen as a matter of expediency by Germany which expected to fight a war with the Soviet Union in 1943. However the events of 1939 to 1941 led Germany to attack the Soviet Union much sooner. Operation Barbarossa began on June 22, 1941. By June 27, 1941, Guderian's tanks had advanced the 200 miles to Minsk and 300,000 prisoners had been taken. By July 16, they had advanced another 200 miles and were at Smolensk, taking another 200,000 prisoners. At this point, they were 200 miles from Moscow. They had plenty of time to make decisive gains before the start of the Russian winter.

However, Hitler and the German High Command then disagreed on strategy. The High Command wanted to continue the attack (in a north-east direction) for Moscow on the assumption that the main Soviet armies would be brought to the defence of Moscow, and defeated there. Hitler wanted to attack Leningrad (which was to the north-west) and Stalingrad (which was to the south-east) on the assumption that the destruction of these cities named after such important Communist leaders would be the end of Bolshevism.

The resulting arguments led to time being wasted, a division of forces, and attacks in all three directions. It wasn't until October that the main attack on Moscow was renewed, and not until December that the German Army reached Moscow. By then, the Russian winter had started, the Soviet commanders had prepared their first major counteroffensive, and the German Army was forced to retreat, having failed to achieve any of its objectives.

The following year, 1942, saw a limited German offensive in the South of the Soviet Union. It began on June 28 with Rostov, the first major objective, being captured on July 23. Hitler then divided his forces, with one army under Kleist aimed at the oil fields of Caucasia, and the other under Paulus aimed at Stalingrad. The double objective and the resulting division of resources were to lead to defeat.

The available manpower and fuel resources were insufficient to achieve both objectives.

22.6 The Pacific Ocean

In the early part of the 20th century, Japan's objective was predominance in Asia. It was militarily successful in wars with Russia, Korea and China. However, by the 1930s, it hadn't achieved its objective. Soviet Russia was getting stronger. And half of the Japanese army was tied down by growing Chinese resistance. US influence in Asia was growing. The colonial powers of Britain and the Netherlands still controlled huge areas of Asia.

The war in Europe offered Japan a chance to achieve its objectives. In September 1940, it joined with Germany and Italy in the Tripartite Pact, hoping to neutralise its conflicts with the Soviet Union (which had a non-aggression pact with Germany), paralyse US influence, and exploit the colonies of the European powers. During the next year its opponents (the Allies) increased diplomatic and trade pressure with the result that war became increasingly likely.

Japan's military strategy was based on control of the seas. Provided it could control the Western part of the Pacific Ocean it could achieve its objective of predominance. The Eastern part of the Pacific Ocean is almost devoid of islands (hence air and sea bases) so any attack on Japan from that direction would be difficult.

In 1941, the Allies had about 300,000 troops in Asia. They were widely dispersed, had little combat experience, and were supported by obsolete planes. The Japanese army alone was over 1,000,000 strong. It was well-equipped and had been battle-hardened in China. The Japanese expected it to achieve victory quickly. Then a defensive ring would be built, from Burma in the west to the Gilbert Islands in the east, to keep out the British and Americans.

The Japanese attacked Pearl Harbor on December 7, 1941. The US fleet there was destroyed. By June of 1942, Japan controlled most of the Western Pacific. Its control extended from the Aleutian Islands in the North, down past the Kuril Islands, parts of China, Korea, Indochina, Siam, Burma, Malaya, the Dutch East Indies, Borneo and New Guinea to the Solomon Islands. Its control of the surrounding seas allowed it to move troops and resources from one country to another. This made it difficult for an enemy to bring together the forces that could start to take back the conquered territory. And made it almost impossible to attack Japan by air.

To defeat Japan, the US developed a strategy to first destroy Japanese naval supremacy in the Western Pacific and then make use of US air power. This strategy took the strategy of "control of the seas" one step further to include control of the air over the sea. On June 4, 1942 a Japanese force led by 4 aircraft carriers attacked Midway, one of the few islands in the Pacific it didn't control. The US Navy, having broken the Japanese Navy's code, was waiting for them and Japan lost all 4 aircraft carriers. Midway was saved from invasion, and from then on, the Japanese were on the defensive and the initiative passed to the US. American forces moved across the Pacific to Japan in a series of battles (Gilbert Islands,…, Saipan,…, Iwo Jima,… Okinawa) following the same outline strategy. The battles took place within range of existing American air bases and were in places suitable for runways and anchorages. The attack would begin with a heavy air attack to destroy the defending Japanese air forces. Then a heavy air attack would bombard Japanese troops. US aircraft carriers would prevent Japanese reinforcements. Landing craft would bring US troops ashore. After fierce fighting they would take control. Engineers would land to build runways and port facilities. The next attack would be prepared.

22.7   Lessons Learned

Lessons can be learned from the above examples of the application of different military strategies, and related success and failure factors (Fig. 22.2). They are a useful input when developing strategies for PLM.

22.7.1   History Repeats Itself

In the brief descriptions of military strategy given above there is a certain amount of repetition. In completely different eras and geographical locations, countries have had similar objectives and strategies. For example, both England and Japan had strategies based on "control of the seas". Both Germany and the US had strategies based on "control of the air".

22.7.2   Over Time, Strategies Change

As the environment and the resources change, strategies change. A strategy that may succeed at one time and in one place may be disastrous under other conditions. It is sometimes said that Generals prepare to fight the last war. This can be seen in

history repeats itself
over time, strategies change
a strategy can be offensive or defensive
a small range of simple strategies
the choice of strategy depends on the objectives
there's a hierarchy of strategies
it's dangerous to change strategy during implementation

Fig. 22.2 Lessons learned

France in the First World War where the French generals' desire to attack stemmed from Napoleon's strategies. But the conditions created by the development of machine guns and artillery meant that a defensive strategy was appropriate. By the time of the Second World War, the value of defence had been understood and the Maginot line created. However, the resources available had changed again, and a strategy based on defence led to a French defeat a few weeks after the start of the German offensive.

It can also be seen how the Prussian and German strategists ranged from defensive to offensive strategies through the 19th century in response to the changing environment.

22.7.3 Offensive or Defensive Strategy

A strategy can be offensive or defensive. In most cases it seems that an offensive strategy is necessary. There are occasions, though, such as in the First World War, where a defensive strategy based on blockade, and sapping the strength of the enemy, is successful.

22.7.4 Small Range of Simple Strategies

Potential strategies are shown in Fig. 22.3. There are strategies of control, and there are others ranging from offence to defence.

These strategies all appear simplistic and are described in a few words. This is because strategies have to be simple. Otherwise, few people will be able to understand them. And even fewer will be able to implement them.

control of the seas
control of the air
control of a land region
attack in overwhelming strength
attack with overwhelming speed
destroy the enemy's will to fight
divide the enemy's resources
cut the enemy's communication lines
cut the enemy's supply lines
siege
blockade
impregnable defence

Fig. 22.3 A small range of simple strategies

22.7.5   Strategy Depends on Objectives

The choice of strategy depends on the objectives. There's always a choice of possible strategies. No strategy is going to be right under all conditions. The only way to judge whether a strategy is right or wrong, is whether or not it results in the objectives being met.

22.7.6   Hierarchy of Strategies

Countries have a hierarchy of strategies. A country will have a strategy for a particular battle. There will also be a strategy for a series of battles, such as those of the US in the Pacific after Midway. At the same time, the US was also fighting in Europe so had a strategy there. The strategies in Europe and in the Pacific fitted into an overall strategy.

22.7.7   Danger of Change During Implementation

It's dangerous to change strategy during implementation. Once the decision has been taken to select a particular strategy, it's dangerous to organise or use resources differently. Lee's attack at Gettysburg didn't correspond to the South's strategy of defence, and led to the South's defeat. The hesitation of the German Army in front of Moscow led to the eventual attack taking place in much worse conditions in the Russian winter of 1942. It also gave the Soviet Union the time to regroup its forces.

22.8   Principles of Military Strategy

Military strategy has been studied for thousands of years to understand the "rules" for successful war. Commanders and military observers have tried to identify strategic constants. These are principles of strategy that remain valid despite technological and environmental change. One of the earliest attempts was Sun-tzu's 13 principles of strategy written down in "The art of war" about 400 BC. Sun-tzu stressed the importance of taking account of political considerations. Many of his ideas were used more than 2000 years later by the Chinese communist armies.

By the 1980s, the Soviet, UK, and US military were more or less agreed on the 11 principles of military strategy shown in Fig. 22.4.

objective	keep the basic objective uppermost in mind. Don't be distracted by less important matters
offensive	a defensive strategy is sometimes appropriate, but in the long run, victory can only be achieved with an offensive strategy
unity of command/co-operation	modern warfare brings together different types of forces (army, navy, air force). To succeed, they have to work together under a unified command
concentration of force/effort	in battle, concentrate forces and aim them against an enemy weak point
economy of force/effort	use minimum force to achieve an objective. Any additional force is wasted
manoeuvre/flexibility	the strategy shouldn't be rigid. It should allow different options to be followed depending on the evolution of events
surprise	aim to outwit the enemy, striking when and where least expected
security	take action to prevent the enemy achieving surprise
simplicity	complex strategies aren't well-understood, don't get properly implemented, and lead to defeat
maintenance of morale	one's own forces may be defeated if their morale, or the morale of their civilians, is low
administration	a successful result in battle or in war requires enormous administrative and logistic support

Fig. 22.4   Eleven principles of military strategy

22.9   Manufacturing Strategy

The history of war goes back thousands of years, providing many examples of strategy. Another area where examples of strategy are numerous is in manufacturing operations. Like armies, manufacturing organisations need a strategy to meet their objectives, and to manage and use their resources. The latter include people, machines, methods, materials and money.

For thousands of years, progress in increasing manufacturing productivity was slow. However, a few hundred years ago, mechanisation made possible a leap forward. The machines introduced in the Industrial Revolution led to an organisation of work that differed from the previous approach. Adam Smith in "The Wealth of Nations" (1776) described the new system in a pin factory, "One man draws out the wire; another straights it; a third cuts it; a fourth points it; a fifth grinds it at the top for receiving the head; to make the head requires two or three distinct operations; to put it on is a peculiar business; to whiten the pin is another; it is even a trade by itself to put them into the paper; and the important business of making a pin is in this measure divided into about 18 distinct operations".

Workers were assigned to a particular position at which they carried out a specific task. The owner supervised the workers making sure they worked at the pace of the machines. This led to a division of labour between the owner and the workers. The owner couldn't watch over all the workers all the time, so a hierarchy of supervisors and managers was developed.

In the 19th century, machine tools changed the environment again. They enabled strategies of mass production with the characteristics shown in Fig. 22.5. In mass production, tasks can be performed by unskilled workers, often immigrants or agricultural workers leaving the land, since much of the skill is in the machine and the organisation. Manufacturing enterprises grew to such a size that a large hierarchy of supervisors and managers became necessary. The increasing size and complexity of operations called for a large management staff including accountants, engineers and personnel managers.

high volumes
mechanisation
organised material flow through various stages of manufacturing
sub-division of labour
low skill level of workers
managerial staff with specialised skills
simplification and standardisation of common parts to allow long production runs of parts that can be fitted to other parts without time-consuming adjustment

Fig. 22.5 Characteristics of mass production

The next step, introduced to manufacturing by Henry Ford, was the assembly line. Its concepts had been developed in the meat-packing industry in Cincinnati and Chicago, where overhead trolleys moved carcasses from one stationary worker to another. Each worker did one task, at a pace dictated by the line, minimising unnecessary movement and increasing productivity. Ford applied these methods to the manufacture of cars, reducing the price of cars, bringing it within reach of more people. According to Ford, the assembly line was based on the planned and continuous progression of a commodity through the shop, the delivery of work to a worker (instead of leaving it to the worker to find it) and an analysis of operations into their constituent parts.

Frederick Taylor brought a scientific approach to these principles. A new discipline, industrial engineering, appeared. Taylor broke each job down into its constituent parts, analysed them to find out which were essential, and timed the workers with a stopwatch. With superfluous motion eliminated, the worker, following a machine-like routine, became much more productive.

However, in the years after the introduction of scientific management, its disadvantages, due primarily to neglecting the human element, began to appear. Elton Mayo, a social scientist, carried out experiments at the Hawthorne plant of the Western Electric Company in Cicero, IL to see how changes in lighting affected productivity. He found that productivity rose even when lighting conditions didn't change. Just by involving the workers, a new attitude was created. This result led to strategies of worker involvement.

Mass production increased the trend to an international division of labour. Factories often needed raw materials from other countries. Saturation of national markets led to a search for customers overseas. Some countries became exporters of raw materials and importers of finished goods, while others did the opposite.

The introduction of computers in Manufacturing in the mid 20th century led to strategies of Shop Floor Automation (NC machines, CNC machines, robots, and Flexible Manufacturing Systems). It also led to the introduction of MRP and ERP systems for planning and control of manufacturing and logistics.

In the 1960s and 1970s, Total Quality and Just in Time (JIT) strategies were introduced to cut out waste in Manufacturing. Stocks were reduced, and non-value-adding activities eliminated. Assembly lines were simplified by focusing on a particular product line. Later in the 20th century, these ideas were extended, and Lean Manufacturing strategies were developed.

The skills needed by assembly-line workers are easily acquired. Standards of living in many developing countries exporting raw materials are so low that wages can be kept below those of already industrialised countries. As a result, developing countries can adopt strategies of industrialisation and export of manufacturing goods. In response, manufacturers in developed countries outsource, getting parts made in low-cost countries. In the early 1990s, original equipment manufacturers (OEMs) in the electronics industry faced pressure to get products to market faster than their competitors. They took to outsourcing in a big way, with parts or whole products made or assembled in developing countries. This started with outsourcing of printed circuit board assembly to electronics manufacturing services (EMS) providers, and eventually led to an EMS industry which offers design, manufacturing and related services to the OEMs.

The logical ultimate in the evolution of strategies seems to be the re-configurable Lights-out Factory producing customised products in a batch size of one. This implies elimination of all manual labour and the introduction of flexible manufacturing and assembly machines with automatic controls providing accuracy and quality beyond human skill levels.

From the above, it can be seen that, as in the military environment, when resources and technologies change in the manufacturing environment, strategies also change.

22.10 Company Strategy

Both military and manufacturing strategies change in response to the changing environment of resources and technologies. The strategies that companies adopt are also subject to change. Two main strategies have been used by companies to meet their objectives. One of these is the low-cost, "cost leadership" strategy. The other is a high-value strategy based on differentiation.

A cost-leader aims for the lowest product cost in a particular industry. This usually requires a high market share and a high volume of standard products. It implies substantial capital for large continuous-flow production runs and facilities. By selling a low-cost product in large numbers, the costs of product development and manufacturing equipment are spread over a large number of products and become relatively insignificant. Usually it's the manufacturing cost that's most important, so this type of company focuses on reducing the cost of manufacture. This implies strong abilities in facility engineering, manufacturing engineering and purchasing.

High-value differentiation strategies are based on having a product or service that differs significantly (for example, by virtue of its design, or technology, or customer service) from those of competitors. Higher prices can be charged because of the uniqueness of the product and the few available alternatives. To make the product special usually requires skills in identifying customer needs, and in defining the product correctly.

Other strategies include "niche", "trend-leader" and "follower". A niche strategy serves a particular market segment, or particular type of customer, or particular geography, or particular part of a product range. Within a given niche, a company can hope to succeed with either a cost-leadership or a differentiation strategy.

A company with a trend-leader strategy will constantly innovate in an attempt to lead the market and be the first to produce a particular product or service, and gain the associated benefits. This type of leader is unlikely to be a cost-leader, due to the difficulty of getting products to market first. Instead, revenue is generated from sales to customers who are anxious to be "early adopters", and are willing to pay the additional costs this entails. This strategy requires good product development skills so that a market-leading product can be brought to market quickly.

A "follower" is a company that enters the market when the leader has moved on to the next generation of products, or when the leader can be attacked through cost or quality features. A follower could aim to be a cost-leader. The follower doesn't aim to sell to one of the few early adopters of the product (who often represent less than 10 % of the market) but aims to sell to the main market (the other 90 %). For a follower, it's less important to have skills to develop new products than to be able to understand and improve what has already been developed. This calls for skills in reverse engineering and in reducing product costs.

The above description of strategy may seem theoretical. In reality, the strategies of many companies don't fall nicely into one of the above categories. Many companies pick and mix, taking elements of different strategies to create their own strategy. Recent years have seen the introduction of new strategies such as "low-cost variety", "fast response time", "partnering", and "process-based" strategies (rather than product-based strategies) such as "capabilities-based competition" and Concurrent Engineering. The driving force behind many of these new strategies was Japanese companies using manufacturing excellence to gain competitive advantage. They put new concepts into production quickly, reduced manufacturing times to the minimum, and continuously pumped out new and innovative products. Manufacturing and engineering were equals with marketing and finance in the eyes of top management, and considered essential in the process of developing strategy. The performance of Japanese companies showed that activities in the product lifecycle can provide a competitive advantage. For example, a company which is better at developing new products and services can use this advantage to gain market share. While competitors are busy developing the same abilities, the leading company introduces new products and features faster, and also develops new abilities. When a competitor reaches its targeted level of improved competence, the leader is ready with a newly developed advantage and the competitor is again behind. It spends money to build competence which doesn't provide the needed return on implementation, because the environment has changed.

Strategies such as "fast response time" have been introduced because, as a result of technological advances and changing customer behaviour, products have increasingly short lives. To make money on a short-life product it's important to bring it to market quickly and give it the longest life possible. This also means that product offerings will be fresher. And the latest technology can be included because less time passes between definition of the product and its arrival on the market. Less time in development means less labour and less cost. The company responds quicker to customers, gets more sales, and sets the pace of innovation. A company like this is going to need a strategy that allows it to develop new products quickly, and get them into production quickly, to change production volume quickly as demand builds up, and to switch to production of other products when demand drops.

"Partnering" is often driven by the need for innovation and the limited resources available for developing new products. Partnerships between companies allow greater value and features to be offered to customers, while allowing each partner to concentrate efforts on things they do well.

A "capability" is a clearly-identified and well-defined set of business processes. Capability-based companies achieve competitive success by making their key business processes (the ones that make them leaders) as effective as possible.

"Process-based" strategies are based on the belief that "we know how to do things well". It's not the particular product that counts, but the successful way it can be got to customers. This type of company needs to have a good understanding of its processes, and the ability to adjust them to handle different products.

According to the Bible, "What has been will be again, what has been done will be done again; there is nothing new under the sun." This can be applied to strategy development. Company strategies bear similarities to military strategies. New business strategies draw on elements of old business strategies. The Napoleonic strategy of focusing resources and attacking on a weak point in the enemy line can be compared to a "niche" strategy of a company, focusing resources on a particular part of the market. The military strategy of "Attack with overwhelming speed" corresponds to the "fast response time" business strategies.

Strategies for PLM can be expected to share characteristics with military, company and manufacturing strategies.

22.11 Principles of Business Strategy

Just as there are "principles of war" that can help in the development of military strategy, there are "business principles" that can help in the development of business strategy. For example, Peters and Waterman identified common attributes of excellent companies in their 1988 book "In Search of Excellence" (Fig. 22.6).

| a bias for action |
| close to the customer |
| autonomy and entrepreneurship |
| productivity through people |
| hands-on, value-driven |
| stick to the knitting |
| simple form, lean staff |
| simultaneous loose-tight property |

Fig. 22.6 Common attributes of excellent companies

Although the context is completely different, there are similarities between these attributes and the military principles. "A bias for action" can be compared to "offensive". "Stick to the knitting" can be compared to "concentration of force/effort".

Understanding strategies in other environments helps get an understanding of strategies for the activities of the product lifecycle. For example, how could Ford's assembly lines be translated from the production environment to the product development environment? How could Just In Time strategies be applied in the Imagination phase of the product lifecycle? How would the military principle of "unity of command" translate to the organisation of the activities of the product lifecycle? If Elton Mayo found that worker involvement increased productivity, what would be the effect of increasing customer involvement in the product lifecycle?

22.12   Importance of Strategy

Strategy may seem intangible and irrelevant for some people. And they may think that strategy development is unnecessary. If so, they should look at the effects of the strategies chosen by some military and business leaders.

The French strategy in the First World War led to the death of about 1.5 million French soldiers. Another 4 million suffered injury.

In the 1980s, Switzerland's Swissair was considered one of the world's leading airlines with an excellent reputation for quality service. Switzerland doesn't belong to the European Union, and Swissair feared it would be excluded from the European internal market. In the mid 1990s, it developed a strategy to ensure its place in Europe by buying stakes in several European airlines. These weren't such good performers. After the events of September 11, 2001, airline passenger numbers and revenues dropped sharply. Swissair filed for bankruptcy with about $10 billion of debt. It had 70,000 employees.

22.13   Principles of Strategy

22.13.1   Principles of Military Strategy

From study of the military environment, a set of military principles was developed (Fig. 22.7).

These principles help in the definition of military strategies.

Principles of Military Strategy
objective
offensive
unity of command/co-operation
concentration of force/effort
economy of force/effort
manoeuvre/flexibility
surprise
security
simplicity
maintenance of morale
administration

Fig. 22.7 Principles of military strategy

22.13.2   Company Principles

From study of the business environment, a set of attributes of excellent companies was developed (Fig. 22.8).

They help in the definition of business strategies.

As Fig. 22.9 shows, although the principles address the different domains of war and business, there are several similarities.

22.13.3   PLM Principles

From study of the product lifecycle environment, we developed a set of PLM principles that can be used to help in the development of PLM Strategies (Fig. 22.10).

Company Principles
a bias for action
close to the customer
autonomy and entrepreneurship
productivity through people
hands-on, value-driven
stick to the knitting
simple form, lean staff
simultaneous loose-tight property

Fig. 22.8 Common attributes of excellent companies

Principles of Military Strategy	Company Principles
offensive	a bias for action
simplicity	simple form, lean staff
concentration of force/effort	stick to the knitting
manoeuvre/flexibility	simultaneous loose-tight property

Fig. 22.9 Similarities between the military and company environments

Principles for PLM Strategy
focus on the Product
involve the Customer, listen to Product Feedback
remember the Planet and Mankind
simple slim-line organisation
highly-skilled people
use of modern technology
coherent PLM Vision, Strategy and Plan
continually increase sales and quality, reduce time cycles and costs
watch the surroundings
maintain security

Fig. 22.10 Principles for PLM strategy

Again, although the domain addressed is different, it can be seen that there are similarities between these principles and those applied in the military and company environments. For example, the PLM principle of "watch the surroundings" has a direct military parallel.

The above principles of PLM Strategy will be addressed in more detail in the following chapters. Some will be described briefly below.

22.14   Implications of Principles

Some of the implications of "Focus on the product" are shown in Fig. 22.11.

Some of the implications of "Involve the Customer, listen to Product Feedback" are shown in Fig. 22.12.

Some of the implications of "Remember the planet and mankind" are shown in Fig. 22.13.

Some of the implications of "simple slim-line organisation" are shown in Fig. 22.14.

Some of the implications of "highly-skilled people" are shown in Fig. 22.15.

Some of the implications of "watch the surroundings" are shown in Fig. 22.16.

Some of the implications of "maintain security" are shown in Fig. 22.17.

a Chief Product Officer (CPO) with unity of command over the product
five-year Product Plan and Strategy
platform products and derivative products
part re-use
integrated Product Portfolio

Fig. 22.11 Implications of "focus on the product"

get Product Feedback
listen to the Voice of the Product
involve the customer in product development
listen to the Voice of the Customer
use technologies such as RFID

Fig. 22.12 Implications of "involve the customer, listen to product feedback"

investigate opportunities of sustainable development
investigate opportunities resulting from environmental requirements
investigate opportunities resulting from ageing populations in industrially developed countries
investigate opportunities resulting from large populations in developing countries

Fig. 22.13   Implications of "remember the planet and mankind"

simple organisational structure
simple, clearly defined processes across the product lifecycle
product-focused organisation
Product Lifecycle Owner
simple product lifecycle methodology
cross-functional teamwork

Fig. 22.14   Implications of "simple slim-line organisation"

hiring good people
training
multi-cultural workforce
need for generalists and specialists
need for soft skills and hard skills
career paths
skills matrix

Fig. 22.15   Implications of "highly-skilled people"

watch the surroundings, that's where the customers are
watch the surroundings, that's where the competitors are
watch the surroundings, that's where most new trends and new technology are found
watch the surroundings, that's where the danger is lurking

Fig. 22.16   Implications of "watch the surroundings"

maintain security in bars and trains. Potential customers may be close by, and overhear the details of how you hope to fix the problems with that product they were going to buy
maintain security in restaurants and planes. Competitors may be close by, and learn of your new products and pricing strategies.
maintain security in chat-rooms and e-mails. You don't know who may be reading what you write
maintain security in buildings. Your competitors may be eavesdropping from outside
maintain security when travelling. Your flight may be delayed and your luggage searched
maintain security in Information Systems. Competitors and other organisations may attack with viruses, worms, hacking and spying programs

Fig. 22.17   Implications of "maintain security"

22.15   Coherent PLM Vision, Strategy and Plan

The PLM Vision provides a Big Picture of the future environment, and the expected performance and behaviour. It provides a picture to guide people in the choices they have to make during strategy-setting and planning of resources, priorities, capabilities, budgets, and the scope of activities.

The future PLM Strategy defines how resources will be organised to achieve the objectives. It defines policies for the management and use of these resources.

Once the PLM Vision and the PLM Strategy are defined, it will be possible to develop an Implementation Strategy to achieve them. And the planning of detailed implementation activities can start.

PLM plans address all the components of PLM such as product data, equipment, human resources, applications and processes. Individual projects are identified and planned. Their objectives, action steps, timing and financial requirements are defined. The relative priorities of projects are understood. When planning is complete, implementation can take place.

The end result of the chain from business mission and objectives through PLM Vision, strategy, plan and implementation is that the PLM organisation behaves in such a way that the company meets its objectives.

22.16 Continually Improve

Some of the implications of "continually increase sales and quality, reduce time cycles and costs" are shown in Fig. 22.18.

A company should proactively aim to increase sales. The opposite approach is to set out to reduce sales. This is likely to send the wrong signals to employees and customers. Without the pressure to improve, employees will spend more time on internal politics, angling for promotion and more office space. Customers will assume the company is on the way out of the market, and can't be relied upon for long-term service. If a company can't see opportunities in its existing markets, it must enter new markets with innovative products and services. These should make extensive re-use of existing parts and information.

A company should aim to increase product and service quality. The concepts of TQM should apply both at the level of the whole company and at the level of the product lifecycle. The company must have a culture, attitude and organisation that allows it to provide, and continue to provide, its customers with products and services that satisfy their needs. The culture requires quality in all aspects of operations, with things being done right the first time, and defects and waste eradicated.

A company should have a bias for cycle time reduction. Cycle time has become a key competitive parameter. Reduced lead times open up new market opportunities and improve profits. They reduce market risk by reducing the time between product specification and product delivery. The sooner that customers use a product, the sooner their feedback can be incorporated in a new, improved version. Getting a

| increase product sales |
| increase sales of services |
| increase product quality |
| increase service quality |
| reduce time cycles throughout the product lifecycle |
| reduce costs throughout the product lifecycle |

Fig. 22.18 Implications of "continually increase sales and quality, reduce time cycles and costs"

product to market early will mean that more people will buy it during the early stages of its life. This is because there will be less competition. Slower competitors won't have got their products to market.

Short cycles provide an opportunity to gather a bigger share of the market by being first. In addition to a higher market share, early introduction of a product means a company can ask a higher price. This is possible both because it will be seen as a new and better product, and because there will be less competition from lower-cost products. So, by getting to the market before competitors, a company can have its products on the market longer, and increase its market share, revenues and profitability.

Early introduction also means the company will get the best customers, the ones who will pay more to get the product early. Not only will they pay more, but this kind of customer will also be back for more, or other, products.

Another reason for reducing cycle time is that, in fast-changing technological and consumer environments, sales revenues get eroded because products become obsolete sooner. The reduced time between product launch and product retirement erodes sales revenues. Since this phenomenon of earlier product retirement depends on factors beyond a company's control, the only way a company can lengthen a product's life is to get it to market earlier.

Cycle time reduction leads to faster increase of development experience. Because development cycles are shorter, there will be more of them. So a company will go up the experience curve faster. Which means it can make its products even better and faster.

Another advantage of a short development process is that, as well as finishing development earlier, it will also be possible to start development later. And, starting development later than competitors, means the customer's requirements should be understood better and should be less likely to change. So the faster developer will face less risk.

And, by starting later, it's possible to exploit the latest advances, most recent technologies, and newest styles and fashions. Bringing products to market quickly means that less time passes between definition of the product and its entry to the market. So product offerings will be fresher and the latest technology can be included.

Short cycles are ideal for companies wanting to offer customised products. Because the development process is clean and short, it will be easier and cheaper to adjust to special orders, so they'll be delivered on time and on budget. The company will be seen as a leader in innovation, and customers will want to buy from it again and again.

Over time, the advantages that result from reducing cycle time will build up. During the time that the competitors of a fast developer are busy trying to develop the same abilities, the fast developer will introduce new products and features, and will also develop new abilities. When a competitor reaches the level of competence it thought it needed to compete effectively, the fast developer will be ready with a newly developed advantage and the competitor will be behind again.

A company should have a bias for cost reduction. With average manufacturing wages in the United States many times greater than those in China, it can be expected that product costs will continue to drop, and that product development, manufacturing and support costs will continue to drop.

As well as being quicker, a shorter development process also costs less. Many companies find that if they reduce development cycle time by 40 %, development cost is reduced by 10 %. Doing things quicker (provided they are being done right) means less effort is needed.

Chapter 23
Getting Executive Support

23.1 Getting Started

Currently, nearly all companies have some components of PLM in place. If they didn't, they would have gone out of business. Typically, in the past, though, these companies managed the product in different unconnected ways at different times in the lifecycle with different approaches, processes and applications. These included Product Portfolio Management, Product Data Management, Configuration Management, Product Recall, Customer Complaint Management, Product Warranty Management and Engineering Change Management. These companies didn't have a "PLM" that managed a product continuously and coherently throughout the lifecycle. Products were managed in one way in early stages of the life. Then in a different way during their development. Often the company didn't manage the product during its use, and partially or totally lost control of the product. Sometimes the company managed the product again when the product was due for disposal. Sometimes it didn't.

Because there wasn't a single, or a coherent, approach or technique or solution or application in a company to manage products across the lifecycle, many problems occurred. With the applications, information and processes spread out between different functional organisations, nobody had a full overview of the product. Product developers could see product details in their CAD system, but had no idea if customers really needed these details or even if these products were being sold. Meanwhile Product Managers looked at Sales figures, but didn't have access to the fine details of products and features, so couldn't see how these were related to sales results. Executives received good-looking Word and PowerPoint reports, but had no access to the underlying data that would help them take better decisions. Quality problems communicated by product users were dutifully logged in Quality databases. But often, the developers of new products couldn't access these databases, and the same problems were designed into new products. Many things were done separately, in separate departments or functions across the lifecycle.

© Springer International Publishing Switzerland 2016
J. Stark, *Product Lifecycle Management (Volume 2)*,
Decision Engineering, DOI 10.1007/978-3-319-24436-5_23

Perhaps companies didn't manage the product as well as they could have done, but, of course, to some extent they managed it. Some managers made sure that products were sold, making money for shareholders, and enabling employees and suppliers to be paid. And in other parts of the organisation, other managers made sure that new products were developed and brought to market. Various elements of PLM were being done departmentally. To improve productivity, many companies automated some of their product-related activities long ago, creating Islands of Applications and Data. However, they were rarely able to leverage these investments to achieve the expected improvements in business results. Even so, they continued to look for ways to move forward, with the result that thousands of companies of all sizes are now either considering investing in PLM or in expanding their existing PLM implementation. There are hundreds of companies offering application software that they describe as PLM software, and many of these companies have annual revenues of over $1 billion. There are thousands of companies offering consulting and system integration services in the area of PLM. There are many conferences on the subject. International research projects have been launched into PLM subjects. The subject of PLM is mentioned in the mainstream press as well as in industry journals and technical publications. Numerous articles are written about PLM and its role in managing the entire product lifecycle and the extended virtual enterprise.

CIOs read in IS publications that PLM is the final strategic building block for their enterprise application architecture. They see that CIOs in other companies are looking to combine collaborative Web platforms and integrated enterprise applications to support better the thrust to bring competitive products to market faster. Product Managers read that the product is once again at the heart of business strategy, and PLM will enable a quantum leap in product innovation. They attend conferences and hear Corporate Innovation VPs explaining the benefits of PLM and the need to act before it's too late. Business Process Managers read in Quality journals that PLM is the final plank of the Business Process Framework and that, to earn Quality awards, they should define and deploy lean PLM processes. CEOs read in the business press that PLM will help increase revenues and earnings by bringing better products to market faster, and extending the lives of mature products. Many people in all types and sizes of company are now aware that PLM is on the way (or has arrived), and ask what they should do about it. The answer seems to be, in theory at least, that the benefits of PLM will be reaped after the development of a PLM Vision, Strategy and Roadmap, and the deployment of PLM.

23.2   Not so Easy

The theory may be correct, but it can be difficult to put into practice. PLM in one company is often very different from PLM in other companies. The PLM Initiative of one company is likely to be very different from that of others. It's not a case of "one size fits all". There may, for example, be differences due to the different span of activities in different companies, to the different focus of PLM in different industries, to differences in company size, to different levels of PLM Maturity, and to different reasons for starting a PLM Initiative.

23.3 Different Span of Activities

PLM, and the PLM Initiative, in one company may be different from PLM in other companies because of the different span of activities. One company may just provide design services, and focus on the development phase of a product. In its view, the main activity of PLM may be the use of 3D CAD applications. Another company, such as an aircraft manufacturer, may be involved with its products across their entire lifecycle, which could be more than 50 years. In addition to IS applications, it may have a much wider scope of PLM, also including business processes and data management applications.

23.4 Differences Between Industries

Currently, PLM is being used in a wide range of industries. It's used in discrete manufacturing, process manufacturing, distribution and service industries, as well as in research, education, military and other governmental organisations (Fig. 23.1). There are many differences between these industries, and they have different needs and priorities. As a result, although PLM is used in many industries, it's implemented and used differently in different industries.

For example, in the automotive sector, companies must bring innovative new products to market frequently. They must also cut costs and improve productivity. Product Development is seen as a key activity to achieve these targets. As in the aerospace industry, the definition of product geometry data has high priority in automotive companies. Collaborative Product Development plays an important role because of the high level of outsourcing and offshoring. Other components of PLM help maximise the reuse of components, parts, and assemblies. Automated workflows speed up processes such as Production Part Approval Process (PPAP) and Advanced Product Quality Planning (APQP), and ensure compliance. European Directives are leading car companies to manage the end of life of the product better. Companies across all tiers of the automotive industry implement PLM to speed time

aerospace	apparel	automotive	beverage	chemical
consumer goods	construction equipment	defence	electrical engineering	electronics
financial services	food	furniture	life sciences	machine tool
machinery	medical equipment	mechanical engineering	petrochemical	pharmaceutical
plastics	plant engineering	rubber	shipbuilding	shoe
software	transportation	turbine	utility	watch

Fig. 23.1 Industries using PLM

50% faster product development
greatly reduced data transfer time
reduced time to communicate changes from development to manufacturing
standardisation of product development processes across multiple sites
improved collaboration with partners
improved management of variants
50% decrease in quotation time
reduced document control costs
increased outsourcing to low-cost suppliers

Fig. 23.2 Typical benefits of PLM in the automotive industry

to market, reduce costs and increase new business achievement rates. Typical benefits are shown in Fig. 23.2.

Companies with long lifecycle products, such as aircraft and power plants, focus on configuration management to support future access to data about the products. These products are often highly complex, with electronic, software and electro-mechanical components. There are regulatory requirements for data retention and auditing. In aerospace companies, Configuration Management plays an important role. Workflows speed up design reviews and change management. Collaborative Product Development is important in this industry as development work is often shared between several companies in different countries. Conformance with European Aviation Safety Agency (EASA) and Federal Aviation Administration (FAA) requirements is needed.

In many high tech industries, companies aim to be market leaders by bringing innovative new products to market before competitors. They need short development cycles and maximum reuse of existing parts. Typical results with PLM are shown in Fig. 23.3.

In the consumer electronics industry, the focus is on managing the BOM across the Extended Enterprise. Companies have to take account of fast-changing global and local trends. There's an increasing need to meet environmental regulations and compliance requirements such as those resulting from the Restriction of Hazardous Substances (RoHS) and Waste Electrical and Electronic Equipment (WEEE) Directives introduced by the European Union. The RoHS regulations, for example, require electronics companies to provide proof that they have complied with regulations limiting the amount of six hazardous materials, including lead, in their products.

In industries such as industrial equipment, factory automation and heavy vehicle, reliability is important for customers. Products are often complex and

reduced product development time
greatly reduced product change cycles
improved document management
reduced change management headcount
increased outsourcing
enhanced history tracking
global accessibility to product data

Fig. 23.3 Typical benefits of PLM in high-tech industries

change cycle reductions
reduced scrap and rework costs
reduced time to volume production
reduced time for generation of Bill of Materials
management of customer-specific products

Fig. 23.4 Typical benefits of PLM in industries with engineered-to-order products

engineered-to-order. Configuration Management is a key issue. Typical results with PLM are shown in Fig. 23.4.

In the pharmaceutical industry, the focus is on discovering new compounds, and managing clinical trials. Idea Management is important, as is conformance with regulations. Typical results with PLM are shown in Fig. 23.5.

In the chemical industry, conformance with REACH, the EU regulatory framework for the Registration, Evaluation and Authorisation of Chemicals is needed.

In the Nuclear Power industry, safety and security are all-important. Regulations depend on the country. In the US for example, regulations are set by the Nuclear Regulatory Commission. In the UK, it's the Nuclear Installations Inspectorate.

The medical equipment industry needs to bring innovative products to market rapidly and demonstrate compliance with Food and Drug Administration (FDA) regulations requiring correctly controlled documents, drawings, and data management procedures.

In the fashion industry, time-to-market, fast response to change, and collaborative working between designers in one country and factories in others are all important.

Companies in the utility sector have to meet stringent environmental regulations. With many small subcontractors involved in developments, exchange of product data between different applications is a key issue.

Thus, although companies in different industries have similar objectives for PLM, the exact requirements may differ. PLM isn't "off-the-peg", "one size fits all". The functionality and implementation priorities depend on the market needs and objectives of the company. The general definition of PLM is "PLM enables a company to manage a product across its lifecycle, from cradle to grave, from the very first idea for the product all the way through until it's retired and disposed of." This applies for all products from companies of all types and sizes in all industries.

In the pharmaceutical industry, for example, the definition becomes "PLM enables a pharmaceutical company to manage all its pharmaceutical products across their lifecycles, from cradle to grave, from the very first idea for a product all the way through until it's retired and disposed of." For tyres, the definition becomes

more new products
reduced product development cycle time
extended product lifecycles
reduced document control costs
improved product data visibility

Fig. 23.5 Typical benefits of PLM in the pharmaceutical industry

"PLM enables a tyre manufacturer to manage all its tyres across their lifecycles, from cradle to grave, from the very first idea for each tyre all the way through until it's retired and disposed of."

23.5   Different Reasons for PLM

Although it's possible to describe the levels of PLM maturity, many Middle Managers are unsure of their current PLM status, and of how to proceed with PLM. They can see the potential for major benefits, but find it difficult to know where and how to achieve them. The reasons to implement PLM differ from one company to another, and depend on the particular position and objectives of the company. Middle Managers may see opportunities in many of the following areas, but it may not be easy to find the best path forward.

23.5.1   Cost, Quality, Time, Business Process Improvement

Many managers can see opportunities, with PLM, for cost and time reduction, and quality and business process improvement. They often see cost reduction as an important reason for introducing PLM (Fig. 23.6).

Quality Improvement is also an important reason for managers thinking of introducing PLM. They look to PLM to improve quality in many areas (Fig. 23.7).

Time Reduction is another important reason for managers introducing PLM. They see opportunities throughout the product lifecycle (Fig. 23.8).

product development costs
direct material costs
warranty costs
prototyping costs
validation costs
personnel costs
inventory costs
production costs
service costs
Information System costs

Fig. 23.6 Potential sources of cost reduction with PLM

improve conformance with customer requirements
reduce product faults in the field
prevent recurring product problems
reduce manufacturing process defects
reduce the number of returns
reduce the number of customer complaints
reduce errors, rework and wasted efforts

Fig. 23.7 Potential sources of quality improvement with PLM

reduce time to market
reduce time to volume
reduce time to value
reduce time to profit
reduce issue resolution time
reduce project times
reduce project overrun time
reduce engineering change time
reduce cycle times

Fig. 23.8 Potential sources of time savings with PLM

improve business decisions
improve visibility over the supply chain
increase visibility into manufacturing operations
improve risk management
reduce engineering changes late in the lifecycle
ensure compliance with standards and regulations
provide traceability
manage product portfolios
analyse product information across the product lifecycle
provide feedback from each phase of the lifecycle
enable better management of outsourced tasks

Fig. 23.9 Other potential improvements with PLM

Business Process Improvement is an important reason for introducing PLM. In many companies, managers are looking at streamlining and harmonising processes such as New Product Development and Product Modification. When companies reengineer processes they have the opportunity to identify the most effective way to work, remove waste activities and get Lean. The introduction of PLM provides an opportunity for them to define and implement the best product-related processes across the product lifecycle.

Companies are also looking to PLM to help in other areas. PLM is so pervasive in a company that it can provide benefits in all sorts of activities, including those shown in Fig. 23.9.

23.5.2 *Innovation*

Thanks to globalisation, companies now have the possibility to sell their products and services world-wide. But they also now have competitors from all over the world. This increased competition means they have to develop better products, develop them faster and develop them at lower cost.

Product innovation is becoming a prime concern for many companies. Company leaders are often frustrated by the low level of product and service innovation in their companies. They want managers to turn on new revenue streams and ramp faster. They want to get increased revenues sooner. They're looking for PLM to increase the innovation rate without compromising creativity or quality.

23.5.3   Compliance

Middle managers are faced with an increasing number of regulatory requirements. These are often voluminous and liable to frequent changes. Just managing the regulations and linking them to different products and services in different countries is a time-consuming task. PLM provides product developers and compliance specialists with rapid access to the right information.

Regulators need proof that their requirements have been met. The proof comes in the form of documents. These documents are managed in PLM. They include documentation of product characteristics, documentation of analysis of the product, and documents concerning tests of the product. Other documents, for example, process descriptions, describe the way that work is carried out. The templates, results, process descriptions and workflows necessary to demonstrate compliance can all be managed within the overall PLM environment.

23.5.4   Mechatronic Products

Many companies develop mechatronic products. These products contain mechanical, electrical, electronic and software modules. Companies usually develop mechanical, electrical and electronic components in a similar way, with similar processes and applications. However, in the past, the processes and applications used for software development have generally been very different. Using two separate sets of processes and applications creates all sorts of problems. It can lead, for example, to customers receiving control software that doesn't correspond to their product hardware. Middle managers look to PLM to provide a better way to manage mechatronic products.

23.5.5   Collaboration

Many companies have moved away from the model of a single product development department and a single manufacturing location. There can be various reasons for this. They include globalisation, a need to shift work offshore to low-cost countries, and a desire to work with the best people, wherever they're located. However, relocating R&D activities changes the organisation of work. New approaches are needed to manage and work effectively in the new environment of networked and fragmented research, development and support. PLM enables integration of the design chain (internally, and externally with suppliers), to achieve Global Product Development. It enables integration of the supply chain (internally, and externally with partners) to achieve Global Manufacturing.

Web and collaborative technologies that support the PLM activity enable both research and development to be carried out in a well-managed way in multiple locations. They enable product developers, sales people and service workers to interact with customers and partners on a global basis. They allow product development and support to occur on a 24/7 basis. Team members can be based anywhere yet work together in spite of space, time and organisational differences. They don't need to be co-located. PLM enables them to achieve use and re-use of common parts, worldwide engineering change management, and global information exchange, synchronisation and interoperability.

23.5.6   Intellectual Property Management

Product data/information (product know-how) is one of the most valuable resources in a company. It's an increasingly valuable resource for corporate growth, and must be kept secure. PLM provides the "Intellectual Property Vault" for protection in the face of increasing global competition and the potential risks from terrorism and economic espionage.

23.6   Limited Headway

23.6.1   Middle Managers

There are many reasons for moving ahead with PLM. However, due to the enterprise-wide scope of PLM, it can be difficult for middle managers to start activities on PLM. Often, they don't have the required authority or responsibility. In addition, they're usually already overloaded with other activities and projects that have higher priority and are already running. Frequently, the result is that they make little or no headway with PLM. This can have negative effects (Fig. 23.10).

| decisions about next steps for PLM are delayed |
| achievements of PLM benefits is slow |
| frustration of product developers and product managers |
| problems arising with partners wanting to move ahead faster |
| the company falling behind competitors |

Fig. 23.10 Potential effects of making limited headway

23.6.2   Executives

Due to the enterprise-wide scope of PLM, it's at the level of the VP (or business executive of similar rank and power) that action has to be taken if the expectations of PLM are to become reality.

However, in today's highly competitive global environment, many business executives feel that they're already overloaded with responsibility and work. Perhaps they have been given additional responsibilities extending beyond their usual areas. For example, they may have been tasked with integrating newly acquired companies, or with overseeing operations in Brazil, India, Russia, South Africa or China. They may be involved in other projects, such as headcount reduction and the introduction of lean techniques. With little time available, they may not want to get involved with a subject such as PLM that can seem unclear in both scope and potential benefit.

Another reason that executives may not be convinced that they should invest time and effort in PLM is its enterprise-wide character. This may lead executives to look at PLM and decide that it doesn't lie in their particular domain of responsibility. A CIO may get the impression that PLM is mainly an issue for Product Managers and Product Development Managers. But Product Managers may see PLM as being mainly a question of application systems, so lying in the IS area.

Another issue is that many experience-hardened business executives are sceptical of claims for new breakthrough approaches and technologies. They may see PLM as just one more breakthrough among the many that are touted. It may be difficult to convince them that it will bring success.

Many executives are looking for short-term improvements with impact on the financial figures in the next quarter. They're likely to consider that PLM doesn't fall in that category.

In many organisations, there's not yet a corporate plan or funding for PLM. There's no PLM budget, and executives haven't been assigned to PLM, or set an annual target for PLM. As a result, none of the executives may feel any responsibility for PLM.

23.7   Company Dilemma

A dilemma arises in many companies as people see the need for, and opportunities of, PLM yet don't see the expected resulting action. On one hand, there's a feeling in the company that PLM should be implemented (Fig. 23.11). On the other hand, due to various concerns (Fig. 23.12), there's little progress with PLM.

the product is at the heart of business strategy. PLM enables a quantum leap in product innovation
PLM can meld collaborative Web tools and enterprise apps in a push for market-leading products
PLM is the final strategic building block for the CIO's enterprise application architecture
PLM enables information automation and system integration with accurate and timely product data
PLM enables benefits for the 80% of the product-data consumer-base outside the R&D Department
PLM is the final plank of the Business Process Framework
PLM is a keystone activity of the Lean Enterprise
PLM is part of the foundations of the Extended Enterprise
PLM increases earnings, getting better products to market faster, extending lives of mature products

Fig. 23.11 Reasons for implementing PLM

business executives are already stretched with other tasks
there isn't a clear vision of PLM for people to aim at
the company is waiting for market improvement before investing in new initiatives
the company is busy with other projects
headcount reduction has resulted in a lack of resources
PLM responsibility isn't defined
PLM doesn't fall nicely into an individual department's scope
PLM may look too strategic and long-term

Fig. 23.12 Factors holding back PLM progress

23.8 Personal Dilemma

In this situation, with PLM looking strategic, but not being acted on by high-level executives, middle managers face a dilemma. Should they try to do something about PLM, or should they forget about PLM and carry on with "business as usual"?

If they do try to do something about PLM, they may well be seen later as having been instrumental in helping the company achieve major benefits through use of PLM. They may enable the company to seize new opportunities and solve long-running problems. Of course, on the other hand, if they try to do something about PLM without support from above, they could expose themselves to criticism for not doing what they've been told to do. They could be blamed for not following the plan prepared by their boss. Even worse, they can be accused of lowering morale and productivity by making unnecessary suggestions for change.

Sometimes they start to make a list of reasons to justify why they don't need to do anything about PLM (Fig. 23.13).

there are already many projects running in the company
PLM isn't the only issue in today's global industrial environment
the company has slimmed down. There aren't enough people for a PLM project
few people have a broad enough overview to lead a PLM project
it's not clear who should be responsible for PLM
business executives are already stretched with other tasks
the CFO has put new initiatives on hold
middle managers are already stretched with other tasks
middle managers don't have the authority to launch company-wide PLM activities
people enjoy fire-fighting, the present environment. Why rock the boat?
many people can't see the potential improvements with PLM

Fig. 23.13 An initial list of reasons justifying a lack of PLM action

managers of projects that overlap with PLM will fight it. They want to keep their projects
PLM will be massive, but it's not clear exactly what it is, or what its scope will be
people talk of PLM in different contexts. This is confusing
people who don't know about PLM find it difficult to understand how it can help them
executives don't understand enough about products to see the need for change
the company is focused on short-term payback. PLM looks long-term for the CFO
PLM looks confusing and difficult to succeed with
without a dominant vendor driving the PLM market, it may be unwise to start with PLM

Fig. 23.14 More reasons to justify a lack of action

Then they go back to work on everyday business. At the back of their mind, new entries for the list appear. After a while, they go back to the list, and add a few more reasons (Fig. 23.14).

Then they go back to work on everyday work. After a few weeks, they begin to think about PLM again, and find some more reasons for the list (Fig. 23.15).

Having made such a list, the manager realises that it might be better to try to do something about PLM. Otherwise, they could be accused of being negligent. Or of not offering the company the opportunity to make major gains through the use of PLM. Of course, the manager may then think that PLM will come one day anyway, and for the moment it's probably not required, as top management hasn't asked for it. And of course, they can comfort themselves with the thought that there's no way they can do it on their own, so they might just as well wait until their boss tells them to do something about it. And of course, if they did try to do something about PLM, they would expose themselves to criticism for doing something that wasn't in their job description. So they may think that the best way forward is to get on with that small improvement project which was planned the previous year, even though it probably won't lead to significant results.

23.9   Going Nowhere

These dilemmas have arisen for many managers in many organisations. They lead to a repetitive situation of the type shown in Fig. 23.16.

PLM isn't in the company's annual plan or budget
there's a lack of methodologies for the implementation and operation of PLM
there's a lack of documented PLM Best Practice
the CIO may be concerned about expensive integration

Fig. 23.15 Even more reasons to justify a lack of action

in previous years, the company had many performance improvement projects, for example, to implement new application software, define business processes, and take on board Concurrent Engineering
in spite of all the past projects, there's a problem related to some of the company's products. There's discussion about how to solve it. The usual way to solve it would be to launch an improvement project
when middle level managers start looking at the details of the proposed improvement project, they see many causes for the problem. And these involve several processes and several departments
they realise that what's needed is some kind of overall joined-up PLM approach that addresses the problem in a wider context of many applications, processes, and techniques
they think about starting a project to develop an overall PLM Strategy
they look round the organisation for someone to lead such a project, but find that, after all the downsizing, offshoring and outsourcing, nobody has the time to do it
they look outside the company, and are quoted more than $50k by consultants for a PLM Strategy
they discuss if they really should spend $50k on a voluminous report, or if they should invest in licenses for a new application that will make everyday work easier
they decide to buy the new licenses and start the improvement project, even though they think it would be better to address the problem in a project with a wider scope
they continue to think about how to find the resources to develop a PLM Strategy
while thinking about this, some more product-related problems (such as lack of product innovation, product configuration errors, field failure reports being lost) occur, and get their attention
when they look at these problems, they see that these problems don't have a clearly-defined stand-alone scope, but involve several processes, several applications and several departments
this confirms the feeling that what's really needed is some kind of overall joined-up PLM approach
however, another review of availability shows there's nobody available to lead an initiative, and none of the business executives have been given the responsibility for PLM.
they start some more small projects to address the latest product-related problems

Fig. 23.16   A repetitive situation

23.10   Examples of the PLM Dilemma

The situation described in the previous section may seem absurd, but it arises in many companies. And it can continue for a long time before a true PLM activity is started. Here are some examples that I've experienced.

Company A, in process manufacturing, had been working for several years to deploy a cross-functional product development process. Asked about PLM, they replied that the CFO had said they would have to complete that deployment before starting an initiative in the area of PLM.

Company B, in consumer electronics, had recently launched a corporate effort to redefine all process maps to take account of globalisation, the new ERP application and the Web. It was a major effort, and executives were wary of starting a parallel PLM initiative. They said the company couldn't handle two major initiatives at the same time.

Company C, in the telecomms sector, was in a phase of merger and restructuring in response to global changes in that industry. The main priority was to get the existing Technical Information Systems, which were based on different architectures, databases and applications, and were on several different continents, to work together. This was a massive task and used all available resources. Nobody in the IS organisation had the time to work on PLM.

Company D, in the automotive sector, was proud of its application of CAD, CAM, CAE, PDM and Digital Manufacturing, but was faced by many problems in the area of Software Configuration Management. They wanted to solve that specific problem before starting a project with a scope as wide as PLM.

Company E, in the aerospace industry, had several overlapping improvement projects on subjects that fell into the area of PLM. Some people had proposed consolidating these projects into one PLM project. But the managers of the overlapping projects claimed that would slow down progress. Although the Engineering VP was supportive of a PLM project, the CIO and the Quality VP were opposed.

Company F, in the mechanical engineering sector, was in a phase of rationalising existing Information Systems. PLM was seen as something fuzzy that couldn't be pinned down. They decided they would look at it when they had a clearer understanding of their new system architecture.

Company G, in the machine tool sector, wanted to reengineer its approach to product development to take better account of customer requirements. It didn't want to address Information System issues. Due to the cost of the ERP project, the CEO had forbidden any customisation of enterprise applications.

Company H, in the pharmaceutical sector, had hundreds of R&D projects running. One was a high-profile project to find a way to give management an overview of the current status of all R&D projects. That project had top priority, and no resources would be put into new projects until it had succeeded. PLM was on the back burner.

Company I, in the electronics sector, was reviewing, again, its Engineering Change process. Some of the people in the project thought the problem wasn't the change process, but the product structure. They wanted to take a more global approach to the problem. But the project charter didn't allow for that.

Company J, in the financial services sector, felt that it had taken a piecemeal approach (Fig. 23.17) to its product-related applications and processes in the past, and thought it was missing something. It was looking to solve that issue by bringing together all available resources for an ERP project.

Company K, in the plastics processing industry, had decided to stop all improvement projects until its markets started growing again. PLM was considered unimportant (Fig. 23.18), and was on hold.

Company L, in the power equipment industry, having recently terminated major projects to harmonise application systems and improve business processes, was running a product structure optimisation project to enable more modularity and easier configuration for sales over the Web. Until that was completed, it would be difficult to start another project addressing the product.

Company M, in the medical equipment industry, had recently acquired a company making software for its products. It was looking to see how best to integrate operations and offer integrated solutions to its customers. As PLM wasn't in the annual plan, it wasn't addressed.

Company N, in the heavy vehicles industry, was struggling to find a way to deliver highly customised products with a Configuration Management application nearly 30 years old. There were several reasons why it wasn't easy to move forward. One problem was that the IS VP, the Engineering VP and the Marketing VP all claimed that PLM wasn't their responsibility.

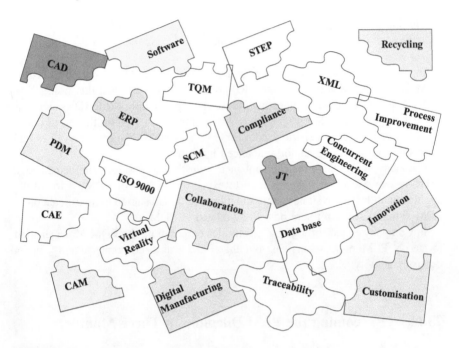

Fig. 23.17 A piecemeal approach

Fig. 23.18 Some people underestimate PLM

Company O, in the electronics industry, finding that software was becoming a major part of its product, was looking at ways to integrate the development, purchasing and management of mechanical, electronic and software components. That project was called 2020 Vision, and PLM may be included among its objectives.

In Company P, in the electronics industry, the PDM Manager tried to start a PLM project, but was told that the ERP project team already had that task.

In Company Q, in the engineering industry, the provider of the CAD system was restructuring its portfolio. There was a discussion as to whether this would be a good time to change to a single vendor for CAD and PDM. The subject of PLM had been sidelined until the vendor announced its plans.

Company R, in the medical appliance industry, successfully implemented a PDM system. However, when it tried to expand the scope of PDM, it found the application didn't have all necessary functionality. It had started to investigate other PDM systems. It wasn't sure how PLM related to PDM.

In Company S, part of a global electronics corporation, the PDM Manager tried to start a PLM project but was told that PLM was a corporate activity, not a company activity.

23.11   Overcoming the PLM Dilemma in Three Months

Managers in many companies face a dilemma over PLM. On one hand, it's clear that PLM makes sense and that it's gaining in importance and acceptance. On the other hand, it's not clear what to do about it, or who should take action.

However, it's clear that, at some stage, the person who will have to take action is a top-level business executive with the authority and responsibility to address a subject that's enterprise-wide and addresses products, processes and applications. Someone who's responsible for ensuring the company improves business performance and makes money for shareholders.

And it's clear that the action will include the launch of a PLM Initiative, the development of a PLM Strategy and the deployment of PLM.

And presumably it's clear that before the top-level business executive can launch the Initiative, someone else will have to explain the case for PLM to them, very clearly and concisely, and in language they understand.

And, presumably, knowing how things work in many companies, that explanation will be in the form of a PowerPoint presentation which will be prepared and presented by one or more people who report to that executive. And the objective of the presentation will be to help senior executives take action.

The content of the presentation could include the points shown in Fig. 23.19.

The presentation could take about 1 h. It could include about 20 slides (Fig. 23.20).

After initial discussions between middle managers, the possibility of making such a presentation can be discussed with a key executive. A draft presentation can be built. The subject can be discussed again with the executive, this time with the help of the slides. More feedback will help improve the presentation. Other people

PLM manages the product all the way across its lifecycle. (There's been nothing available to do this in a coherent way, and that's caused problems)
PLM provides visibility about what's happening to the product across the lifecycle. (It will be clear what's happening with products and projects)
PLM gets products under control across the lifecycle. (Which means that executives will be in control, face less risk and have more influence)
in the past, products were to a certain extent managed across the lifecycle, so a lot of the components needed for PLM already exist. (Which means that PLM doesn't involve starting from new, but building on what already exists)
the benefits of PLM are measurable and visible on the bottom line. (Typical targets for PLM are to increase product revenues by 30% and to decrease product maintenance costs by 50%)
PLM is holistic. (PLM doesn't just address one resource, and improve use of that resource while reducing the effectiveness of other resources)
there's currently not an off-the-shelf solution for PLM. (Which means that each company must define its own solution for PLM)
with PLM, one person will be responsible for all the products, which will be visible and under control. (Instead of having unclear multiple responsibilities)
the company should launch a PLM Initiative. (PLM enables the company's product-related objectives to be achieved)

Fig. 23.19 Potential contents of the presentation

Title of the presentation	1 slide
Contents of the presentation	1 slide
Objective of the presentation	1 slide
This is PLM	8 slides
PLM: our benefits and opportunities	4 slides
Three ways to move forward with PLM	3 slides
Ten step approach to PLM Launch	2 slides

Fig. 23.20 Structure of the presentation

Meet with the executive	Month 0
Create draft presentation	Month 0
Show presentation to executive	Month 1
Improve the presentation	Month 1
Present the presentation again to the executive	Month 1
Discussions with other executives	Month 2
Define and launch PLM Initiative	Month 3

Fig. 23.21 Timeline for preparing to launch the PLM initiative

will be invited to join the discussion. Before long, the executive will be making the presentation to other executives, and the company will be on the way to PLM.

Discussing and creating a presentation doesn't take long. The timeline for the above activities could be similar to that shown in Fig. 23.21.

23.12   The PLM Initiative

PLM Initiatives will be different in different companies because the companies are in such different situations (Fig. 1.32).

As a result, there isn't a single, off-the-shelf, PLM Initiative that will fit everybody. Without knowing the exact situation within a company it's not possible to know what it should do. This can be demonstrated by considering two companies

(Company A, Company B) of similar size and supply chain position supplying similar products to similar OEMs in their industry. Company A reports that it's reduced its Engineering Change time by 80 % by implementing a new PLM software application. What reduction do you think Company B can achieve by implementing that PLM application?

The answer, of course, is that it's impossible to give a meaningful answer. What really happened in Company A? Is the 80 % reduction due to implementing an application or improving the processes? Was the process previously manual or already automated? Does the reduction apply to all products or just to one? Does it apply to all sites or just one? And how does the environment in Company B relate to that in Company A? Has Company B already implemented that PLM application? In which case it may already have achieved a 90 % reduction in Engineering Change time.

Although there are thousands of different PLM Initiatives in thousands of companies, there are often some features common to their Initiatives. As a result, although each company has to build its own PLM Initiative, it can draw on experience from other companies.

Without knowing the details of a particular company's PLM initiative, it's clear that a PLM Initiative will last several years, and that it isn't realistic to expect that everything will be done at once. As a result, it's useful to develop an Implementation Strategy and a PLM Plan to identify what should be done. And to prioritise the order in which these things should be done.

For some activities, it's clear that they can't be done together. For example, after PLM has been successfully deployed, it must be maintained. It can't be maintained before it's been deployed. However, for other activities, it may be less clear in which order they should be carried out. For example, it may not be clear if a process should be improved before it's automated, or if it should first be automated. And then improved once the automated process has been used and understood.

Because a PLM Initiative addresses so many components such as products, processes, people, data, and information systems, it may not be clear initially how a company can handle such a huge project. It's even possible that someone will suggest that it's not possible to manage an activity with such a wide scope. They may suggest cutting off a piece of PLM and focusing on that one piece. For example, some people might want to focus on the Product Definition phase of the product lifecycle, and ignore the other phases. Others may want to focus on IS. By focusing their resources in one area, they may hope to get a better understanding that will lead to faster progress and better results.

However, the danger of initially restricting the scope in this way is that it may result in the loss of many of the potential benefits of PLM. It's by bringing together, and joining up, previously disparate and fragmented activities, applications and processes, that PLM overcomes the many problems that result from the old unconnected approach. Cutting off a piece of PLM runs the risk of leading to a new fragment with similar problems to those of older fragments.

There are alternatives to reducing the potential size of PLM by cutting off a piece of PLM and addressing it separately. The first step should be to look for a structure and organisation for PLM that will help simplify its understanding and implementation. When this has been achieved, the many opportunities within the scope of PLM can be prioritised. And an implementation roadmap built up from manageable pieces.

Full achievement of PLM can be expected to take a lot of effort and a long time. That's normal. PLM is a major business activity running across the complete product lifecycle and the Extended Enterprise. PLM has a wide scope, and the PLM environment is complex. To achieve PLM will require a lot of effort over many years. In theory, it might be possible to do everything in one project, but a single, huge, multi-year project is likely to end in disaster. In reality, it's better to run a formal PLM Initiative containing many smaller, shorter, more focused projects. Without a formal PLM Initiative, there's the danger that some important activities will slip out of view, some won't occur, some activities will overlap, the results of some activities will conflict, and some important decisions won't be taken. The end result is likely to be project failure, or downgrading of objectives. Many of the potential benefits of the PLM Initiative will be lost.

A PLM Initiative, made up of many individual projects of various sizes, can be compared to a Development Program, with multiple development projects, that's set up to develop a series of related products. There's a need for leadership of the overall Program, but equally importantly, each of these projects will have its own project leader, objectives and tasks. In the case of the PLM Initiative, the leaders of the individual projects report to the PLM Initiative Manager.

In theory, a PLM Initiative can be led by anyone who can run a complex, cross-functional project. In practice, it's good to have a leader who has experience with the company's products at different phases of the lifecycle, who can handle the cross-functional aspects of PLM, and who has experience of managing the various components of PLM, such as applications, processes, data and work methods.

23.13   PLM Initiatives, from Strategic to Tactical

The scope of PLM is very wide. The range of possible PLM Initiatives is also very wide. The PLM Initiative of a particular company may depend on a range of factors such as its existing PLM status, its financial health, its competitive environment, and its available management skills. The result is that the PLM Initiative it launches may fall anywhere in the range between "supremely strategic" and "totally tactical" (Fig. 1.35).

It's important to make clear to everybody concerned just what the PLM Initiative is expected to achieve. There's a danger that people will expect strategic results from a tactical approach and a tactical investment. This issue can be illustrated by reference to Fig. 1.34 that shows the results of one of our surveys into different types of approach.

The results make sense. They show that major gains come from long-term strategic approaches, not from short-term tactical projects. However, they go

Fig. 23.22 From vision to
implementation

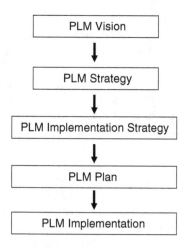

against the philosophies of "getting something for nothing" and "getting something for nothing, fast".

For a truly enterprise-wide PLM Initiative (Fig. 23.22), the first step may be to develop and communicate a Vision of the proposed new environment, including a future PLM Strategy, so that everyone knows where they're going. The step after that will be to define an Implementation Strategy to achieve the PLM Vision. Then an Implementation Plan has to be developed. Once the plan has been implemented, the benefits can be harvested.

However, if the focus is departmental cherry-picking, then the Vision, and even the Strategies, may not be needed. In all cases though, a plan will be needed to show what has to happen, when it should happen, and who does what to make it happen.

The exact details of a PLM Initiative will be different in every company because all companies are different. Different companies have different products, different positions in the supply chain, different management styles, different business processes, different information systems, and different objectives. As a result, their PLM Initiatives will be different. Some companies will have already taken initial steps on the road to PLM, and the first steps they need to take now will differ from those of a company that's only just starting out towards PLM.

Usually the first steps in a PLM Initiative are to understand what it means for the company, and how PLM can be achieved. Different companies focus on different parts of the product lifecycle, and many different scopes and approaches for the Initiative are possible.

In some companies, a Feasibility Study will be carried out to find out which type of approach (Fig. 1.36), and which level of response, is appropriate.

The results of the study should be documented in a Feasibility Study report (Fig. 1.38).

The Feasibility Study may lead to identification of the need for some specific actions (Fig. 23.23).

Fig. 23.23 Actions identified
in a feasibility study

better understand the product lifecycle
better understand activities and processes across the lifecycle
use a PLM phase/gate methodology across the product lifecycle
define the roles in the product lifecycle
train people to work effectively in a lifecycle environment
define information needs across the lifecycle
create a glossary of words used frequently in PLM
manage product development projects better
position and quantify each product in the lifecycle
define end-of-life needs
define product grouping
maintain and reuse product development knowledge
use a Product Data Management system across the lifecycle

Other companies may feel that a Feasibility Study isn't necessary. For some companies, the first step in the PLM Initiative could be to carry out a pilot implementation of PLM software. They may think it's best to get hands-on experience as soon as possible, and then, on the basis of that experience, they'll start the planning and communication activities.

In other cases, companies will prefer to start by planning what they're going to do, make everyone aware of what's happening, and then start looking at different possibilities for PLM. One of these could be software-related, but alternatively they might decide that the first step is to train their product teams.

A PLM Initiative is a major improvement activity that's likely to result in activities and changes that affect many people throughout the company. In most companies it takes a long time and a lot of effort to bring about change. This is likely to be the case for the changes related to PLM.

Recognition of the need for a clearly defined and professionally managed change activity is a key feature of successful change. It's helpful to understand that change is a major activity in its own right, and is a project in its own right, with its own objectives, activities, tools, techniques and metrics. In successful change, communication, learning and reward systems are given a high priority. Communication is a necessary element of a change project, but alone it's not enough. Communication and learning are good and necessary, but they're not enough. Communication, learning, and new reward systems are all important factors in successful change.

23.14   Understanding the Way Forward

Experience shows that it can take longer to make progress with PLM than expected. Often, one of the reasons for this is a need to broaden the understanding of PLM issues among business executives. Another is the difficulty of identifying the best approach to PLM and justifying the business case.

The way forward will be different for different companies. They'll be approaching a PLM Initiative from different starting positions (Fig. 23.24). As a result they'll have different questions about PLM.

| looking at PLM for the first time |
| having understood PLM, creating a PLM business case |
| expanding an implementation from PDM to PLM |
| responding to competitive pressures demanding improved performance |

Fig. 23.24 Different starting positions before a PLM initiative

| what is PLM? |
| how and where should we start with PLM? |
| how can we improve chances for success with PLM? |
| what should our PLM concept include? |
| where does PLM fit with other initiatives in our company? |
| our CAD and ERP system vendors have different PLM concepts. Who's right? |

Fig. 23.25 Initial questions about PLM

23.14.1   First Time Entry

A company looking at PLM for the first time may have many general questions. Examples of this type of question are shown in Fig. 23.25.

23.14.2   PLM Business Case Creation

Another company may be creating a business case for PLM. It's likely to have some very specific questions, such as those shown in Fig. 23.26.

A company at this stage could be looking for the best way to develop a justification of the PLM Initiative (Fig. 23.27).

23.14.3   PDM to PLM Expansion

Another company may be intending to evolve from departmental use of a PDM system to a strategic approach to PLM. It could have the type of questions shown in Fig. 23.28.

| what should we include? |
| how can we quantify the value? |
| what figures are realistic? |
| how do we calculate ROI? |

Fig. 23.26 Questions about a PLM business case

Phase	Type of Savings	Value of Cost Savings	Type of Gains	Value of Revenue Gains
Imagine	Cost reductions (manpower, fees)	Comparatively low	New products and services	Very High
Define	Cost reductions (project, manpower)	Comparatively low	Better products and services	High
Realise	Cost reductions (material, manpower)	Medium	Fast availability of customized products	High
Support/ Use	Cost reductions (material, manpower, warranties)	Medium	Upgraded/ extended products & services	Very High
Retire/ Recycle	Cost reductions (manpower, fines)	Medium	Material reuse	Medium

Fig. 23.27 Justification of a PLM initiative

what do we do next?
where can we gain the biggest benefit?
how can we stop struggling with multiple CAD and PDM systems?
how can we automate our business processes?
how should we build a data model for the lifecycle?
how do we handle multiple applications resulting from acquisitions?
how can we get our support costs under control?

Fig. 23.28 Questions related to expansion from PDM to PLM

23.14.4 Competitive Pressures

Another company may be facing business drivers demanding much greater effectiveness and efficiency. It could have the type of questions shown in Fig. 23.29.

23.15 The 10 Step Approach to PLM Launch

To answer questions such as those mentioned above, we developed the Ten Step Approach to PLM Launch. The overall intention was to answer the questions, overcome any issues, and enable faster PLM progress. The approach offers

how can we grow at 10% per year?
how can we compete against low-cost producers?
how can we produce more products faster?
how can we support products world-wide?
how can we identify more great products?

Fig. 23.29 Questions oriented to improved business performance

| build a business case for PLM and get management buy-in to proceed |
| uncover hidden needs and opportunities for PLM beyond the obvious |
| identify the best PLM approach aligned with business objectives |
| clarify the scope of PLM |
| gain clearer understanding of the ROI potential of PLM |
| define and prioritise a clear PLM Roadmap |
| implement PLM quickly and cost effectively, avoiding pitfalls |
| improve overall PLM success |

Fig. 23.30 Aims of the ten step approach

1	PLM Status Review, Data Gathering
2	Executive PLM Education and Awareness
3	Best Practice Positioning
4	PLM Concept Generation and Analysis
5	PLM Scope Definition; Roadmap and Plan Generation
6	Business Benefits and Business Case Development
7	ROI Calculation
8	Management Report Preparation
9	Executive Presentation
10	Executive Decision Support

Fig. 23.31 The ten steps of the ten step approach

companies a structured way to determine opportunity and problem areas that can be addressed by PLM. It's based on our experience of many companies in many industry sectors. Its aims are shown in Fig. 23.30.

The aims correspond to those of many companies that we work with. Based on these needs, we developed the ten steps of the approach (Fig. 23.31).

Experience shows that these ten steps help in understanding how PLM can be applied to a business most effectively, and in getting executive approval for the PLM initiative to proceed. The approach has been used in many companies, at different stages of PLM progress, in many industries. It's been found that the ten steps make it clear to everyone involved what has to be done.

In a medium-sized company, a typical project will run six weeks (Fig. 23.32), a very cost-effective six weeks compared to the months or more of time and expenses that can be saved down the road.

Clearly-defined deliverables for each step help show how the project will proceed, and make sure that key findings and proposals are captured and retained (Fig. 23.33). For example, the deliverable from the "PLM Status Review, Data

	Step	Wk 1	Wk 2	Wk 3	Wk 4	Wk 5	Wk 6
1	PLM Status Review; Data Gathering						
2	Executive PLM Education and Awareness						
3	Best Practice Positioning						
4	PLM Concept Generation and Analysis						
5	PLM Scope Definition; Roadmap and Plan Generation						
6	Business Benefits & Business Case Development						
7	ROI Calculation						
8	Management Report Preparation						
9	Executive Presentation						
10	Executive Decision Support						

Fig. 23.32 The ten steps planned over six weeks

Step	Main Deliverable
PLM Status Review, Data Gathering	a report on the as-is situation, and expectations for the to-be situation
Executive PLM Education and Awareness	a PowerPoint presentation addressing potential benefits and opportunities of PLM
Best Practice Positioning	improvement opportunities, strengths and weaknesses
PLM Concept Generation and Analysis	a report on potential PLM concepts, and reasons for the choice of a particular concept
PLM Scope Definition; Roadmap and Plan Generation	PLM Scope; PLM Roadmap; Plans
Business Benefits and Business Case Development	a report on expected costs, benefits, value and ROI
ROI Calculation	a realistic calculation of Return on Investment
Management Report Preparation	a Management Report and a presentation
Executive Presentation	full understanding of the PLM proposal
Executive Decision Support	a Go/No Go decision

Fig. 23.33 The deliverables from each of the ten steps

Gathering" step includes an overview of the current situation. Much of this will be in the form of text, but it will also include numerous tables, lists and graphics such as histograms, pie charts and radar charts to help visualise why certain recommendations are warranted.

The deliverables from the "PLM Concept Generation and Analysis" step include, for each concept or option: a description; the benefits; the strengths and weaknesses; other issues; main activities; elapsed time; manpower requirements; costs; risks.

The deliverables from the "Management Report Preparation" step are a comprehensive report and an accompanying PowerPoint presentation that can be presented to executives.

23.16 Results of Use of the Ten Step Approach

The following examples show the benefits achieved by some companies that followed the approach.

23.16.1 Understanding and Quantifying Options

This company wanted to understand and quantify the different options that had been suggested with 2D and 3D CAD, PDM, workflow management, BOM Management, product development process improvement, and a new development methodology. The 10 Step Approach showed that there were three main options, and highlighted their different costs and benefits. In particular, it showed that the benefits of the low-cost option would be negligible, yet the other options would require significant management involvement. This led the company to appoint a PLM VP to drive the PLM initiative forward and achieve maximum benefit.

23.16.2   *Managing the Post-acquisition Situation*

As a result of an acquisition, this company had different CAD and PDM systems, and different product-related processes and methods, at different sites. It wanted to identify the best solution and understand the associated implementation tasks and costs. The 10 Step Approach showed many additional issues and opportunities that hadn't been addressed, and led to a common PLM Strategy for all sites.

23.16.3   *From PDM to PLM*

This company had identified the need for a PDM system, and wanted help with specification of PDM system requirements, short-listing, benchmarking, project planning, cost evaluation and ROI (Fig. 23.34). The 10 Step Approach showed the need for PLM. It simplified the project and led to faster implementation.

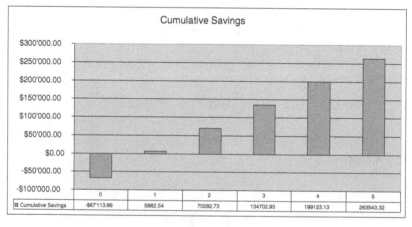

Fig. 23.34 Showing the value and ROI of the project

23.16.4 Getting Started with PLM

This company had identified the need for a PLM solution. Management initially wanted help with identification and detailing of different PLM concepts. The 10 Step Approach led into support for selection and implementation of the corresponding PLM applications.

23.16.5 Engineering Change Management

This company wanted help with the definition and automation of its Engineering Change process. The 10 Step Approach showed that the source of the problems it hoped to overcome with the EC project was outside the process. The proposed EC project was doomed to failure. The project was redefined to enable an increase in business value.

23.16.6 Identification of Benefits and Risks

This particular company had started a project to select and implement a PLM solution. As the size of the potential benefits became clear, management asked for external support to validate the findings and identify potential risks. The 10 Step Approach quantified a realistic ROI. Risks were identified, classified and quantified. A risk management approach was implemented.

23.16.7 Two Proposed Solutions

This company had received a proposal from its CAD application vendor for a PLM solution, and a very different proposal for PLM from its ERP vendor. The 10 Step Approach highlighted the differences between the two proposals and showed how they related to business objectives. This allowed the CFO to launch an opportunity study to show which approach would be best for the company.

23.16.8 Common Benefits

The above examples show how the 10 Step Approach to PLM Launch tends to broaden and deepen a company's understanding of PLM. It raises the level of awareness among executives, eventually leading to a PLM Initiative of greater benefit to the company.

Activity	1	2	3	4	5	6	7	8
Clarify the PLM Vision, including the future PLM Strategy								
Detail the Current Situation								
Detail the Future Situation								
Develop the Implementation Strategy								
Revisit the Roadmap								
Describe the Next Steps								
Develop the Plan for the Next Steps								
Report the Progress								

Fig. 23.35 The plan for next steps

23.17 Plan for Next Steps

The plans generated in Step 5 of the Ten Step Approach include a detailed plan for the next steps. In addition to the planned activities shown in Fig. 23.35, this will also show deliverables, participants and responsibilities.

Chapter 24
Developing PLM Vision and PLM Strategy

24.1 Deliverables of Vision Development

The three deliverables of the activity of developing PLM Vision and PLM Strategy are shown in Fig. 24.1.

24.1.1 The PLM Vision Report

It's useful to define the shape of the Vision report before starting to develop the Vision. Team members will then be aware of what they have to achieve. The typical content of the PLM Vision report is shown in Fig. 24.2.

For several reasons (Fig. 24.3), the Vision should be documented in a formal report.

24.1.2 PowerPoint Vision Presentation

The main characteristics of the PLM Vision should be documented in a PowerPoint presentation. It will be easier and quicker for people to understand this presentation than for them to read the report.

24.1.3 One-Page PLM Vision Overview

To help communicate the Vision, a short, easy-to-understand Vision Statement should be developed. A single sheet of paper is an ideal communication medium.

© Springer International Publishing Switzerland 2016
J. Stark, *Product Lifecycle Management (Volume 2)*,
Decision Engineering, DOI 10.1007/978-3-319-24436-5_24

Fig. 24.1 Three deliverables
of vision development

| a formal PLM Vision report |
| a management presentation |
| a one-page summary for the desktop |

Fig. 24.2 Typical structure
of the Vision report

Title : A Vision of our PLM Activities

Table of Contents

Executive Overview

Section 1 - The Company
1.1 Company objectives and strategy
1.2 Key success factors for the company
1.3 Key issues : markets, customers and competitors
1.4 Key issues : products

Section 2 - PLM Project Progress
2.1 Recommendations from the PLM feasibility study
2.2 Critical issues for PLM

Section 3 - The PLM Vision
3.1 The environment (past, current and future)
3.2 Driving forces
3.3 Description, scope and objectives
3.4 Products and customers
3.5 Framework and components
3.6 Relationship with other activities
3.7 Finance

Section 4 – The PLM Strategy
4.1 Candidate Strategies
a.   brief description of the selected PLM strategy
b.   analysis of the selected PLM strategy
c.   strengths and weaknesses
d.   response to opportunities and threats
e.   fit to company strategy
4.2 Detailed description of the PLM strategy. Organisation and policies
a.   products and services
b.   portfolio
c.   customers
d.   activities, processes
e.   equipment and facilities
f.   human resources
g.   technology
h.   methods
i.   data
j.   information systems
k.   standards
l.   relationships with other activities
m.   interfaces
n.   operations
o.   metrics
p.   planning and control
q.   quality
r.   finance
Section 5 - Implementation strategy
Section 6 - Outline implementation plan/roadmap
a.   major projects
b.   project objectives, timing, resources, costs, benefits, priorities,
organisation
Section 7 - Outline first year plan for the PLM Initiative

Appendices
1. Team Members
2. Interviews

A very concise description of the Vision, of between 40 and 80 words, can be laid
out on one page. It should be displayed on desks and walls. This will help keep the
essential elements of the Vision in everybody's minds.

to distribute the Vision for review and correction
to distribute the Vision to increase awareness
to reduce the risk of the Vision being forgotten
to reduce the risk of misunderstanding
to reduce the risk of distortion in support of particular interpretations
to communicate the Vision

Fig. 24.3 Reasons for developing a formal Vision document

24.2 The Visioning Process

The development of a PLM Vision requires the identification and consideration of many factors. Some of these will be related to competition, some to customers, some to technology. When we work with companies to help them develop an Vision we spend a lot of time getting them to identify the factors related to PLM. We help them to develop the Vision, but we don't develop it for them. By making them go through the Visioning process for the future environment in detail (in the same way they go through the Visioning process for their vacation in detail), we get them to understand the issues and become responsible for the result. In this book, it's difficult to go through this process in an interactive way. But, as you're reading this, you can try to identify factors that are relevant to your company but not included here.

24.2.1 First Thoughts for the Vision

A first attempt at identifying factors that may have an impact on PLM might lead to the list shown in Fig. 24.4.

These factors are obviously important when developing a Vision. They strongly affect business performance. A meaningful Vision must take account of them. You have to understand these factors, develop a coherent Vision of the future that takes account of them, and then work out corresponding plans and budgets.

At this stage, some people may feel they've done enough visioning, and will ask if it wouldn't be better to start with a budget and discuss how it could be spent.

After some thought though, most people will realise it makes sense to continue the visioning process. As they continue to think about the factors impacting the Vision, they might want to extend the list with the factors shown in Fig. 24.5.

Fig. 24.4 Initial result of listing factors impacting PLM

customers
suppliers
competitors and other market players
environmental, geopolitical and other developments
technological developments
national, international, and industry standards
key market and customer requirements

Fig. 24.5 More factors
impacting PLM

future product and market scope, emphasis and mix
desired position in value chain. OEM or supplier
sources of competitive advantage
company Vision, strategy, and organisation
company leadership
expected company and business development
key business success factors
projected company resources and capabilities
relationships with the world outside the product lifecycle

Fig. 24.6 Other factors
potentially related to PLM

the PLM Vision, strategy, plans and organisation
borders, boundaries and themes of the Vision
business philosophies, culture and leadership
choice of metrics
interdepartmental communication
approach to project funding
cost justification
improvement initiatives
the management of change
forces for change
quality
standards
processes
computer systems and applications
best practices
human resources
Information

As they go deeper into the subject, they might propose another set of factors (Fig. 24.6).

By now, the list is getting quite long, and people can see that the development of a PLM Vision has to address a lot of issues. The Vision must provide all the answers to questions that may be asked about these issues. Figure 24.7 shows that all sorts of questions may be asked about these factors.

24.2.2 Horses for Courses

The answers to all the questions that can be asked about each of the above factors during the development of the Vision are going to be very company-specific. In addition to the company objectives, they will depend on many factors (Fig. 24.8).

The industry sector is an important characteristic. The PLM activity of an automotive supplier making a high-volume, standard product will be different from that of an aerospace company with a make-to-order product which differs from one customer to the next. The organisational structure of the company is also an important characteristic. There'll be many differences between companies that organise by cross-functional project teams, and companies that have a strong departmental focus to their organisational structure. The culture of the company is another important characteristic. In some companies, top management takes all decisions, and bottom-up initiatives are discouraged. In others, management actively tries to get decisions taken by people lower down the hierarchy. Sometimes

How will we structure the lifecycle?	Which activities will be in PLM?
Which products/services will we provide customers?	Where will we make money?
Will we be a product manufacturer or a distributor?	Which activities will be outsourced?
Will we be a company that just develops new products?	What should our main activity be?
Will we be a contract manufacturer?	How big will our product portfolio be?
Will our only customers be OEMs?	How many new products each year?
Will our only customer be the government?	Will we be an environmental leader?
How are the sectors we compete in evolving?	What are the key issues in these sectors?
What will be the critical success factors for the future?	Which new markets could we enter?
Can we get 75% of revenue from products < 5 years old?	Can we make cash cows out of mature products?
What new opportunities does the Web give us?	What new technologies provide opportunities?
What's the worst problem we've had with a product?	What lessons can we learn from that?
What are Marketing's criticisms of Engineering?	How can we learn from them?
What changes might affect us?	Who will be our customer?
What will be the needs of the customer?	What will be value to the customer?
Where will the customer be?	What will the customer buy?
How will the customer buy?	What's the best way to respond to competition?
Which companies could we work with?	What laws and regulations could affect us?
How should we measure our performance?	How should we be judged?
Do we need more research effort and facilities?	How can we improve innovation?
How can we improve time to market?	Which new technologies should we be using?
What type of potential employee will be available?	Do we need new offices/office layouts?
How can we best position ourselves for the future?	Which activities should be outsource?
Should we diversify or buy a competitor?	How can we manage cross-functional activity?
What will our business development strategy be?	If our main market collapses, how do we survive?
Which of our products are real turkeys?	Why did we develop them?
Why didn't we stop the development earlier?	Do we need new offices/office layouts?
Can we develop products for developing countries?	Where do we want to be strong?
Where is it acceptable to be weak?	What might happen and be an opportunity?
What might happen and be a threat?	How could we be more competitive?
What happens if we are more responsive?	What happens if we reduce time to market?
What happens if we increase our range of skills?	What happens if we reduce costs?

Fig. 24.7 Questions to test out the draft PLM Vision

the company's industry sector
the company's organisational structure
the company's culture
the company's resources
the current approach to product development and support
the geographical location of the company's units
the size of the company

Fig. 24.8 Factors impacting a company's vision

the situation isn't clear, with management claiming people are empowered, while those lower down the hierarchy claim they aren't. The current approach to product development and lifecycle support will have major implications for the approach in the future. Without some extremely good reasons, major changes to people's behaviour are difficult to bring about. Due to all these differences, it's unlikely that two companies will have exactly the same Vision. And it's unlikely that a company can find an off-the-shelf PLM Vision.

24.2.3 Executive Input

Having got this far, it will probably be clear that only a few people will be able to answer some of the questions that are raised during the development of the Vision and Strategy. In particular, few people may know about the business objectives of the company as they relate to PLM. Often, such information can only be obtained from top management. Sometimes, the information will be available because the company has already defined its PLM strategy. In other cases, top management won't be able to give the answer directly. It will then be necessary to piece it together from information that's more readily available, such as the company strategy, and the individual IT, R&D/Engineering, Manufacturing, Marketing, Innovation, Customer Support and Sustainability strategies.

The relationships between business objectives and PLM may also become apparent from the issues raised, and the concerns expressed, by top management when discussing PLM and related subjects. There may be many objectives, but it's important to identify and confirm the four or five factors, and business metrics, that are most important for management.

There could be a need to reduce lead times significantly, to improve product quality, or to increase revenues. There could be specific issues that have to be avoided, or relationships with powerful customers that need to be improved. There may be the intention to suppress some product lines, or to develop new, or improved, products. There could be plans to change the way clients and markets are addressed, or the way work is carried out with development and support partners. Management may want to focus PLM on reducing product cost. Management may want to introduce major business programs to improve Supply Chain Management or to achieve Lean Manufacturing. They may want to introduce techniques such as Design for Manufacture and Design for Sustainability. Management may have specific targets in mind, for example to reduce the lead time in the engineering department by 50 %, and reduce recalls by 80 %.

The business objectives provide a clear business focus for the PLM Project. This will help greatly and, in particular, it should prevent drowning in the sea of information that will be produced during development of the Vision and Strategy. Without the business objectives, it's only too easy to produce technical findings that are of no benefit to the business. With the business objectives, there are clear targets in sight, providing focus and priorities. Knowledge and understanding of the business objectives helps balance the perhaps divergent needs of technical wishlists for improved management of engineering drawings, CAD data, engineering documents, lifecycle processes and product workflow.

If possible, the information obtained from management should be quantified. If the information is quantified, it will have more meaning. And can be used later both as a target and as a measure of progress. It's not enough to know that profitability and market share must be increased, some quantification is needed. There are hundreds of ways that qualitative objectives like these can be met. Without quantified targets, it's not possible to differentiate between them. Once management has set the targets, the team developing the Vision will be able to differentiate between possible solutions. It may even be found that a particular target isn't related to PLM.

The factors of most importance to management should be very closely linked to business strategy. These top-level businesses issues are just as important for the future of PLM as the technical requirements. If a particular product or process is to be abandoned or outsourced, then there may be no reason to address it further.

Its exclusion could significantly change requirements. A company intent on market dominance through lowest unit cost may have very different requirements from one that will manufacture high-performance products to specification in a speciality niche market. Performance-driven companies emphasise continually evolving designs, whereas lowest-cost manufacturers will want to freeze product designs.

The business objectives, which are an important input to strategy development, might just be expressed as growth, innovation and profitability. There could be a need to reduce lead times significantly, or to improve product quality and reliability. There could be specific problems that have to be avoided, or relationships with powerful customers that need to be improved. There may be the intention to withdraw some product lines, or to develop new, or improved, products and services. There could be plans to change the way customers and markets are addressed. Product development costs and product costs may be too high. Support staff costs may need to be reduced.

24.2.4 Looking at Competitors

Another important input for the Vision is information about competitors, and their use of PLM. Ideally, the information about competitors should address a wide range of subjects (Fig. 24.9). The description of the competitive environment should also cover the subjects shown in Fig. 24.10.

The following types of questions need to be asked (Fig. 24.11).

Fig. 24.9 Information about competitors

their company strategy and management
their approach to PLM
their approach to management of change
their products and their lifecycles
their activities and processes for the product lifecycle
the people who work in the lifecycle
information technology
information
practices
performance metrics
quality
standards
finance

Fig. 24.10 Information about the company's environment

geopolitics
economics
business issues
customers and suppliers
legislative issues
environmental issues
standards

Fig. 24.11 Questions about competitors and the environment

Who are the competitors? What are they doing?
What are their key success factors?
Where are competitors based? Why?
What's the competition good at doing?
Where are our competitors strong? Where are our competitors weak?
Who could be our competitors in the future?
What's the current PLM strategy of major competitors? Why?
What could we learn from the competition?
What are competitors planning to do in the future?
How are our competitors organised for lifecycle activities? Why?
Which technologies are most important to competitors?
Which product technologies will become key?
Which process technologies will become key?
What products compete with ours, and what technology do they use?
What services do competitors offer?
What national and international laws, standards and regulations affect us?
What factors in the market place will have most impact on us?
What's happening that could be an opportunity for us?
What's happening that could be a threat for us?

24.2.5 With Executive and Competitive Input

Answering some of these questions, asking others, expanding and regrouping the factors can soon lead to an impressive collection of factors, as shown in Fig. 24.12. In practice, we often build this up using Post-its and flipcharts.

24.2.6 Review

To see if the list is sufficiently complete, it's useful to test it with some more questions (Fig. 24.13). If it's complete, then these questions won't lead to additional factors being added to the list.

The above set of questions is far from complete. However, it shows the range of questions that could be asked. The more questions that are asked at this stage, the more robust the Vision will be.

24.3 Haziness of the Vision

Creating a Vision that meets the business objectives and the requirements of the people carrying out activities in the product lifecycle isn't as simple as it looks. First of all, nobody knows exactly what the Vision will look like, so it's impossible to provide a detailed description of what to aim for. Secondly, there's an infinite number of possible Visions.

As there are an infinite number of possible Visions, it will be best to identify several potential scenarios, and show why one is preferred to the others. It's usually best to investigate three or four scenarios. Each scenario should be described in detail along with its strengths and weaknesses. This helps get an in-depth

Product Lifecycle			
portfolio management	release	technical publications	outsourcing
requirements definition	change management	process planning	assembly
requirements management	process definition	installation	manufacturing
product portfolio, today	process measures	field service	overhaul
product portfolio, in five years	process ownership	help desk	disposal
system engineering	process management	NC programming	upgrading
idea management	elimination of waste	lifecycle support	refurbishment
design	process improvement	maintenance	recycling
development	supplier involvement	repair	research

Company	Products	People
company strategy	future product & market scope	knowledge
company mission and objectives	future emphasis & mix	training
company philosophy and beliefs	product range, customisation	professionalism
company vision, strategy, and plans	type of product, prototypes	rewards and remuneration
company organisation	incorporation of new technology	skills and skill transfer
company leadership	product/ process multi-technologies	enthusiasm
departmental strategies	simplicity/complexity of design	ambitions
expected company development	standard material, parts & modules	expectations
key business success factors	product cost, lead times	beliefs
projected company capabilities	product idea, product definition	involvement in change
company targets	embedded software	career development
company expectations	product portfolio	job content
company alliances and partners	customer requirements, feedback	culture
partnering strategies	sustainable development	facilities
outsourcing strategies	services provided on products	offices
product range	level of functionality	hire and fire
innovation management policy	volumes, costs	roles and role models
sales targets	internet-enabled?	titles and job descriptions
alliance management	impact on society	mobile devices

Departments	Product development	Performance metrics
interdepartmental communication	leadership	lead times
understanding of other's needs	objectives, vision, strategy	financial ratios
trust of other departments	structure and organisation	% spend on training
conflict resolution	capacity, location	customer service metrics
feedback	project management technique	customer satisfaction
role and rank within the company	orientation	first pass yield
relationships with the outside world	management structure	engineering productivity
supplier and customer involvement	management skills	value-added
	management training	% cross-functional working
Management of Change	attitude to risk /risk management	on-time delivery ratio
improvement projects	breakthrough/ update approach	benchmarking
approach to implementing change	outsourcing	continuous improvement

market	customers	standards	PLM	information	Knowledge
competitors	location	company	definition	structure	knowledge management
customers	number	industry (FDA)	scope	communication	technical publications
progress	type	type	description	management	marketing publications
globalisation		national (DIN)	understanding	flow	sales brochures
stockholders		global (ISO)	benchmarking	meetings	on-line sales catalogues
banks	practices		processes	data bases	electronic communications
venture funds	techniques		activities		web
	methodologies		tasks		intranet
	best practices				intellectual property issues

quality	technology	information systems	society	finance
TQM	product technology	infrastructure	environment	capital
cost of quality	process technology	applications	sustainability	cost structure
prevention	technology watch	ERP, CRM, SCM	pollution	project funding
assurance	awareness	CAD, CAE, CAM, PDM	global warming	cost justification

Fig. 24.12 Factors impacting PLM

What's the overall business objective?	What's the overall business strategy?
What's the overall basis for competition?	What are the activities of the product lifecycle?
What does PLM do for competitive advantage?	What could PLM do for competitive advantage?
Who are internal customers of lifecycle activities?	Who are external customers of lifecycle activities?
What are the needs of the customer?	What has to be done to satisfy the customer?
What's value to the customer?	What are key elements of customer service?
Where are the customers?	How do the customers buy?
How do we measure customer satisfaction?	What are our products and services?
What's unique about our products?	Where are we strong?
Where are we weak?	Where are we strongest?
What are the key skills?	What's the key know-how?
What's the key resource?	What's the key activity?
What's the critical technology?	What's unique about our technology?
Are our people dynamic and innovative enough?	What are the key issues we face?
What are the key metrics of the lifecycle activities?	What are we good at doing?
What could we be good at doing?	What are we not good at doing?
What and where are the current lifecycle resources?	How are they organised?
Why are they organised like this?	What are their roles?
Who is responsible for the lifecycle activities?	What are our core capabilities?
What key capabilities of the lifecycle are exploited?	Which key lifecycle capabilities could be exploited?
Which products and services are the most profitable?	Who are the most important customers?
What do the most important customers buy?	Is competition price sensitive?
How are we perceived by our customers?	How do we communicate with our customers?
How successfully have we met objectives in the past?	What prevented us from meeting our objectives?
How do we view our current performance?	What were our past strategies?
How successful were they?	How were they developed?
How were they communicated?	How were they implemented?
What were our past plans?	How successful were they?
What assumptions have turned out to be invalid?	What have been recent performance changes?
What went wrong recently? Why?	What went right? Why?
What has recently changed? What are implications?	What's currently changing? What are implications?
What are basic beliefs of the people in the lifecycle?	What are the unwritten rules across the lifecycle?
What are the fundamental assumptions?	What gives us most problems?
What should we do differently?	What stops us from doing it?

Fig. 24.13 Questions to test the range of factors

understanding of a proposed solution. Often it's by trying to understand the strengths of one scenario that the weaknesses of other scenarios become apparent.

24.4 Vision Structure. Slicing and Dicing

It's relatively easy to make long lists of factors that could influence the activities of the product lifecycle. It's more difficult to create a concise Vision from them. However, listing the factors is the first step. Then they can be put in the Trash Can, where they can be shaken, rattled and rolled.

Instead of having hundreds of individual factors to consider, wouldn't it be easier to think about a smaller number? Could all those factors be put in a small number of groups? Perhaps they can be put into three or four groups, such as people, processes and systems? Let's see. A human resources group could bring together things like culture and leadership (Fig. 24.14).

No, that doesn't seem to have created a vision. Maybe it would be better to chunk all those factors differently. What about men, machines, methods, materials and money? Or perhaps that long list of issues could be sliced into management, customer, worker, partner and competitor sections? The management group could include corporate strategies, departmental strategies, customer and supplier

culture	customer and suppliers	new technology awareness	time performance
leadership	partnerships	new technology uptake	schedule performance
organisation	departmental metrics	communication of information	product cost
philosophies	organisation	standards	product functionality
professionalism	philosophies	suppliers of services	customisation
department's role	quality		quality
department's rank	activity role and rank		service
skills	standard practices		customers
	standards		technology
	suppliers of services		

Fig. 24.14   Four groups of factors

Global	globalisation will continue ; global R&D and global supply management will be needed
Time	product development times will be cut by a factor of 4 ; product introduction rates will double
Growth	a company has to aim to grow sales at least 10% per year, else it stagnates
Quality	quality will be a must ; ISO 9000
Environment	ISO 14000 will be a must
Costs	costs will be continually driven down by competitive pressure across the globe
Customers	customers will want customised products ; each country will have special needs to be met
Technology	new technology will lead to major changes and new opportunities

Fig. 24.15   Another attempt at creating a Vision

relationship, finance, initiatives, leadership, organisation and philosophies. And perhaps there should be a "Market" group, of customers and potential customers, suppliers, competitors, regulatory authorities and government. But wouldn't it be good to have something about global forces? Let's try again (Fig. 24.15).

No, still no vision appearing. However, at least it's clear that there are many ways to slice and dice the factors, and that it's useful to try several arrangements to see how things relate together. The structuring process is very important because it helps bring the Vision into existence. By repeatedly stirring the information in the Trash Can, and then allowing it to settle, the "cream" will eventually rise to the top in the form of the Vision. It's a long process that involves repeatedly turning over and restructuring a lot of information from many sources. Care has to be taken to avoid jumping to conclusions before all the information has been fully processed.

24.5   A Five-Step Process

With the development of the Vision under way, it's now time to turn to the Strategy. Developing and implementing a PLM Strategy is a five-step process (Fig. 24.16).

Step 1	the information with which the strategy will be developed is collected and assembled
Step 2	several potential strategies are identified, formulated and described in terms of resources, and the organisation and policies to be applied to the resources
Step 3	potential strategies are evaluated and tested, and the most appropriate strategy is selected and detailed
Step 4	the chosen strategy is communicated to everyone affected by it
Step 5	detailed planning is followed by implementation

Fig. 24.16   Five steps of strategy development

24.6   Step 1: Gathering Information

A very good understanding of the activities and the resources in the product life-cycle is needed to develop the PLM Strategy. This understanding must be based on factual information, not on guesses and opinions. The required information includes the Vision and associated targets, customer and innovation objectives, details of resources, capabilities and the environment, strengths and weaknesses, opportunities and threats. This information should be available after the Vision-related activities described in previous sections.

24.7   Step 2: Identifying Strategies

In the second step of strategy development, several potential strategies are identified, formulated and described in terms of the organisation and policies to be applied to the resources. It's always useful to identify and describe several possible strategies. This will improve the chances of finding the best strategy since the most obvious strategies aren't necessarily the most appropriate.

24.7.1   Resources

The resources that can come into play in the product lifecycle, and therefore should be addressed in the PLM Strategy, are shown in Fig. 24.17.

24.7.2   Strategy

A range of strategies will be possible in the PLM environment, just as there are many possible strategies in the military and business environments (Fig. 24.18).

Fig. 24.17 Resources in the lifecycle

facilities such as offices, manufacturing plants, service centres
equipment such as computers, machine tools, assembly line machinery
people and their skills, morale, know-how
supplies such as raw materials and energy. Waste materials
finance, for example, for capital investment and for financing projects
reputation with potential customers and employees
time, in the form of response time, development time, time-to-market
knowledge, information, data, documents.
information systems
processes
working methods and techniques
communication in terms of equipment. Communication with customers
command and control in terms of management of lifecycle activities
space. Favourable locations
alliances with other organisations, companies, government, etc.
standards
existing products
customers

military environment	business environment
control of the seas	cost leadership
control of the air	differentiation
control of a land region	niche
attack in overwhelming strength	leader
attack with overwhelming speed	follower
destroy the enemy's will to fight	low-cost variety
divide the enemy's resources	fast response time
cut the enemy's supply lines	partnering
siege	process-based

Fig. 24.18 Strategies in military and business environments

Some candidate strategies for PLM will be apparent from the organisation's current and previous strategies, from those of competitors and other market players, and from reading about military and business strategies. In some cases, a candidate PLM strategy will be closely linked to the company's business strategy. For example, a cost-leader will be more interested in reducing manufacturing costs than developing market-leading functionality for its products. This has follow-on repercussions on PLM Strategy. Sometimes a completely new strategy will be developed. Often this will come from the "Trash Can" approach of continually sifting through information until something useful rises to the surface. In many cases, a candidate strategy will be developed by a "pick and mix approach" to strategies found by the above methods.

The strategy chosen for PLM has to meet the objectives of the company. As each company will have a different objective, as well as different resources and a different environment, the strategy a company develops will be different in some respect from that of any other company. It's to be expected that the PLM Strategy selected will contain some elements of one (or more) of the basic strategies, and also have some additional elements.

The hierarchy of strategies can make the identification and understanding of strategy elements very difficult. For example, going back to the military environment, the US had a strategy in the Pacific after Midway, and at the same time was fighting in Europe, so had a strategy there. The strategies in Europe and in the Pacific fitted into another, overall strategy. There can be many answers to the question "What was the US strategy?" since there were strategies in different areas and at different levels. For companies, and for PLM activities, there can be similar confusion.

For example, a company may have a strategy of "customer focus". Because the PLM activity is an important activity, it will probably be expected to adopt this strategy. Company management may also set "customer focus" as an objective for the product development organisation. The product development organisation may respond to this objective with a strategy of "customer focus" or another strategy such as "fastest time to market". The product development organisation may also decide to adopt a strategy of "customer focus" to its internal customers within the company, for example to F&A and Service, and include their representatives in project teams. The result of so many "customer focus" strategies may be confusing.

However, confusion needs to be avoided. The PLM Strategy must be very clear and simple, because if it's confusing and complex it will fail.

24.7.3   Strategy Elements

The name given to a strategy has to be meaningful and self-descriptive. For example "control of the seas". At a lower level than the strategy itself are "strategy elements", addressing particular resources and activities, which also need to have simple names and clear descriptions.

"Customer focus" is an example of a strategy element that could be identified in the second step of PLM strategy development. Then it would have to be described in the context of the organisation, activities and resources of the lifecycle. It would soon be realised that "customer focus" isn't sufficiently descriptive or wide-ranging to build a strategy for the product lifecycle. For example, it says nothing about products or human resources. "Customer focus" is more a PLM principle than a strategy. It's generally agreed to be "a good thing".

Usually a strategy can't be based on just one strategy element, one improvement initiative, or one resource. It's not enough to claim "Lean" as the PLM Strategy, or to claim that the choice of a particular location for the PLM organisation is the strategy. PLM Strategies aren't one-dimensional. Several strategy elements need to be combined to develop a particular organisation's strategy. It may appear that all elements should be needed, but in practice, organisations have limited resources so can't do everything. An attempt to do everything would lead to confusion, and probably nothing would get done. As a result, choices have to be made and a clear strategy has to be created.

Examples of the types of strategy elements that can be proposed are shown in Fig. 24.19.

Fig. 24.19 Examples of strategy elements

```
customisation capability. Customers can configure products
the highest functionality products and / or services
the most robust product or service
the most sustainable products. Perpetual recycling
the best processes across the lifecycle
environmental-friendly products and processes
fastest time to market
market-leading hi-tech products
bundled solutions, rather than individual products
long-life. Buy once, use forever
most mobile products and services
lowest-cost competitor
best service over the lifecycle
the safest products
maximum re-use of parts
the widest range of products and services in the market
the least variation between products of succeeding generations
products that are easy to integrate into solutions
reduce to core product development activities
lean organisation
most appreciated by customers
lowest product development costs
partnering
most skilled workforce
```

The exact meaning of a strategy element will differ from one company to another. For example, the strategy elements of "fastest time-to-market" and "lowest-cost competitor" could both be implemented in many ways. "Fastest time-to-market" could be implemented by building up a pre-defined stock of solutions, by increasing the number of engineers, or by shortening the product development process by removing non-value-adding activities. "Lowest-cost competitor" could be implemented with cost reduction programs, capital expenditure cuts, headcount reductions, or by improving the effectiveness of the product development process. The criteria for selecting strategy elements, and deciding how they'll be implemented, will be made clear to some extent by the objectives provided by the business strategy, and to some extent by the application of PLM principles.

The strategy development activity aims to find the most suitable way to carry out the activities of the product lifecycle and meet the objectives with the limited resources available. It may well be that there's no strategy that allows the PLM activity to meet the objectives with the resources available. In this case, either the objectives, or the resources, need to be changed. It is, of course, much better to find this out during strategy development than by failing to meet the objectives.

Business managers shouldn't ask PLM to aim for the highest functionality product, the fastest time to market and the lowest costs. In practice, this is likely to be impossible. Just as in war, commanders who don't have very clear objectives and feasible strategies won't succeed. By describing, and then analysing, strategies this can be found out before the battle.

Strategy elements can be described in exhaustive detail. Some of the key points are shown in Fig. 24.20.

Each strategy element has to be described in detail. These details will probably be different for each company. Similarly, the strategy elements may be interpreted differently in different companies.

The detailed descriptions of each of these strategy elements will differ from one company to another. They may be lengthy. This can be illustrated with "re-use".

The first time that information, or a part, is created it goes through a long process involving a lot of quality checks and validation. It's then used, and any remaining errors are uncovered and corrected. By the time someone wants to re-use it, it will be more or less bug-free. Re-use of existing information and parts doesn't require any development time. It can help reduce development cycles. By re-using existing designs, less investment will be needed for the development of new designs. As there will be fewer designs, and fewer changes to designs, the overhead costs associated with managing, storing, copying and communicating designs can be reduced. The development cycle will be shorter if existing information is re-used. Design and verification time will be reduced. Less engineer cost and less overhead cost will be incurred. There'll be less need for simulation. There'll be less scrap. Re-use of existing information with known characteristics should result in an optimised design. These individual benefits will all contribute to a reduction in the overall cost of the product. Reuse savings come from reductions in design and

Fig. 24.20 Key points of
strategy elements

```
customer involvement
          customer involved in product development and support teams
          frequent communication with the customer
          in-depth customer surveys
          specifications negotiated and optimised with customers
          development and support work on customer sites
          online feedback from customer use
customisation capability
          perceived by customers as excellent at customisation
          able to identify customisation requirements quickly
          able to identify customisation requirements correctly
          able to carry out requested customisation quickly
the highest functionality products and / or service
          perceived by customers as the market leader
          market expectation of continued functionality leadership
          unique product and / or service
          technological leadership
          premium features
the most robust product or services
          product has long mean time between failure
          product requires little maintenance
          extensive simulation and testing
          simple product or service with small number of parts
          product certification demonstrating robustness
the best processes
          implement the best development process and methodology
          able to do tasks well
          customer belief in process competence
          process certification
fastest time to market
          projects started early with up-front effort profile
          information shared as early as possible
          simulation rather than test
          information re-used
value-adding PLM
          move up the added-value chain
          focus on key value-adding activities
          excellent reputation for key activities
          activities that add little value are outsourced
lowest-cost competitor
          simple product or service
          standardisation
          re-use
```

engineering costs, reductions in prototype and testing costs, and reductions in field service costs.

Re-using information avoids the cost and time of repeating a task that has already been carried out. When information is created, it goes through a process including specification, development, production, test, modification, use, and storage. The process includes quality checks to ensure that the information is correct. It includes decision points where choices are made as to the best way to proceed. The overall process takes a certain amount of time, and costs a certain amount of money. At the end of the process, the information exists and is correct. If exactly the same information is created again from scratch, the process will be repeated. It will take about the same amount of time to create the information as it did the first time, and it will cost about the same. However, instead of recreating the information from scratch, the information created the first time can be recalled and re-used. This information is known to be correct and usable. The process of recalling it should take very little time and money. It's much quicker and cheaper to re-use information than to re-create it.

Similarly, if an existing part is re-used there's no need to go through a wasteful process of re-developing the part, re-developing the process to make and support it, simulating performance, re-developing tools, and waiting for prototypes.

By re-using an existing part, time and money is saved, and quality is guaranteed. Also, since a new part number isn't needed, overhead isn't increased, there's no need to extend part files, there's no need to find additional storage space, and there's no need for additional working capital for holding costs. Because there are fewer parts, it's easier to locate existing parts, so access times can be kept acceptable. The exact figure depends on the specific industry and part, but the creation of a single new part, including all the activities and overhead it engenders, is generally put at several thousand dollars. If creation of a new part can be avoided, time and money are saved.

It's useful to document the description of the strategy element so that everyone can understand what's proposed, and can contribute to improving it.

24.7.4 *Implications of Strategy Elements*

Different companies will develop different strategies. These strategies may have very different implications for the resources used in the activities of the product lifecycle. Consider the two strategies of "value-adding lifecycle" and "the lowest-cost product".

"Value-adding lifecycle" focuses on the key lifecycle activities that differentiate a company from its competitors. This often means outsourcing lifecycle activities that aren't strategic. The main activities of an organisation with a "value-adding lifecycle" strategy could be defining customer requirements, system engineering, simulation, and management of suppliers. Everything else, including design and drafting of components and parts, and manufacturing and assembly, could be outsourced. The organisation would employ highly experienced, competent and creative engineers. It might use a technique such as Quality Function Deployment to capture the Voice of the Customer. Its key application systems could be used for solids modelling and simulation of the system design.

A "lowest-cost product" strategy could focus on Value Analysis, simplifying the product, reducing the number of parts, using standard parts, using cheaper parts, and reducing machining and assembly time. The main activity of such an organisation would be to work at the level of detailed components and parts. That's just the opposite of the organisation described above. Highly creative engineers and system engineers wouldn't be needed to make detailed changes. QFD would be less important since the focus wouldn't be on developing something the customer might want in the future. The focus would be more on reducing the cost of something that has already been developed for the customer. The key CAD system might well be a simple PC-based drafting system. The most appropriate use of simulation could be in simulating the manufacturing process, not in simulating the product.

There are similar implications for other strategy elements. Some examples of the type of action that each strategy element might lead to are shown in Fig. 24.21.

Fig. 24.21 Actions related to
strategy elements

```
customer focus
        improve customer relations
        implement customer satisfaction metrics
customisation capability
        define and improve processes
        use QFD
        exploit technological base better
        implement CAD/CAE/CAM
        implement simulation tools
the most robust product or service
        use Taguchi methods
        use Robust Engineering
        use DFA/DFM/DFX
the best process
        carry out process mapping
        define and improve processes
fastest time to market
        build on platform products, modular design
        use Rapid Prototyping
        re-use
value-adding PLM
        use System Engineering
        outsource
lowest-cost competitor
        remove non-value-adding activities
re-use
        build on platform products, modular design
        use Group Technology
        use CAD and PDM
the widest range of products and services
        build on modular designs
        have a wide range of skills
automation
        use CAE/CAD/CAM, PDM, simulation tools
        use NC machine tools
standardisation
        build on modular design
        use Group Technology
flexibility
        have a simple clear process
        have a skilled workforce
minimise costs
        eliminate non-value-adding activities
```

24.8   Step 3: Selecting the Preferred Strategy

In the third step of strategy development, potential strategies are tested, and the
most appropriate strategy is selected and detailed. It will be useful to investigate
three or four alternative strategies. This should lead to an in-depth understanding of
the possible strategies. The strengths and weaknesses of a particular strategy often
become clear when examining the strengths and weaknesses of other strategies.

The analysis of the different strategies is often called SWOT analysis. The
acronym stands for strengths, weaknesses, opportunities and threats. These are the
four factors that have to be described and compared for each of the possible
strategies.

The SWOT analysis for PLM is carried out after identifying several potential
strategies. The questions are aimed at finding out which strategies are suitable, and
which of the suitable strategies is the most appropriate (Fig. 24.22).

Although the same SWOT approach will be used, the SWOT analysis carried out
for a particular organisation will differ from that carried out in any other organisation. The detailed situation in different companies will be different.

Fig. 24.22 Questions in
SWOT analysis

Does this strategy meet the objectives?
Is it in line with overall company objectives?
Will it enable us to achieve the mission?
How does it address key issues?
How does it relate to key success factors?
How will we measure progress towards specified targets?
What would be the key metrics?
How long would it take to implement?
What are the overall costs?
Do we have the financial strength to do this?
What are the benefits to the organisation?
Do we have the resources to do this?
Can we find/afford missing resources?
Does this strategy play to our strengths?
Does it address our weaknesses, and help us against threats?
Does it allow us to take advantage of opportunities?
What will be the impact on customer responsiveness?
Would this strategy save time?
What would be the effect of this strategy on quality?
What would be the effect of this strategy on product costs?
What would be the effect on product development cost?
What would be the effect on product support cost?
What are the risks associated with implementing the strategy?
What are the risks associated with not implementing it?
What are the technological implications of the strategy?
What are the human resource implications of the strategy?
What practices are implied by the strategy?
What are the implications of the strategy on use of information?
What would be the effect of this strategy on customers?
What would be the effect on partners and suppliers?
What would be the effect on management and on our workforce?
What would be the effect on competitors?
What would be the effect on regulators?
What would this strategy do for our products and services?
How would we sell our products/services under this strategy?
Can we implement the organisation to do this?
Do we have the managerial ability to do this?
How does this strategy differ from our current strategy?
Does this strategy make it easier to develop new products?
Does this strategy offer something our competitors can't do?
Do we have the resources for strategy implementation?
How does this strategy make life easier for our competitors?
Will we face new competitors if we select this strategy?
Does this strategy allow us to respond to opportunities?
Does this strategy really correspond to what we are trying to do?
Will this strategy allow us to portray ourselves as we desire?
Would this strategy work for anyone - what's special for us?
What are the implications for facility location?
What are the implications for investment, profit and loss?
With this strategy, could we adapt if the environment changes?
With this strategy, could we adapt if demand changes?
Does this strategy provide a basis for long-term competition?

The SWOT analysis should be as complete as possible, looking into all areas of relevance to the organisation. A partial list would include customers, competitors, suppliers, technology, products, costs, cycle times, quality, information, computer systems, management of change and human resources.

As the analysis is carried out, it may be found that important information is missing. If so, the necessary time should be taken to find or develop it. It's important that the analysis is carried out on the basis of facts, not guesses and opinions.

After the SWOT analysis has been carried out in the third step, one strategy must be selected. It should then be documented in detail.

24.9   Step 4: Communicating the Strategy

In the fourth step of the five-step strategy development process, the chosen strategy is communicated to the people who will be affected by it, or involved in its implementation. Communication of the strategy is essential. A strategy is useless unless the people who are going to be involved are fully aware of it, can understand it and can implement it.

The name given to a strategy must be so short and simple that everyone (not just strategy developers in company headquarters) can understand it. In PLM, everything is simple, but even the simplest thing is difficult to achieve. If what appears simple is actually difficult, what appears difficult will be well nigh impossible.

Different strategies will be selected by companies in different circumstances. In some cases, a strategy with a name such as "control of the products" may be suitable. Customers will be happy to know that the products are under control. Internally, the company, with its products under control, will be able to reduce costs and time cycles. In other circumstances, for example, for a supplier of commodity, low-tech components to OEMs, a "fast–follower" strategy could be suitable, watching competitors closely, but not investing heavily in research.

A strategy has to be realistic. It has to be expressed in language that everyone can understand. There's no sense in including motherhood statements, or in claiming to greatly improve performance overnight.

It's important to have general agreement on vocabulary. Otherwise, if the members of a group of people are asked to describe the strategy they'll all give different answers. Before communicating the strategy, make sure that everybody understands the meaning of the terminology.

Before disseminating the strategy, test it out on a few people who are typical of the audience. If they don't understand it, then make it understandable before communicating it more widely.

No-one is going to be interested in a PLM Strategy that doesn't seem to involve or concern them. Make sure people realise the relevance of the strategy for them as individuals.

As most people are very busy with other things apart from PLM Strategy, there's a good chance that they won't take it in the first time that they hear or see it. The strategy message needs to be repeated. Repeat it on different occasions and in different ways, such as in formal and informal presentations, meetings, intranets and newsletters.

Decide who the strategy needs to be communicated to before starting to communicate it. Product developers, field engineers, customers, company management, suppliers, regulatory organisations, other companies?

Communicate the strategy to everyone involved with PLM. The more they understand it, the more chance it has of being implemented successfully. Make sure people understand why it's important to understand the strategy. Make sure they understand why the strategy is necessary for the organisation. Make sure people can see where it takes them as individuals.

The PLM Strategy development process will probably lead to a new PLM Strategy, different from what went before, implying many changes.

24.10 Implementation Strategy

The Implementation Strategy (also known as the Deployment Strategy, or the Change Strategy) shows how to get from the current use of PLM resources to the future use of PLM resources (Fig. 1.44). There are many ways to get from the current use of PLM resources to the future use of PLM resources. These options should be documented and analysed. Again, SWOT analysis is appropriate. The best option is selected. This is the Implementation Strategy. It should be documented in detail and communicated.

24.11 The PLM Plan

The Vision Report should contain a PLM Implementation Plan showing how the strategy will be achieved over the length of the total implementation (for example, five years) and a more detailed Plan for the first year. The structure for the Plan should follow that for the PLM Vision, making it easy for managers and others to see how all the issues are linked, and will be addressed. As an example, the Plan could address the elements shown in Fig. 24.23.

Different views of the Plan should be prepared, with different levels of detail. The first view should be a block diagram only showing which types of activity will take place in each year (Fig. 24.24). This level of detail may be sufficient for company management.

Other views of the plan will show more details of the activities. They'll be needed for people who participate in, and manage, the activities.

1	Start-up
2	Management of the Initiative
3	Organisational structure (across the extended enterprise)
4	Processes
5	Organisational structure (internal)
6	People
7	Information
8	Working methods
9	Information systems
10	Accompanying Actions

Fig. 24.23 Possible elements of a PLM implementation plan

	Year 1	Year 2	Year 3	Year 4	Year 5
Start-up					
Project Management					
Organisational structure (extended)					
Processes					
Organisational structure (internal)					
People					
Information					
Working methods					
Information systems					
Accompanying Actions					

Fig. 24.24 Block diagram representation of the PLM plan

24.12 PLM Vision and Strategy Team

The PLM Vision and Strategy development activities described in the previous sections should be carried out by a Team, not by one person. However, the PLM Report should be written by one person, so that it's clear and coherent. Otherwise it could be just a jumble of ideas written by different people with different styles.

The PLM Vision and Strategy should be developed by a Team with members representing top management, functional organisations, and product organisations. The Team that creates the Vision should be representative of all the functions in the product lifecycle. It needs to include high-level managers and low-level workers. Membership should also be arranged to allow the presence, if required, of representatives from groups such as customers, contractors, and partners. It's a good idea for the Team to "walk the product lifecycle". The Team should actually walk around all those places where activities happen along the product lifecycle. That way they'll get an idea of what really happens in the lifecycle. They may be surprised to see the conditions in which some activities are carried out. Inappropriate locations, with broken-down facilities, insufficient space, outdated equipment and poor security, may keep costs down, but may not be conducive to high quality behaviour and an interest in the customer.

Top management participation is needed to ensure that the strategic business focus is taken into account. Upstream of the PLM Vision is the company's business strategy. This will probably have been communicated in a relatively concise form, behind which lies a mass of detailed information and decisions. In many organisations, only company managers, perhaps supported by staff groups, have a sufficiently broad view of the current and future situation to be able to pass on the full message.

The PLM Vision and Strategy Team should include competent and powerful individuals from all the functions that participate in the product lifecycle. These functions could include R&D, design engineering, manufacturing engineering, marketing, sales, F&A, manufacturing planning, manufacturing, logistics, maintenance and end-of-life. It will also be useful to include in the Team an individual who will represent the interests of people and organisations outside the company, such as suppliers and customers, who come into contact with the product or product data.

The aim of being multi-functional isn't achieved if some departments propose lightweight Team members whose questions and opinions will be ignored. Similarly, there's no point in including busy departmental managers on the Team if they intend to delegate their role to junior staff.

Positive Team membership characteristics include involvement, commitment, support, hard work, an open mind, a good knowledge of a function, the power to make changes happen, the ability to avoid pulling rank, and a good understanding of the fact that nobody knows everything.

Techniques, tools and resources that the Team can use to help them develop the Vision include brainstorming, scenario building and comparison, simulation, Delphi techniques, experience, intuition, questionnaires, surveys, benchmarking, and process analysis.

Development of the PLM Vision requires a lot of effort from functional managers and product managers. They have in-depth knowledge of the business. They know about the past, and have expectations for the future. They're the people who will be most affected by the future PLM Strategy. They have to make sure a realistic strategy is developed, and take responsibility for it. They may be responsible for starting and encouraging the process of developing the PLM Vision, making sure that the strategy that's developed is relevant to the company's business strategy, and making sure that everybody who will be affected by the strategy is involved in its development. Once a strategy has been agreed, they'll be responsible for communicating and implementing it.

Some people who work in the low-level activities of the lifecycle should be included in the Team. They can provide both their own input and that of their colleagues. Functional managers and product managers should ensure that everybody in the organisation is aware that the Vision is being developed, and that their input is welcome. Time should be set aside to tell them why it's necessary to develop a Vision, and what will be done after it has been developed. Involving everybody in the process increases the knowledge and experience base on which the Vision is built. It may generate ideas that would otherwise have been ignored. It helps make people aware of the key issues the organisation is facing, and the various possible solutions. Once the Vision has been developed, it will be easier to communicate it to people who have participated in its creation and are aware of its importance.

To develop the PLM Vision and Strategy deliverables, the Team will have to interview many people in many activities, and collect details of the various resources they use. As it probably won't be possible for all Team members to be involved in gathering the information to answer each question, it will be best for them to work in small groups of 2, 3 or 4 people. The person from the function most directly related to a particular question should be involved in finding the answer, but so should people from other functions. A Team member from the upstream source of the information, and a Team member from a downstream user function are suitable candidates. This reduces the risk of a biased answer, and helps Team members gain cross-functional knowledge.

The Team is jointly responsible for the results it produces. At the end of the process, all Team members should understand all the answers, understand what they imply, and agree with the findings to be reported. If this is the case, a good information base about the product lifecycle will have been built, and consensus achieved. This will be the ideal starting point for the next activities.

Top management should define the objective of the Team, give responsibility and authority to the Team leader, and inform interested parties throughout the company about the Team's activities and its objectives. In particular, top management should inform managers of functions where product data is created, or used. They should make it clear to these managers that the Team leader has authority to ask questions and to ask for details of documents, applications and activities related to products. In turn, these middle managers should inform their subordinates what's happening, and ask them to provide the necessary information.

The Team leader is given authority and responsibility for the Team's activities by top management. The Team's activities are likely to be time-consuming. So it's unlikely that the leader will be a top manager. More likely, the leader will be a middle manager reporting directly to a top management PLM sponsor and champion. As the team's activities are going to be cross-functional, and involve working at many levels in the company, the Team leader should be picked with care.

The Team leader should have the authority and the responsibility to carry out the required activities. If however, it appears that progress isn't fast enough, or is being opposed by particular interest groups, the Team leader should be able to go back to top management for guidance and support.

24.13   Ramping up PLM Knowledge

PLM is a relatively new subject, and few, if any, of the individuals in the Team will know much about it. Some of the Team members may lack some knowledge, or may not be used to working in a team. It may be necessary to provide appropriate training and/or education.

The Team members may be able to attend an introductory PLM course so that they can find out about some of the basics of PLM, and learn together some of the jargon and vocabulary of PLM. Useful information on PLM can also be gathered from books, journals and web-sites, from conferences and seminars, from demonstrations by vendors of PLM applications, and from visits to other companies using PLM.

Getting a good understanding of the environment in which the PLM activity takes place requires understanding competitors, other players in the market, and the overall position of the industry. In some cases, competitors may be amenable to an exchange of information about PLM.

The initial understanding and awareness of PLM shouldn't be restricted to the Team, but spread as widely as possible. Management should be kept up-to-date, as should potential PLM users. The more that management knows about PLM, the more supportive it will be of the Team. The more that the users know about Team progress, the more supportive they'll be, and the less likely they'll be to start independent competing activities and overlapping projects.

Although the members of the Team, and other members of the company, will learn much from their participation in the various activities, it will be helpful if they can receive some basic information early on from a neutral, experienced PLM expert. Otherwise, it's only too likely that they'll go round in circles for a considerable time, with each member of the Team becoming more and more closely attached to the idea that PLM exists mainly to solve his or her everyday problems. Some will see PLM as being only a solution to CAD data management problems. Some will see it as a Bill of Materials project. Some will see it as being the answer to configuration management and traceability problems. Some will see it as a way of making sure that their favourite procedures are implemented. As time goes by, each one will become more and more convinced that they alone are right. The intervention of a neutral, experienced expert at an early stage can prevent this negative and resource-wasting state of affairs arising.

Chapter 25
Example of a PLM Vision

25.1 Overview

The purpose of the PLM Vision is to communicate to many different people an overview of what PLM is and will be, why it's important, and how it will be achieved. As they may not understand instinctively what "PLM" is, it may be useful to communicate PLM in terms of concepts that they do understand. Such as products, applications and processes. These can be thought of as components of PLM. It's easier to ask questions, and provide answers, about the components than about PLM. For example, what structure should our products have? How many products should we have? Which applications will we need for PLM? Which business processes correspond to PLM?

The three components mentioned above (products, applications and processes) aren't sufficient to describe PLM completely. They don't address, for example, the people in the organisation, or the organisation itself. The following list of 12 PLM components is more complete (Fig. 25.1). Any question about PLM in the future should be related to one of these components.

Ignoring part, or all, of any of these components in the Vision would weaken the Vision. It would lead to questions about its validity. Ignoring part, or all, of any of these components in the deployment of PLM would weaken the overall ability to manage products across their lifecycles.

25.2 Overview of the PLM Vision

The following paragraphs describe a concise PLM Vision for a particular company. For some people, such as the CEO, the level of detail they provide might be sufficient. Other people, such as programmers of application program interfaces, might find that they don't have enough detail.

© Springer International Publishing Switzerland 2016
J. Stark, *Product Lifecycle Management (Volume 2)*,
Decision Engineering, DOI 10.1007/978-3-319-24436-5_25

Fig. 25.1 Twelve
components of the vision

| Vision, Strategy, Plan, Metrics |
| Products |
| Customers |
| Organisation |
| Management, Control, Visibility |
| Lifecycle and Processes |
| Collaboration |
| People and Culture |
| Data, Information & Knowledge |
| Facilities, Equipment, Applications & Interfaces |
| Mandatory Compliance, Voluntary Conformity |
| Security and Intelligence |

The global surroundings exert all sorts of pressure on the company. Examples include: the environmental movement; globalisation; global competition; lean manufacturing techniques; the fluctuating value of the dollar; China's manufacturing industry; product liability problems; changes in the age-distribution of populations; the information society and rapidly advancing technology.

According to company plans, the company has three strategic objectives. They are innovation, growth and profitability. The company will be a product developer and manufacturer. It will sell products worldwide in more than 100 countries. Last year it sold 3 million units; in 5 years the target is to sell 3.7 million units. In the last 5 years, the company introduced 10 new products; in the next 5 years it plans to introduce 30 new products. These will be market-leading products for the consumer market. Company R&D expenditure is planned to rise six points to 4.0 % of revenues. Over the next five years, the number of repeat customers should increase by 15 %. The company aims to cut costs by 3 % per year.

Our products are expected to become more complex. More technologies will be integrated. There'll be an increase in the percentage of electronic and software components. More options will be available. The products are likely to be specific to each customer. The products are expected to be more intelligent, capable of providing feedback about status, use and location. Product leasing is expected to expand at the expense of outright purchase. As a result of global pressures and of company plans, the company won't be able to work in the future as it did in the past. Products will have to be developed, supported and managed in a different way.

The best way to do this will be with a cradle-to-grave, PLM, approach to products. We'll have a single set of integrated optimised lifecycle activities from the first idea for a product through to its retirement.

The focus of PLM is our products. There's nothing more important in the company's future than its products, and the way they'll be developed and used. Without those products, and their customers, there'll be no revenues. The objective of PLM is to maximise the value of the company's products across their lifecycles. PLM aims to increase continually the value of the company's Product Portfolio, and its revenues and profits. It aims to bring profitable products to market quickly. It aims to increase product quality and reduce product-related time cycles and costs. PLM success will be measured in four areas: financial performance; time reduction; quality improvement; business improvement. PLM brings together everything to do with the product, including product development, product recall and product liability. It manages, in an integrated way, the parts, the products and the product portfolio. It manages the whole product range, from individual part through individual product to the entire product portfolio. With PLM, all the product-related issues are united under the PLM umbrella and are addressed together in a joined-up way. The approach is holistic. PLM is seen as the way to address all the product-related issues.

We need to get a full understanding of our customers and the way our products behave. We need to understand the nature of the future market out there. We need to understand what customers really want, and understand what their usage of our products is telling us. As a result of our PLM approach, customers will be proud and pleased users of our products.

We'll be fully responsible for our products. Products and processes will respect the environment. The company's products are at the heart of the PLM Vision. There'll be a modular range of products, based on platform products. We want to maintain some manufacturing facilities for strategic activities. However, as much as 85 % of manufacturing will be outsourced. Outsourcing and offshoring will be balanced between dollar, euro and yuan regions. Major assemblies will be defined, but detailed design will be outsourced. Twenty main partners will provide more than 60 % of parts. These partners will be closely integrated, and will be expected to carry out a large part of the development work. They'll be involved as early as possible in development work. As well as participating in the industry e-marketplace, we'll extend our existing suppliers' portal. And we'll integrate the purchasing information system with the ERP system. Customers will be able to buy our products 24/7/365 from an e-catalogue and our e-shops. Product support activities will be distributed across the world, wherever customers are located. There'll be a small but very important core product support organisation providing leadership and direction.

Many of the companies that participate in our extended enterprise will be on other continents. Many of them will be at multiple locations. Nevertheless, the PLM structure and working methods of our extended enterprise will enable products to be successfully managed throughout their lifecycle. Knowing the structure of the extended enterprise will enable the objective, content and location of each activity along the product lifecycle to be defined. The processes that make up the overall lifecycle will be defined and documented. Processes will integrate international and industry standards, and ensure required compliance. Then, on the basis of their skills, knowledge and competence, we'll assign the people, and build units, teams and groups to do the tasks. We'll probably have several types of functional, product and project organisations of people.

People need to have the right skills, be properly trained, communicate well, and understand where they fit in the process. Otherwise, they won't contribute effectively to the lifecycle. A significant investment in training will be needed. The percentage of female and minority employees will be increased. They'll help us understand our customers better. The best way of working will be identified. People will use the most appropriate methodologies, working methods and techniques.

After the processes and tasks are clear, information requirements will be defined. With the work definition and information requirements clear, information systems (and the interfaces between them) will be identified to help people carry out their everyday work.

25.3   More Detailed Overview

This section provides more detail about each of the twelve components shown in Fig. 25.2.

The Vision, Strategy, Plan and Metrics are integrated. Clear metrics will enable fast implementation of PLM and easy identification of the results of deployment. They'll enable comparison between the Vision and the deployment.

There's an objective to increase continually sales and quality, and reduce time cycles and costs.

Focused on our products, we'll increase revenues with an on-going stream of innovative new products. The value of current and future products will increase each year for the benefit of customers, employees and shareholders. Products will be platform-based. There'll be a high level of part reuse. The exact configuration of each product will be known at each stage in its lifecycle.

There'll be more customers. The products will increasingly satisfy our customers, who will acquire and use more products.

The company's organisation will be both customer-focused and product-focused. A simple slim-line organisation will be built around Product

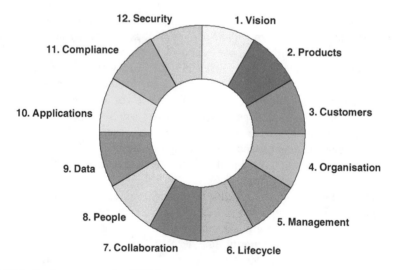

Fig. 25.2 Twelve components of PLM

Family Teams focused on product families. A Chief Product Officer will be appointed with responsibility for all products.

PLM is the management system for all products. It will provide full control and visibility over products throughout the lifecycle.

Products are managed across their lifecycle through five phases: imagination; definition; realisation; support; retirement. A single set of coherent industry-standard business processes will be used across all sites.

Products are managed across the lifecycle in a collaborative, extended enterprise environment.

People will work in a team environment. On average, people working in the PLM environment will receive 10 days of training per year. Enterprise Change Management will be an important activity supporting the deployment of PLM.

A single set of coherent industry-standard documents will be used throughout the extended enterprise. The right data will be available to the right people, wherever they are, at the right time. A full and up-to-date electronic definition of each product (the "Digital Product") will be available at each stage of the lifecycle.

Appropriate facilities and equipment will be used throughout the product life-cycle. Wherever possible, industry-standard applications and interfaces will be used. Applications such as Automated Product Idea Generation, Virtual Engineering, Digital Manufacturing, Collaborative Product Support and Computer-Aided Recycling will support each stage of the lifecycle.

All relevant mandatory regulations will be identified and integrated into processes and automated workflows. All relevant voluntary guidelines will be identified, reviewed, and where appropriate, integrated into processes and automated workflows.

Access control in PLM will ensure that availability of data corresponds to the company's security policies. All security-related activities will be identified and integrated into processes and automated workflows. All intelligence-related activities will be identified and integrated into processes and automated workflows.

The following sections describe each component of the Vision in more detail.

25.4   Vision, Strategy, Plan, Metrics

There's a document describing the PLM Vision, a document describing the PLM Strategy, a document describing the Implementation Strategy for PLM, a document describing the Plan for implementation of PLM, and a document describing the PLM Metrics.

The PLM Vision, PLM Strategy, Implementation Strategy, PLM Plan and PLM Metrics are closely integrated.

The Vision is a major input for development of the Implementation Strategy. In turn, this is a major input for development of the plan. The Plan drives the deployment. The success or failure of the deployment is measured with the metrics.

The PLM Vision provides a Big Picture of the expected PLM environment five years from now. It describes the performance and behaviour that's expected at that time. It helps people take decisions when setting strategy. It helps plan resources, capabilities, budgets and the scope of activities. The Vision will include target values for key metrics.

The PLM Strategy reflects the organisation's overall business strategy, and defines how PLM resources will be organised.

The PLM Plan shows the detailed activities and resources needed to achieve the Vision. The PLM Plan initially addresses a five-year timeframe. It's updated annually. The Plan addresses all the components of PLM such as human resources, applications, and processes. Individual projects are defined and planned in terms of objectives, action steps, timing and financial requirements. The relative priorities of projects are described.

The planned PLM deployment will take five years. However, it's expected that, once it's complete, there'll be numerous opportunities to extend and optimise PLM. During and after deployment, progress is measured against the PLM Metrics. PLM success is measured in four areas: financial performance; time reduction; quality improvement; business improvement.

25.5   Products

This component of the Vision is addressed in five parts.

25.5.1   Product Focus

The product is the focus of PLM (Fig. 1.17). This is a basic belief of PLM. Without its products, and the related services, the company wouldn't need to exist and wouldn't have customers. Focused on its products, the company generates revenues from an on-going stream of innovative new products. Great products make it the leader in our industry sector. Great products lead to great profitability.

25.5.2   Product Portfolio

There's a document defining the Product Portfolio. The Product Portfolio contains both the existing products and those under development.

The Product Portfolio plays a key role in PLM. It's structured into product families, product lines, products, platforms and modules. It contains actual and forecast sales figures, costs and Intellectual Property values across the lifecycle for current products and for products in the pipeline.

There's a document showing the current status of the Product Portfolio.

The products within a product family are similar. They're created with modules built on a basic platform. There are similarities between their specifications, features, parts, modules, drawings, manufacturing processes, assembly techniques, distribution channels and end-of-life. As a result, there's no need to re-invent the wheel for each new product in a family.

Products are structured with great care. Products are structured so that the individual requirements of specific customers can be met with little additional effort. Products are structured so that modules adding little value can easily be outsourced or offshored. Modules are structured so that it's easier to offer customised products. Functionality has been developed, so that, on option, each product can have a unique identifier making it traceable throughout its life.

At any given time, several generations of a product family are under development. Once the first generation of the family has been developed, the cost and time to develop succeeding products decreases. The second generation re-uses 75 % of the parts used in the first generation, greatly reducing product and process design and verification effort.

25.5.3 Five-Year Strategy and Plan

There's a five-year strategy for the Product Portfolio. It shows how resources in the product lifecycle will be organised to achieve the company's product objectives. It defines policies for the management and use of resources in the product lifecycle. For example, it defines policies for platform products, modular products and part re-use.

There's a five-year plan for the Product Portfolio. It shows how existing products are expected to be upgraded and how new products will be introduced. It shows how and when platforms evolve, how modules are introduced, and how parts and modules are re-used. The five-year plan helps decision-taking in related areas such as outsourcing, offshoring and Supply Chain Management.

The objective of continuously increasing the value of the Product Portfolio, and increasing product value and revenues, drives product innovation. Necessity is the mother of invention. When existing markets offer little further opportunity, new markets are entered with innovative products that make extensive re-use of existing parts and information.

25.5.4 New Technologies

The Product Portfolio shows the introduction of new technologies into products.

Technology roadmapping and product roadmapping help identify and prioritise new opportunities.

A detailed review was carried out to identify duplicate and similar parts. These were then eliminated. The number of modules has increased, but the number of individual parts has been greatly decreased. The reuse frequency of both modules and the remaining parts has increased.

A detailed review was carried out to identify recurring design problems. Action to eliminate these was taken at the source of the problem. Costs due to product recalls, failures and liabilities have been reduced by 50%.

Existing products were reviewed. Products with low and declining sales for which no feasible upgrade path was possible were retired, licensed or sold. An upgrade plan was developed for the other existing products. As a result, they've all been upgraded within the last five years. Revenues from these products have increased by 25%. Revenues from new services on these products have risen by 50%. All product versions currently on the market are less than five years old.

New products are now introduced at twice the rate of five years ago. 90% of a new product is recyclable.

Sales of the company's products, and related services, have increased at an average 10% per year.

Products from acquired companies were integrated into the Product Portfolio within three months of the acquisition date.

More than 99% of products didn't require major maintenance in their first three years of use.

An increasing proportion of revenues is coming from customised products, providing services to support product use, and refurbishing existing products. Additional revenues are also coming from new environment-friendly products, and from providing, and taking financial and environmental responsibility for products produced in low-cost countries.

Fig. 25.3 Reporting progress with products

Among the new technologies of particular importance are those which can be used to provide, automatically, detailed direct feedback, "in-use information", from the product. They include RFID, mobile telephony and the Web. The "Voice of the Product" is considered to be as useful as the "Voice of the Customer". It provides important data about the way that products are actually used by customers. It gives us the opportunity to understand how our products behave.

25.5.5 *Progress with Products*

Targets are needed to measure the success of PLM deployment.

The "report" in Fig. 25.3 might be written five years after the PLM Initiative is started.

25.6 Customers

This component is addressed in four parts.

25.6.1 Customer Focus

PLM puts the focus on the product, but it doesn't forget the customer. The company's revenues come from customers purchasing products. Customers buy great products. Customer satisfaction can only be achieved by providing a high-quality product. A company wouldn't get a sale without a competitive product, even if it had all the knowledge in the world about its customers. And what the customers had dreamed, imagined, thought, scribbled, twittered, blogged and said.

To achieve great products, it's important to listen to customers, and to understand how they use products. It's important to take account of what customers say about products, and what potential customers say about a competitor's products. It's important that many customers choose the company's products rather than those of competitors. It's only if enough customers buy the company's products that, at the end of the financial year, revenues will exceed costs. If this isn't the case, then before long, we'll go out of business.

25.6.2 Voice of the Customer

It's important to listen to the customer. It's important to make sure that the customer's demands, expectations, requirements and wishes are reflected in the product. Listening to the customer is a key part of the product development and support process. It's also important to watch how customers use products and see how they adapt them to improve performance. The customer can often provide a lot, if not most, of the product specification.

It's good to listen to the Voice of the Customer. But it's even better to involve the customer in product development and support activities. That enables customer knowledge, experience, behaviour and requirements to be taken into account during the entire product lifecycle, and not just during product development. The best way to take account of them is by directly involving the customers all the way through, from the beginning of the product lifecycle to the end.

25.6.3 Customer Involvement

Involving customers early in the product lifecycle reduces development time and costs. It brings new ideas and potential problems to light quickly, avoiding expensive rework. Involving customers throughout the lifecycle helps avoid the discovery of problems when it's too late to avoid their effect.

Customer Surveys are carried out to discover what customers are thinking about existing products and future products. And, using technologies such as RFID and the Web, information is exchanged directly with customers using the product.

Five years ago, we had customers in 10 countries. We now have customers in 55 countries. Of these customers, 50% are now in the US, 30% in Europe and 20% in Asia Pacific. Our objective is to have 40% of the customers in the US, 35% in Europe and 25% in Asia Pacific. The number of customers has increased by 25%. The number of repeat customers has increased by 40%. The number of customer complaints has decreased by 50%.

Fig. 25.4 Reporting progress with customers

Getting feedback from a customer at the actual time of use provides even more valuable information than a survey form.

Customer representatives are involved in product development and support teams. Office space is made available for them so they can work on-site to develop components to their specifications. Joint reviews of field reports are held with customers to get their reactions to existing products. They help us understand the performance of re-usable components of existing products. This provides a mass of information about customer requirements, views, experience and wishes.

Working closely with customers helps ensure that our products really meet customer requirements. Usually these close relationships with customers prevent errors occurring, so reduce costs. They're also a source of innovative new ideas. And they also save time because people only work on what's really required. We don't want to work on activities that are of no interest to customers or even to potential customers, so don't add value.

25.6.4   Progress with Customers

Targets are needed to measure the success of PLM deployment.

The "report" in Fig. 25.4 might be written five years after the PLM Initiative is started.

25.7   Organisation

This component is addressed in three parts.

25.7.1   Product Family Teams

To meet customers' requirements for better products, we'll deploy PLM. We'll have several types of functional, product and project organisations of people. However, the organisation will be primarily product-focused and built around Product Family Teams (PFT) focused on product families. Gone, with Product Focused Teams, are numerous layers of hierarchy, large staff groups and managers supplying conflicting information. These would make it difficult for company management to understand what's really happening and what actions will lead to improved performance.

Each PFT will focus on a single product family, so it will focus on a single product lifecycle. This will result in a better understanding of the activities related to that product family. It will be easier to see how they can be improved and where most value can be added. The process will be continuously improved. Information flow will be organised to meet the needs of the process. The PFT will become more and more competitive. Its products will be closer and closer to customer requirements.

Because they'll focus on one product family, the people in the PFT will know their products in depth. Through training and experience, they'll know how to make a valuable contribution to the product. Everyone will learn about the process. It's important that they know who does what, what has to be done, where everything is, and how things are organised.

A PFT will focus on one product family. It, and its partners in our extended enterprise, will install and optimise the best equipment for that family, and cost-justify it over several generations of the product, not on just one product. Because the PFT will focus on one product family, planning and scheduling will be easier. There'll be no need to switch resources between projects addressing completely different products. It will be easier to plan to have the right resources available when they're needed. It will be easier to plan ahead, since so much will already be known about the next generation of the product.

25.7.2   PLM Responsibilities

In the PLM organisation, responsibility for the company's products will be with the Chief Product Officer (CPO). We have one manager responsible for other key parts of the business. For example, finance is the responsibility of the Chief Financial Officer (CFO), information the responsibility of the Chief Information Officer (CIO). The CPO will have the responsibility to develop the five-year strategy and plan for products and product-related policies (for example, for platform products, modular products and part re-use), and to achieve the targets. The CPO will report to the CEO, as do the CFO and the CIO. Product Family managers will report to the CPO.

The Chief Information Officer will be responsible for the IS architecture and infrastructure that enables effective product lifecycle management. Aligning information technology decisions with the PLM needs of the business will help drive the company forward.

The Product Data Management (PDM) Manager will be responsible for product data, and will report to the CIO.

The PLM Initiative Manager will lead the PLM Initiative with the aim of introducing highly effective and successful PLM. The PLM Initiative Manager will work closely with the CPO and the CIO.

Just as Process Owners are responsible for individual processes in a company, the Product Lifecycle Owner will be responsible for the maintenance and improvement of the overall product lifecycle.

25.7.3   Product Development and Support Methodology

The product development and support methodology will be defined in detail. Everyone will be trained in its use. They'll work in harmony across the lifecycle. People will understand the tools that the company has chosen to work with, not only so that they can make best use of them, but also to understand what their team colleagues are doing.

25.8   Management, Control, Visibility

This component is addressed in three parts.

25.8.1   Management

PLM will be the management system focused on the product.

Product Portfolio Management will enable the company to manage its products for the medium-term and the long-term. The process of Product Portfolio Management will be defined.

Project Management and Program Management will enable the company to manage its product-related projects for the short-term and the medium-term. The processes of Project Management and Program Management will be defined.

The PLM Plan will show management the detailed PLM deployment projects. Individual projects will be defined in terms of objectives, activities, timing and financial requirements.

With PLM, top managers understand and can formulate the need for effective product lifecycle management. They define the key metrics and strategies.

25.8.2 Visibility

PLM will give visibility about what's happening over the product lifecycle. It will give managers visibility about what's really happening with products. Before, managers were often faced by an opaque mountain of unclear and conflicting information. PLM will offer managers an integrated view over all product development projects, and over the market performance of existing products. It will provide them with the opportunity to manage better. Based on valid information, they'll be able to take better decisions.

25.8.3 Control

PLM will enable the company to be in control of its products across their lifecycle. It will get our extended enterprise under control. It will reduce the cost of new products and the time it takes to bring them to market. PLM will enable the company to be in control of its product development projects. Modifications to products will be implemented faster. Part reuse will be increased. Mastering the activities in the lifecycle will make it easier to provide reliable products and to sell related services. Better control over the lifecycle will provide better assurance about environmental impacts. It will be easier to take account of potential risks. Being in control will open up new opportunities. With population levels rising fast in many countries around the world, there'll be many customers in distant locations, but PLM will keep control over the products they use, however far away they may be.

25.9 Lifecycle and Processes

This component is addressed in eight parts.

25.9.1 Phases of the Product Lifecycle

The product lifecycle is defined as having five phases: imagination; definition; realisation; support; retirement. It's recognised that, for users of the product, there are also five phases in the product's lifecycle: imagination; definition; realisation; use (or operation); disposal (or recycling). The first three phases are the same for the company and the user, the last two are different.

25.9.2   Management of the Product Lifecycle

The Product Lifecycle Owner has responsibility for defining and maintaining an effective product lifecycle, including the definition of the details of the lifecycle structure.

There's a document describing the lifecycle structure.

25.9.3   Life Cycle Design and Analysis

Life Cycle design and analysis will play an increasing role. All issues related to a product's life will be considered at the outset, including those involving the product once its useful life is over. Life Cycle analysis will be carried out over the complete cradle-to-grave lifecycle including analysis of use of raw materials, production methods and usage/disposal patterns.

25.9.4   Lifecycle Modelling and Analysis

The product lifecycle will be modelled and analysed to identify where most value can be added, and where waste can be reduced. Opportunities will be found in the early phases of the lifecycle to increase the speed of generating ideas, translating them into products, launching new products, and generating revenues and profits. Opportunities will be found in the mid-life phases of the lifecycle to ensure sales of a product are as high as possible, for example by extending the life of patents, and protecting the customer base against competitors. Opportunities will be found at the end-of-life phases of the lifecycle to increase sales with upgrades, or to exit the product graciously with product retirement, licensing or sale.

25.9.5   Process Definition and Automation

Clearly-defined, coherent, well-organised processes across the product lifecycle lie at the heart of effective PLM. These processes will be waste-free and low-cost. They'll enable concurrent involvement by people in different functions and locations. They'll be well-documented. Otherwise it will be difficult to improve them further. The key roles in the processes will be identified and described, along with the corresponding task and information characteristics. People in many different companies working in different places round the world may take these roles. Hundreds of people may be directly involved in these tasks. The process needs to be explained to them, with regular refreshment. To avoid confusion, the message needs to be very clear.

A clear, standard process architecture will enable coherent working across the product lifecycle. There's a document describing the process architecture across the lifecycle. A common harmonised version of each process in the product lifecycle will be used on all sites.

Relationships between the processes in each phase of the product lifecycle will be defined. Relationships between the processes in different phases of the product lifecycle will be defined. Relationships between processes in the area of Product Lifecycle Management, and those in other areas such as Supply Chain Management and Customer Relationship Management, will be defined. When possible, process steps will be automated in workflows. The workflows will be consistent with the process definitions. When possible, appropriate methodologies and working techniques will be defined for each process step.

25.9.6 Standard Lifecycle Processes

The company will define standard processes, standard data and standard applications that it, and its many suppliers, customers, and partners in the extended enterprise, can use to save time and money. Without such standards, each interface between different processes and applications would be a source of chaos, would add costs, and would slow down the lifecycle activities.

25.9.7 Standard Lifecycle Methodologies

Without a standard product development and support methodology, it's unlikely that people are going to be able to work in harmony across the lifecycle. A well-defined methodology lets everybody know exactly what's happening at all times, and tells them what they should be doing. It defines the major lifecycle phases and explains what has to be done in each phase. It shows how the phases fit with the company organisation and structure. It shows the objectives and deliverables at the end of each phase, and the way that phases connect together. It shows which processes, applications, methods, techniques, practices and methodologies should be used at which time in each phase. It shows the human resources that are needed, identifying the type of people, skills, knowledge, and organisation. It shows the role and responsibilities of each individual and the role of teams. It shows the role of management, project managers, functional reviewers and approvers. It describes the major management milestones and commitments. It describes the metrics used in the process.

> The lifecycle architecture was defined and applied. A lifecycle-wide process architecture has been defined and applied. The number of different, site-specific, variants of what should be the same process has been reduced by 50%. The target is to implement a common harmonised version of each process in the product lifecycle across all sites. There's a common harmonised Product Change Management process.
>
> The number of process steps that have been automated in workflows has been increased by a factor of four. There's still a long way to go. Initially, different sites had very different processes and applications. A lot of harmonisation was needed before it made sense to introduce automated workflows. After reviewing quality problems, feedback processes were defined and introduced to ensure effective feedback of information from product users to product developers. Processes have been reviewed and upgraded with activity steps that ensure and demonstrate compliance with regulations.

Fig. 25.5 Reporting progress with processes and the lifecycle

25.9.8 Progress with Lifecycle and Process

Targets are needed to measure the success of PLM deployment.

The "report" in Fig. 25.5 might be written five years after the PLM Initiative is started.

25.10 Collaboration

Collaborative technologies support PLM activities across the lifecycle, enabling work to be carried out in our extended enterprise in a well-managed way in multiple locations. Activities take place on different sites, and information is available on different sites. Team members may be based anywhere yet work together in spite of space, time and organisational differences.

Increased collaboration requires changes to behaviour and methods of working. It calls for increased trust, communication and commitment.

The "report" in Fig. 25.6 might be written five years after the PLM Initiative is started.

> There's a document describing the Collaboration Strategy. The Collaboration Strategy defines how, and under which conditions, collaboration takes place in the different stages of the product lifecycle.
>
> Collaboration policies were defined, a collaboration guideline was produced, and training was carried out. Processes were modified. Technical infrastructure (such as a collaboration portal and project work areas) was defined and implemented to enable use of a range of technologies. These include webcasting, podcasting, videoconferencing, audio conferencing, web conferencing, automated information feeds, twittering, collaborative blogging, electronic whiteboards, discussion groups and collaborative content co-authoring.
>
> Standard processes, standard data and standard systems are used in the collaborative extended enterprise environment.
>
> To enable effective working in the Extended Enterprise, a small brochure, "Collaboration Capability and Competence", was produced and made available to potential partners and customers. It describes the resources and skills available for product-related activities, as well as those supporting potential collaboration.

Fig. 25.6 Reporting progress with collaboration

25.11 People and Culture

This component is addressed in four parts.

25.11.1 Team Culture

The Product Family Teams will have a team culture, with people from different functions working together and in parallel. Team members will come from many functions such as marketing, regulatory, design, service, manufacturing engineering, test, quality, purchasing and disposal. They'll work together, sharing information and knowledge, and producing better results faster than they would have done if operating as individuals with limited specialist knowledge in traditional departmental or functional organisations. Their composite knowledge of design, processes, materials, manufacturing, recycling, quality, regulations and customer requirements will be applied to develop the best definition of the product and its manufacturing, support and disposal processes.

Team members will think about the product across its lifecycle. They'll have a clear view of the status of a product. Engineers designing a product will take account of how it will be manufactured. And how it will be disassembled and recycled. The recycling specialists will keep up-to-date with environmental laws. They'll keep development engineers informed. Together, they'll work out how to design products that can be disassembled quickly, and how to re-use parts in new products. With each new product, they'll further extend their knowledge for application on future products. Together they'll take better decisions, reducing rework, bottlenecks and waiting time. As a result, products will get to market sooner, products will fit better to customer needs, costs will be reduced and quality will be improved.

Working together, team members will use a common, shared vocabulary and standardised data definitions. Information will be well-organised and shared, so information access will be fast. Understanding of information will be improved as fellow team members will be available to provide guidance and more details.

The improved communication among team members will help reduce unnecessary product changes. It will help increase downstream awareness early in the development process. Team members from upstream functions will have a better understanding of downstream reality, resulting in a reduction in problems for downstream functions. The reduction in changes will result in less rework, and in a reduction in the overall product development and support cycles. The reduction in changes will reduce the burden on the product change system.

25.11.2 Skilled, Competent People

Although the company will have a focus on the product, this doesn't mean that human resources will be ignored. The development and support of high quality products will only be possible with highly-skilled, well-trained, highly-motivated people. The workforce will be one of the key components of successful activities across the product lifecycle.

There'll be a need for a variety of people with different types of skills for the activities across the product lifecycle. Even though teamwork will be common, there'll still be a need for people with very specialised individual skills, for example, to carry out analysis work requiring in-depth knowledge and long experience in interpreting results.

Alongside the need for specialists will be a need for multi-skilled generalists. These are people with skills in several areas. Generalists will have a different role from specialists. They'll lead teams, act as an interface between specialists, and provide the link to customers and suppliers. Everyone, specialists as well as generalists, will need the hard skills required to work with the company's applications, processes, and methodologies. People will also need soft skills, such as the ability to work in a team, and the ability to communicate well with their colleagues. They'll need to be able to work with people who come from other functional, cultural and national backgrounds. They'll need to be adaptable, and open to new ways of working, new ideas and new challenges.

There'll be a document describing the skills required of the Product Family Team, a document describing the skills of the Product Family Team, and a document describing the training plan for the Product Family Team.

25.11.3 Quality Culture

There'll be a culture of quality both at the level of the whole company and at the level of the product lifecycle. The company has, and will maintain, a culture, attitude and organisation that allows it to provide, and continue to provide, its customers with products and services that satisfy their needs. This requires quality in all aspects of operations, with things being done right the first time, and defects and waste eradicated.

Although PLM has a focus on the product, people working in the product lifecycle will have a customer orientation. They'll know that, unless customers buy their products, the company will go out of business.

With PLM, people think of both profit and the planet. In addition to financial issues, they take account of non-financial issues such as the environment, social issues, health, education and sustainable development.

The number of people using the PLM infrastructure of processes and applications has increased by 100%. In core competence areas, headcount has increased by 10%. In areas outside core competence, headcount has been reduced by 30%, mainly through outsourcing of certain activities. Training of people in core competence areas has increased by 50%.

Fig. 25.7 Reporting progress with people

25.11.4   Progress with People and Culture

Targets are needed to measure the success of PLM deployment.

The "report" in Fig. 25.7 might be written 5 years after the PLM Initiative is started.

25.12   Data, Information and Knowledge

This component is addressed in six parts.

25.12.1   Clean, Standard, Process-Driven Data

Throughout the product lifecycle, product information is all-important. It's all that people can work with when the product doesn't physically exist in their environment. Product data is a strategic resource. Its management is a key issue. It needs to be available, whenever it's needed, wherever it's needed, by whoever needs it, throughout the product lifecycle. Working closely together, Product Family Team members will use a common, shared vocabulary and standardised data definitions for the PLM environment. To save time and money, team members will want to work together using standard processes, standard data and standard applications. They'll also want to work with standard processes, standard data and standard applications with their suppliers, customers, and partners in our extended enterprise environment. Without standards, each data interface between different processes, applications and documents is a potential source of errors, adding costs, and slowing down activities. Once industry-standard processes are clear, their data and document requirements can be defined, and document definition, use and exchange agreed. A single common standard template for each document can be introduced across the extended enterprise.

Feedback about the use of one generation of a product helps improve future generations of the product. Information from product use will be used in product development.

There'll be a document describing the common, shared vocabulary and standardised data definitions for the Product Family Team, and a document describing the documents used by the Product Family Team.

25.12.2   Digital Data

All information will be converted to digital form so that it can be used, managed and communicated effectively. Correct and up-to-date digital data about the Product Portfolio, existing products, and products under development, are needed for short, medium and long-term decision-making. Digital product data will flow smoothly through the lifecycle, and will be available when and where needed.

There'll be a document describing the data model and flow.

25.12.3   Data Management

A Product Data Management (PDM) system will provide people in the product lifecycle with exactly the right information at exactly the right time. Having digital product data under PDM control will help achieve the objectives of improved product development and support. With PDM, it will be much quicker and easier to access, retrieve and reuse product data. The PDM system will manage all data defining and related to the product across the product lifecycle from initial idea to retirement. It will provide controlled access to correct versions and configurations. It will enable tracking of product configurations.

25.12.4   Legacy Data

The different types of legacy data will be identified. Policies will be defined for managing them and, where possible, for eliminating them.

25.12.5   Data Exchange

A review will be made of the need for different data formats. Where these are found to be necessary, standard approaches will be implemented for data exchange.

25.12.6   Progress with Data, Information and Knowledge

Targets are needed to measure the success of PLM deployment.

The "report" in (Fig. 25.8) might be written five years after the PLM Initiative is started.

More than 99.9% of data in use are in digital form. The number of different versions of document templates across sites has been reduced to one for each template. All sites use the same document template for Product Change Management. In 85% of cases, duplicate data (such as duplicate part descriptions) have been eliminated to leave a single clean data element. The target is still 100%.

Fig. 25.8 Reporting progress with data

25.13   Facilities, Equipment, Applications, Interfaces

This component is addressed in four parts.

25.13.1   Facilities

The most appropriate facilities will be used across the product lifecycle. Sometimes they'll have been purchased by the company, sometimes leased. Often they'll be the facilities of other organisations in the extended enterprise environment.

25.13.2   Equipment

The best equipment will be used across the lifecycle. Advantage will be taken of modern computer-controlled technology, such as 3D printers that produce accurate physical parts directly from a CAD model.

Because a Product Family Team focuses on one product family, the best equipment can be purchased, installed and optimised. It can be cost-justified over several generations of the product family.

Manufacturing and maintenance equipment will be simulated and optimised before use. Simulation will help study the performance of a plant before it's been physically built or implemented. Computer-based simulation is low-cost and effective. It uses the models designed in the computer that would normally be the basis for building the plant. It makes it easier to evaluate before implementing. It allows errors to be identified and corrected before they are implemented. Models can be built, tested and compared for different concepts. "What-if" analysis can be carried out. Recommendations for improvement can be made.

25.13.3   Application Standardisation

Application programs, such as Automated Product Idea Generation, Virtual Engineering, Digital Manufacturing, Collaborative Product Support, and

Computer-Aided Recycling, are used in the corresponding phases of the product lifecycle. Application programs also manage the Product Portfolio.

We'll save a great deal of time and money by the use of standard processes, standard data and standard applications. Without such standards, each process or application interface would be a source of problems. Without such standards, duplicate applications are a source of waste. Applications will be harmonised over all sites, and across the lifecycle. There'll need to be very good reasons to have, for example, different CAD applications, or different versions of the same CAD application, on different sites. Such differences can be a barrier to communication and progress.

There'll be a document describing the application architecture.

25.13.4 Interfaces

PLM applications contain important product information that must be made available to other enterprise applications such as ERP, CRM and SCM. PLM applications need to have access to information that's managed in other enterprise applications.

Interface programs are costly to develop and maintain, error-prone, and potential breakpoints impeding smooth process and information flow. As a result, all interfaces will be reviewed frequently and their existence questioned.

The target is to eliminate 20 % of interface programs per year.

25.14   Mandatory Compliance, Voluntary Conformity

This component is addressed in two parts.

25.14.1 Mandatory Compliance

PLM supports our activities to meet mandatory compliance requirements of international and industry regulations in areas such as health, safety and environment. It helps maintain documentation in required formats, and provides an audit trail showing actions taken.

There's a document describing the compliance requirements.

25.14.2 *Voluntary Conformity*

PLM allows the company to do more than just comply with regulations and laws. It allows us to go further, and demonstrate our beliefs in the importance of the environment, social justice, health, education and sustainable development. PLM enables voluntary compliance with recommended practices and guidelines in these areas. PLM enables us to act responsibly and address the effect of policies for sustainable production and consumption of existing and new products. Voluntary conformity can improve financial performance. Sustainable development and environmental needs represent major business opportunities for faster growth and profitability through improved current products and services, and innovation of new products and services. PLM lets us take advantage of voluntary self-regulation initiatives and use them to build new markets.

25.15 Security and Intelligence

This component is addressed in two parts.

25.15.1 *Security*

PLM provides security in the face of increasing global competition and the potential risks from terrorism and economic espionage. Product information is an increasingly valuable resource for corporate development and must be kept secure.

PLM helps provide security in Information Systems, protecting against viruses, worms, hackers, pirates, hijackers, phishers, denial of service attacks and spying programs. It enables consistent corporate-wide action to keep up with evolving systems, networks and changes in staff behaviour.

We must also maintain security in areas where people may be less vigilant, such as in telephone conversations, chat-rooms, collaborative workspaces, e-mails, and use of portable computers and similar lap-top and hand-held devices.

We'll implement procedures and take measures to keep information secure in our buildings. We know that competitors may be eavesdropping from outside to steal trade secrets. Similar rules will be in force in bars and restaurants. We know that competitors at the next table may be listening for high-value details of new products among the high volume of low-value corporate gossip. And there'll be procedures in force for travel. We know that luggage may be searched for any potentially valuable financial or technical information, such as designs, patterns, plans and procedures.

There's a document describing the Security policy.

25.15.2   Intelligence

PLM will help the company to carry out three important intelligence activities. PLM will support managers in the development of strategies and plans. It will provide valuable information about customers and competitors, reducing the risks associated with decision-taking. PLM will help monitor trends in new technology that could affect the company's future. It will enable rapid extraction and analysis of information from electronic files, synthesis of knowledge by data-mining and content-understanding techniques, collation of facts and inferences, and removal of repetitive and irrelevant information. It will allow candidate decisions to be presented on the basis of risk and reward, and will identify and prioritise decisions and strategies. PLM will enable the implementation of counter-intelligence measures to protect against economic espionage.

There's a document describing the Intelligence policy.

25.16   Linking PLM Drivers to PLM Benefits

The PLM Vision helps show the link between PLM drivers and PLM benefits (Fig. 25.9).

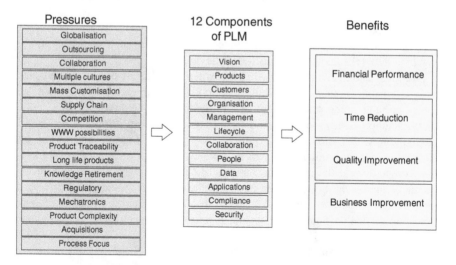

Fig. 25.9 PLM drivers, vision and benefits

Chapter 26
The Current Situation

26.1 Some Questions

When a company decides to describe the current situation of its PLM environment, there are usually a lot of questions (Fig. 26.1).

Questions like these are helpful. As Rudyard Kipling put it:

I keep six honest serving-men
(They taught me all I knew);
Their names are What and Why and When
And How and Where and Who.

You can learn a lot from good questions if you take the time to answer them.

Another frequently asked question runs along the lines, "Why bother to look at the current situation? I'm interested in how to work better in the future, not showing how we do things badly today".

OK, but your home would be "better" with a swimming pool, a sauna and a helicopter landing pad on the roof. Shall we start building? Or would you like someone to check first if the walls would support the extra weight? In other words, would you get some information about the current situation?

If you don't know the current situation, you may be missing key information that you need before making your proposal for the future situation. If you don't know what you have, or what the problems are, it's going to be difficult to know what you're going to improve, or fix, and why. There are probably things you need to remove before you add new things. There could be many things that work very well. You may not want to change them, because changing them might impair performance, not improve it. If you don't know the current situation, you may miss easy improvement opportunities. If you know what you do badly, you can make sure you don't do it again in the future, and don't propose the same wrong things for the future.

And, to successfully implement change, you need to communicate it to, and convince, many people. You need to communicate a clear documented message. If

© Springer International Publishing Switzerland 2016   509
J. Stark, *Product Lifecycle Management (Volume 2)*,
Decision Engineering, DOI 10.1007/978-3-319-24436-5_26

what's the scope of this work?
how do we do this?
why are you guys doing this?
what do we have to produce?
where do we start from?
how far do we go?
who's doing this?
what comes after?
when do we do it?
how do we report it?

Fig. 26.1 Questions about the description of the current situation

you can't even explain how things are today, it's unlikely that anyone's going to believe your suggestions for the future.

26.1.1 Starting Point

The starting point for the description of the current situation results from the objectives set by top management. This information may be given directly by management, or may need to be pieced together from information which is available elsewhere, such as in the Vision, in a proposal for this activity of documenting the current situation, in business objectives, or in annual plans and targets.

From the objectives, the scope of the project to describe the current situation can be defined. Once the scope is clear, project activities can be defined. However, before starting any of these project activities, the scope and proposed activities should be discussed and agreed with management.

26.1.2 Scope and Perimeter

The subject matter of the description of the current situation of the PLM environment could include all of the following components: enterprise structure; product structure; product data; business processes; PLM applications; PDM systems; techniques; facilities and equipment; metrics and people.

However, a project to describe the current situation in a particular company often only involves describing a few parts of the situation, not everything. In theory, perhaps it would be best to look at everything. But there could be several reasons why this option isn't appropriate. A company may not have the resources, or the time, to look at all areas. Or, it may already have looked at some areas. Or, it may be that top management wants to limit the description to a few areas.

In our experience, every company looks at something different. For example, one of our customers makes a more-or-less standard product. During the last decade, the number of customers grew rapidly, with many of them asking for small customisations to the product. The result was that, instead of having a single standard BOM, the company had very many slightly different BOMs. The objective set by top management was to identify modular BOMs. For that company, the description of the current situation was focused on product structure, product data structures, and data objects.

However, another customer, as a result of acquisitions, had several different PDM systems in different product divisions. The objective of top management was to decide whether they needed to have these different PDM systems, or could use the same application everywhere. In that case, PDM systems were obviously addressed in the description of the current situation. And so were product structures, product data structures, business processes and PLM applications.

The top management of another customer wanted to enable its product development unit in China to collaborate with units in other countries. The description of the current situation addressed PDM systems and other PLM application such as CAD/CAE/CAM applications.

Another company wanted to find out why it was getting so few new products to market. The description of the current situation focused on development and change processes, product structure and product data.

26.1.3 Project Team

The make-up of the team that documents the current situation will depend on the scope of the areas to be described. Usually there's a team member from each of the specific areas to be described. Sometimes these team members are known as departmental specialists, sometimes as functional experts, sometimes as subject matter experts. Some will come from departments such as Marketing, R&D, Production and Support. A Product Manager is often included in the team. And usually there are team members with cross-functional views of the product lifecycle. Examples are a planner, someone from IS, someone from the organisation that manages product data and documents, and someone from the quality or process organisation.

The project team should include people who are knowledgeable about the company, respected, and have time to participate in the project. It's not useful to include departmental bosses on the team if they don't have the time to participate, and intend to delegate their role to junior staff. Similarly, it's not helpful to include lightweight team members whose opinions and efforts will be ignored.

Top management should select the project team leader with care. Project activities are likely to be cross-functional, and involve working at many levels in the company. The project team leader needs to be a good manager, knowledgeable, diplomatic and respected. Top management should give the project leader the authority and the responsibility to carry out the project. However, if it appears that the project isn't moving forward quickly enough, or is being opposed by particular interest groups, the project leader should be able to go back to top management for guidance and support.

Top management should inform interested parties throughout the company about the project and its objectives. In particular, managers of functions in the scope of the project should be informed that the project team leader has the authority to ask questions, and to ask for details about the current situation. In turn, these middle managers should inform their subordinates about what's happening, and ask them to provide the necessary information.

In some PLM Projects, "top management" will be represented by one top manager. In other cases, there may be a Steering Committee, including several top managers, for the project. Sometimes there will be a Task Force, or a Tiger Team, of top managers to drive progress.

PLM is a relatively new subject. Probably, few, if any, of the individuals in the project team will know much about it. Defining the current situation of the PLM environment isn't a frequent activity. Again, few, if any, of the individuals in the project team will know much about it. It may be necessary to provide appropriate training and/or education. Some basic information can be provided right at the beginning of the project by a PLM consultant with experience of describing the current situation in other companies.

The size of the project team depends on many parameters such as the scope of the project and the size of the company. We've worked with project teams of more than 50 people. And we've worked with teams of 2 or 3 people. As for the length of such projects, in one case the description of the current situation was completed in 3 weeks. In other cases, we've been involved in projects that have run for more than a year.

Positive team membership characteristics include willingness to work hard, involvement, support of fellow team members, an open mind, a good knowledge of at least one function, the ability to stop talking and listen, and a good understanding of the fact that nobody knows everything. It's important to be able to distinguish between "this is the current situation" and "this is how I'd like it to be". Similarly, it's good to be able to distinguish between "this interests me" and "this is important for the company". It's also important to be immune to the Stockholm syndrome transposed to the PLM environment. It's only too easy to support the interests of product and service suppliers rather than those of the company.

26.1.4   Improvement Opportunities

Another frequently asked question is "should we look for improvement opportunities when documenting the current situation?" From our experience, the answer is, "No". It's difficult enough just to correctly document the current situation. And since the identification of improvement opportunities is a major undertaking in itself, trying to do the two together usually results in neither being done well. So we recommend doing them separately. However we do suggest that any potential improvement opportunities that appear, or are mentioned, while finding out about the current situation, should be documented.

Another question that often arises is whether the activity of documenting the current situation should include an analysis of the current situation. In our experience, the two activities should be carried out separately. First the current situation should be documented, reported to management, and agreed. Then the analysis can be started. Both tasks require significant effort. There's a danger that if analysis starts before agreement has been reached about the current situation, even more effort will be needed. Analysis can take a lot of time. If there's disagreement about the underlying current situation, it can take even longer. However, improvement suggestions made by people in the product lifecycle, for example during interviews, should be included in the report of the current situation. Management may then decide to take short term actions to address them. And, once documented, the improvement suggestions will be available for consideration in later phases.

26.1.5 Deliverables

One of the dangers when describing the current situation is that important information gets swamped by the huge amount of data that's produced. To avoid this, it's important to define in advance the intended approach and scope (Fig. 26.2).

Defining the deliverables up-front will help define the approach and avoid data overload.

26.1.5.1 Major Deliverables

The deliverables will depend on the scope of the current situation description. As the scope of the description differs from one company to another, the deliverables will also differ. However, there are some deliverables (Fig. 26.3) that we've seen in many descriptions. And there are some deliverables (Fig. 26.4) that we see less frequently and depend on the scope.

| the approach to find out about the current situation |
| the information that's needed |
| the deliverables that will be produced |

Fig. 26.2 Up-front definition of intent

| a description of the scope and the objectives |
| a description of methods used to describe the current PLM environment |
| a description of the project team |
| some examples from the PLM environment |
| a list of any improvements suggested |
| a business case model showing how things work in the current situation |
| some statistics and metrics |

Fig. 26.3 Frequently encountered deliverables

examples of questionnaires used to gather information
a list of people interviewed
a description of product structures
an overall business process diagram
a high-level information flow diagram
a description of each of the major activities in the product lifecycle
flow diagrams for each major activity
a data model
descriptions of documents and applications used in the product lifecycle
examples of reports created in the product lifecycle
examples of metrics in use, such as time per project, or sales per year
statistics, such as the number of new products per year

Fig. 26.4 More specific deliverables

26.1.5.2   Reporting the Deliverables

There are usually two deliverables from the activity of documenting the current situation. They are a detailed report and a management presentation.

26.1.5.3   Report

The project team will collect a mass of data about the current situation. It doesn't need to present all of this raw data in its report. Instead, the project team must distil it into a form in which management can understand it, and can recognise the way in which it relates to the PLM environment, and to the objectives.

The report may be in four parts (Fig. 26.5).

26.1.5.4   Management Presentation

After the project team has described the current situation and created its report, it should outline the main findings in a PowerPoint presentation. The team leader should present this to top management. They will probably ask that it also be made available to functional managers. Some managers will probably want to read the report in detail. It may take several weeks to get the results formally agreed, and the next phase of activity started.

an executive overview
a section describing objectives and approach
a section describing the findings for each component
an appendix containing additional details

Fig. 26.5 Contents of the report

26.1.6 Approach

The methods used to describe the current situation (Fig. 26.6) will depend on the scope.

Whatever the scope, many people will need to be involved. Perhaps, in a very, very small company, one person may know everything about the current situation. In all the companies we've worked with though, nobody has known everything about the current situation. And, to get the complete picture, it's been necessary to involve many people.

26.1.6.1 Questionnaire

One of the ways to understand the current situation is to develop a questionnaire addressing the main topics within the scope. And then get the people who are responsible for the individual topics to answer the questionnaire. A top-down functional decomposition approach can be taken. If for example, several departments are involved across the product lifecycle, then a questionnaire could be created and sent out to the leader of each department. Depending on the answers, additional questionnaires could be developed to get more detailed information in specific areas.

26.1.6.2 Product Path Interviews

Another way of gathering information is to "walk the Product Path", following a typical product through its lifecycle and asking questions where appropriate. This should give a more practical result than use of a questionnaire. This approach can be used to get a good understanding of the current situation of the lifecycle from the points of view of products, processes, people, applications and information. This activity is best carried out by a multidisciplinary group of team members. They can identify the various product paths, the structure of product paths, and their typical activities and events. The project management and product management techniques in use should be identified and documented.

It's to be expected that the team will have to interview many people in many activities, and build up knowledge about many subjects. Due to time constraints, it probably won't be possible for all team members to be involved in every interview.

| creating and sending out questionnaires |
| reviewing responses to questionnaires |
| carrying out interviews |
| documenting real-life examples |
| holding meetings of study groups |
| reviewing documentation |
| modelling and mapping |

Fig. 26.6 Methods for understanding and documenting the current situation

A good approach is for them to work in small groups of 2, 3 or 4 people. The person from the function most directly related to a question should be involved in finding the answer. And so should people from other functions. A team member from the upstream source of the information, and a team member from a downstream user function should be involved. This increases the likelihood of getting a complete, rounded answer.

In general, people working day in, day out, in the product lifecycle are the best source of information about everyday activities, the information that's used in such activities, and short-term improvements. On the other hand, these people often find it more difficult to address longer-term and cross-functional issues. The project team should encourage them to address subjects such as their role, what they create, how they receive work, how they know when to start working on an activity or to change to another activity, the organisational hierarchy, and the release procedures.

26.1.6.3   Study Groups

Study groups may be set up within the project team to address particular issues (Fig. 26.7).

26.1.6.4   Modelling

Models and maps can be developed to help describe the current situation. Some models may already be available. If so, they should be reviewed, and re-used if appropriate. Sometimes a top-down hierarchical approach will be a suitable modelling approach.

The models produced need to be easily understandable for people working in the lifecycle. The level of detail in the model can be progressively increased until they feel that all activities, participants, information use and flow are shown on the model or map. Other people in the lifecycle can be asked to comment on the model. Models produced by people in neighbouring activities can be put together to understand interfaces.

However, since many months can be spent in developing models, the modelling activity should be organised carefully by the team leader. The objectives of modelling, and the expected end-product, must be defined in detail before initiating modelling activities. Modelling can become very expensive and time-consuming if pursued without a clear objective and to a too-detailed level.

| understanding the current product processes |
| understanding the current product realisation processes |
| understanding the current product support processes |
| describing the applications currently in place |
| identifying the PLM approach of competitors |

Fig. 26.7 Typical issues addressed by study groups

Initially, the project team may only need top-level models. Refinement can start later, once a high-level overview has been prepared and agreed, and it's possible to see where further modelling would be most effective.

26.1.6.5 Review

Once the information for the description of the current situation has been obtained and documented, it should be reviewed and cross-checked. In particular, it should be shown to the people who provided it, to confirm that it really represents the current situation. If it doesn't, corrections should be made and missing details added.

26.1.6.6 Plan

The plan for the activity of documenting the current situation (Fig. 26.8) will depend on the scope. A project plan should be drawn up showing the major tasks and milestones. The plan should be agreed with top management before project activities start. The plan should be communicated to everyone involved. It should be kept up to date and visible. Top management, in particular, should know what progress is being made, and be kept aware of any problems that may arise.

Communicating a plan for the next steps in the PLM Initiative (Fig. 26.9) will help management understand how the overall PLM activity is moving forward.

It may seem unnecessary to create a project plan for an activity as basic as documenting the current situation. However, as President Eisenhower put it, "In preparing for battle, I have always found that plans are useless but planning is indispensable." Planning is useful. Thinking about, and discussing, what's needed, and anticipating what could happen, helps to avoid pitfalls, focus on important activities, and to carry them out in the most effective way.

Activity	1	2	3	4	5	6	7	8
Start-up and Training								
Review the Current Situation Documentation								
Get, and structure, high-level data								
Review available data. Identify missing data								
Get additional data								
Review available data. Get any missing details								
Prepare Draft Report								
Prepare Management Presentation								
Present to Management								
Finalise the Report								

Fig. 26.8 The plan for description of the current situation

Activity	1	2	3	4	5	6	7	8
Document the Current Situation								
Finalise the Current Situation Report								
Describe the Future Situation								
Develop the Implementation Strategy and Plan								

Fig. 26.9 Plan for the next steps

26.2 Extended Enterprise

Sometimes, the Extended Enterprise will be in the scope of the activity of describing the current situation. In other cases, it will be out of scope. However, even if it is in scope, the details of the description of the Extended Enterprise will vary from one company to another.

Some companies may be in only one Extended Enterprise. Others could be in a different Extended Enterprise for each product line.

Activities in the description of the current situation could include creating lists of suppliers, partners and customers, and indicating their role in the Extended Enterprise. Organisation charts can be collected and included in the documentation. A description of the departments in scope could be given, along with their roles in the product lifecycle. Documenting the number of people in each organisation could help to increase understanding. In some cases, it's useful to give a brief history of the organisations, showing when they joined the Extended Enterprise, and indicating any specifics, for example, cultural, legal or linguistic characteristics.

Potential metrics depend on the scope and circumstances (Fig. 26.10).

26.3 Product

Products may be in the scope of the description of the current situation. If they are in scope, it's likely that the details of their description will vary from one company to another.

It may be useful to describe the structure of the Product Portfolio, with the products in each segment or family being listed. Product structures can be shown for each family, describing modules and main assemblies as well as lower-level structures. A matrix can be created to show interfaces between modules.

The different types of product can be described. The documents used to describe products can be listed. The metadata describing products in applications and databases can be listed.

the number of Extended Enterprises in which the company operates
the number of organisations in each Extended Enterprise
the percentage of outsourced activities

Fig. 26.10 Examples of potential metrics related to the Extended Enterprise

percentage of products less than five years old
numbers of product lines, products, modules, and components
number of hierarchical levels of product structure
number of generations of product family concurrently worked on
number of complaints per product
numbers of new, modified, retired products per year
level of part reuse
degree of product reliability

Fig. 26.11 Examples of potential metrics related to products

Product naming and numbering schemes can be described.

The product can be shown as it evolves across its lifecycle. Timelines can show how typical products and families evolve. The age distribution of products can be described.

Sales figures and cost structures can be documented for product families and specific products.

Any activities that are specific to particular products should be documented. Any product-related differences between different parts of the PLM environment (for example, sites or product families) should be described.

Potential metrics will depend on the scope and circumstances (Fig. 26.11).

26.4 Processes

Processes may be in the scope of the description of the current situation. If so, an overall process map will help position the individual processes. This may already exist in the form of a Business Process Architecture or Process House. It may be documented in an ISO Manual or a Quality Manual.

Depending on the circumstances and the time available, it may be appropriate to list processes, showing their name, owner, objective, input, output and metrics.

Alternatively, an "Overview Matrix of All Processes" can be created. In the first column, the processes can be listed by name. Other information about each process, such as that shown in Fig. 26.12, can be listed in other columns.

An "Individual Process Matrix" can be made for each process. In the first column, the participants can be listed. In the columns for each "participant row", the

the objective of the process
the process owner
the participants involved in the process
the customer of the process
the event that causes the process to be started
the first action in the process
the last action in the process
the event that happens at the end of the process
the neighbouring processes
the applications used in the process
the procedures that apply to the process
the process metrics

Fig. 26.12 Examples of information documented for each process

activities of each participant can be shown. Then, for each process, work with the process participants to make a map of the process.

Another approach to process description is hierarchical, starting with business processes and going down to individual tasks. At each level, the objective of each activity should be documented. If necessary, the activity can be broken down into its constituent tasks. For each activity, the information input, created, used, and output can be described, as can the sources of information and the definitions of information. The cost of the activity can be documented. The people involved in the activity and their roles can be described. The frequency and execution time of each activity can be indicated. Any applications used in the activity can be identified, their information requirements described, and their interfaces with other applications detailed. Procedures and performance measures associated with the activity can be described.

The engineering change management process is often in project scope. If so, both data creation and change processes may need to be described, clarifying the roles and rights of participants. The time and effort required to carry through changes can be described.

Perhaps some processes that run outside the company will need to be described. And, any differences between the same processes in different parts of the environment should be described.

Potential metrics will depend on the scope and circumstances (Fig. 26.13).

26.5   Product Data

Product data may be in the scope of the description of the current situation.

If product data is in scope, a product data model can be created to show how products are represented by documents and other data.

A list of documents and other data in the product lifecycle can also be provided. The different types of document can be described briefly and their characteristics and volumes noted. The different types of data can be described briefly, with their characteristics and their volumes being noted. For each data type, and each document type, the most important attributes can be documented.

Transition rules between different states of data can be described.

number of processes in the company
number of processes in the scope of PLM
percentage of identified processes completely defined
time to execute a process
cost of process execution
number of times a process is executed annually
number of participants in a process
number of tasks in a process
difference between planned and actual cycle time for a process
number of product development projects abandoned each year

Fig. 26.13 Examples of potential process-related metrics

Data flow diagrams can show how product data flows throughout the product lifecycle. The flow of product data at departmental boundaries, can be described.

The different structures of product data, such as Bills of Materials and parts lists, can be described, as can other associations such as product/drawing relationships.

Owners and users of product data can be described. Access needs, and the rights of users and groups of users, can be described.

Shared and redundant data can be described.

Data standards and templates can be described.

The data security and data integrity situation can be described.

A picture of the current organisation of the company, from the point of view of product data, can be provided. This can show the number of users and their locations, both geographically and functionally, and the way they store and communicate information. It can show where data is created, modified and stored, and how it's communicated and shared.

Potential metrics will depend on the scope and circumstances (Fig. 26.14).

26.6   PLM Applications

PLM applications may be in the scope of the description of the current situation. If so, an inventory of existing PLM applications can be made. It should include all applications related to products across the product lifecycle. As well as applications such as ERP, CAD, CAM and CASE, it should include applications used in analysis, project management, technical publications, documentation management, configuration management, and sales. The applications should be described.

The number and roles of the users of each application can be documented.

An inventory can be made of interfaces between applications. Interfaces and information transfer between applications can be described, as can transfer of data to and from supplier and customer applications.

Any differences between applications in similar situations in different parts of the environment should be described. They can include the use of different versions of the same application, and the use of applications from different vendors to achieve the same result.

Potential metrics will depend on the scope and circumstances (Fig. 26.15).

number of different data types
volume of data of each type
number of documents of each type created annually
number of documents of each type under modification
percentage of data on electronic media
time spent looking for data
number of times that key data is recreated
level of data reuse
quality level of data

Fig. 26.14 Examples of potential data-related metrics

number of different applications
number of vendors providing applications
number of users
investment in PLM applications as percentage of budget
annual running cost as percentage of budget
annual ratio of PLM applications staff to value-adding staff in the product lifecycle
number of interfaces
cost of interfaces

Fig. 26.15 Examples of potential application-related metrics

26.7   Product Data Management

PDM systems may be in the scope of the description of the current situation. If so, an inventory can be made of existing ways of managing product data. These can include PDM systems, manual systems for the management of product data, and other product data management approaches (such as databases, file management systems, other applications).The types of data managed can be documented.

Any other data repositories can also be documented. They may contain product data such as product names, Bills of Materials, manufacturing instructions, technical drawings, product specifications, CAD files, process specifications, quality data and test results.

The current methods of creating, numbering, classifying, storing, archiving, obsoleting and otherwise managing product data can be described.

The cost structure for management of product data can be documented.

The number and roles of people managing product data can be described.

Potential metrics will depend on the scope and circumstances (Fig. 26.16).

26.8   Equipment and Facilities

Equipment and facilities may be in the scope of the description of the current situation. If so, existing equipment and facilities can be listed and described. Their year of installation can be noted. Their location can be documented.

Potential metrics will depend on the scope and circumstances (Fig. 26.17).

the cost of managing product data
the quality level of product data management
the volume of product data managed electronically
the volume of product data managed manually

Fig. 26.16 Examples of potential PDM-related metrics

26.9 Techniques

Techniques may be in the scope of the description of the current situation. If so, there may be two areas to address.

Any special techniques (such as Design for Six Sigma) used at some stage of the product lifecycle can be listed and documented. Their characteristics can be documented (Fig. 26.18).

A list of current corporate or departmental improvement projects that may affect the product lifecycle can be drawn up. The objective, scope, cost and expected effect of each can be described.

Potential metrics will depend on the scope and circumstances (Fig. 26.19).

26.10 People

People may be in the scope of the description of the current situation. If so, there may be several areas to address.

The number and location, both geographically and functionally, of the people working in the product lifecycle can be described.

Organisation charts, job descriptions, skills matrices and training programs can be included in the documentation.

The roles of people in the lifecycle can be described. The PLM-related skills of people in the company can be described.

the annual maintenance spend
the annual investment in new equipment and facilities
average age of equipment
average age of facilities

Fig. 26.17 Examples of potential equipment- and facility-related metrics

date of introduction
objective
scope
role
cost
expected benefits
actual benefits

Fig. 26.18 Characteristics of techniques

the total number of special techniques
the number of new techniques introduced annually
the cost of special techniques
the financial benefit of special techniques

Fig. 26.19 Examples of potential equipment- and facility-related metrics

Any differences between different parts of the PLM environment, such as between different sites. can be described.

Potential metrics will depend on the scope and circumstances (Fig. 26.20).

26.11   Metrics

Metrics may be in the scope of the description of the current situation. If so, existing metrics can be listed and described. They may be related to all sorts of parameters such as lead times, cycle times, raw materials, engineering changes, parts, defects, costs, and data. Other examples are shown in Fig. 26.21. Current and target values can be documented.

Any differences between sites should be documented.

Potential metrics will depend on the scope and circumstances (Fig. 26.22).

26.12   Organisation

Organisation may be in the scope of the description of the current situation. If so, organisation charts can be included for those components in scope, such as products, processes, applications, data and people.

number of people with each particular skill
percentage spend on training
skills available as percentage of required skills
number of people working in the product lifecycle
ratio of design engineers to manufacturing engineers
ratio of design engineers to support engineers
number of product development projects concurrently worked on by individuals

Fig. 26.20 Examples of potential people-related metrics

the number of existing products, assemblies, parts and tools
the annual number of new products, parts and tools
the number of new software versions implemented annually
the number of modifications
the number of new and modified drawings and other documents
the number of drawings and other documents released daily
typical product development project times
typical product support project times
the number of engineering changes
the time taken to process engineering changes
average number of levels of Bills of Materials
maximum number of levels of a Bill of Materials

Fig. 26.21 Examples of metrics

number of metrics
number of metrics in each phase of the lifecycle
number of metrics addressing the entire lifecycle

Fig. 26.22 Examples of potential metric-related metrics

| number of PLM components for which organisational structures are documented |
| number of levels of organisational structure |
| span of control |
| number of functions included in a product team |

Fig. 26.23 Examples of potential organisation-related metrics

A matrix can be created to show, for each PLM component, the related organisational structures.

Any differences between different parts of the environment should be described. Potential metrics will depend on the scope and circumstances (Fig. 26.23).

26.13   PLM Environment Model

If possible, a simple model should be described, showing how the various components and metrics of the current PLM environment fit together.

Chapter 27
Current Situation Examples

27.1 A Fragment

The following is a fragment from the report of the review of the current situation in a company in the process manufacturing sector.

The review showed that, although there are, and have been, many development projects, few new products get to market. Many projects are failing to achieve acceptance in the Final Prototype activity. Even apparently very small projects are taking a long time to come to fruition.

There is no easy way for management to see the status of development projects. Sometimes it is not clear which criteria are being applied to decide if a project can move forward.

There is incompatibility between the IS applications in use, with the result that there is data duplication between applications, and sometimes data are manually re-entered. There have been misunderstandings in projects due to a lack of clear definition of particular words. There is confusion, for example, between product features and product characteristics, and between customer requirements and application requirements.

There is a lack of agreement among the product development team members from different functions. People in Marketing feel that their ideas and opportunities are being lost because R&D has no time to work on new projects. People in Production complain that "R&D's projects" interfere with their plant and production runs, costing them excessive set-up time, unnecessary downtime and reduced yield. People in R&D complain about the overload of work they face, the continual demand for changes from Marketing, the huge volumes of paperwork they have to produce, the numerous meetings they have to attend, and the lack of time to do anything useful for their projects in the lab.

There is no methodology showing product developers what they should be doing at each time during a project.

© Springer International Publishing Switzerland 2016
J. Stark, *Product Lifecycle Management (Volume 2)*,
Decision Engineering, DOI 10.1007/978-3-319-24436-5_27

27.2   Automotive Company

The following fragment of a summary of the current situation in an automotive
company was published in the first edition of "Product Lifecycle Management: 21st
Century Paradigm for Product Realisation".

Current situation

Very few of today's corporate managers understand the requirements for new product
development. They are happy to leave the Engineering function to itself, and let it do what
it likes, provided it doesn't want to spend lots of money. The main criticism that top
management has of Engineering is that products represent the engineers' dreams, not the
customers' requirements. Time and again, new designs are for rugged pick-up trucks,
high-powered sports cars and futuristic luxury models - yet most customers just want a
low-cost reliable car to get to and from work, the mall, and the football stadium.

Top management is frustrated, and talks more and more about out-outsourcing Engineering.
Top management can't understand why the engineers always start their designs with a
blank sheet of paper. Can't they re-use existing parts, or use purchased parts? Why do they
always try to do it all themselves? Can't they go out and see what customers really want?
Can't they listen to the marketing specialists and use the specifications that come from the
market? Can't they make themselves clear when they communicate with the plants? Can't
Engineering understand the difference between lowest initial cost and lowest lifetime cost?
Can't Engineering see that competitors' designs are fresher, have more variety, and are
technically more sophisticated? Isn't it obvious that it's better to take 2 years on a
development rather than 7 years? Can't Engineering understand that if a mid-life
replacement is late, customers won't just wait for it to arrive, they'll go and buy a competitor's product?

Engineering management recognises it has some problems, but knows it has a lot of
solutions. Approval to develop, over the next five years, its proposed New Product
Realisation Process will guarantee quality improvement by an order of magnitude. If top
management would only provide the funding for its 10-year CIE (Computer Integrated
Engineering) project, it will be able to slash lead times.

Engineering management sees the main problems with performance as being related to top
management attitudes and behaviour. Top management seems to have no real understanding of the underlying engineering processes, and seems to run the business on the
basis of a simplistic, top-down, cost-centre view. In this picture the business runs itself, and
top management makes fine tuning through annual "flavour of the year" adjustments. One
year it's Total Quality Management, then it's Customer Focus, and then Logistics
Management, or Cycle Time Reduction. The title is always written with capitals, but even
this doesn't make people think it's important - everybody knows that next year it will have
disappeared.

Another criticism from Engineering is that every time things look bad, top management
"downsizes" across the board. Downsizing by 10 % means reducing headcount by 10 %, so
a certain number of people, regardless of their skills, knowledge, or their role in the
engineering process have to go. Middle management decides who should go and who
should stay, so middle management stays, while design and manufacturing engineers go.

27.3 Engineering Company

The following fragment of a summary of the current situation in an Engineering company was also published in the first edition of "Product Lifecycle Management: 21st Century Paradigm for Product Realisation".

Current situation

In spite of all the effort that top management is putting into the improvement process, the engineering managers around the world feel that the real problem is that no clear direction is being set. There are countless exhortations to work harder, to schedule better and to "do your best". One top manager even spread the message that people weren't expected to work the 40 h in their job contract, but to do 60 h a week. This went down badly with teams trying to introduce Just-In-Time and to reduce cycle times. They preached that wasted effort is the cause of most problems in business processes, and that if it could be removed, things would get done faster yet with less effort. They counselled that rather than working longer hours, people should work smarter. The 60 h week was seen as confirmation that top management had lost control. It was yet another unrealistic target that would distort their efforts to improve the process. Unrealistic targets were often proposed by top management or the sales force, and this gave the impression that development was always late, when it was actually on time compared to its own targets.

An on-going problem is that far too many projects are handled at the same time by a few people, and a lot of time is lost as the effort is switched from one project to another. One year, top management came up with the idea of using a scheduling system on a PC to enable engineers to do more work. Engineering management had explained that scheduling wasn't the problem, but were forced to implement this idea from above. In the meantime, top management still holds projects up by forcing everyone to wait for management decisions that are only made at monthly management meetings.

Engineering management is aware that new product development performance could be better, but they aren't quite sure what to do about it. They know for example that most of the time, 80 % of a new product already exists in other products, but don't know how to access the information or how to reuse it.

Top management is tired by the Engineering organisation's unquenchable desire for high-risk, high development cost projects. The culture of the Engineering organisation doesn't seem to tie in with the rest of the company. The engineers are individualistic and don't even seem to understand the benefits of working in teams. They rarely talk to their colleagues in manufacturing. Top management has tried for a long time to communicate with the engineers, but has given up since the engineers never seem to say anything in management meetings. At times, top management has seriously discussed outsourcing the entire new product development process, and focusing on financing, production and marketing.

27.4 Electronics Company

The following fragment of a summary of the current situation in an electronics company was published in the first edition of "Product Lifecycle Management: 21st Century Paradigm for Product Realisation".

The response to change

The company has made tremendous efforts to change. It has all the latest CAD, CAE and CAT systems. It spent a lot on ERP systems and on getting JIT working. It invested a lot in new manufacturing facilities. One year it did TQM and another year it did SPC.

In spite of all the investment, there has not been enough improvement in performance for the company to remain competitive. The main problem is that its products are consistently late to market, and some 40 % of projects to develop new products fail.

Looking back at some of its recent initiatives, the company found that when changes were made, they were uncoordinated, project-oriented, non-interrelated, and non-sustaining. For example, one VP would push the idea of strategic IS, while another tried to do TQM and SCM, and someone else did fuzzy logic. One VP wanted to build a "lights-out" factory in Silicon Valley. His successor wanted all production and assembly done in the Philippines.

Initiatives were not brought together. Improvement activities conflicted. And, often, by the time initiatives got down the hierarchy to working engineers they had already been watered down, and since the next initiative was known to be on its way, no-one could be motivated to change their behaviour.

Current situation

A lot of effort and money has gone into attempts to change, but the end result has met no-one's expectations, and some people are very unhappy with the results. Top management has come to the conclusion that product development is an unmanageable black box, a Black Hole with a never-ending appetite for dollars. Management thought it had got to the stage of being able to estimate likely underestimates of new product development costs and cycles until they saw some of the software that was developed to go in the products. 90 % of the time this was more than 3 months late. It was always full of errors, and 40 % of the effort went into fixing the bugs.

Top management believes that developers don't actually understand the business environment and don't want to communicate with anyone else in the company. They seem to be incapable of teamwork.

Product development managers know they waste a lot of time. They know that sometimes they put too much effort on the wrong project and don't get the expected payback for their investment. However, they believe the real problems in the company are caused by top management. Top management responsibilities change frequently. Top managers know they'll change job before initiatives and projects are finished, so they try to get easy short-term success, and leave the long-term hard issues to the next guy. They start something with a bang, and a few months later it disappears without even a whimper.

Product development managers feel that top management is dominated by the bean counters. They believe the financial controller runs the business, putting together plans and budgets in his spreadsheet. They say he does this very well - but there is no link to the customers or the products. And other top managers are so busy looking at the figures he produces they don't have time for customers and products.

Product development managers claim top managers use the wrong measurement systems to judge performance. The main indicator for product development performance is product development headcount. They complain they're rarely involved in decision making.

Product developers claim they are assigned to far too many projects. One was assigned to 15 projects at the same time. So many that he wasn't even sure which projects he was working on. Product developers claim that managers don't define in enough detail what is expected of them in a particular project. Different managers have different expectations, but these are not clearly explained. After many reorganisations there are numerous uncoordinated systems and inconsistent sets of documentation. Product engineers claim it's not surprising that projects fail when it's unclear what the targets are, who should do what, or how it should be done.

27.5 Aerospace Company

The following fragment of a summary of the current situation in an aerospace company was published in the first edition of "Product Lifecycle Management: 21st Century Paradigm for Product Realisation".

The response to change
The company has made tremendous efforts to change. It has been through extensive corporate restructuring activities and has divested some operations. It has started joint ventures and new relationships with new partners. It has tried many new strategies, and is torn between the benefits of focused factory, low-cost, niche, agile, and high-velocity manufacturing. As one of the leading companies in its various markets, it is generally one of the first to develop and use new techniques. It has all the latest CAD, CAE, CAM, aerodynamic and structural analysis systems. It has invested a lot in new plants, introducing new techniques wherever possible. It has invested heavily in TQM, CIM and time-based management. It was heavily involved in CALS activities, was one of the first companies to get involved in GATEC (Government Acquisition Through Electronic Commerce) and CITIS (Contractor Integrated Technical Information Service). It was also a leader in IETMs (Interactive Electronic Technical Manuals).

In spite of all the investment, there has not been much change in performance. Although performance has improved a little, the results are nowhere near as good as expected. Competitors are known to be making much faster progress.

The company has been evaluating its efforts, which have not been so successful, and trying to work out where it went wrong. Looking back it now recognises that the company focus was too far from Engineering. When business conditions were good, top management attention was elsewhere, for example, on Mergers & Acquisitions. Without focus and pressure from top management, Engineering, like other functions, felt no pressure to significantly improve performance. It over-engineered many of the products. Then came the end of the Cold War and the recession, and top management has been so worried about not getting enough work, and wondering which operations to sell off and which to buy from other companies in a similar position, there's been no time to think about productivity improvement.

Without an overall focus, many of the improvement programs that were started have developed a life of their own, and instead of helping to reduce costs, have only increased them. For example, a lot of money was spent on customising the CAD system. This should have been left to the system vendor. Eventually the company decided to change to a system from another vendor, so most of the customising effort was wasted. A lot of money was spent on developing an in-house Product Data Management system. Again this appears to be something that should have been left to the vendor community. The company's mission is to develop aerospace products and services, not to develop software to support product development.

The company has found it very difficult to improve performance within its departmental organisation. Performance improvements are implemented on a departmental basis. Each department is responsible for its own performance, so it does what it can to improve itself.

The result is generally invisible. Even if Marketing could identify which potential customers were going to buy which products on a given day, it wouldn't make much difference. By the time that Engineering has deformed the product specifications, and Manufacturing has made whatever adjustments it deems necessary, and Finance has pushed the price up, the potential customer will already have bought the competitor's product. Even if Engineering buys the most modern CAD technology, it's not going to make much difference. Designing products that customers don't want with a modern CAD system isn't any better than designing products that customers don't want with an old CAD system. More unwanted designs will be produced, creating even more pressure on Manufacturing, and distorting the production plans. Manufacturing's new MRP system would probably be able to handle all the new designs, if only someone knew how it worked with Engineering's new CAD system.

During piecemeal implementation, the departments don't work together. Each does its own thing. The resulting sub-optimisation has little overall effect. Activities involving more than one department are not considered for improvement as it would be impossible to get everyone to agree, so activities like engineering change which involve 16 departments, more than 50 documents, and a 9-month cycle time are not considered for improvement.

Current situation

Top management is concerned that Engineering still seems unable to keep to plan. No sooner is a plan in place than Engineering wants to change it. The different engineering departments seem unable to work together, reports from different departments are often inconsistent, and even when they address the same subject, different departments come up with different answers. There appears to be continual interdepartmental strife, with departments not working together to solve problems. Each has to solve problems from its own viewpoint. They don't share important data (e.g. on customer requirements and competitors) between departments, and don't share reasons for engineering choices with manufacturing engineers.

The engineering function has become very expensive to run, and a major customer for capital investment. In view of its cost, top management is pursuing options to spin it off as a separate company, or to sell it to a competitor. Any increase in its efficiency will have a positive effect on its chances of survival. However, much of the engineering process seems uncontrollable, and engineering management finds it difficult to get productivity up.

Top management has also been looking at setting up an organisation along the lines of Lockheed Martin's Skunk Works or Boeing Phantom Works.

Engineering managers recognise they have frequently missed important deadlines and that some of the big projects have taken too long, for example the one that came in 9 years late. They realise that marketing, engineering and manufacturing processes are changing fast under the influence of new techniques. They know that management processes and organisational structures must change correspondingly. They read about other companies using new approaches to reduce product development time, to reduce batch sizes, to increase quality, and to improve overall productivity of the workforce. When they look at the way their own company is behaving, they see nothing likely to help the company gain or maintain a competitive advantage. They feel they're missing out - but don't know what to do about it.

In spite of top management effort in restructuring, re-engineering and other improvement initiatives, engineering managers feel the real problems are at top management level. They say there are too many people who were once working in government bureaucracies, too many corporate staffers, too many levels of middle managers. They recount countless horror stories highlighting top management's failure to understand the specifics of the business. Although theirs is essentially a long-term business, they say that management is primarily short-term profits-oriented, and unable to define or stick with a long-term view. Because they can't trust top management, they say they always add 15 % to cost estimates, so that when management makes across-the-board cuts they will be able to absorb the cuts and continue with their programs.

The company has made tremendous efforts to change. It has all the latest CAD, CAE and CAT systems. It spent a lot on ERP systems and on getting JIT working. It invested a lot in new manufacturing facilities. One year it did TQM and another year it did SPC.

27.6   Data at the Manufacturing Interface

The fragment in Fig. 27.1 comes from the report of a review of the data received by a company's manufacturing site ("The Site") from the development engineering sites.

Subject	Statement
Accessibility	Data can be difficult to access as there is not a single database for all information. In the extreme case, data must be accessed in five applications, as well as on paper.
Accuracy	Accuracy of data is not known. There is no indication as to whether the information is a rough guess, or 100% accurate.
Creation	The Site creates data using the same applications as the development engineering sites. However, it has its own templates.
Change	The Site may receive data, review and redline it, and send it back. As there is not a common change system across all sites, the next version that The Site receives may take no account of the redline.
Common Definition	Across the sites, there are not common definitions of documents and data elements.
Communication	There are difficulties communicating with design engineering. Frequently, e-mails are not answered. Telephone calls are not always returned promptly.
Compatibility	Data is compatible. The Site uses the same applications (at the same level) as the development engineering sites.
Completeness	Data packages received by The Site are said to be frequently incomplete. A quick review of the most recently received packages showed that about half were incomplete.
Consistency	Data created by the development engineering sites is not consistent. Each has its own approach. Also, with the exception of one development engineering site, data received from each of the other sites is not internally consistent. Different people on these sites use different names, relationships and structures.
Cost	'A single hourly rate is applied to everything done by The Site. There is no differentiation by task.
Security	There is no control over data taken out of The Site. Also, any data can be sent anywhere by e-mail.
Timing	Data received by The Site from the development engineering sites comes in any order and at any time.
Usability	Data received by The Site from the development engineering sites frequently needs to be reformatted before it can be used. A quick review of the most recently received data showed that about 20% needed to be reformatted.
Value	There is no figure for the value of The Site's data.
Version Management	The Site keeps and manages all versions of data that it receives. The development engineering sites use different versioning schemes, complicating The Site's version management activities. For some data elements, it's unclear which version is the Master version.

Fig. 27.1 Review of data at the manufacturing interface

27.7 Product Data Report Table

Figure 27.2 comes from the report of a review of product data throughout the product lifecycle. The subjects to be addressed were grouped into 20 categories. The relative weight (1 = low, 10 = high) of each category was defined. For each category, the current performance (P = Poor, F = Fair, G = Good, V = Very Good, E = Excellent) was determined.

27.8 Current Situation Summary

Figure 27.3 comes from the report of a review of applications, data, processes, people and techniques throughout the product lifecycle.

The criteria reviewed were different for each of these components. In the case of applications, for example, the review addressed the status of the application architecture, the percentage of application needs defined, the percentage of applications implemented, the percentage of application use documented, and the resulting quality.

The relative weight (1 = low, 5 = high) of each criteria was defined. For each criteria, the current situation (1 = Poor, 2 = Fair, 3 = Good, 4 = Very Good, 5 = Excellent) was determined.

Category	Weight	P	F	G	V	E
Data across lifecycle	9			x		
Data archival	3				x	
Data cleanliness	8	x				
Data costs and value	3		x			
Data creation and change	7			x		
Data definition and model	10		x			
Data, digital data	5				x	
Data exchange (internal and external)	6			x		
Data feedback	3	x				
Data, legacy data	4				x	
Data management (applications)	8				x	
Data management (human resources)	5			x		
Data management (processes)	8			x		
Data metrics	4		x			
Data objects completeness	9		x			
Data ownership and responsibilities	10		x			
Data re-use	4		x			
Data security	5			x		
Data use everyday	9			x		
Data users	7		x			

Fig. 27.2 Review of product data

			1	2	3	4	5
Applications	Status of Application Architecture	3			X		
	% application needs defined	2		X			
	% applications implemented	2		X			
	% applications integrated	1	X				
	% application use documented	4				X	
	Quality of applications	3			X		
Data	Status of data model	3			X		
	% data defined	2		X			
	% data integration	2		X			
	% data measured	2		X			
	% data computer-based	4				X	
	Quality of data	2		X			
Processes	Status of Process Architecture	5					X
	% processes detailed	4				X	
	% processes implemented	4				X	
	% processes measured	4				X	
	% processes automated	2		X			
	Quality of processes	4				X	
People	Status of HR Architecture	5					X
	% skills needs known	3			X		
	% hard skills known	5					X
	% soft skills known	2		X			
	% training planned	2		X			
	% training given	2		X			
Techniques	Status of Techniques Architecture	2		X			
	% techniques detailed	2		X			
	% techniques implemented	2		X			
	% techniques integrated	1	X				
	% techniques automated	1	X				
	Quality of techniques	1	X				

Fig. 27.3 Review of PLM components across the lifecycle

Fig. 27.4 Rose diagram of
PLM component ratings

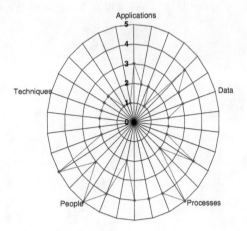

Fig. 27.5 Populations of data creators and viewers across the lifecycle

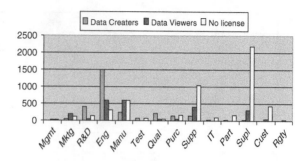

Fig. 27.6 Average time between portfolio entry and product launch

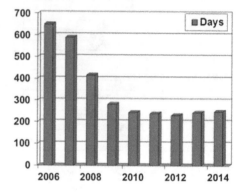

The current situation may also be presented in other ways, such as with a rose diagram (Fig. 27.4), and in histograms (Figs. 27.5 and 27.6).

Chapter 28
Describing the Future Situation

28.1 Some Questions

When a company decides to describe the future situation of the PLM environment, there are usually a lot of questions (Fig. 28.1).

Usually, companies start to describe the future situation after they've described the current situation. In other cases, though, the two activities will run in parallel. And sometimes, the description of the future situation will be carried out independently of the activity of describing the current situation. The order of these two activities depends a lot on the expectations of top management, the circumstances, and the scope of the environment that's addressed.

The description of the future situation will be much more concise than that of the current situation. Since the future situation doesn't exist, it's difficult to describe it in a lot of detail. But that's OK. The intent is only to outline the expected future situation. Main characteristics and desired performance levels can be described, and requirements can be prioritised. However, even though the expected output of this activity is high-level, it can take several months to achieve agreement about the future situation.

We're often asked if the PLM Implementation Strategy should be developed in parallel with the description of the future situation. The logic for developing them together is that it would be a waste of time to take several months to describe a future situation, only to find, when developing the Implementation Strategy, that it's not possible to achieve. However, in our experience, when the two are done together, the future situation isn't defined correctly. Some components and issues are ignored, glossed over or forgotten. Others are excluded because people feel that they won't fit with the Implementation Strategy, even though this hasn't even been defined. In our experience, it's best to first define the future situation. Only when it has been defined correctly and agreed, should development of the Implementation Strategy start.

| how can we know what the future should look like? |
| how can we describe what we don't know? |
| how deep do we go into the details? |
| who does it? |
| how do we do this? |
| what do we report? |
| aren't there many possible future situations? |
| can't we buy a description of the future situation? |
| where do we start? |

Fig. 28.1 Questions about the description of the future situation

28.1.1 Scope and Perimeter

The description of the future situation of the PLM environment, as for the description of the current situation, could include the many subjects in Fig. 28.2.

Usually, however, a project to describe the future situation only involves describing some parts of the situation, or perhaps just one. In theory, it would be best to look at everything. But, as with the current situation, there are often good reasons why this option isn't appropriate. For example, the company may not have the time, or the resources, or the need, to look at all areas.

In our experience, every company looks at something different.

One company we worked with wanted better management visibility into the progress, or otherwise, of product development projects. And they wanted product development time reduced by 15 %. And an effective obsolescence process.

In another company, the PLM project team had been told by top management of the intended growth in revenues for the next 7 years, and were expected to develop and implement a corresponding PLM Strategy. As a result, they wanted to look at the entire scope of PLM.

One customer operates in over 100 countries. Making sure they stay aware of regulations world-wide was a major target. Related issues were to increase local sourcing and to reformulate products to reduce pollution.

Another customer had product developers on many sites working for many OEMs with different CAD/CAM/CAE applications. Top management was looking for a common project management approach across the company. They were hoping to include all development projects in a common portfolio, which would be managed with a common enterprise-wide process supported by a common enterprise-wide information system with a single database. And they wanted a

| enterprise structure |
| product structure |
| product data |
| business processes |
| PLM applications |
| PDM systems |
| techniques |
| facilities & equipment |
| metrics |
| people |

Fig. 28.2 Potential subjects for the future situation

multi-CAD solution enabling them to use resources and skills effectively across multiple projects and sites. And they wanted to improve requirements management, particularly for software development.

Another company, as a result of acquisitions, had several part numbering systems. Their objective was to introduce a common part numbering system across the company.

In another case, a customer wanted to introduce a cross-functional Stage and Gate process and methodology supported by a PLM application. They had several different types of project, including Product Idea Management, New Product Introduction, Technology Development, Product Improvement and Technology Improvement. They wanted all of these project types to have the same overall structure and follow the same overall process, yet include the necessary specifics at the detailed level. One of their targets was to be able to see the status of all projects, and that of the overall Project Portfolio, at any time.

Other targets for the future situation have included increased product traceability, reduced environmental footprint, new product lines, improved feedback about products during middle-of-life, withdrawal of poor-selling products, and increased value at end-of-life.

28.1.2 Project Team

The make-up of the project team depends a lot on the scope and objectives. The team leader for the description of the future situation is often the same person as for the description of the current situation. It's an advantage to have a good understanding of the current situation, and to know how the PLM Initiative has progressed. Often, some of the other team members who participated in the description of the current situation will be in the team. And usually some new faces join the team.

New team members are unlikely to know much about the project. They will need to be trained in the techniques that the team is using, such as process modelling and data modelling. And, to be familiar with the work that has been done, the deliverables that the team has produced should be explained to them. The objectives of the project should be explained to them. They should be informed of the project team's view of PLM. Otherwise they are likely to assume that the team's view of PLM is the same as their individual view.

One or two team members need to build a picture of the overall future situation. Usually, many such pictures, "scenarios", are possible. It takes time, and a lot of information, to build these pictures. In one company we worked with, the project leader was fond of quoting Van Gogh's "first I dream my painting, then I paint my dream", and these team members were referred to as "the dreamers". In other cases, they are referred to as scenarists.

Initially, most of the other team members are involved in activities to create, from various internal and external sources, the information with which the scenarists will build their scenarios of the future situation.

Later, these team members will analyse the scenarios, looking for ways to strengthen them, and for ways to eliminate incoherencies and other weak points.

From the scenarios, an agreed description of the future situation will be developed and documented.

28.1.3   Initial Input Collection

At the beginning of this activity, it's useful to document what's known and what's required. One part of this task addresses the available information, such as the Vision and the description of the current situation. It would be a waste of time and effort to reinvent existing information. Another part addresses the objectives and targets. Are there specific objectives from top management? Or will objectives have to be pieced together from other information? Objectives announced in previous phases of the project may need to be validated. An objective documented during the review of the current situation may have changed since then.

28.1.4   Future Situation Deliverables

One of the dangers when describing the future situation is that people will describe what they understand, and what interests them, but not what's wanted. Describing what they, personally, want is relatively easy. However, it can be a daunting task to describe what others may want, but don't communicate, and perhaps can't even be imagined. To help keep on track, it's useful to define three targets in advance (Fig. 28.3).

Defining the deliverables up-front will help define the approach.

28.1.4.1   Major Deliverables

The deliverables will depend on the scope of the project. As the scope is likely to differ from one company to another, the deliverables are also likely to differ. However, there are some common deliverables (Fig. 28.4) that we've seen in many

| the approach for describing the future situation |
| the information that will be needed |
| the deliverables that will be produced |

Fig. 28.3 Three targets to define in advance

projects. And there will be some specific deliverables (Fig. 28.5) that depend on the company and the project.

The specific deliverables depend on the scope and the methods applied.

28.1.4.2 Reporting the Deliverables

As Fig. 28.6 shows, there are usually two deliverables from the activity of describing the future situation.

28.1.4.3 Report

The report of this activity could be in eight parts (Fig. 28.7).

a description of the scope and the objectives
a description of the methods used to describe the future PLM environment
a description of the project team
a brief overview of the scenarios created and investigated
a brief overview of the expected overall future situation
a brief overview of the expected future situation for each PLM component in scope
a list of expected benefits
a business case model showing how things will work in the future situation
some metrics and their target values

Fig. 28.4 Frequently encountered deliverables

results of analysis of the current situation
input from activities such as benchmarking
a list of people involved
a high-level description of future product structures
a single-page targeted business process diagram
a high-level future information flow diagram
a paragraph describing each major activity proposed for the product lifecycle
flow diagrams for each major future activity
a one-page future data model
examples of reports needed in the product lifecycle
target values for metrics such as time per project, number of new products per year

Fig. 28.5 More specific deliverables

a detailed report
a management presentation

Fig. 28.6 Future situation deliverables

an executive overview
a section describing objectives
a section describing methods used to identify and describe the future situation
a section describing the findings
a section describing the overall picture of the future environment
a section giving details for each of the in-scope components
a section showing how the proposed future situation meets the objectives
an appendix containing additional details

Fig. 28.7 Structure of the report

The report doesn't need to detail all the activities carried out by the team, or to include all the data they've collected. The detailed material can be referenced in the report, and made available if requested.

28.1.4.4 Management Presentation

After the project team has described the future situation and created its report, it should document the main findings in a PowerPoint presentation. The team leader should present this to top management. Usually it will also be presented to functional managers. Some managers will probably want to look at, and discuss, the detailed findings. It may take several weeks to get the results formally agreed, and the next phase of activity started.

28.1.5 Approach

The approaches, or methods, used to describe the future situation will depend on the scope and objectives. Probably some will focus on the company itself. And probably some will look outside the company.

Among the activities focusing on the company itself could be a review of existing project information, an analysis of the current situation, and identification of improvement suggestions from people working in the product lifecycle. Among the activities looking outside the company could be benchmarking, visits to other companies, reviewing maturity models, reading technical literature, and attending conferences.

28.1.6 Plan

The plan for the activity of outlining the future situation will depend on the scope. A project plan (Fig. 28.8) should be created to show the main tasks and milestones.

Activity	1	2	3	4	5	6
Start-up and Training. Review of available data.						
Get information from internal sources						
Get information from external sources						
Review data. Identify and get missing information						
Build scenarios						
Analyse scenarios, identify preferred Strategy and Plan						
Prepare Draft Report						
Prepare Management Presentation						
Present to Management						
Finalise the Report						

Fig. 28.8 The plan for the description of the future situation

Activity	1	2	3	4	5	6
Detail the Future Situation						
Confirm and communicate the Future Situation						
Develop the Implementation Strategy						
Develop the Implementation Plan						

Fig. 28.9 The plan for the next steps

The plan should be agreed with top management before project activities start. It should be kept up to date and visible. Top management, in particular, should know what progress is being made, and be kept aware of any problems that may arise.

It's important to have a plan, even though it's unlikely that activities will occur 100 % according to plan. As Robert Burns put it,

The best laid schemes o' mice an' men
Gang aft agley,
An' lea'e us nought but grief an' pain

Communicating a plan for the next steps in the PLM Initiative (Fig. 28.9) will help management understand how the overall PLM activity is moving forward.

28.2 Internal Input

Among the activities addressing the company itself, rather than other companies, could be a review of existing project information, an analysis of the current situation, and review of improvement suggestions from people working in the product lifecycle.

28.2.1 Existing Information

There'll probably be a lot of information about the future situation in documents such as the Vision, project proposals and annual plans. This information could include lists of planned improvements projects as well as proposals for other changes.

28.2.2 Current Situation Analysis

The description of the current situation will be a source of information for the future situation. A review of the deliverables describing the current situation can be carried out. Objectives of the review could include looking for weak points that can be eliminated, and identifying areas where other improvements can be made.

Depending on the scope of previous activities, the description of the current situation may address several components such as processes, applications, data, and people. In most companies, the report will show many opportunities for potential improvement. There may be some duplication that can be eliminated. Some elements (such as a particular application or process) may be organised differently in different locations. The best organisation could be applied everywhere. Activities may be carried out differently in different locations. Some elements may be missing in some locations. There may be some redundancy and inefficiency in the current lifecycle. Some activities may be illogical, wasteful or in the wrong order. Some administrative procedures may be unnecessary or duplicated. Some information may never be used. People may be doing the same thing in different ways, using multiple versions of the same processes, and multiple applications with similar functionality. There may be different definitions of the same objects and activities. Maybe there's not a standard terminology. Some documents may not be under control. There may be missing information. Some information may be difficult to access. Some processes may be missing. There may not be general agreement about how work should be done. There may be a lack of training. Sometimes, product development and change aren't under control. Gaps, duplication and differences can make it difficult to offer tip-top products, difficult to compare performance across locations, difficult to make worldwide reports, and difficult to move people from one location to another. Training and support may be ineffective because there's a common approach, but people at different locations have different requirements.

28.2.3   Requests from Lifecycle Participants

Another source of information for the future situation is the people working in the product lifecycle. To identify and describe their suggestions, similar techniques can be used as for the description of the current situation. These include questionnaires, interviews and modelling.

28.3   External Input

Among the activities looking outside the company could be review of technical literature, benchmarking, visits to other companies, reviewing maturity models, and attending conferences.

28.3.1   Smart Products

Technical literature, such as magazines and newsletters, is usually full of ideas for the future situation. For example, a lot is written about Smart Products. These are products that can sense and communicate information about their condition and environment. Information about the use of the product can be automatically collected by the product itself. This information can then be transmitted back to the manufacturer and other participants in the product value chain. They can use it in many ways. The information can be transformed into knowledge to support existing products and to help create new high-value products and services. It offers new opportunities to understand the way products are being used and behave.

Smart Product technologies can be used in many industry sectors. For example, a train carriage can send a message to the operator when it needs maintenance, and send a message to the manufacturer to say which parts were over-engineered. At End-of-Life, full information on a car's history will be available, enabling selective component reuse and recycling. During the Middle-of-Life, use of white goods can be optimised to minimise energy loss and pollution. In the Middle-of-Life, key components can be replaced before failure, rather than after failure, avoiding costly out-of-operation time.

28.4   Product Strategy

Since the focus of PLM is the product, another input to the description of the future situation can come from knowledge of other companies' Product Strategies (Fig. 28.10).

28.4.1   Managed Complexity and Change OEM

The strategy of Managed Complexity and Change (MCC) takes an OEM where less competent potential competitors can't follow, leaving them bemused and trailing. This is the typical strategy for an OEM with its roots in high-cost countries (such as the US, Western Europe and Japan) and a desire to provide products worldwide.

Offering complex, frequently updated products through a global capability, the strategy puts the company on a playing field on which few can compete. In this strategy, the OEM will often define major assemblies, then outsource their development and production to Global Complex Assembly Providers.

An OEM with an MCC strategy has to make money for its stakeholders, but it doesn't have to make its product. It can outsource to the best development, manufacturing, sourcing and delivery networks. The OEM focuses on managing its portfolio of products, its product deployment capability, customer requirements,

| Managed Complexity and Change OEM |
| Global Complex Assembly Provider |
| Low-Cost Commodity Supplier |

Fig. 28.10 Examples of other companies' product strategies

product architecture, product specifications, supplier management, system integration, final assembly and customer feedback. The OEM doesn't make commodity parts and assemblies. However, the OEM may develop and produce particularly strategic or complex components, as well as those using new technologies.

28.4.2   Global Complex Assembly Provider

The strategy of a Global Complex Assembly Provider (GCAP) is to provide high-value, complex major assemblies worldwide to OEMs.

To attract OEMs, the GCAP will aim to be a world-leader for particular components, sub-assemblies and assemblies. It will aim to master particular technologies and competencies.

A company with a GCAP strategy will often develop and manufacture some complex components and sub-assemblies, assemble these with parts and sub-assemblies from a Low-Cost Commodity Supplier, and supply the resulting assemblies to OEMs.

A company with this strategy will need to be global to respond to the needs of its customers, the OEMs. It will also need to be financially strong, as OEMs often expect GCAPs to share investment and risk.

28.4.3   Low-Cost Commodity Supplier

The strategy of Low-Cost Commodity Supplier is typical for companies that supply commodity parts and services to Complex Assembly Providers and OEMs. Often these companies operate in low-cost regions.

28.4.4   Product Portfolio and Product Architecture

The strategy of Managed Complexity and Change implies careful definition of the Product Portfolio and the architecture of products. The objective is to be able to launch worldwide, in quick succession, many new products based on a small set of common platforms, modules and interface components.

The basic platforms should change infrequently and be as similar globally as possible. The platform may represent 80 % of the product, the modules 20 %. As much as 90 % of a new product may be the same as for a previous product, and only 10 % different. New modules and low-cost facelifts to existing modules will enable the customer to be presented frequently with a succession of "new" products.

Platforms, modules and interface components need to be designed so that they can be assembled in different ways to give customer-specific products.

The decision criteria for what's in the platform, and what's in the modules and interfaces, depend on the product and market. For example, the platform may be heavy or very large, suggesting local production, whereas modules may be light-weight and easy to air-freight.

Often, product localisation will be achieved with modules. In some cases, the modules may include what the customer sees, while the platform includes what the customer doesn't see. In other cases, the user interface may be in a module that remains unchanged while the platform is upgraded.

Sometimes, installation, maintenance and remanufacturing requirements can affect the product architecture. Specific modules may be developed to enable easy installation and easy maintenance.

Special modules can also be included to protect Intellectual Property and to prevent counterfeiting. Trade secrets may be split between several modules, each of which is produced by a different partner or supplier.

In some industries, the modules may include parts that are fashion-related, seasonal, related to recurring sports events, or changing frequently for another reason.

28.4.5 Beyond the Product

Apart from the product itself, the Product Strategy should also address the product deployment capability. There are many issues about what should be done locally (implying it may be done differently in different locations) and what should be done globally (implying "common to all locations"). Key to success is getting the right balance between centralisation and decentralisation, and between global commonality and local specifics. The point of equilibrium is different for different products and different processes.

Pricing should be in local currencies. Costs need to be compared in a common currency. Product costs should be calculated globally. There may be global recommendations for product pricing, but local product prices need to be based on local market conditions.

Product and process-related regulations need to be gathered and understood locally, but managed globally to achieve synergies.

Applications should be common worldwide, and updated at the same time everywhere.

Business processes for a particular product should be as common as possible. Implementation of processes should start from the top, and work down to the level at which it no longer makes sense to enforce commonality.

To achieve the objective of the basic platform being as similar as possible worldwide, the product definition of the platform is usually carried out by a global team. However, the product definition of the localising modules is better done by local teams with an everyday understanding of local markets. This implies local development activities round the world that are close to customers, even though this may add complexity to the environment. Local teams for marketing and development can be brought together with communications and collaboration technology. As different parts of the world are usually at different parts of the economic cycle, development resources not needed at one location at a particular time can be used to support other locations that do need resources at that time.

Heavy and/or voluminous products and platforms are often manufactured locally to benefit from low-cost manufacturing facilities, and/or to minimise transport problems and/or costs. Some percentage of an assembly or product may need to be produced locally to meet government requirements for product origin.

Light-weight products with high added value, such as software, whisky and wristwatches, may be assembled in one central location, then shipped worldwide for local distribution.

Sales and support are usually best done locally. Local salespeople should have a better understanding of a customer's needs and desires than faraway top executives. Local support teams can provide fast onsite assistance to customers. If they can't solve a particular issue, a more experienced global SWAT (Special Weapons And Tactics) Team can be involved over the Web, or flown in.

Customer requirements need to be gathered locally, but managed globally to achieve synergy. Although managed globally, they need to be available to local teams for new developments and modifications, and to maintain customer relationships.

For a new product, the global aim is often to lower the price compared to the previous product, or to keep it constant while increasing functionality.

For a new product, the global aim is to be able to launch at the same time worldwide. However, the local launch dates should be based on local market conditions.

Whereas a single engineering BOM may be the target for a product worldwide, manufacturing BOMs often need to be plant-specific. Manufacturing equipment, accessories and consumables tend to be different in different plants and different countries.

28.5 Common Steps

Another input for the future situation can be an understanding of the steps taken by companies wanting to develop, sell and support products on a worldwide basis. Their targets include a reduction in costs, and an increase in the number of customers, sales and profits. Quality remains an imperative. They don't expect it will be easy to make the changes necessary to enable global products. They expect it will take a long time. They recognise the need for clear vision, strategy and plans for the next five to ten years. They've recognised the threats and opportunities of the environment, and are taking action to manage their products, to be in control of the products and to reduce the related risks. They're responding proactively. They take the following common steps.

28.5.1 Think Global

The first ground rule for global products is to think global. Aiming for a global product will lead to a different order of magnitude for sales and customers. It's a different ball game, and requires a different mindset.

28.5.2 Understand the Global Market

It's necessary to understand the global market and the customers. The company must offer products that customers in different parts of the world will want and will buy.

28.5.3 Select the Markets

The company must decide in which main geographical markets the product will be available. Until specific markets and products are defined, the discussion of what should be global, and what should be local, tends to be academic.

28.5.4 Product Strategy

Once the competitive battleground has been selected, a corresponding Product Strategy is needed.

28.5.5 Upfront Planning

A lot of upfront planning is needed to define the way that a company will work in the future with its products. It's better to work out first exactly how products will be managed in the future. The alternative of starting projects for individual products without a clear plan is unlikely to succeed.

28.5.6 Prescriptive Approach

A prescriptive approach is required. In other words, the company has to define what must be done. Then everyone has to follow these rules. The opposite approach, "anarchy", is that everyone would decide locally what they want to do. The result would be problems when people in different locations try to work together.

28.5.7   Clear and Common Terminology

A clear and common terminology is needed company-wide. For example, people need to agree on the meaning of words such as product definition, customer requirement, product requirement, and product specification. Another set of words about which people may have a different understanding is version, variant, release, option, model, and revision. These are often used with different meanings, leading to confusion.

28.5.8   Architectures and Models

Architectures and models describing the product and the product deployment capability will be required. In the world of global products, they are needed to describe and communicate how a company's resources are organised.

28.5.9   Digital Product and Digital Manufacturing

Digital models enable faster and lower-cost development, analysis, modification and simulation of products, processes and equipment.

28.6   Benchmarking

Benchmarking can also provide input for the future situation. It can be used to understand how other organisations carry out specific product-related activities, particularly those for which they are believed to be more effective. If there are specific activities that they do better than the company, then these activities could be candidates for improvement in the future situation.

Benchmarking can be applied to any activity. The following example results from a benchmarking activity by a group of companies in the area of Portfolio Management. Portfolio Management includes the activity of sequencing the set of product opportunities available to a company for investment. It identifies the sequence in which the corresponding product development, improvement and phase-out projects should be carried out to provide most value.

28.6.1   Actual Situation

Detailed Portfolio Management reviews were carried out by each company in the benchmarking group. They showed that, with corporate objectives demanding more for less, there was a widespread need to improve the overall return on the investment in product development and improvement projects. There were too many projects relative to the available resources. As a result, many projects were being staffed with fractional resources. The reviews showed a lack of consistency between the evaluation methods used to value different projects. Different managers developed and used different types of business cases, making it difficult to compare different projects and to identify the relative priorities of projects. The result was product development times of many years, and about half of the projects being cancelled before new product commercialisation, even though significant resources had already been invested. It was often found that product developers were good at researching and inventing new technology and ideas. However, they were less good at focusing on projects that would create business value for the company. Product developers would keep poorly performing projects alive long after they should have been killed. These projects continued to eat resources that would have been better used on more valuable opportunities.

28.6.2   Improvement Objective

The companies wanted to be sure that their product development resources were focused on the best set of projects. They wanted to be sure that these projects were aligned with business objectives. After identifying and reviewing best practice companies, the benchmarking group estimated the improvement potential. For one company, this implied reducing project cycle times by about 45 %, and increasing successful project completion rates by about 30 %. Another objective was to be able to compare quantitatively the value of each project in the portfolio, so that decision-makers would have clear visibility of the situation and could make informed decisions.

28.6.3   Action

To improve the situation, it was clear that a Portfolio Management process should be defined, implemented and used for all projects in the Portfolio. A related application would be implemented to support the process.

Among the requirements for the Portfolio Management process was a need to track the value of the overall portfolio and of individual projects. Another was to enable understanding of the upper and lower limits of a project's value. This implies

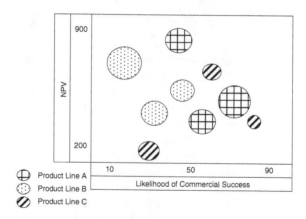

Fig. 28.11 Likelihood of success (x), NPV (y), resource cost (bubble size)

understanding the variables that can lead to different values. Another requirement for the process was a feedback loop to address non-conformance and to derive more benefit in the future from real-life experience.

One requirement for the application was the use of templates to provide a standard basis for understanding, evaluating and comparing projects. Another was the output of standard reports that give executives information such as the degree of portfolio alignment with strategic targets, the expected portfolio return, the mix of short-, medium-, and long-term projects, project risk vs reward, and the mix of risk across all projects in the portfolio (Fig. 28.11).

Other requirements included support for performance metrics. These included the Net Present Value (NPV) of projects launched during a particular time period, cycle time, and the number of projects for which premature termination occurred.

28.6.4 Results

Figure 28.12 shows typical results after implementation of the new process and application.

standardised evaluation of the economic value of projects
a consistent, comparable basis to assess risk and success
improved understanding of risk and uncertainty
increased transparency and credibility of project information
decision-taking based mainly on project value
updating of changes to project value as project conditions changed
early identification of projects failing to achieve the targeted value
a doubling of the value of the projects worked on
a doubling of the number of projects terminating successfully
a doubling of the number of projects terminated early
a 50% increase in the number of projects started
a 40% decrease in project cycle time

Fig. 28.12 Typical results of implementation

28.6.5 Lessons Learned

The benchmarking group found that creating a Product Portfolio requires top management involvement and agreement, cross-functional involvement and enterprise-wide involvement. One of the main benefits of Portfolio Management is to know that the best set of projects has been selected. The Portfolio Management activity must be driven by a clear business strategy. The Portfolio Management process must fit smoothly with other company processes.

For the portfolio management process to be successful, the most important characteristics of development projects must be identified. They must have the same definition everywhere in the company. Otherwise it won't be possible to aggregate projects. All projects should be in the same portfolio, although they may aim to "grow revenues" or "cut costs". Projects of both types may use the same resources and lead to a similar result. Projects need to be categorised. Otherwise, important projects can get lost in the mass of less important projects. And, without categorisation, projects with very different characteristics may be compared, leading to conclusions that make no sense.

A top manager should own the Portfolio Management process. The Portfolio Management process should include activities to provide praise for good performance, and to provide ways to learn from problems. The activity of understanding the upper and lower boundaries of a project's value was found to be beneficial because it encourages identification and understanding of project risks. Explicitly managing project risks helps product developers to increase project value. It's important to develop rules about the criteria for decisions to "continue the project" or to "kill the project" before these decisions are on the verge of being taken.

Although clear definition of the process itself is all-important, a Portfolio Management application is needed to help enforce use of the process. A secure Web-based application can enable both data security and wide data visibility.

Other lessons learned from the benchmarking activity are shown in Fig. 28.13.

28.7 Towards the Future Situation

The many activities addressing internal and external sources of information related to the future situation will create a lot of data. From this data, the scenarists will pick and mix to build and describe the scenarios for the future situation.

| a Portfolio Management process can lead to changes to the organisation chart |
| when implementing new processes and applications, remove previous processes and systems |
| good portfolio decisions can only be taken on the basis of high-quality data |
| standard templates should be introduced for frequently used project documents |
| metrics are needed to track portfolio progress and project progress |

Fig. 28.13 Lessons learned

Each scenario should be analysed in detail, looking at all aspects, including potential costs and benefits. This will help increase understanding of its strengths and weaknesses. Often it's by trying to understand the strengths of one scenario that the weaknesses of other scenarios become apparent.

The strong points of a scenario will be built into other scenarios. Scenarios with weak points will be reinforced. Eventually a preferred scenario for the Future Situation will emerge.

As well as documenting the proposal for the Future Situation, the project team needs to document the major differences between the most likely scenarios. This information should be included in the report.

The description of the future situation of the PLM environment will depend on the scope, but could include, for example, information about components such as business processes, a product data model and PDM systems. The information for each component should be documented.

28.8 PLM Environment Model

A business case model should be prepared to show how the various components and metrics of the PLM environment will fit together in the future situation.

Chapter 29
PLM Implementation Strategy and Plan

29.1 Some Questions

When a company starts to develop the PLM Implementation Strategy and Plan, there are usually a lot of questions (Fig. 29.1).

If the company is following the path of PLM Vision, PLM Strategy and PLM Plan, then, by the time it gets to this stage, it will have a strategy. It will have defined the PLM Strategy. The PLM Strategy shows how PLM resources should be used in the future.

But the PLM Strategy isn't the same strategy as the PLM Implementation Strategy.

The PLM Implementation Strategy shows the activities that have to be carried out to get from the current use of PLM resources to the future use of PLM resources.

The PLM Implementation Strategy is likely to be very different in different companies. Their current situation is different, their future situation is different. The scope of activities considered is likely to be different. And there are many ways to get from the current situation to the future situation. So, it's to be expected that each company will create a different Implementation Strategy. In turn, it's to be expected that each company will create a different Implementation Plan which will be built up of manageable and prioritised sub-projects.

It's best to develop the PLM Implementation Strategy when both the current situation and the future situation have been defined. However, if an appropriate PLM Implementation Strategy can't be found to meet the desired future situation, then it may be necessary to loop back and review the future situation. This can lead to overlap between activities.

We're often asked "should we develop the Implementation Strategy in parallel with the future situation?" In our experience, the two activities should be done in series. First describe the future situation. Then develop the Implementation Strategy.

© Springer International Publishing Switzerland 2016
J. Stark, *Product Lifecycle Management (Volume 2)*,
Decision Engineering, DOI 10.1007/978-3-319-24436-5_29

Fig. 29.1 Questions about
the implementation strategy

what do we have to do?
we know where we want to be. But how do we get there?
there must be many plans. How do we find the best?
when should we make the plan?
in which function should we start implementation?
when do we get the benefits?
how can we plan what we've never done?
don't we already have the strategy?
should we outsource?
should we prototype?
should we go step-by-step or go for a Big Bang?
should we make small improvements or a quantum jump?

The likely result of doing the two in parallel is that the future situation won't be described fully. Instead, before its description is complete, some initial implementation activities will be started to address "high-priority" issues "requiring immediate solution".

In such an exciting fast-moving environment, nobody will have the time or the desire to complete the description of the future situation. However, after a while, it will be realised that an important part of the future situation was forgotten. And the "high-priority, immediate solution" issues won't have been solved because they weren't investigated in enough detail. The plan will be changed quickly to take account of what was forgotten. The changes to the plan won't be communicated properly. Before long, the project will be out of control.

29.1.1   Starting Position

The starting point for the development of the Implementation Strategy and Plan is usually that both the current situation and the future description have been described and documented.

Before starting to develop the Implementation Strategy, it's useful to recap what's available. There should be a lot of foundation information available including objectives and targets, a PLM Vision, a PLM Strategy, a description of the current situation, and a description of the future situation. It's on the basis of these that the Implementation Strategy will be developed.

29.1.2   Scope

The scope of the Implementation Strategy and Plan is usually the same as for the descriptions of the current situation and the future situation. However, something may have changed during those activities, so it's best to review the scope before going further. It could include some or all of the subjects shown in Fig. 29.2. If it appears that there's been a change of scope, then it's best to review this with top management before starting to define the Implementation Strategy.

Fig. 29.2 Resources
potentially addressed by the
implementation strategy

| enterprise structure |
| product structure |
| product data |
| business processes |
| PLM applications |
| PDM systems |
| techniques |
| facilities and equipment |
| metrics |
| people |

29.1.3 Approach

Many activities take place as the Implementation Strategy and Plan are developed. Depending on the situation, some of them may be carried out in series, some in parallel, and some may be repeated several times.

29.1.3.1 Basic Understanding

To be able to develop the Implementation Strategy, it's necessary to have a basic understanding of both the current situation and the future situation.

People who participated in the activities of describing the current situation and the future situation will have that understanding. People who didn't, will need to be brought up to speed. New team members need to be trained in the techniques that the team is using. And, to be familiar with the work that has been done, they should be taken through the deliverables that the team has produced. The objectives of the project should be explained to them.

29.1.3.2 Other Inputs

In addition to information related to the current situation and the future situation, special requirements from top management may have to be considered in the strategy development process. Another input that should be taken into account is the need for a Change Strategy. A PLM Initiative usually results in many changes being proposed. However, it's not easy to successfully bring about many changes. Change is difficult, time-consuming and costly. There are three important tools of change. They are communication, learning and reward systems. They are often ignored in a PLM Initiative but, without them, the desired change is unlikely to occur.

29.1.3.3 Gap Analysis

A starting place for developing the Implementation Strategy is to understand the gap between the current situation and the future situation. It will be helpful if the future situation was outlined using the same structure as that used for the documentation of the current situation. That will make it relatively easy to compare what is needed with what exists. For example, the description of the future situation may call for a single PDM system, but there may be multiple PDM systems in the current situation. There may be no obsolescence process in the current situation, but an obsolescence process may be required in the future situation.

The gaps should be listed and described in a Gap Description Matrix (Fig. 1.47).

29.1.3.4   Top-Down and Bottom-up

The gaps need to be addressed both top-down and bottom-up to get to the Implementation Strategy. Both approaches are needed to anchor the strategy in reality.

29.1.3.5   Top-Down

With the gaps between the current situation and the future situation identified, the top-down approach aims to describe, in just a few words, how the future situation can be achieved. This is a broad-brush approach aiming to identify main groups of actions.

29.1.3.6   Gap Elimination Proposals

With the gaps between the current situation and the future situation identified, the bottom-up approach approach looks for ways to close them. Several ways should be proposed to eliminate each gap. They should be described in a Gap Elimination Matrix (Fig. 1.48). Each activity should be described, clarifying its scope, the constituent tasks, the resources it needs, its deliverables, the gap(s) it eliminates, and its relationships with other activities. Often, it's more effective to close a group of gaps than to close them one by one. Gaps should be grouped, and ways proposed to address each group. The different ways to eliminate gaps, and groups of gaps, should be examined. Proposals for corresponding projects can be made.

29.1.3.7   Project Proposals to Scenarios

Once the project team has completed the above steps, it should be in a position to identify potential solutions, or scenarios. There's an infinite number of strategies, and the likelihood of finding the most appropriate at the first attempt is low. It will be best to identify several potential scenarios. They should show different high-level ways to reach the overall future situation and different ways to reach the future situation for each component. Each scenario should be described in detail.

The strengths and weaknesses of each scenario should be described. This helps get an in-depth understanding of each proposed solution. Often it's by trying to understand the strengths of one scenario that the weaknesses of other scenarios become apparent.

The project team should consider all aspects in the scenarios including costs (Fig. 29.3), benefits and risks.

29.1.3.8 The Proposed Strategy

Analysis of the scenarios leads to a preferred Strategy. This should be documented in the report.

The report could be in six parts (Fig. 29.4).

29.1.3.9 Detailing the Implementation Plan

The Implementation Plan should address the long term and the short term. For the long term, it provides management with the information necessary to understand activities, resources and timelines. The more specific the plan is, the better. It should define an overall implementation timetable. It should show how the PLM implementation will be split into manageable phases. The plan may be a phased plan, for example with Phase 1, Phase 2 and Phase 3. Or the intention may be to have a PLM Program, with several projects. Or the intention may be to have a PLM Project, with many sub-projects.

Each project in the plan should be detailed (Fig. 29.5).

| purchase or lease of equipment, facilities and applications |
| maintenance, installation, and expansion |
| planning |
| training |
| development of procedures and documentation |
| standardisation |
| facility and system management and support |
| security |
| management roles and responsibilities |
| work methods |
| potential re-assignment of roles and responsibilities |

Fig. 29.3 Examples of costs to be considered for each scenario

| an executive overview |
| a section describing objectives and approach |
| a section describing the scenarios examined |
| a section describing the proposed Implementation Strategy |
| a section describing the proposed Implementation Plan |
| an appendix containing additional details |

Fig. 29.4 Contents of the report describing the implementation strategy

Fig. 29.5 Information
needed about each project in
the plan

purpose
scope
objectives
timing
investments and other costs
relationships with other projects
priorities
expected benefits
participants
other resources

The short-term plan should show management which actions need to be taken initially. The plan is more likely to be accepted if it includes some actions that will lead to short-term savings and other short-term benefits.

29.1.4   Project Team

The make-up of the project team depends a lot on the scope and objectives. The team leader for this activity is often the same person as for the descriptions of the current and future situations. It's an advantage for the team leader to have a good understanding of the current situation and the future situation. Often, some of the other team members who participated in the descriptions of the current and future situations will be in the team.

New team members should be trained in the techniques that the team is using. The deliverables that the team has produced, and the objectives of the project, should be explained to them.

The scenarists who developed the scenarios for the future situation may also be in the team. They can now be involved in developing scenarios for the Implementation Strategy. A project planner should be included in the team, and will be involved in developing the Implementation Plan.

29.1.5   Deliverables

The activity of defining the Implementation Strategy and Plan usually results in a detailed report and a management presentation.

29.1.5.1   Reporting the Deliverables

The report will describe the strategy. It will include plans at different levels of detail and addressing different time periods.

29.1.5.2 Report

The project team will produce a lot of paper during this activity. This will include a description of gaps identified, analysis of gaps, identification of scenarios, analysis of scenarios and suggestions for plans. Only some of this needs to be included in the report. It should be distilled into a form in which management can easily understand it. It should be easy to see what is going to be done, and how it will result in the objectives being met.

29.1.5.3 Management Presentation

After the project team has developed the Implementation Strategy and Plan, and created its report, it should document the main findings in a PowerPoint presentation for top management. Probably this presentation will also be made to functional managers. Some managers may want to look at the detailed findings. It may take a few weeks to get the results formally agreed.

29.1.6 Plan

The plan to develop the Implementation Strategy and Plan will depend on the scope. A project plan should be drawn up showing the major tasks and milestones (Fig. 29.6). The plan should be agreed with top management before project activities start. It should be kept up to date and visible. Top management should be informed of progress, and kept aware of any problems that may arise.

Communicating a plan (Fig. 29.7) for the next steps in the PLM Initiative will help management understand how the overall PLM activity is moving forward.

Activity	1	2	3	4	5	6	7	8
Provide training as required								
Review current and future situations								
Identify and analyse Gaps								
Make Gap Elimination Proposals								
Create Project Proposals								
Build scenarios								
Analyse scenarios, identify preferred Strategy and Plan								
Prepare Draft Report								
Prepare Management Presentation								
Present to Management								
Finalise the Report								

Fig. 29.6 The plan for developing the implementation strategy and plan

Activity	1	2	3	4	5	6	7	8
Develop the Implementation Strategy								
Develop the Plans								
Communicate and Confirm								
Launch initial Implementation Activities								

Fig. 29.7 The plan for next steps

in which order should components be addressed?
what size should proposed activities be?
is a prototype appropriate?
how long should a project be?
how do we choose initial activities for the first phase of implementation?
in which part of the company should initial activities take place?

Fig. 29.8 Frequent questions about implementation

29.2   Influencing Factors

In most of the companies that we have worked with to develop the PLM Implementation Strategy, there have been similar questions about implementation issues (Fig. 29.8).

29.2.1   Order of Components

Questions are often asked about the order in which a company should address PLM components such as processes, data and applications. The answer depends on the situation in the company. Sometimes it will be appropriate to address processes first, sometimes data, sometimes products. In a start-up situation, the product data, which identifies the product, needs to be identified first. Then the process to create it can be defined. In a mature company, though, with the product data defined, it's likely to be the process which needs to be addressed first with a view to improvement. Probably there will have been many uncoordinated changes and additions to the process over the years, and the potential for improvement is likely to be high. In a start-up situation, after the product data and process have been defined, then an application can be selected to support the process. In a mature company, with the product data and process defined, the functionality of a new application could enable improvement of the process.

In general, a balanced approach, depending on the current situation, is best. For example, some work should be done on processes, and then some work should be done on product data. Then a pause to review the results. Then some more process work, and then some more work on product data. The two will be implemented side-by-side in the effort to achieve the objectives.

29.2.2   Prototype

Sometimes, it's clear what needs to be done to go from the current situation to the future situation, but the company doesn't have enough experience to know in which order, or how, to do this. In such cases, a well-planned prototype can be a useful tool for increasing understanding, and learning enough to be able to take better decisions.

Prototyping should be a short-term activity focused on producing useful results. However, it can appear expensive if it involves significant up-front expenditure in training and other organisational activities, and longer-term benefits are excluded from a cost-benefit calculation.

Care needs to be taken to ensure the prototype takes place in conditions that represent normal use. It needs to be well-planned. The objective and deliverables must be clearly defined before the prototype starts. Care needs to be taken to ensure that it isn't hijacked by a few individuals with a special agenda.

The experience gained from the prototype must be communicated regularly, otherwise it will be of little value.

29.2.3   Bite Size

Companies can choose to move forward towards the future situation with a step-by-step approach or a Big Bang.

A step-by-step approach moves the company forward from the current situation in clearly-understood increments. It has the advantage that each step is so clearly understood that it should succeed. And, as each step is implemented, people can see and appreciate the progress. Should any problems occur, they can be resolved before they get out of hand. On the other hand, the approach runs the risk that the gains due to many small steps will never add up to a significant benefit. And that the project will be stopped before all steps are completed. Or will be stopped even before all steps are started.

A Big Bang approach aims to implement all the changes at the same time. It has the advantage that, by addressing all changes at the same time, a great improvement can be made quickly. On the other hand, failure will lead quickly to a great disaster. Although a Big Bang approach may appear quick and simple, the result is generally the opposite. A successful Big Bang only comes after a very long preparation and planning phase, during which many people lose interest and motivation because of the lack of results.

29.2.4   Starting Activities

The decision as to where the initial implementation of PLM should take place will vary from one company to another. In some companies, one function is seen as a strategically important function, and able to act fairly independently. Initial activities could be targeted on this function.

In other cases, the decision can depend on the size of the gap between the current situation and the future situation. It can also depend on previous experience with similar projects.

Starting the project with a short, simple activity that will soon show success can lead to increased project acceptance. It's helpful to be able to show successful initial activities in a project.

29.2.5   Simple High-Level Message

Whatever the details of the Implementation Strategy and Plan, it's always good to be able to communicate a simple high-level message.

One company we worked with called its initiative CHAIFA (Commonise, Harmonise, Align, Integrate, Fill, Add). This communicated the six main elements of the strategy and their relative priority. Achieving a common approach across the company, wherever possible, had the highest priority. CHAIFA was the concise high-level message. At lower levels there was a mass of detail about projects addressing many PLM components at many sites.

Another company focused on four main elements (Fig. 29.9).

Activities addressing the focus on value included partnering with risk-sharing assembly providers, and outsourcing or off-shoring of low-value activities. Unification activities included integrating independent national company organisations, and integrating unconnected applications. An example of harmonisation was to implement the same version of a CAD application on all sites. Alignment included adjusting workflows in an application so that they were in line with the steps defined in the corresponding business process.

Fig. 29.9 Four key deployment elements

Value
Unification
Harmonisation
Alignment

Fig. 29.10 Elements
frequently missing in 2015

| Global Product Roll-out |
| Compliance Management |
| Intellectual Property Management |
| Global Requirements Management |
| Product Portfolio Management |
| Innovation Management |

29.2.6 Extending Deployment Capability

Often, the first step when extending the deployment capability is to include elements that are missing in the company, but exist and are in use in similar companies. The elements that are missing evolve over time. Figure 29.10 shows examples of elements frequently missing in 2015.

The second step when extending the deployment capability is to include new elements as they become available. Typically, such elements appear on the market at least five years before they achieve industrial strength. Technology Watch and Road-mapping help to identify them and to track their emergence.

Chapter 30
PLM Action

30.1 Some Questions

When a company starts to implement the PLM Implementation Strategy, there are usually a lot of questions (Fig. 30.1).

If the company is following the path of PLM Vision, PLM Strategy and PLM Plan, then, by the time it gets to this stage, it will have an Implementation Strategy and an Implementation Plan. The latter will show the planned activities, their sequence and their timing.

30.1.1 Starting Position

At the beginning of the implementation activity, it's useful to check that all the deliverables expected to exist at this time are available. If not, corresponding activities to create them may need to be planned. It's also useful to review and confirm whatever plans are in place. This is because the situation may have changed in some way since the plans were made. For example, the availability of people may have changed. Or, other events in the company may have impacted some aspect of the PLM project. As a result, before starting detailed implementation activities, it's best to review the situation with the project's sponsors. If the plan no longer makes sense, then, with top management guidance, it should be modified. Perhaps the scope of some activities, or the order of activities, may need to be changed.

© Springer International Publishing Switzerland 2016
J. Stark, *Product Lifecycle Management (Volume 2)*,
Decision Engineering, DOI 10.1007/978-3-319-24436-5_30

Fig. 30.1 Questions about
implementation

what do we do now?
where do we start?
when do we start?
do we have enough funding?
who manages this?
who does all the work?
how do we involve people outside the PLM Team?
does this project ever end?
how do we organise this?
how do we know if we're making progress?

30.1.2   Recap

It's useful to review the project documents that are available at this stage. There
should be a lot of information about the project including objectives and targets.
There could be information about PLM Vision and PLM Strategy. There should be
a plan showing detailed sub-projects. There could be descriptions of the current
situation and the future situation. This information shouldn't be forgotten. Much of
it will be useful for implementation activities.

30.1.3   Differences Between Companies

In our experience, the detailed actions in the Implementation Plan differ significantly from one company to another. Perhaps this is to be expected, as companies
are in such different situations. The scope of activities considered in their PLM
projects is often different. Their current situations are different, their future situations are different. And, even if they were the same, it's likely that different
companies would have created different Implementation Plans with different
activities and different timings. There are many ways to get from the current situation to the future situation.

However, although, in their details, the activities that are planned to take place
will be different in different companies, there are also likely to be some similarities.

30.1.4   Roles, Responsibility, Involvement

Up to this stage of a PLM project, most of the work will usually have been done by
PLM Team members. However, in the implementation phase, an increasing proportion of work and responsibility should be taken by the managers and users who
work in the product lifecycle and will be impacted by the implementation.

Nevertheless, the implementation phase of the PLM project should be run under
the overall authority of the project team. The team should retain responsibility for

key issues such as the implementation plan and budget, process definitions, data models, departmental reorganisation, and training.

There are often issues around the involvement, in the PLM Initiative, of people who work with the product at some stage of the lifecycle, but aren't in the PLM Team. Their input will be needed because they are the only ones who know the details of what happens in their part of the product lifecycle. However, their managers may claim that these people are working full-time on "everyday activities", and have no time to work on activities for the PLM Initiative. If this happens, top management involvement is often necessary to avoid the Initiative grinding to a halt.

One of the major responsibilities of the project team at this stage is to make sure that the implementation takes place within the agreed budget and time limits. The project team is also responsible for ensuring close integration between organisational and technological activities. This is a testing time for the project team. Everything they have done so far will be to no avail unless real benefits are now produced for the company and for people in the lifecycle.

The team members may feel submerged under a huge load of detailed everyday tasks. However, they mustn't forget that the objective of a PLM project is to increase company revenues and portfolio value. Until a project can demonstrate it's done this, it hasn't succeeded, and it's certainly not finished.

30.1.5 Project Team and Others

The team leader for the implementation activity is often the same person as for the descriptions of the current and future situations, and for the development of the Implementation Plan. It's an advantage in this phase for the team leader to have a good understanding of the current situation and the future situation.

The make-up of the project team depends a lot on the scope and objectives of implementation. Often, some of the team members who participated in the descriptions of the current and future situations will be in the team.

At implementation time, the workload usually increases significantly, and additional team members are needed. New team members should be trained in the techniques that the team is using. The deliverables that the team has produced should be explained to them. The objectives of the project should be explained to them.

People who participate in the project, but are not in the PLM Team, also need to be trained in the techniques used in the Initiative. Otherwise, it may be difficult for them to participate effectively.

30.1.6   Actions and Reports

Associated to the planned activities should be clearly defined targets and useful, measurable deliverables. Each activity should result in a detailed report which may be presented to the project's sponsors and other members of top management. It's important to define the deliverables in detail before starting the activity. This will help avoid the activity going off course. It will also make it easier to know if the activity has reached the required target.

30.1.7   Reporting Progress

Progress needs to be distilled into a concise form so that everyone involved in the project can understand it quickly. Management needs to see what progress has been made, what's going to be done next, and how this will result in achievement of the objectives.

30.1.8   Balanced PLM Action

The many components of PLM (Fig. 30.2) need to be addressed together. If only one is addressed, then the initiative becomes unbalanced.

For example, a PDM system can only be addressed for a short time before questions arise about the definition, structure and status of the product data that the application will manage. And questions will then arise about the processes and activities to create and change this data. And these questions will lead to other questions concerning changes to the organisational structure to enable the most effective application of the processes.

PLM stretches across many functions. It affects many people. If the project is to succeed, the organisational impact will be wide. A clear understanding of the organisational issues is needed. These range from basic issues such as training, through the development of working procedures, to standards and policies addressing the use of PLM, and major functional reorganisation.

Fig. 30.2 Components of PLM

| enterprise structure |
| product structure |
| product data |
| business processes |
| PLM applications |
| PDM systems |
| techniques |
| facilities and equipment |
| metrics |
| people |

30.1.9 Plan

Due to the wide scope and many components of PLM, the implementation is likely to be carried out in phases. As a result, there's usually a set of plans addressing different time periods.

A top-level phase plan should show the entire project over its expected duration (Fig. 30.3).

Another plan will show the major tasks and milestones of the current phase (Fig. 30.4). The plan should be confirmed with top management before project activities start. It should be kept up to date and visible. Top management should be kept aware of progress, and of any problems that may arise.

Creating and communicating a plan (Fig. 30.5) for the next steps in the PLM Initiative will help management understand how the overall PLM activity is moving forward.

Phase Activity	Y1	Y2	Y3	Y4	Y5	Y6	Y7
Prepare Phase 1							
Execute Phase 1 activities							
Prepare Phase 2							
Execute Phase 2 activities							
Prepare Phase 3							
Execute Phase 3 activities							
Prepare Phase 4							
Execute Phase 4 activities							
Prepare Phase 5							
Execute Phase 5 activities							

Fig. 30.3 The planned phases of the project

Activity	M1	M2	M3	M4	M5	M6	M7	M8
Detail the plan for Phase 1 activities								
Manage the Phase 1 activities								
Carry out activities related to product structure								
Carry out activities related to processes								
Carry out activities related to product data								
Carry out activities related to PDM								
Carry out Portfolio Management activities								
Finalise deliverables. Prepare report								
Report Phase 1 activities								

Fig. 30.4 The plan for the first phase

Activity	Q1	Q2	Q3	Q4	Q5	Q6
Execute Phase 1 activities						
Describe Phase 2 activities in detail						
Develop the plan for Phase 2						
Agree and communicate the plan for Phase 2						
Execute Phase 2 activities						
Describe Phase 3 activities in detail						

Fig. 30.5 A plan for next steps

30.2   Forewarned Is Forearmed

Because PLM initiatives are important, it's sometimes assumed that once they're started they'll succeed. However, more than 50 % of them fail. All sorts of things can go wrong. Key people leave. Important decisions are taken later than expected. Unexpected roadblocks and bottlenecks appear. People don't deliver what they promised.

Understanding the factors that affect the success of the Initiative helps avoid project failure.

30.2.1   Looming Failure

PLM initiatives fail for all sorts of reasons. Many overrun, many don't meet business objectives. A survey among PLM users and potential users showed a variety of reasons for lack of success (Fig. 30.6).

But in spite of the potential risks and pitfalls, implementation of an effective PLM activity will eventually take place. Forewarned and forearmed, pitfalls can be avoided.

failing to start the initiative properly. The initiative didn't have an agreed objective, an agreed scope or agreed funding. It started, and soon stopped
failing to define needs correctly. To avoid "paralysis by analysis", needs of people working in the lifecycle weren't investigated
failure to provide promised resources, with the result that key activities weren't carried out, and the project didn't progress
failure to train participants in the Initiative. Not knowing how to work, they failed
failure to clarify ROI. As the ROI wasn't clear, the project wasn't launched properly
failure to understand how the environment should change. There was a belief that new applications would automatically provide the required results. They didn't
failing to maintain focus. The initial focus changed to implementing an application. It was implemented. There were no real benefits. But lots of unforeseen costs and problems
failing to keep track of progress. Everything in the project was said to be fine. Until the project collapsed. By which time hundreds of thousands of dollars had been wasted
failure to accept an honest ROI calculation showing a low ROI. The expected benefits were exaggerated until an acceptable ROI was found. But, unrealistic, it couldn't be achieved
failing to get commitment from top management. They let the project start, but stopped it when it didn't lead to rapid improvement
failing to involve middle management. The initiative faded away when department managers felt that the results of the project would cause them to lose power
failing to involve all participants. R&D chose a particular application. Other departments refused to use it. They hadn't been consulted about the choice
believing in miracles. Management hired a management guru for the initiative. Once the contract was signed, the guru disappeared, replaced by inexperienced MBA students
failing to clarify responsibilities. Management hired a consultancy for the initiative. Roles and responsibilities were unclear. Key activities weren't assigned. The project flopped
believing in miracles. Management was assured by an application vendor that all needed functionality would be in the next release. It wasn't, and the initiative targets weren't met
failing to involve all participants. Decisions were taken by four self-proclaimed specialists. When the solution was implemented, it was rejected by people who hadn't been consulted

Fig. 30.6 Some reasons for failure of a PLM initiative

30.3   Recommendations for Project Managers

The environment in which a PLM Initiative takes place is usually complex and difficult to manage. Many projects fail. Many projects swing between elation and despair (Fig. 30.7).

In a PLM project, there are always many issues to address. There's always something new happening. There's always something changing. There's always something happening that could result in failure to meet the plan.

30.3.1   Wait and Think

There are so many issues for the PLM Project Manager to address, that "wait" and "think" are among the best recommendations for success. Before rushing off to fight the first fire that shows up, PLM Project Managers should pause. They should make sure they really know where they are, and where they're going. They'll never succeed if they don't start from the right place. And, they won't succeed if they don't know where they're going. They can test their readiness with the questions in Fig. 30.8.

PLM Project Managers should be able to answer "Yes" to every question in Fig. 30.8. If they can't, they should take some time to plan what's really needed for the next few months. That's more important than trying to solve whatever appears

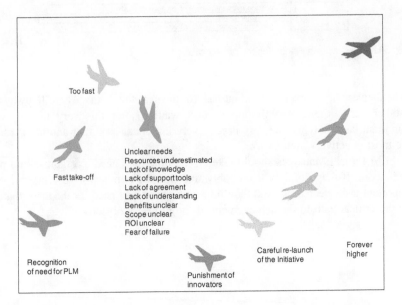

Fig. 30.7 Ups and downs of a PLM initiative

Objectives	
	are they clear?
	are they documented?
	is there consensus about the objectives?
Targets	
	are there clear quantitative targets to aim for this year ?
	are there approximate quantitative targets to aim for next year ?
Plan	
	is it documented?
	is it agreed?
	has it been communicated?
	is it a plan for several years?
Tasks	
	are they clearly defined?
	are you sure they aren't open to misinterpretation?
	are you sure no-one has redefined them without telling you?
	are you sure the tasks don't conflict with other people's tasks?
	are you sure there are no unbudgeted tasks?
People	
	are the people assigned to the project full-time?
	are you sure team members don't have other things to do?
	will team members work hard?
	are you sure team members won't back off when the going gets tough?
Dependencies	
	can you really rely on the people you depend on for success?
	are you sure people will give you the resources they've promised?
	are you sure you're not dependent on the success of other projects?
Leaders	
	are you sure that support of project sponsors won't waver?
	do your sponsors have enough power for you to win through?
	are you sure you're not getting conflicting signals from management?
	are you sure that management expectations aren't too high?
Budget	
	is your budget clear and agreed?
	is the money really there?
	do you have enough money in case something goes wrong?
Scope	
	is it clear which products are in the scope of the project?
	has the scope of the processes to be managed been clearly defined?
	are you sure you can handle the scope of processes?
	has the scope of the product data to be managed been clearly defined?
	are you sure you can handle the scope of product data?
Training	
	is everyone sufficiently aware of the need for PLM?
	has everyone had enough training?
	are you sure that no-one will cut your training budget?

Fig. 30.8 Readiness questions for the project manager

to be the very urgent, very important task for the next few hours. There'll always be another very urgent, very important task waiting after this very urgent, very important task. Solving one very urgent, very important task after another is rarely the most effective way to work.

PLM Project Managers should take all the time they need to organise and plan their work. If they do, they'll probably find that the number of very urgent, very important tasks decreases. And they'll have the time they need for the tasks that are on the critical path to the achievement of the project's objectives.

define the business benefits of the PLM project very clearly
define the financial objectives of the PLM project
put the focus on achieving the benefits. Keep the focus
include people from all parts of the product lifecycle in the project team
select a project leader who can handle the cross-functional aspects of PLM
commit the project leader to the project for the long term
judge the project leader and the project team on the business results of PLM

Fig. 30.9 Guidelines for success

30.3.2 Focus on Benefits

A PLM implementation project can only be considered as successful if it meets its objectives. Achieving the objectives may take a long time. During that time, unexpected events and special interests will tend to drive it off course.

To maintain the right PLM focus over the long term, it's helpful to follow the guidelines in Fig. 30.9.

The objectives of the PLM Initiative need to be expressed in financial terms. Build the project team around people in the product lifecycle who are looking to improve product development, management and support performance. They'll want PLM to perform well because their personal performance will depend on its performance. Focus the project on results of PLM, not on secondary or tertiary issues, such as application selection. Don't let people think that the project ends as soon as applications have been chosen. Make sure they're aiming for productive use of PLM and the resulting financial benefits.

Initiatives in areas like PLM take a heavy toll. Managing an initiative to change a cross-functional environment isn't easy. In addition to good advice about the project, some personal advice will also be useful (Fig. 30.10).

30.3.3 Customer Care

The scope of a PLM Initiative is so wide that it will come to the attention of people throughout the company. Although there will be a single "official" description of the objectives of the Initiative, it's likely that many individuals will have their own unspoken targets (Fig. 30.11).

be proactive. Sometimes you'll be wrong, but on average you'll succeed 7 times out of 10. If you're not proactive you'll never succeed
decide what you believe, and go for it. Don't waste your time hesitating and shilly-shallying
focus on activities linked to business objectives. If you succeed you'll be rewarded. If you fail, you'll have learned something useful for your next job
aim for business performance improvement, not application implementation
don't waste your time criticising people around you. It doesn't add value for you or the initiative
don't work on the same subjects as people 10 years younger. They do it cheaper and better
don't work on the same subjects as people in low-cost countries. They do it for less
don't be an information receiver and re-transmitter. Be an information creator and transmitter

Fig. 30.10 Survival rules

CEO	increase shareholder value. Clarify responsibility for products
CFO	clarify financial performance of products. Reduce risks and waste
CIO	provide the best IS support to PLM, a major business activity
CPO	clarify the situation of all products and parts. Improve portfolio value
Product Manager	clarify the situation of products. Improve profitability. Reduce risk
CAD/PDM Manager	clarify tasks, roles and responsibilities

Fig. 30.11 Potential targets of key individuals

CEO	launch the PLM Initiative. Ensure support by all managers
CFO	ensure the Initiative is funded. Participate in the Initiative
CIO	train, and provide, staff for PLM and the PLM Initiative
CPO	provide staff and input to the PLM Initiative
Product Manager	support the PLM Initiative. Clarify position of products
CAD/PDM Manager	support and participate in the PLM Initiative

Fig. 30.12 Involvement from key individuals

Understanding the targets of each individual helps understand how they can best participate in, and support, the PLM Initiative (Fig. 30.12). Often, they will be very helpful once they understand how PLM will help them to achieve their targets.

30.3.4   Do's and Don'ts

Finally, in this section of recommendations for project managers, Fig. 30.13 shows some do's and don'ts.

30.4   Actions

The following sections look in detail at some frequent actions related to products, processes, product data and portfolio management. In our experience, the activities in these areas tend to be similar in many companies.

On the other hand, activities in other areas, such as applications, techniques, equipment, human resources and organisational issues, tend to be more specific to each company. Activities in these areas are briefly addressed below.

30.4.1   Company-Specific Actions

In the area of PLM applications, we see a range of activities. They range from high-level activities such as the definition of an overall IS architecture for PLM, to low-level activities focused on details of one particular application.

Techniques are often specific to a particular industry sector, or to a particular position in the supply chain. In many companies, there's a strong focus on one

Do take account of organisational and cultural issues	Don't believe that buying PLM applications will automatically lead to successful PLM performance
Do remember that enterprise-wide introduction of PLM is lengthy, costly and cross-functional	Don't expect a low-cost local solution to give the same results as an enterprise-wide solution
Do run PLM as a cross-functional, high-level activity reporting directly to top management	Don't expect a few engineers and programmers to find a good solution without management input
Do take the time to understand the holistic character of PLM	Don't try to split off and solve one component of PLM before understanding all components
Do start by trying to understand the business objectives and how they can be achieved	Don't start by trying to model all the product data flows
Do consider the many forms of product data	Don't only address CAD data, or BOMs, or menus, or installation manuals
Do take the opportunity to improve the product workflow	Don't automate activities that add no value
Do look outwards and take account of customers, suppliers and competitors	Don't just look inwards and focus on internal activities. Don't take a NIH attitude
Do remember that different people have different objectives for PLM	Don't believe that everyone wants PLM for the same reason
Do remember the value of product data as a corporate asset. Manage it as other assets	Don't focus on reducing the cost of PLM applications. Focus on increasing portfolio value
Do remember that only top management has the power to modify the organisational structure	Don't imagine you can do it alone, bottom-up, from a single functional area
Do involve top managers	Don't only talk to PLM Team members
Do remember the need for training and education	Don't think training is not needed because a PLM application looks costly
Do implement by steps, within an overall plan	Don't try to do everything in one Big Bang, mega-project
Do work hard to understand the lifecycle processes	Don't think you can manage product data without understanding the processes where it's used
Do involve people at all levels, and from all lifecycle functions, from the beginning	Don't try to go it alone. If you do, it won't be long before your project is stopped
Do make one person responsible, really responsible, for the PLM initiative	Don't try to share the responsibility, or give it to a committee
Do remember that there is a high risk associated with a PLM initiative	Don't assume a PLM Initiative is easy
Do remember the only constant is change. Be prepared for it	Don't expect everything and everybody to stand still while you look for a solution
Do ensure a focus on areas of maximum value-added, wherever they may be in the lifecycle	Don't focus on what looks easiest. It may not lead to increased value
Do ensure people have the time they need to learn the skills they need for the Initiative	Don't assume that everybody knows everything about PLM
Do ensure there is a common understanding of PLM, the problems and the opportunities	Don't let everybody pull the Initiative in different directions

Fig. 30.13 Do's and Don'ts

technique. However, we don't see many companies focused on the same technique. Each company seems to have its own preferred technique, or techniques. And, only very occasionally, do we see an effort to integrate and align several techniques.

Often, facilities and equipment aren't considered to be part of a PLM Initiative. When they are included in the Initiative, activities are often focused on acquisition of new equipment for a particular phase of the lifecycle.

In many PLM Initiatives, organisational issues aren't considered in the initial plan. However, as the project evolves, it becomes apparent that organisational changes will be necessary to achieve more effective PLM. In some cases, the resulting approach to the required activities is bottom-up. In others, it's top-down.

Similarly, in many PLM Initiatives, human resource issues aren't considered in the initial plan. Again, it's only as the project progresses that the need to understand, manage and ensure improvement of human resources becomes apparent. In some cases, this leads to a broad range of activities related to training and education. In others, it leads to a single specific activity such as the development of a skills matrix.

30.5   Product Structure

In most PLM projects, there will be a sub-project addressing improvement of product structures. However, in most companies, prior to the PLM project, activities in this area will have been infrequent. As a result, it's likely the project team members won't have experience in this area, and won't know what to expect.

30.5.1   Method

We take an eight-step approach (Fig. 30.14) to sub-projects addressing product structures. We've developed it during many product structuring projects.

30.5.2   Tools

We use tools such as Excel and PowerPoint to document the current and future situations of product structures. These tools may not appear to be highly sophisticated, but they have the advantages of being widely available, usable by most people, and understandable by most people. A hidden advantage of their lack of sophistication is that they force people to think about structures to reduce the amount of work. With PowerPoint, it's usually possible to represent structures of 50 parts. As for Excel, we've used it in projects involving products and systems that contained thousands of parts.

There are, of course, more sophisticated applications that include specific functionality for Product Structure Management. However, they often require additional training and licensing. And, if someone doesn't understand the basics of product structure management in Excel, they're unlikely to understand it in a more sophisticated application.

1	Prepare	Write down the scope and objectives. Plan the expected activities, taking care to include activities such as planning, communicating, reporting, interviewing, documenting, presenting and sustaining
2	As-is	Understand and document the as-is situation of product structures. Document objectives, performance measures, problems, requirements
3	To-be	Define 3 or 4 options for the to-be situation. SWOT to get the best
4	Strategy	Identify several potential strategies. SWOT to get the best
5	Plan	Develop an implementation plan for an initial project, and for further rollout phases
6	Communicate	Communicate a compelling case for action
7	Implement	Start small. Aim for early success. Check results against targets. Measure benefits. Communicate success
8	Sustain	Before the initial project ends, prepare follow-on activities. When the initial project comes to an end, start follow-on activities

Fig. 30.14 Eight step approach

30.5.3 Different Product Structures

One of the first things we do in a sub-project addressing product structure is to get people to think about products in different ways. We find that most people in a company have a very individual and limited view of the company's products. They frequently work with a single view, day in, day out. We try to get them to think of other views of products, for example, a bottom-up view and a top-down view (Fig. 30.15).

Another way to think about products is in their various manifestations across the lifecycle (Fig. 30.16).

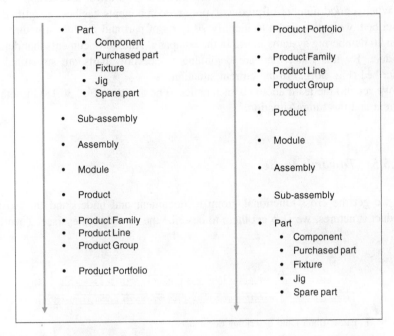

Fig. 30.15 Part to portfolio, portfolio to part

Idea	Development	Realisation	Support	Retirement
acquired	developed	finished	catalogue	scrap
discovered	purchased	semi-finished	export	salvage
existing	re-used	raw	repair	re-cycle
			replacement	re-use
	standard	packaged		
	customised			
	product	tool		

Fig. 30.16 Part layout across the lifecycle

30.5.4 As-is Situation

To help launch the description of the as-is situation, we get a cross-functional group of people from the company to indicate 20 different product types across the lifecycle. They might start off by listing a spare part, an acquired part, new parts, old parts, parts exported to different countries, purchased parts, and assemblies.

The number of different types of product in a company differs widely between companies. So, if people can indicate 20 different types, we get them to identify another 20. And so on. We ask them to find an example of each type of part, and show it to the group. This is helpful as there are usually some people in the group who have never seen some of these product types.

Then we get the group to do the same exercise with lower-level components.

When a good number of product types and components has been identified and described, we get the team to identify 20 types of part and product structure. And then 20 numbering systems in use in the company. And 20 documents that define a product. By this time, they are beginning to see how parts are structured and described (Fig. 30.17) in the current situation.

We remind the team about characteristics to be avoided (Fig. 30.18) in both the current and the future situation.

30.5.5 Towards to-Be

As we get the cross-functional group to document and understand the existing product structures, we also get them to describe the problems they face. Then they

| part number | revision index | part name | part type | effectivity | alternate |

| product description | product requirements | technical requirements | product specification |

Fig. 30.17 Part attributes and describing documents

Overproduction	anything that is produced, but isn't needed by a customer, adds to costs and wastes time. For example, unnecessarily large batches. Unwanted parts have to be stored, financed, managed, removed from the store
Excess	any transportation of parts and products beyond that needed to produce and deliver to the customer wastes time, effort, money, and runs the risk of damage or loss
Transportation	developing unnecessary functions and features takes time and money, but the customer won't pay for them
Over-processing	the materials, time and effort used for parts and products that are put in inventory create costs. But no payment is received from the customer
Inventory	movement of people adds no value to the product
Movement	defects lead to rework, extra costs, quality inspectors, upset schedules
Defects	parts, products and people waiting for the next activity are incurring costs

Fig. 30.18 Seven wastes to be avoided

Overproduction	Making batches of parts and products larger than customer needs
	Not reusing an available BOM, but creating a new one
	Creating a new part even though the part already exists
Excess transportation	Some of the parts going to the assembly area, but then being moved elsewhere until the other parts are ready
	Shipping the order to the customer for installation, then shipping it offsite until the problems are fixed
Over-processing	Products with components and features the customer doesn't use
	Creating unnecessary intermediate assemblies
	Spending too much time maintaining unwieldy BOMs
	Making a part that's unnecessarily precise
	Developing software that has too much functionality for the customer
Inventory	Inventory of raw material
	Inventory of components
	Just-in-case inventory on customer site
	Inventory of finished products
Movement	Travelling to a customer site to fix problems
	Sending the installation team to the customer site for installation, but the parts aren't all there for installation
	Going to meetings and finding we don't have the customer data to be discussed
Defects	Defects leading to rework and quality inspectors
	Incorrect product structures leading to recalls. Product liability costs
	Entering the wrong part number in a BOM
	Creating conflicting product structures in different departments
	Including a part in a module that doesn't match with a mating part in the other module
	Out-of-sync BOMs in different parts of the company
	Nonconformity between order and delivery
Waiting	Waiting until we transform the order into the right BOMs
	Waiting for installation to be complete before the customer pays
	Waiting for Engineering to finish the design
	Waiting for machine set-up
	Waiting for the rest of the data packet to get to the machining cell

Fig. 30.19 Examples of problems to be avoided

list things that have gone wrong, and things that should be avoided in the future (Fig. 30.19).

30.5.6 To-Be

The team documents the results of the above work in a report. The as-is product structures and their problems are documented and analysed. The objectives for future product structures are documented. They should be known from the objectives of the PLM project. If not, they can be interpreted from the project objectives, and the resulting suggestions discussed with the project's sponsors.

The results of the sub-project are usually presented to project sponsors. This will help keep them informed of progress. And it will give them an opportunity to comment on the findings.

Next, the cross-functional team designs the product structures for the future. The team should include people from different functions such as engineering, logistics, production, sales and support. This enables inclusion of different views and requirements such as those of a customer, a design engineer, a cost engineer, a manufacturing engineer and a support engineer.

Several options for improvement are proposed, investigated and compared by the team. Their suggestions for the future situation, including the expected benefits, are documented and presented to management.

30.5.7 Benefits

The benefits of implementation of the resulting product structures are often impressive. In some cases, the time to create or modify a structure was reduced from several weeks to a few hours. In other cases, the complexity was reduced by over 80 %.

30.6 Processes

In most PLM projects, there will be a sub-project addressing business process improvement. Again, though, this type of project is infrequent in many companies. Many of the members of the team may never have participated in such an activity. As a result, it may be difficult for the project team to know what to plan and how to go about their work.

30.6.1 Similarity

There is often similarity between a sub-project addressing business processes in one company, and a sub-project addressing business processes in another company. For example, there may be a similar need to document processes. Similar process mapping tools may be used. Similar processes may be addressed. Similar performance improvements may be targeted. Similar automated workflows may be possible.

30.6.2 Similar but Different

Although there may be similarity between sub-projects in different companies addressing processes, there may also be differences. One of the reasons for this is that the relative importance of processes depends on product strategy.

An OEM with its roots in high-cost countries, and a desire to provide products world-wide, will often have a strategy of Managed Complexity and Change. The OEM focuses on managing its portfolio of products, its product deployment capability, customer requirements, product architecture, product specifications, supplier management, system integration, final assembly and customer feedback. In this environment, the focus of the sub-project may be on processes such as managing outsourcing, requirements management, project management, portfolio management, change management, and product customisation.

The strategy of a Global Complex Assembly Provider is to provide high-value, complex major assemblies worldwide to OEMs. It will often develop and manufacture some complex components and sub-assemblies, assemble these with parts and sub-assemblies from a Low-Cost Commodity Supplier, and supply the resulting assemblies to OEMs. In a company with such a strategy, the focus of the process sub-project may be on the processes of proposal management, technology innovation, intellectual property management, product modification and change management.

The strategy of Low-Cost Commodity Supplier is typical for companies that supply commodity parts and services to Complex Assembly Providers and OEMs. In these companies, the focus of the process sub-project may be on the processes of order management, product costing, product modification, and reverse engineering.

30.6.3 Method

For sub-projects addressing processes, we take an eight-step approach similar to that used for product structuring. We've developed it based on experience gained in many process mapping, process definition and process re-engineering projects (Fig. 30.20).

30.6.4 Tools

We use tools such as Excel, PowerPoint and Visio to document the current and future process situations. These tools may not seem highly sophisticated. However, they're widely available, usable by most people, and understandable by most people.

1	Prepare	Write down the scope and objectives. Plan the expected activities, taking care to include activities such as planning, communicating, reporting, interviewing, documenting, presenting and sustaining
2	As-is	Understand and document the as-is situation. Document the specifics of the As-is product lifecycle processes, activities and steps. Document input and output information. Document users and use of the information. Document objectives, performance measures, problems, requirements. Identify problems and weaknesses holding back performance. Identify waste. Identify the causes
3	To-be	Define 3 or 4 options for the to-be state. SWOT to get the best
4	Strategy	Identify several potential strategies. SWOT to get the best strategy
5	Plan	Develop a detailed implementation plan for an initial project and for further rollout phases
6	Communicate	Communicate a compelling case of success
7	Implement	Start small, get some success. Check results against targets. Communicate success
8	Sustain	When the initial project ends, start the planned follow-on activities

Fig. 30.20 Eight step approach to process improvement

There are, of course, more sophisticated applications that include specific functionality for process mapping and improvement. However, they often require additional training and licensing. And if someone doesn't understand the basics of process mapping in Visio, they're unlikely to understand it in a more sophisticated application.

30.6.5   As-is Situation

We get a cross-functional group of people from the company to document the current processes. The group should include people from different functions such as engineering, production, sales, support and end-of-life. This enables inclusion of different views and requirements such as those of a customer, a design engineer, a cost engineer, a manufacturing engineer and a support engineer.

Often the company will have defined a multi-level process architecture. That can be a good starting point for further documenting the processes. Sometimes a company will have carried out value stream mapping. That can also be a good starting point for documenting the as-is situation. And its results can be a useful input to the description of the current situation.

It's relatively easy to map processes in PowerPoint, starting from the top level (Fig. 30.21), and working down. For example, the sub-processes making up, at the next level down, the "Product development processes" in Fig. 30.21 can be shown on another slide. They could include clarification of requirements and

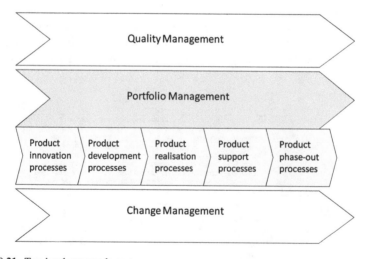

Fig. 30.21 Top-level process layout

specifications, development of concepts, selection of the preferred concept, detailing of the selected concept, and definition of a prototype. And then, at the next level down, the sub-processes making up, for example, "development of concepts", can be shown on another slide.

A similar approach can be taken with Excel. In this case, though, instead of laying out processes across the slide, they can be listed vertically in the cells of a column. There is more room available with Excel, so the process name can be included in an adjacent cell. Other cells across the row can be used to document information such as the process owner, the participants in the process, the input, the deliverables, the workload, the value, and the average time spent in the process.

With a top-down approach, each process can then be divided into sub-processes, and further detailed.

30.6.6 Towards to-Be

We get the cross-functional group to document the existing processes and the as-is process structure. As they do this, they identify related issues, and document the problems they face. They draw up a picture of things that have gone wrong, and things that should be avoided in future (Fig. 30.22).

30.6.7 To-Be

The as-is processes and their problems are documented and analysed. The objectives for future processes are known, or developed, from the objectives of the PLM project.

When the as-is process has been fully understood, the results of the above work are described in a report.

The report from the process sub-project is usually presented to project sponsors. This keeps them aware of progress. And it gives them an opportunity to respond to the findings.

The next step for the cross-functional team is to identify possibilities for improvement of the as-is situation. These may result from eliminating waste, simplifying some processes and activities, and restructuring other processes and activities. Other improvements leading to an improved to-be situation can be made by introducing new technologies, applying software to automate mundane tasks, and taking advantage of best practices.

We get the team to identify several different options for improvement. We assign each option to a different team member, and get them to improve "their" option. Usually this leads to improved options. Then we reassign each option to a different team member, and repeat the process.

Overproduction	Creating a process description from an empty sheet, not a template
	Creating information in a process that already exists in another process
	Repetition of the same activity in different processes
	Repetition of the same activity in the same process
	Defining processes without target values or measurement points
	Creating a duplicate value stream
	Doing more work on a project than is required, slowing other projects
Excess transportation	Pushing data to a distribution list, even to people who don't need it
	Multiple hand-offs of data in a process
	Excessive delegation
	Automated workflows with unnecessary steps
	Automated workflows that include steps only rarely relevant
	Changing processes frequently
Over-processing	Too many sign-offs in process activities
	Making glamorous presentations for a Steering Committee
	Approving documents that have already been approved
	Reviewing results that have already been reviewed
Inventory	Defining processes before they are needed
	Defining processes that create information before it's needed
	Creating an inventory of waiting projects
	Atomisation of data. Producing data bit by bit
	Defining processes with unnecessary storage activities
Movement	Travelling to off-site meetings
	Process steps requiring search for drawings and other information
	Clicking though multi-level menus and lists
	Travelling to a customer site to fix problems
	Switching between multiple projects
	Going to status update meetings
Defects	Forgetting to develop a risk management activity in a process
	Not developing a portfolio management process
	Making mistakes in processes
	Making mistakes in process documentation
	Making processes without quality control
	Documenting processes so badly they're impossible to understand
	Not documenting application workflow in a process description
	Not documenting a process
	Not updating a process description when the process changes
Waiting	Preparing training material long before people will be trained
	Waiting for sign-offs
	Waiting due to bottlenecks
	Waiting due to serial flow, instead of parallel flow

Fig. 30.22 Examples of activities to be avoided

The resulting preferred option for the future situation, and its potential benefits, are presented to management.

30.6.8 Benefits

The benefits of the new processes are often impressive. Cycle times have been reduced by up to 90 %. Significant reductions, sometimes as much as 50 %, were made in workload. And visibility of process activity was greatly increased. Sales have been increased because of faster, more accurate response to customers. Other improvements have included improved compliance and less rework cost.

30.7   Product Data

In most PLM projects, there will be a sub-project addressing product data and Product Data Management.

30.7.1   Method

We take an eight-step approach to product data (Fig. 30.23). It's similar to that shown for product structuring. We've used it in many sub-projects addressing product data.

30.7.2   Tools

We use tools such as Excel and PowerPoint to document the current and future situations of product data. These may not seem to be highly sophisticated. However, they're widely available, usable by most people, and understandable by most people.

There are, of course, more sophisticated applications that include specific functionality for data modelling. But, they often require additional training and licensing. And if someone doesn't understand the basics of data and document management in Excel, they're unlikely to understand it in a more sophisticated application.

30.7.3   As-is

We get a cross-functional group of people from the company to document the use, flow and structure of product data throughout the lifecycle. We start off by asking

1	Prepare	Write down the scope and objectives. Plan the expected activities, taking care to include activities such as planning, reporting, interviewing, documenting, presenting, communicating and sustaining
2	As-is	Document the as-is situation of data. Document its users and its use, flows, types and structures. Document objectives, performance measures, problems, requirements. Document PDM systems. Document other PLM applications that manage product data. Identify data-related problems and weaknesses. Identify the causes
3	To-be	Define 3 or 4 options for the future situation. SWOT to get the best
4	Strategy	Identify several potential strategies. SWOT to get the best strategy
5	Plan	Develop an implementation plan for an initial project, and for further rollout phases
6	Communicate	Communicate a compelling case for improvement
7	Implement	Start small, get some success. Check progress against targets. Communicate success
8	Sustain	When the initial project ends, start previously planned follow-on activities

Fig. 30.23 Eight step approach

them to identify the types of entities in the PLM environment that are represented by the data. This usually leads to a list such as that shown in Fig. 30.24.

Then we ask them to identify potential locations of the data describing these entities. This usually leads to a list such as that in Fig. 30.25.

Then we get the team to list, for example, all the paper documents in the PLM environment. Excel is an ideal support tool for this activity. Different types of information about each document can be entered column by column. Examples are owner, type, title, and creation date. The potential states of data can be defined. They may include in-process, in-review, released, and obsolete.

Then we get each team member to show the other team members some examples of the documents they work with daily in the product lifecycle. We ask them to include a best practice example, and a worst-case scenario. The latter often lead to consternation, as the team sees examples of documents with multiple document numbers, handwritten changes written on computer-generated documents, content that no-one can understand, and missing information.

Once the team members have a good understanding of the data and documents in the PLM environment, we get the team to create a simple data model in PowerPoint showing how entities such as products and documents are related.

30.7.4   Towards to-Be

In addition to documenting, understanding and detailing the existing data and documents, we get the cross-functional group to describe the problems they face with product data. As they continue to document the as-is structure and document the problems they face, they can draw up a picture of things that have gone wrong, and things that should be avoided in future (Fig. 30.26).

Fig. 30.24 Examples of entities in the PLM environment

| products |
| components |
| processes |
| documents |
| equipment |

Fig. 30.25 Locations of data

| paper documents |
| electronic documents |
| processes |
| databases |
| metadata descriptions |
| applications |
| interfaces |
| directories |
| files |
| desks |
| legacy systems |
| archives |

Overproduction	Producing drawings of a family of 12 parts, when a customer asks for 2
	Creating data/documents that will never be used
	Recreating data/documents that will never be used
	Recreating data/documents that exist elsewhere
Excess transportation	Pushing information to everyone on a list, even if they don't need it
	Unnecessary transportation of documents around the shop floor
	Excessive use of cc: on e-mails with long attachments
Over-processing	Too much detail in a report
	Too much detail on a design drawing
	Creating data that has no effect on the product
Inventory	Building an inventory of unused information
	Creating speculative part numbers that are never used
	Piling up drawings waiting to be signed
	Making stacks of reports waiting to be read
	Creating data just-in-case
Movement	Walking around the office looking for drawings and other information
	Travelling to a customer site to collect data
	Attending status update meetings
	Attending status review meetings
Defects	Writing reports that contain errors
	Making mistakes on drawings
	Entering incorrect numbers and incomplete information
	Including opinions and guesses, instead of facts, in a report
	Using an erroneous data model
	Using erroneous data exchange/translation applications
	Transferring data through bug-ridden interfaces
	Creating incorrect relationships in an entity-relationship model
Waiting	Waiting for sign-offs
	Waiting to read a report
	Waiting for a drawing that is in a bottleneck
	Waiting due to serial, rather than parallel flow

Fig. 30.26 Examples to be avoided

30.7.5   To-Be

When the as-is situation of data has been fully understood, the results of the above work are described in a report. The as-is data and its problems are documented and analysed. The objectives for the future situation are documented.

The results from the sub-project describing the as-is situation of data are usually presented to project sponsors. This will help keep them informed of progress. It will also give them an opportunity to give their feedback on the findings.

We then get the team to address improvement of the as-is situation. Many types of improvement can be proposed and evaluated. We get the team to propose several options. There may be an option focused on improving the quality of product data. Another option could be to improve the activities addressing product data. There can be suggestions to clean the data, removing redundant data and correcting incorrect data. Other improvements can be suggested, including the introduction of new technologies, such as a PDM system, and taking advantage of best practices. Individual options can be combined and reworked to get the best result.

The options for improvement are investigated and compared by the team. Their suggestions for the future situation are documented and presented to management.

30.7.6   To-Be Data Model

We get the team to develop the basics of a data model for the future situation. Fortunately, it doesn't take long to develop such a basic model. The intention is only to highlight the main entities and their relationships, and to identify the main attributes for each entity. We don't expect the team to develop a complete, fully-detailed data model. This can be done later with the involvement of other people who have the corresponding specific skills and experience.

30.7.7   PDM

In a PLM project, as well as a sub-project addressing product data, there will often also be a sub-project addressing PDM. An early task in this activity is to define the steps towards selection of an application (Fig. 30.27). Once these are clear, a detailed plan can be made for the corresponding activities of the team.

Sometimes we help the team to create the RFQ, starting with a structure such as that shown in Fig. 30.28.

We help the team to review responses to the RFQ, and to identify the most promising application. Sometimes we then help the team to define a benchmark to better understand the differences between candidate applications.

Fig. 30.27 Steps towards selection of a PDM system

increase understanding of PDM
define requirements for product data management
identify potential solutions and applications
send a Request For Quotation to selected vendors
review replies to the RFQ
make a shortlist of most likely solutions
carry out a benchmark of short-listed solutions

Fig. 30.28 Example of RFQ structure

objectives
scope
processes
infrastructure
applications
interfaces
information warehouse
information structure management
workflow management
system administration

30.7.8   Benefits

Documenting the as-is situation shows some of the problems with data. In one case, several data items were found to have an annual decay rate of over 40 %. In another case, merging data from three sources, more than 20 % of documents were found to have duplicates. About 11 % of data had some kind of inaccuracy. In another case, more than 15 % of relationships showed some kind of inaccuracy. Often it's found that data and documents are not re-used across the lifecycle, but that the information they contain is re-entered in a later phase.

The benefits achieved with sub-projects to improve product data and product data management have been impressive. Performance measures such as data quality and reuse of data have been improved by more than 30 %. Other examples of benefits achieved include a reduction in data entry cost of more than 10 %, and a reduction in data management costs of 15 %.

30.8   Portfolio Management

In some PLM projects, there'll be a sub-project to address Portfolio Management.

The product portfolio plays a key role in PLM. Because PLM aims to maximise the value of the portfolio, management of the portfolio is one of the most important activities in PLM.

Portfolio Management includes activities to sequence the potential product opportunities. These activities identify the sequence in which product development, improvement and phase-out projects should be carried out to provide most value. Chapter 26 included examples of reports showing project-focused information (Fig. 30.29).

Portfolio Management also includes activities addressing current products. These activities lead to another set of reports. They show information such as the expected age of products over the coming years (Fig. 30.30), the planned future of current products (Fig. 30.31) and the planned source of future products (Fig. 30.32).

We find that participants in a Portfolio Management sub-project, such as Portfolio Managers and Product Managers, usually have a very clear idea of what they want to achieve. For example, they want Portfolio Management to help them derive new product strategies, enhance traditional risk/return calculations, and visualise the result and return of different product-related projects. In global markets with, on one hand, vendor consolidation, and on the other hand, many start-ups, they see things changing fast. They know what reports they need, and they expect

| likelihood of success of a project compared to its NPV |
| cumulative cost and value of projects |
| customer view of project innovation compared to the internal view of risk |

Fig. 30.29 Examples of project-focused reports

Fig. 30.30 Planned age of
products over the next 7 years

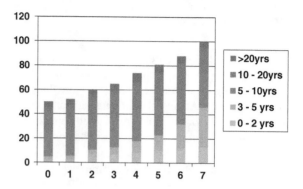

Fig. 30.31 Planned future of
current products

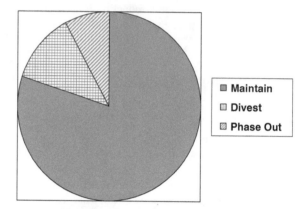

Fig. 30.32 Planned source of
future products

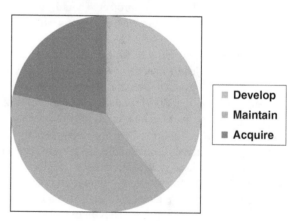

portfolio reviews at least every month so that they can keep up-to-date with the situation.

However, in our experience, many companies don't have the underlying data required to support such activities. In Portfolio Management, without good input data and assumptions, results may look very precise and meaningful, but be of little or no value.

The first step in these sub-projects is usually to carry out a review of the current Portfolio Management process and identify areas of weakness. This usually highlights the lack of high-quality data. And it often shows that the processes to create such data don't even exist.

Based on the results of the review, a plan is then made to create, or improve, the processes. Usually, this will also lead to production of the required data. In other cases, additional activities may be planned to address the lack of high-quality data.

30.9 PLM Environment Model

In some PLM projects, there'll be a sub-project addressing the development of a model of the product lifecycle environment. This type of model is intended to show how the components of the PLM environment are related, and how changes to the components can affect performance measures.

30.9.1 Evolution

Over the years, the focus of leading-edge companies has changed. In the 1970s, the focus was on individual applications. In the 1980s, it moved to product data and PDM. In the 1990s, it moved to business processes. At the beginning of the century, there was a strong focus on products and product structures.

By 2010, the focus of leading-edge companies was changing to the development of a model of the product lifecycle environment. With the basics of PLM in place, they wanted a way to better understand the PLM environment, and to be able to see how resources should be addressed to improve performance. They wanted to develop a model to show, for example, whether the company would achieve a greater benefit from implementing a particular plan to improve processes, from implementing a specific set of actions to improve product data, or from purchasing a particular PLM application.

30.10 Initiative Progress Reporting

With so much to do in a PLM project, PLM team members sometimes forget the objectives of the project. They concentrate so much on short-term tasks, requiring completion in a few days or weeks, that the overall objectives disappear over the horizon.

Fig. 30.33 Progress towards targets

	Target	To Date
Rate of introduction of new products	+100%	+25%
Revenues from extended product life	+25%	+4%
Part reuse factor	7	2.2
Costs due to recalls, failures, liabilities	-75%	-5%
Development time for new products	-50%	-5%
Cost of materials and energy	-25%	-4%
Product recyclability	90%	12%
Product traceability	100%	22%
Lifecycle control	100%	14%
Lifecycle visibility	100%	18%
Revenues from new services on existing products	+40%	0%

However, the sponsors of the Initiative are not so interested in the results of day to day tasks. They want to see progress towards the targeted objectives (Fig. 30.33).

The team should develop procedures to capture, at regular intervals, the data from which such information can be prepared.

30.11   PLM Review

Occasionally, for example every 9 months, there should be a formal review of the progress that's being made towards meeting the targets. The review should lead to a report sufficiently detailed to keep management informed of progress. Deviations from plans should be noted and explained. Proposals for changes to plans should be documented and discussed with management.

Figure 30.34 shows some of the possible areas for review. There are so many possible areas to review that, in many PLM projects, a different subset of these areas will be considered in each review.

There are several reasons for asking these questions (Fig. 30.35).

benefits	budget	business processes	communication	configuration control
contingency planning	costs	data model	deliverables	eng. change mgmt.
external resources	end of life	governance	implementation plan	lifecycle range
mgmt. of change	people	PDM functionality	PLM applications	portfolio value
process architecture	metrics	project documents	product structure	project manager
project objectives	project team	quality	requirements	responsibilities
risk management	roles	scope	strategy	targets
task definition	time	training	vision	workflows

Fig. 30.34 Possible areas for a PLM review

identifying and quantifying progress towards targets
communicating progress towards targets
making the status of the progress visible
getting a feel for any slippage between project plans and reality
in the event of major problems, putting the project back on the right path

Fig. 30.35 Reasons for a review

One advantage of a regular review is that it keeps the focus on business benefits. Unless the focus is maintained on the expected benefits of a PLM project, it's unlikely that they'll be achieved. There are always many opportunities for a PLM project to drift off on a different tack that will take it far from the targeted destination.

Another advantage of a regular review is that it helps keep PLM on management radar screens. It's only too easy for management to agree to a large investment in PLM, and then, faced with everyday fire-fighting activities in other parts of the company, forget about the PLM project, assuming subconsciously that it's making good progress.

The problem with this approach is that there's a 50 % chance the project won't be making good progress. About five out of 10 PLM projects fail to meet their targets.

30.11.1   Project Progress

In 50 % of cases, the project will be on the right track, and making good progress. In such cases, the review will show that it should continue. A few improvements and adjustments may need to be made.

However, in other cases, it will become clear that the project isn't making good progress. In such cases, it will probably be necessary to take a major decision about the future conduct of the project. There'll be various options (Fig. 30.36).

The first option is unlikely to work. First of all, management probably doesn't have the time to provide the required support. And, without more management involvement, it's not likely the team is suddenly going to change its behaviour and act differently.

The second option may work, but it's unlikely to work. It's rarely the case that a project only goes off course because of the project leader.

The third option will lead to another failure unless a lot of work is also done to identify the causes of failure. Usually these are related to corporate and departmental culture. Unless they're identified and resolved, they'll still be there to cause project failure for the new team.

| leave things as they are, hoping that the project leader and the project team will be able to put things right with some help from management |
| allow the project to continue, but with a new project leader |
| restart the project with a new project leader and/or a new project team |

Fig. 30.36 Future options for the project

30.12   External Audit

It's often helpful to get an external expert to audit PLM progress (Fig. 30.37). Such an audit shouldn't be seen as a criticism of the project team, which is probably doing a great job and making a lot of progress. Instead, it should be seen as a way of stepping back from everyday tasks, and an opportunity to look again at the big picture. The resulting report should help with planning and moving forward.

The scope of the audit may be similar to that of the review mentioned above (Fig. 30.34). It may include general questions such as those shown in Fig. 30.38.

30.13   PLM Thought and Action

So long as there are products, and companies wanting to improve performance, so long lives PLM and the desire to improve PLM performance.

an independent assessment of project progress
an opportunity to communicate progress
expert advice concerning the project's next steps
an opportunity to involve management
high level of acceptance of results by management
an opportunity to focus again on project objectives
support and communication of next steps

Fig. 30.37 Advantages of an audit by an external expert

is the project scope clearly defined? Has any scope creep been reflected in the plan?
were expected benefits identified at the start of the project? Has progress to benefits been measured and reported?
is there a plan showing current activities, and their objectives, resources, deliverables and reports? Is the plan up to date and realistic?
is time on the project recorded correctly?
are the objectives, deliverables and reports clearly defined and documented.
are requirements well documented? Is there a tracking system for requirements and changes to requirements?
are costs being tracked properly? Have cost variations been measured and reported?
is there a Quality plan in place? Are there Quality checkpoints? Has there been an independent Quality review?
are there enough resources on the team? Are team members working full-time on the project? Does the project team have the right tools and skills?
are roles and responsibilities clearly defined? Do they correspond to reality?
is there a training plan? Have team members been trained?
is there a change strategy in place? Has communication been extensive? Have changes to reward systems been addressed?
have all key stakeholders been identified and involved in developing the plan?
are external partners being used? Are there contracts to retain key external partners?
is there an up to date risk management plan? Are regular risk reviews undertaken?
is there an issue escalation process? Are there contingency plans?
is the business providing the necessary level of support for the project? Is there sufficient executive support for the project?
is there effective sponsorship? Is there an effective guiding group? Is there an effective project driver?
are project documents well organised. Are they under version control? Are they well-structured and understandable?
are there agendas for meetings? Are there meeting minutes?
is there a project glossary?

Fig. 30.38 Some general questions for a PLM audit

Running around like a chicken without a head, trying to implement all sorts of new PLM techniques and applications, without thinking about an overall vision and approach, is unlikely to lead to success.

But, sitting in an ivory tower, dreaming of visions and strategies, isn't going to lead to success either.

Between 2000 and 2015, we helped more than 50 companies with their PLM projects. In several cases, we worked with them for more than five years, sometimes at the strategic level, sometimes down in the details of process steps, data relationships and customisation of application functionality. Based on this experience, it's clear that, for PLM success, both thought and action are needed. As Henri Bergson wrote, "Think like a man of action, act like a man of thought."

Appendix A
PLM and Big Data

A.1 This Chapter

This chapter looks at the role of Big Data in the world of PLM. The concept of Big Data appeared in the 1990s, although many companies didn't start to address the subject until later. Big Data, and the related Analytics, offer companies opportunities to rapidly manage and analyse huge volumes of data and gain deep insights. The first section of the chapter introduces Big Data in the PLM environment. The second section clarifies the meaning of Big Data. The next section gives examples of Big Data across the product lifecycle. The fourth section of the chapter describes why it's important to understand Big Data in the PLM environment. Section 5 describes opportunities available with Big Data, as well as the expected benefits and value. The following section identifies application areas of Big Data in each phase of the product lifecycle. Section 7 of the chapter outlines issues that arise around Big Data. The two following sections address typical issues and success factors of Big Data projects. The final section addresses the relationship between a Big Data project and the PLM Initiative.

A.2 Introduction to PLM, PDM, Product Data and Big Data

Product Lifecycle Management (PLM) is the business activity of managing, in the most effective way, a company's products all the way across their lifecycles.

Product data is the data that defines and describes a product. Product data is needed to develop, produce and support a product throughout its lifecycle.

A PDM system is a very specific type of PLM application. It has the primary purpose of managing product data.

Big Data is the name given to a collection of technologies, resources, activities and opportunities including:

- extremely large volumes of digital data
- the collection and management of this data

© Springer International Publishing Switzerland 2016
J. Stark, *Product Lifecycle Management (Volume 2)*,
Decision Engineering, DOI 10.1007/978-3-319-24436-5

- the provision of this data in raw or aggregated form
- Analytics, the computerised analysis, or "mining", of this data
- predictions made by a company based on the analysis
- decisions taken by a company as a result of the analysis
- value added for the company as a result of the predictions, decisions and/or analysis.

A.3   Introduction to Big Data

The concept of Big Data appeared in the late 1990s. Before then, people thought of data mainly in terms of kilobytes, megabytes and gigabytes. At that time, there wasn't a lot of data in the world, just a few billion GB. As Fig. A.1 shows, a GB is a billion bytes, so a few billion GB is a few exabytes, a few quintillion bytes.

Ten years later, there was a lot more data in the world, a few trillion GB, a few zettabytes. And by 2015, there were a few tens of zettabytes, with a few quintillion being added each day.

"A few quintillion bytes" is a lot of data to add each day. It's a few billion GBs. It's coming from all sorts of sources (Fig. A.2).

Sometimes an individual instance of data from one of these sources of data may be quite small, sometimes it will be much larger (Fig. A.3).

Just as much data can result from data with small instance sizes as from data with large instance sizes. It depends on how many instances occur. For example, the

Unit of Measure	Abbreviation	Number of bytes		
kilobyte	kB	10^3	Thousand	1000
megabyte	MB	10^6	Million	1000000
gigabyte	GB	10^9	Billion	1000000000
terabyte	TB	10^{12}	Trillion	1000000000000
petabyte	PB	10^{15}	Quadrillion	1000000000000000
exabyte	EB	10^{18}	Quintillion	1000000000000000000
zettabyte	ZB	10^{21}	Sextillion	1000000000000000000000
yottabyte	YB	10^{24}	Septillion	1000000000000000000000000

Fig. A.1 Units of measure

Fig. A.2 Sources of Big Data

Social Media	Sales information
Blogs	Advertising feedback
Communities	Websites
Search engine queries	Companies' computer systems
E-mails	Industrial equipment sensors
Tweets	Simulation and manufacturing activities
Consumer behaviour data	Test activities
SMS	On-board product sensors

Fig. A.3 Different sizes of instances of data

Type of data	Typical instance size
A part number	a few bytes
A search engine query	a few bytes
A sensor reading	a few bytes
Web form comments	a few bytes
Social media comments	a few hundred bytes
An e-mail	a few kB
A PowerPoint presentation	a few MB
A CAD file	a few MB to a few GB

Wikipedia	More than 100 billion page views each year
Facebook	About a billion users each day
LinkedIn	About 360 million registered users
Search engines	More than 1 trillion searches each year
Twitter	500 million tweets each day
Point-Of-Sales terminals	About a billion
Credit/debit card transactions	Hundreds of billions each year
E-mails	More than 100 billion per day
YouTube	More than a billion minutes of video uploaded per month
SMS	About 600 billion sent per month
Websites	About a billion
Blogs	More than 500,000
A long-distance flight	A petabyte of sensor data
A long distance car trip	A terabyte of sensor data
Shop floor measurement data	Billions of files per month

Fig. A.4 Estimates of volumes in the Big Data world (2015)

volume of a billion search engine queries could be less than that of one large CAD file. The volume of a million e-mails could be less than that of one large PowerPoint file.

The term "Big Data" refers to the huge volumes of data, measured in petabytes, exabytes and more, coming from the sources mentioned above. Figure A.4 gives some examples of the volumes of some of these sources in 2015.

A.3.1   Three Contexts of Big Data

It can be seen from Fig. A.4 that there are three main contexts of Big Data:

- Commercial Big Data (such as Point of Sales data)
- Social Media and General Internet Big Data
- Product Big Data.

A.3.2   Commercial Big Data

How much commercial Big Data is there? A billion PoS terminals (10^9), averaging a thousand transactions per day (10^3), and a transaction size of a thousand bytes (10^3) would generate a few petabytes (10^{15}) per day.

Most companies don't have millions of customers. A very large company with a few tens of millions of customers (10^7), and a million bytes of data (10^6) about each one, would have a few tens of terabytes (10^{13}) of customer data. Most companies would have less.

So it looks as if, in 2015, Commercial Big Data is measured in terabytes and petabytes.

A.3.3   Social Media and General Internet Big Data

How much social media and general internet (e.g., search engine queries, e-mails) Big Data is there?

100 billion (10^{11}) emails with an average size of a few tens of kilobytes (10^4) would generate a few petabytes of data (10^{15}) per day.

So it looks as if, in 2015, Social Media and General Internet Big Data is measured in terabytes and petabytes.

A.3.4   Product Big Data

Much of the Big Data in the industrial context is generated by sensors. As an extreme example of Big Data in the industrial context, the particle collisions in CERN's Large Hadron Collider (LHC) generate about one petabyte of data per second. Fortunately, not all of this, only about 25 petabytes per year, needs to be stored.

It seems as if a few petabytes of Commercial Big Data are generated each day, and a similar volume by the LHC each second. And there are 86,400 s in a day!

In an industrial plant, the sensors may "sense" (be read) less frequently than in the LHC. But even if the sensors on each machine in a plant are read 4 times a second, the plant could generate 24 MB of data per minute. If each sensor on a locomotive is read 10 times a second, the locomotive could generate 24 MB of data per hour. If each sensor on a wind turbine is read 400 times a second, the turbine could generate 10 GB of data per day. If a long distance car trip can generate a terabyte of data (10^{12}), a million such trips (10^6) can generate an exabyte of data (10^{18}), a quintillion bytes. (And it's estimated that there are more than a billion cars in the world.)

A.3.5   Big Data and Analytics

Whatever the structure, volume and source of Big Data, there's so much of it that it wouldn't make sense for a human to try to read it and make sense of it. Instead, computer programs work on it. Sometimes they aggregate data from different sources. Sometimes they slice and dice data from a single source. At their heart are algorithms analysing the data (numbers, texts, photos, videos, locations, times, preferences, age, salary, investments, purchases, gender, race, religion) and searching for patterns, correlations, meaning and other valuable information.

These algorithms could be looking for different types of things. As US Secretary of Defense Donald Rumsfeld put it in 2002, there are:

- known knowns (the things we know that we know)
- known unknowns (the things we know that we don't know)
- unknown unknowns (the things we don't know that we don't know)

Organisations could develop analytics for all these cases.

Being personal, sometimes the analytics may look at all the data that exists about you to try to work out what you're going to buy next. Or which ads to show next on your screen. Or, based on their algorithms' conclusions about what you are likely to buy, which product special offers to send to your smartphone as you enter a mall.

Or they may be trying to work out which product, that doesn't currently exist, you'd like to buy. And sending their findings to companies around the world.

In another context, they could be working out the best time to service a machining centre, taking account of plant load and mean time between failure.

A.4 Big Data in the Product Lifecycle

Big Data is created and used across the product lifecycle (Fig. A.5).

Examples of Sources of Big Data Used	Lifecycle Phase	Examples of Big Data Created
Scientific articles, technical articles, research papers, industry journals, patent information, standards information, industry guidelines, best practices, forums, consumer blogs, PoS data, government rules and regulations, questionnaire answers, purchasing behaviour, problem reports, market data,	Imagine	
parts catalogues, material catalogues, theses, industry blogs, social media, supplier data, specialised websites, equipment catalogues, user communities, industry communities, product feedback (Voice of the Product), etc.	Define	Test data, analysis and simulation results
Specialist technical communities, specialist websites, supplier data, etc.	Realise	Machine data, quality data, manufacturing process data, sensor data, test data
User communities, specialist technical communities, social media, PoS data, etc.	Support. Use	Voice of the Product, maintenance data, on-board system data, operating data, sensor data, IoT data, field service data, CRM data
Specialist technical communities, etc.	Retire. Dispose	Voice of the Product

Fig. A.5 Examples of Big Data in the PLM environment

A.5   The Need to Understand Big Data in PLM

It's important for those in a company's PLM Initiative to understand about Big Data, because it's likely there'll also be a Big Data project in the company. And there can be overlap.

There are two main types of overlap between product data and Big Data.

A.5.1   Overlap in Content

Firstly, there can be overlap because some of the product data that will be stored in a PDM system will be Big Data (Fig. A.6). Companies that have invested heavily in PDM systems to manage their product data may question if their investment can also address Big Data, or if further investment will be needed.

Product data is the data that defines and describes a product. Product data is needed to develop, produce and support a product throughout its lifecycle.

The classification of data as product data or Big Data isn't clear. One person's product data can be another person's Big Data. And today's Big Data could be tomorrow's product data. The amount of overlap between a company's product data and Big Data will also depend on how the company defines them, and how it classifies data. Unless care is taken, this overlap may lead to confusion and waste.

Figure A.7 shows three categories of data that defines or describes the product.

"Company product definition data" is usually created, owned and structured by the company owning the product. It's relatively low-volume, and easy for a person to find and understand. It's usually created by the company's systems. Companies

Fig. A.6 Overlap in content

Company product definition data	Company product description data	External product description data
Part names	Test equipment data	Point of Sales data
Product numbers	On-board sensor data	Tweets
CAD files	Shop floor equipment data	E-mails
....	....	Blogs
....	....	....

Fig. A.7 Three types of product-related data

Fig. A.8 Overlap between
projects

usually think of this type of data as product data and put it in their PDM system. Because the company defines the structure and format of this data, it can write programs to analyse it. As the company has created this data, it should be confident of its quality. That's not to say that it's necessarily 100% clean. There can also be a lot of "dirty data" in a company's PDM system. And there can also be problems when data from different sources are aggregated, e.g. if data doesn't have exactly the same meaning in different sources.

"External product description data" usually isn't created, owned or structured by the company owning the product. It's very high volume, and almost impossible for a person to interpret. It's not created by the company's systems. Usually this data doesn't define a product, but describes it. Because the company doesn't define the structure and format of this data, it will have problems to write programs to analyse it. Companies usually think of this type of data as Big Data.

"Company product description data" is usually created, owned and structured by the company. It's relatively high-volume and difficult for a person to understand. Usually, companies don't put all this data in their PDM system. Instead, they put some of it in their PDM system, but leave most of it in the database of the PLM application that created it. Because the company defines the structure and format of this data, it can write programs to analyse it.

A.5.2 Overlap in Activities

As well as overlap in content, there's also likely to be overlap between the activities of a PLM Initiative and the activities of a Big Data project (Fig. A.8).

A.6 Big Data, Big Value, Huge Opportunity

There's a need for Big Data. And there are many opportunities with Big Data and Analytics (Fig. A.9).

There's a need for Big Data and Analytic technologies because it's difficult to process the huge volumes of data mentioned in previous sections effectively with the technologies previously available.

Manage the huge volume of data that's available
Process high volumes of data effectively
Make data visible, usable and understandable
Make some sense out of all that data
Create previously unattainable insights
Use the data to make predictions faster
Use the data to take decisions faster
Use the data to take action faster
Increase efficiency
Reduce costs
Increase revenues

Fig. A.9 Needs and opportunities for Big Data

There are opportunities for Big Data and Analytic technologies because the huge amount of data from various sources can be analysed with the aim of getting more financial benefit for the company.

A.6.1 Typical Benefits of Big Data

There are many potential benefits of Big Data (Fig. A.10).

Companies create and receive huge quantities of data every day. Some data may come from operations and maintenance. Some may come from sales and service situations. Once all the data has been organised, analytics offer fact-based insight into the entire product lifecycle.

Analytics offers opportunities to leverage data and deliver deep insights for product managers. The company can better understand products, and predict what customers want next. Tailored products can be configured to meet the expressed and unexpressed desires of customers.

A.6.2 The Value of Big Data

Figure A.11 shows ten sources of value for Big Data. These are the areas in which companies typically look to justify investment in Big Data technology.

These sources exist in all industries.

Better understand customer requirements	Customise your communication
Better understand customer feelings about your products	Customise your services
Find out how people might react to your future products	Take better decisions
Use the data to develop new products and services	Get the customer viewpoint of product
Get the customer viewpoint of competitor's products	Identify risks before they become issues
Get the customer viewpoint of missing features	Test your hypotheses
Use customer data to improve service quality	Segment your markets even finer
Uncover hidden patterns and correlations	Bring valuable insights to light
Spot potential disasters before they happen	Get to the root of a particular problem
Help specify new products and services	Drive efficiency

Fig. A.10 Examples of benefits of Big Data

1	Identify and develop new products/services
2	Identify and develop better products/services. Better serve existing customers. Reduce customer churn. Understand field problems faster. Reduce service costs.
3	Better predict/forecast the needs for products/services
4	Have all information available when needed. Make better decisions
5	Identify quickly any lack of compliance. Resolve it by building compliance into workflow
6	Improve business processes, use less resources, reduce process execution time
7	Maintain equipment/machines better, leading to longer life
8	Maintain equipment/machines better. Avoid equipment/machine downtime and idle time as this can be costly
9	Identify optimum operating conditions, improve efficiency. Reduce waste and pollution. Reduce costs. Maximise utilisation.
10	More finely segment the market, better match product/service supply to customer demand. Eventually customise products and services to individual customer needs.

Fig. A.11 Sources of Big Data value

For example, in the pharmaceutical industry, companies can work with Big Data (molecules, clinical trials, patient data) to identify markets for new drugs for unmet health needs.

In production environments, companies can analyse Big Data (operating data from machines) to minimise downtime and maximise throughput.

In the power generation industry, wind turbine blade pitch and yaw can be adjusted to whatever is optimum, taking account of factors such as wind speed, grid demand, and the supply from other turbines on a wind farm, and the supply from other sources.

In the air transportation sector, companies can use Big Data to minimise unloading/loading turnaround time of trucks and trains. Just reducing waste by a few per cent can have a significant impact.

In the air transportation sector, aircraft engine performance can be optimised to achieve, for example, reduced fuel consumption and reduced aircraft landing delay. Again, reduction of waste by a few per cent can have a significant impact.

A.7   Application Areas of Big Data Across the Lifecycle

Big Data and Analytics can contribute across the product lifecycle. Figure A.12 shows some examples. For example, at the beginning of the lifecycle, Big Data and Analytics can improve innovation, and help get a better understanding of customer needs. Later, they can improve product development, improve realisation across the supply chain, and improve service. At the end of the lifecycle, they can reduce environmental impact.

Phase	Big Data Use
Imagine	to identify new products and services to anticipate future needs of customers
Define	to better understand customer behaviour to tune existing products and services to help shape new products to reduce iterations in product development to customise products to individual customer requirements to target existing customers with add-on products and services to optimise component performance
Realise	to launch preventive equipment maintenance to get insights that weren't previously available to detect defects in manufacturing to monitor product data quality to drive efficiency across the Extended Enterprise to identify likely problems and prevent them happening
Support. Use	to provide intelligent recommendations to customers to identify counterfeits to reduce operating costs to highlight hidden product problems to understand what led up to a product problem to be able to see tell-tale signs of a problem before it occurs to offer preventive maintenance to optimise product presentation
	to identify the best supply chain to foresee machine wear and tear to improve energy efficiency to customise the supply chain to predict wear to track product behaviour to customise support to detect faults to monitor product degradation to identify fraud to monitor product quality to review product status
Recycle	to minimise waste, and to optimise recovery to reduce environmental impact to predict remaining component lifetimes to process components and assemblies effectively

Fig. A.12 Examples of Big Data use across the lifecycle

A.8   Typical Issues with Big Data

Some of the many questions raised about Big Data are shown in Fig. A.13.

Other issues with Big Data are shown in Fig. A.14.

The CIO organisation in many companies is already overloaded. It may not have the resources to address Big Data and Analytics.

Big Data and Analytics technology is relatively new. A company may find itself pioneering relatively new technologies and approaches.

Big Data comes from many sources. Many of these have data related to several subjects. A company may not have people with the skills to understand each source.

It can be difficult to understand, from so much data from so many internal and external sources, which is most relevant.

Big Data is easy to collect from multiple sources. However, it also needs to be managed properly. Big Data governance needs to be defined.

There's a danger that with so much data being available, it may be possible to link data to particular individuals. This loss of privacy could affect a person's private life. Within a company, it could be used to identify high and low

Volume	There's so much Big Data. How can we handle it? Where can we store it?
Variability	Different programs use different words with different meanings. How can Analytics understand them?
Variety	Big Data comes from so many sources, each with its own format. How can we read them all?
Visualisation	How can we present so much data in an understandable way?
Velocity	There's so much data coming all the time. How can we process it quickly enough?
Value	How can we find anything useful in all that data? How can we show there's any value in Big Data?
Verification	With such huge volumes, how can we check the data and the sources?
Veracity	How can we know that all this Big Data is reliable? Can we trust it?
Vulnerability	How can we know that criminals and competitors aren't manipulating Big Data before we get it?

Fig. A.13 Typical questions with Big Data

Fig. A.14 Other issues with Big Data

Multiple contexts	High-precision meaningless results
Many types of Big Data	Immature Big Data technology
Lack of skills	Lack of experience in Big Data management
Lack of resources	Data security
Data collection difficulties	Data storage
Varying/unknown data quality	Insufficient occurrence of data points
Displaying results	Multiple languages
Incorrect hypotheses	Unclear data lifecycle
Multiple meanings/interpretations	Unclear, fragmented data ownership
Data manipulation	Lack of data structure
Self-inflating results	Conflicting sources of Big Data
Scalability	Availability
Lack of governance	Pressure on the CIO organisation
Pressure on people	Threat to individual privacy
Bugs in algorithms	Viruses in algorithms
Inflexible algorithms	Poorly-performing algorithms

performers, and result in the latter being fired, perhaps on the basis of low quality data, unsound hypotheses and bugs in analytic software.

Analytic results may look highly precise and meaningful. However, if the underlying hypotheses are wrong, the results may look great, but be meaningless and/or dangerous. And, if they're based on insufficient data points, they can also be dangerous.

In some cases, very high availability of data and analytic results is important, for example, for aircraft and surgical equipment. Loss of a few fractions of a second could be fatal.

Scalability of Big Data solutions will be important for many companies. They will want to start with a low volume of Big Data, then scale up to very high volumes of Big Data.

Data can easily be manipulated. Websites and user profiles can be created with false information that is then copied, republished and re-tweeted by other sites and people, eventually gaining credence. Applications can be developed to automatically write negative comments and blogs about a company's products, or give false information about potential customers' requirements for a new product.

A.9 Typical Issues with Big Data Projects

Companies have run many Big Data projects. There's a lot of experience about the issues that can cause problems (Fig. A.15).

Fig. A.15 Typical issues
with Big Data projects

Ignoring existing databases
Ignoring other on-going data-related projects
Putting Big Data technology before business needs
Lack of prioritisation of opportunities
Lack of business case
Unclear business value
Unclear ROI
Lack of governance for Big Data
IS-run project, not business-relevant
Not identifying the right Big Data to work with
Not modelling the data landscape
Not modelling the business processes
Not embedding Big Data in everyday activities
Not going beyond a pilot

Sometimes the Big Data project team is so excited by the great new opportunities of Big Data that they don't even think that the subject may already have been addressed, at least partially, in the company. And then they waste time recreating the wheel.

Or they may fail to see that a similar project is already running in another part of the company.

There's so much Big Data out there, and so many things you could do with it, that it can be difficult to identify the right data and work out how to use it best. It's important to identify and describe the opportunities, then prioritise them. Is the company's objective to customise products to meet the desires of each individual customer? Or is it to show website ads customised to different customer profiles? There are so many things you could do, that, unless you establish clear priorities, you could finish up doing nothing useful. Lack of prioritisation often results from a lack of an agreed business case. In turn, lack of a business case often results from not having a clear understanding of the business value.

Letting the IS Department run the Big Data can be risky. They may put technology issues before business needs. Until the company defines its business needs, the IS Department can't know what technology it will need.

A.10   Big Data Success Factors

Companies have run many Big Data projects. Lessons can be learned from these projects and success factors identified (Fig. A.16).

As in any project, it's important to identify and appoint the Project Sponsor. And then appoint the Project Leader and team members. The objectives and rules for the Big Data project have to be defined. Related projects should be identified. The best way to work with them should be agreed.

Fig. A.16 Success factors for
Big Data projects

Clarify project governance
Understand Big Data and Analytics
Collaborate with related projects
Focus on business targets
Identify strategic areas
Identify changes to business processes
Embed potential benefits in the business
Clarify Big Data governance

Big Data and Analytics may be new subjects for many people. Some education and training will probably be needed. It's important to get a good shared understanding of Big Data before proposing solutions.

It's easy to capture lots of Big Data, much more difficult to do something useful with it. Before coming up with solutions and recommendations for new applications, it's important to identify the business benefits that can be achieved. And then estimate their value. It's only then that executives can decide how much they can invest in Big Data.

Once the business benefits have been identified and agreed, the next step is to understand how Big Data and Analytics will be embedded in everyday activities. Big Data and Analytics aren't needed if they're not going to be used. And, if they are going to be used, some business processes will have to be changed to include them. The business processes that will be affected should be identified. The way that they will be adapted to take advantage of Big Data must be detailed and documented. The corresponding education and training activities should be defined.

It's important to define and document the rules, regulations, procedures and policies that will be applied to Big Data. It's important to train people about them. Without clear documented governance, anything can happen. Without good governance, the countless risks and potential problems lurking in the Big Data environment might suddenly become a big issue.

The data model for Big Data should be defined, detailed and documented. It has to fit in the Enterprise data model.

There need to be rules about accessing, retaining, reading, writing, and communicating data. There are data privacy, security, compliance, quality, and accuracy questions to address. There need to be policies about the data sources to access, what data to use, which data to keep, and how long to keep it.

When the business benefits have been identified, and it's clear how everyday activities should be adapted, then look for applications that can support the activities and help you achieve the benefits. Perhaps the company's PDM system can be used for some of the Big Data. Once changes have been made to business processes, and technology and applications implemented, more training will be needed.

A.11   Big Data and the PLM Initiative

There's so much overlap between Big Data and product data that a Big Data project in a company shouldn't be run separately from the PLM Initiative.

Some joint activities should be organised. Otherwise it's likely that there'll be duplication of activities and confusion. For example, the Big Data project may think that the PLM Initiative is responsible for a particular type of data. Whereas the PLM Initiative may think that type of data is the responsibility of the Big Data project. The result could be that nobody looks at that type of data, and opportunities are overlooked.

The area of "Company product description data" is often a source of overlap. It includes data such as test data, data from products' on-board sensors, and shop floor equipment data.

To keep the Big Data project and the PLM Initiative aligned and in phase, the joint activities should involve members of both the Big Data project team and the PLM Initiative team. It's also useful to include the PLM Initiative leader in the Steering Group for the Big Data project. And to include the Big Data project leader in the Steering Group for the PLM Initiative.

Appendix B
PLM and the Internet of Things (IoT)

B.1 This Chapter

This chapter looks at the role of the Internet of Things from the PLM viewpoint. The concept of the IoT emerged in the 1990s, although many companies didn't start to address the subject until later. The Internet of Things, and the related Smart Products, offer companies and their customers many potential benefits. The first section of the chapter introduces the Internet of Things in the PLM environment. The second section looks, in detail, at its components. The next section describes why it's important to understand the IoT in the PLM environment. The two following sections describe the opportunities and potential benefits of the Internet of Things. Section 6 of the chapter describes the potential impact of the Internet of Things in each phase of the product lifecycle. Section 7 outlines issues that arise around the IoT. The two following sections address typical issues and success factors of IoT projects. The final section addresses the relationship between an IoT project and the PLM Initiative.

B.2 Introduction to PLM and the IoT

Product Lifecycle Management (PLM) is the business activity of managing, in the most effective way, a company's products all the way across their lifecycles.

The Internet of Things is the name given to a collection of technologies, resources, activities and opportunities including:

- the Internet, a communications network
- various electronic devices such as transmitters and sensors
- data transmitted over the Internet
- products
- activities across the lifecycle of these products
- value added for the product manufacturer and/or user.

© Springer International Publishing Switzerland 2016
J. Stark, *Product Lifecycle Management (Volume 2)*,
Decision Engineering, DOI 10.1007/978-3-319-24436-5

B.3   Nothing New Under the Sun

The individual components of the IoT, such as the Internet, and the various devices and other products, have existed for many years.

B.3.1   The Internet, a Communications Network

Research into the Internet, which is a communications network, started in the 1960s.

Before the Internet was developed, many other types of communication networks already existed. They included road and rail networks, telephone networks, and industrial process control networks. These were often limited in geographical spread, proprietary and analogue. They had their own nodes, links, layers of protocols, transmission mechanisms and addressing systems.

The Internet is an open, global, standard network. It uses TCP/IP protocols, structured into protocols at the application layer, the transport layer, the Internet layer and the link layer. In its early days in the 1970s it was seen as a network of computers that could be used to transfer data between intermediate nodes (such as computers) and terminal nodes (such as a computer, user screen or printer). Each node had a unique Internet Protocol (IP) address. With the IPv4 protocol, addresses were 4 bytes (32 bits) long, for example, 127.168.0.1. Use of four-byte addresses limits IPv4 to about 4 billion addresses. The successor protocol, IPv6, uses 128-bit addresses. World IPv6 Launch Day was 6 June 2012.

The "Internet" of the Internet of Things is the same Internet that supports the World Wide Web and e-mail.

B.3.2   IoT Devices

Just as there's nothing new about the Internet component of the Internet of Things, there's nothing new about the devices it uses. For example, thermostats which can be used to control the temperature of a room, or a refrigerator, have been in use for a long time. Aircraft flight recorders have been used since the mid-20th Century to sense and record various parameters about flight behaviour.

In the Internet of Things (Fig. B.1), the "thing" typically includes several devices (Fig. B.2):

- a receiver and transmitter for communication with the Internet
- a sensor to sense some characteristics of the thing or its environment
- a memory to store data

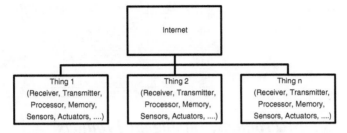

Fig. B.1 Things on the Internet

Fig. B.2 Some IoT devices

activator	receiver
actuator	sensor
display	storage
GPS receiver	switch
memory	tactile sensor
meter	thermostat
microphone	transceiver
motor	transmitter
power module	vision system
processor	voice synthesiser

Fig. B.3 Examples of characteristics sensed and measured

movement	light
speed	electricity
acceleration	magnetism
pressure	temperature
stress	strain
proximity	sound
location	patterns
vibration	humidity
touch	tilt

Fig. B.4 IoT devices and Smart functions

sensors	"seeing", "feeling", "reading", monitoring
a voice synthesiser	"speaking"
motors	moving
GPS	locating
displays	showing information, reporting
microprocessors	"thinking and calculating"
memory	remembering information
a memory	self-identification
a receiver	receiving information over the Internet
a transmitter	sending information over the Internet

- a processor
- operating software
- an actuator to make the thing do something.

The devices sometimes just carry out simple functions such as measuring a temperature and transmitting it across the Internet.

Depending on the type of sensor, many characteristics can be sensed and measured (Fig. B.3).

Sometimes the devices are combined to carry out smart functions (Fig. B.4).

Fig. B.5 Examples of smart products

Smart phone	Smart thermostat	Smart early warning system
Smart car	Smart watch	Smart advertising equipment
Smart machine	Smart wind turbine	Smart electricity meter
Smart tractor	Smart domestic robot	Smart braking system
Smart switch	Smart energy grid	Smart vending equipment
Smart light bulb	Smart person (cyborg)	Smart traffic control system
Smart home	Smart industrial robot	Smart clothes - wearables
Smart mattress	Smart transportation	Smart monitoring system
Smart tractor	Smart health monitor	Smart security system
Smart camera	Smart car park	Smart building system

B.3.3 Smart Products, Intelligent Products

Smart, or Intelligent, Products are products that, in addition to their primary functionality, have smart functionality to decide or communicate about their situation or environment. Almost all products can become Smart Products (Fig. B.5). A washing machine has primary functionality to wash clothes. A smart washing machine, equipped with a scanner, can read the labels on clothes, and select the most appropriate washing and drying cycle. Smart labels in transparent foil around meat products can change colour from blue to red when the temperature rises above the safety limit. A smart lawn mower can be programmed to cut the grass for you. Its sensors see if there are any obstacles, identify the height of the grass, and switch on its motors to go down the garden and cut the grass. A smart microwave oven can identify the food to be cooked, then set the timer and the temperature. A smart water softener can identify the hardness of incoming water, and treat it as required by its hardness and the intended use. With the addition of some devices, almost any product can become a Smart Product. Many more examples could be given, there are many opportunities.

Just as there's nothing new about the Internet component of the Internet of Things, or the device component, there's nothing new about Smart Products. According to the first edition of this book, published way back in 2004, "Intelligent clothes will change performance as the weather changes and the wearer's mood changes."

B.3.4 Data Transmitted Over a Network

In the IoT, products are connected to the Internet. Data representing control commands may be sent to the product's devices, for example, to switch the product on or off. Performance data and feedback information may be transmitted by the product's devices. There's nothing new in the concept of sending data to, receiving data from, and controlling objects over a network. SCADA (Supervisory Control And Data Acquisition) systems, used in industrial process control, do just this. They appeared in the third quarter of the 20th Century.

B.3.5   Something New Under the Sun

If there's nothing new about the idea of controlling objects over a network, or the Internet component of the Internet of Things, or the device component, or Smart Products, is anything about the IoT new?

What is different with the IoT is that the communications network that is used is the Internet. It's an open, standard, low-cost network that, in theory, is available everywhere on Earth. Products can be connected to it, becoming connected products. Then they can be used anywhere and controlled from anywhere.

Back in the 1970s, the Internet was seen as a network of computers that could be used to transfer data between intermediate nodes and terminal nodes (such as a computer, user screen or printer). It wasn't seen as a network over which almost any of the trillions of products in the world could be controlled. With IPv4 this wouldn't have been possible, as IPv4 is limited to about 4 billion addresses. The successor protocol, IPv6, uses 128-bit addresses. 2^{128} is more than a trillion trillion trillion, so IPv6 has enough addresses for quite a lot of products.

B.4   The Need to Understand the IoT in PLM

At first glance, there may not seem much in common between PLM and the Internet of Things. However, the two subjects are closely related, as the "Things" are products. And the devices are also products.

There are many areas of overlap between PLM and the IoT. First of all, there's overlap in content. This can be illustrated by reference to the PLM Grid (Fig. B.6), which shows the resources which can be organised in different ways to manage a product across its lifecycle.

For example, a lot of product data will be needed to describe the IoT "thing", and its devices. This will need to be stored in the company's PDM system. Other PLM applications used by the company, such as CAD, will be needed to develop the thing. As people in a company may have no experience of developing a connected product, they'll need to learn new skills. A new way of working, such as system engineering, may be needed. New PLM applications, not previously used by the company, such as ECAD, may be needed. Business processes will have to be redefined to address all the new activities related to connected products. New equipment may be needed, for example to assemble connected products. Changes will be made to the company's organisation to include new roles addressing connected products. Management will set objectives for the IoT, and put metrics in place to track progress.

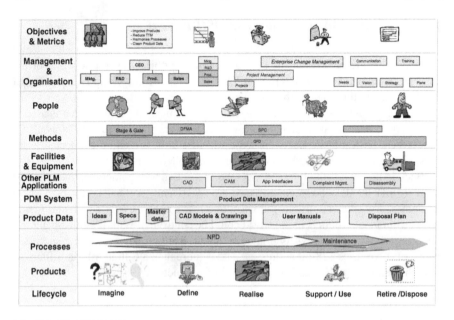

Fig. B.6 The PLM Grid

Fig. B.7 Overlap between a
company's projects

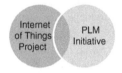

There's also likely to be overlap between the project set up to address IoT and
the PLM Initiative, as they both address the company's products (Fig. B.7).

B.5   The Opportunity of the Internet of Things

B.5.1   Financial Opportunity of the IoT

Various estimates have been made of the size of the various IoT-related markets.
It's expected, for example, that by 2020 there will be a few tens of billions of IoT
devices, with a market size of a few hundred billion dollars.

B.5.2   Strategic Opportunity of the IoT

Before the IoT, many companies lost touch with their product once it went out the factory gate. Once the product was at a customer location, it was difficult for a company to keep control of it. Until a product fault or breakdown occurred, often nothing was heard about the product, nothing was known about its performance. In the pre-IoT environment, it was easy for companies that develop, produce and support products to lose control over a product.

With the IoT, companies can keep in contact with their products across the lifecycle. The Internet of Things enables companies to see their products across the lifecycle. The IoT gives them the possibility to receive information from the product, and control it, even if it is far away. The IoT offers numerous opportunities to get a better understanding of the way products behave over their lifecycle.

The product can create its own onboard electronic log, noting data such as test results, operating conditions, performance, component use and changes, age, date and time, location, etc. All this information can be transmitted periodically (on demand, or continuously) back to the product manufacturer.

With the IoT, companies can offer new services on top of existing products. For example, a vending machine manufacturer can inform soft drinks companies when the stock of drinks in the machine is running low. Machine tool manufacturers can predict when a device at a customer site will fail and propose maintenance activities to avoid problems. This will prevent costly machine downtime and loss of production. Improving security, products, ranging from ships through aircraft to consumer products, can be tracked and unusual situations notified.

Companies can get the product to report back about how it's working. They can listen to the Voice of the Product. Data sent by a product during its operation by a customer can be used in product development to help improve future generations of the product.

With the IoT, companies can also offer new products and services. For examples, new wearable health products can measure blood pressure, glucose levels, pulse, cardiac electrical activity, brain wave patterns, etc., and report results over the Internet to a health service provider.

B.6   Potential Benefits with the Internet of Things

The Internet of Things offers benefits to both product manufacturers and product users. The IoT offers many potential benefits for the manufacturer of the product (Fig. B.8). Some benefits arise when the product is outside the company, from where the product can send data to the company's product developers, maintenance engineers, machine operators and sales force. For example, from data sensed and

Fig. B.8 Benefits for the
product manufacturer

Innovate great new products and services
Understand customer needs better
Enhance existing products with value-adding services
Add value throughout the product lifecycle
Save time
Increase efficiency
Optimise resource usage
Reduce costs across the product lifecycle
Increase revenues

Fig. B.9 Benefits for the
product user

Reduce usage costs	Operate in adverse conditions
Reduce human errors	Operate remotely
Get new products and services	Optimise maintenance
Get improved products and services	Get real-time data
Sense unplanned conditions	Carry out transactions
Get customised services	Reduce risks
Optimise performance	Monitor activities
Prevent misuse	Monitor resource usage
Analyse data	Remember preferred settings
Inform about unexpected conditions	Remember planned operation schedules

communicated by product devices, the company can understand how the product behaves when it's being used by the customer. From analysis of this data, the company can identify innovative new products and services to better meet the customer's requirements.

Other benefits arise when the product is inside the company. For example, product components, sub-assemblies and assemblies can be fitted with transmission devices so that they can be tracked during the realisation phase to improve quality, reduce risks, and optimise manufacturing and assembly times.

The IoT also offers many potential benefits for the user of the product (Fig. B.9). The user can benefit from new and improved services on new and improved products, with new and improved features and functions.

The user can operate the product remotely, and in adverse conditions. For example, with the IoT, on a cold snowy day, on your commute home, you can, from your smartphone switch on the oven and the heating, and switch off the intruder alarm system. As you drive up your property's access road, you can open the garage door and avoid ruining your shoes walking through snowdrifts.

The product user can monitor product performance from afar. For example, from your office you can check the operation of your home's intruder alarm system. Activities can be monitored to reduce risks, for example sensing unexpected operating conditions and responding quickly.

With the IoT, the user can easily provide the manufacturer with feedback such as improvement suggestions at specific operating times.

At the end of life of a product, the data in its log can be reviewed to identify which components can be reused, which can be remanufactured, which should be recycled, which should be disposed of. This can reduce costs and improve compliance with environmental rules and regulations.

B.7   IoT Impacts Across the Lifecycle

The Internet of Things has impacts across the product lifecycle (Fig. B.10).

In the innovation phase, companies can imagine new business models, new products and new services.

In the definition phase, companies can benefit from better understanding of customer needs to develop new products and services. They can also improve existing products and services. Data fed back from products in the field can be fed into the development of the next generation of products. Product developers can review feedback information about reliability problems, failure rates and customer complaints corresponding to specific usage conditions and product performance levels. They can analyse information about typical component lifetimes, average repair and replacement rates, disposal costs, and actual disassembly costs and times. All of this information can be used to improve future components and products.

Data about the product's performance reliability, availability, maintainability and safety during realisation and use can be used to avoid repeating errors, and reduce development time and effort. Designs that have been successful in existing products can be reused. Reusing successful designs should reduce development time. And due to use of the proven design, there should be fewer breakdowns and changes during product operation. Increased customer satisfaction should result from a more reliable product.

In the realisation phase, for example on the company's shop floor, if a component is incorrectly positioned for assembly, its position can be automatically corrected. Components can be tracked through manufacturing and assembly so that anything that goes wrong can be corrected quickly.

In their use phase, products can be tracked throughout their lifetime. Immediate support can be provided if needed. The IoT offers opportunities to get a better understanding of the way products behave over the lifecycle. This can be used to optimise use and maintenance activities of the product. For example, the collection

Lifecycle Phase		
Imagine	Imagine new products and services	Imagine new business models
Define	Develop better	Improve customer interaction
	Better understand customer needs	Make product modifications based on customer usage data
Realise	Anticipate machine failures	Track inventory
	Improve uptime	Track products
	Minimise machine downtime	Reduce energy costs
	Reduce downtime costs	Optimise manufacturing parameters
	Reduce shutdowns	Correct the position of a part
	Track parts in assembly	
Support. Use	Optimise vehicle trajectories	Reduce fuel costs
	Understand customer usage	Propose additional services
	Link insurance premiums to behaviour	Monitor product use
	Track luggage	Enable remote service and repair
	Get real performance data from a product	Enable remote control of the product
	Prevent spare part counterfeiting	Enable remote upgrade of the product
	Enable remote monitoring and service	Track products
	Monitor product condition	Trigger maintenance work
	Monitor the product environment	Switch from fixed schedules to preventive maintenance
Retire. Recycle	Treat waste environmentally	

Fig. B.10 IoT impacts across the lifecycle

and analysis of data from a fleet of commercial vehicles enables a different maintenance schedule for each vehicle based upon its specific usage conditions. Engine components can be monitored for fatigue. Compared to maintenance at fixed intervals, this can lead to a reduction in downtime and maintenance costs. The number of component breakdowns can be reduced.

Maintenance practices for all sorts of household appliances, ranging from water heaters to refrigerators, can be improved, shifting from intervention after breakdown to predictive maintenance. As a result, there should be fewer component failures, reduced maintenance intervention, and reduced maintenance costs. The IoT also enables remote upgrade of a product's control software. This reduces downtime compared to the traditional approach of taking the product to a service centre for an upgrade.

In the retire/recycle phase, products and components can be tracked, identified and treated in an environmentally correct way. Based on information about production dates, maintenance dates, repaired parts, replaced parts and usage conditions, better decisions can be taken about which components to dispose of, which to remanufacture, and which to reuse.

B.8   Typical Issues with the IoT

There are many potential issues with the IoT (Fig. B.11).

IoT technology is at an early stage. It's unclear how it will develop. Currently there are few standards. There's a risk of manufacturers and users finding themselves with many incompatible devices with poor user interfaces. The devices can produce an overload of data that has no meaning, or is difficult to exploit. There are security issues, as devices may be hacked. There are data ownership issues, with data being sent across the Internet between many devices, many computers and many customers, suppliers and manufacturers. There are data privacy issues, with users sending confidential information to products in their homes, and devices sending confidential information back to them. As devices increasingly communicate with each other, without any human interaction, these issues of data security, ownership and privacy will become even more complex.

Another issue is that the Internet of Things is the Internet on which government agencies, crime syndicates and brilliant students hack and snoop tirelessly with state-of-the-art technology. They read everything, hunting for valuable information. Product manufacturers and users may be concerned about important information being read by the unscrupulous. The latter may find out about problems with the product, and pass the information on to competitors. Or they may blackmail the company, threatening to reveal the information to consumer organisations or to government agencies.

Standards	Trust	Integration with other systems
Skills	Theft	Proprietary architecture
Knowhow	Data storage	Proprietary protocols
Data security	Data processing	Limited scalability
Hackable old systems	Flood of devices	Limited customisation
Data privacy	Flood of different interfaces	Limited communication
Legal liability	Flood of data	Increased waste
Regulations	Flood of meaningless data	Difficult recycling
Company structure	Flood of useless data	Intellectual property management
Safety	Flood of transmitted data	Operating performance
Cost	Diversity of devices	Unreliable connectivity
Unfriendly user interfaces	Difficult set-up	Lack of governance model

Fig. B.11 Typical issues with the Internet of Things

B.9   Typical Issues with IoT Projects

There are many potential issues with IoT projects. They arise in many areas (Fig. B.12).

Most existing products weren't designed to be connected. It may be more difficult and costly to adapt them than expected.

Most people don't have experience of innovating or developing connected products. It may take more time and expense than expected for them to learn the appropriate skills. It's difficult to get people to change. It's very difficult to get them to change if nobody can tell them where they're meant to be going or why they're meant to be going there.

Among the strategic issues in the project, should the company change its business model? Which opportunities should it pursue? Among the organisational issues that may arise in the project, how should the company reorganise to take advantage of the IoT? How will it address systems engineering of mechatronic products? Will it do everything in-house, or will it add new design and supply chain partners? Another issue is that the IoT may be seen as a technical topic for engineers to play with in their sandbox, not as a business subject requiring executive involvement.

The IoT offers many opportunities, but the first that appears may not be the best. It may take a long time to find the best opportunities. The project team needs to take the time to find the best opportunity, and not be rushed into something quick and dirty. The IoT project members may feel that they should look for an approach in which the company does everything itself. But they shouldn't forget to look also at alternatives involving design and supply chain partners.

Fig. B.12 Typical issues with IoT projects

Lack of people with IoT skills	Over-optimistic timeline
Lack of people with IoT understanding	Ignoring OCM
Difficulty of getting people to change	Unclear targets
Managing new systems (e.g., ECAD)	IoT icebergs
Changing, unclear business requirements	Immature technology
Forgetting the lifecycle (maintenance, waste)	Poor project leader
Lack of support from the business	Scope change
Lack of executive commitment	No KPIs for success

Fig. B.13 Success factors for
IoT implementation

| Build new and evolving IoT skills |
| Manage organisational change |
| Involve management in IoT decisions |
| Focus on the business, not IoT technology |
| Understand opportunities, generate and filter ideas |
| Clarify the business model |
| Define the best product/service combination |
| Collaborate with lead customers |
| Develop the corresponding IoT design/supply chain |
| Identify and fix bugs before full-scale launch |

B.10   IoT Project Success Factors

Some of the factors that lead to success can be identified from existing IoT projects
(Fig. B.13). In 2015, the IoT is a new area for many company and people. This has
many effects.

Education and training are key to success. People who know little about IoT, and
have no experience with it, are unlikely to succeed. Management involvement is
necessary to set objectives and point people in the right direction. Team members
may be expected to learn all about IoT, but that's only part of what's needed. They
also need to know the targets for the project.

It's important to clarify the company's IoT business model. If there's no change
from the existing business model, perhaps some opportunities are being missed. It's
important to review the possibilities for developing a new design/supply/mainte-
nance chain for connected products. Once solutions have been identified, and
implementation started, it's important that bugs, snafus, gaffes and goof-ups are
identified and fixed before full-scale launch.

B.11   IoT and the PLM Initiative

Often, an Internet of Things project in a company will be run separately from the
PLM Initiative. And usually the IoT project won't be led by the person who leads
the PLM Initiative.

However, there's a lot of overlap between the IoT and PLM. This overlap should
be taken account of when organising the corresponding projects.

Joint activities should be carried out between the IoT project and the PLM
Initiative. The first of these should be to identify and address the areas of overlap.

To keep the IoT project and the PLM Initiative aligned and in phase, the joint
activities should involve members of both the IoT project team and the PLM
Initiative team. It's also useful to include the PLM Initiative leader in the Steering
Group for the Internet of Things project. And to include the Internet of Things
project leader in the Steering Group for the PLM Initiative.

Bibliography

Burden R (2003) PDM: Product Data Management. Resource Publishing

Camp R (1989) Benchmarking: the search for industry best practices that lead to superior performance. Quality Press, UK

Christensen C (1997) The innovator's dilemma: when new technologies cause great firms to fail. McGraw-Hill, New York

Cooper R (1986) Winning at new products. Rinehart and Winston of Canada, Holt

Cooper R (2001) Portfolio management for new products. Basic Books, New York

Crnkovic I (2003) Implementing and integrating product data management and software configuration management. Artech House Publishers, Norwood

Deming WE (1982) Out of the crisis. Institute of Technology

Griffin A (2007) The PDMA ToolBook 3 for new product development. Wiley, Hoboken

Hammer M (1993) Reengineering the corporation: a manifesto for business revolution. HarperCollins Publishers, New York

Khan K (2012) The PDMA handbook of new product development. Wiley, Hoboken

Kotter J (1996) Leading change. Harvard Business School Press, Boston

Pine J (1992) Mass customization: the new frontier in business competition. Harvard Business Review Press, Watertown

Porter M (1985) Competitive advantage: creating and sustaining superior performance. Free Press, New York

Project Management Institute (2013) A guide to the project management body of knowledge. Project Management Institute, Newtown Square

Sandberg S (2013) Lean In: Women, Work, and the Will to Lead. Knopf

Stark J (1988) Managing CAD/CAM: implementation, organization, and integration. McGraw-Hill, New York

Stark J (1992) Engineering information management systems: beyond CAD/CAM to concurrent engineering support. Van Nostrand Reinhold, Hoboken

Stark J (2004) Product Lifecycle Management: 21st century paradigm for product realisation. Springer, London

Stark J (2007) Global Product: Strategy, Product Lifecycle Management and the Billion Customer Question. Springer, London

Sutton R (2014) Scaling up excellence: getting to more without settling for less. Crown Business, USA

Watts F (2011) Engineering documentation control handbook: configuration management and product lifecycle management. William Andrew, Norwich

Womack J (1990) The machine that changed the world. Free Press, New York

Index

© Springer International Publishing Switzerland 2016
J. Stark, *Product Lifecycle Management (Volume 2)*,
Decision Engineering, DOI 10.1007/978-3-319-24436-5

CPSIA information can be obtained
at www.ICGtesting.com
Printed in the USA
LVHW07*0545240518
578250LV00001B/8/P

9 783319 244341